Lecture Notes in Computer Science 11558

Commenced Publication in 1973
Founding and Former Series Editors:
Gerhard Goos, Juris Hartmanis, and Jan van Leeuwen

More information about this series at http://www.springer.com/series/7407

Florin Manea · Barnaby Martin ·
Daniël Paulusma · Giuseppe Primiero (Eds.)

Computing with Foresight and Industry

15th Conference on Computability in Europe, CiE 2019
Durham, UK, July 15–19, 2019
Proceedings

 Springer

Editors
Florin Manea
Christian Albrechts University of Kiel
Kiel, Germany

Barnaby Martin
Durham University
Durham, UK

Daniël Paulusma
Durham University
Durham, UK

Giuseppe Primiero
Università degli Studi di Milano
Milan, Italy

ISSN 0302-9743   ISSN 1611-3349   (electronic)
Lecture Notes in Computer Science
ISBN 978-3-030-22995-5   ISBN 978-3-030-22996-2   (eBook)
https://doi.org/10.1007/978-3-030-22996-2

LNCS Sublibrary: SL1 – Theoretical Computer Science and General Issues

This Springer imprint is published by the registered company Springer Nature Switzerland AG
The registered company address is: Gewerbestrasse 11, 6330 Cham, Switzerland

Preface

Computability in Europe 2019:
Computing with Foresight and Industry
Durham, UK,
July 15–19, 2019

The conference Computability in Europe (CiE) is the annual flagship conference of the Association CiE, a European association of mathematicians, logicians, computer scientists, philosophers, physicists, biologists, historians, and others interested in new developments in computability and their underlying significance for the real world. The Association CiE promotes the development of computability-related science, covering mathematics, computer science, and applications in various natural and engineering sciences, such as physics and biology, as well as related fields, such as philosophy and history of computing.

CiE 2019 had as its motto "Computing with Foresight and Industry" and was the 15th conference in the series. The conference was organized by Algorithms and Complexity in Durham (ACiD), a research group in the Department of Computer Science of Durham University. The 14 previous CiE conferences were held in Amsterdam (The Netherlands) in 2005, Swansea (Wales) in 2006, Siena (Italy) in 2007, Athens (Greece) in 2008, Heidelberg (Germany) in 2009, Ponta Delgada (Portugal) in 2010, Sofia (Bulgaria) in 2011, Cambridge (UK) in 2012, Milan (Italy) in 2013, Budapest (Hungary) in 2014, Bucharest (Romania) in 2015, Paris (France) in 2016, Turku (Finland) in 2017, and Kiel (Germany) in 2018. CiE 2020 will be held in Salerno (Italy).

The CiE conference is the largest annual international meeting focused on computability-theoretic issues. Its proceedings are published in the Springer series *Lecture Notes in Computer Science* and contain the best submitted papers, as well as extended abstracts of invited, tutorial, and special session speakers.

The CiE conference series is coordinated by the CiE Conference Series Steering Committee consisting of Alessandra Carbone (Paris), Gianluca Della Vedova (executive officer), Liesbeth De Mol (Lille), Mathieu Hoyrup (Nancy), Natasha Jonoska (Tampa FL), Benedikt Löwe (Amsterdam), Florin Manea (Kiel, chair), Klaus Meer (Cottbus), Mariya Soskova (Sofia), and ex-officio members Paola Bonizzoni (Milan, President of the Association CiE) and Dag Normann (Oslo).

The Program Committee of CiE 2019 was chaired by Daniël Paulusma (Durham University, UK) and Giuseppe Primiero (University of Milan, Italy) and consisted of 30 members. The committee selected the invited and tutorial speakers and the special session organizers, and reviewed the submitted papers.

The Program Committee invited four speakers to give plenary lectures at CiE 2019: Felipe Cucker (City University of Hong Kong, SAR China), Ursula Martin (University of Oxford, UK), Sonja Smets (University of Amsterdam, The Netherlands), and Linda Brown Westrick (Penn State, USA). The conference also had two plenary tutorials, presented by Markus Holzer (JLU Giessen) and Assia Mahboubi (University of Nantes).

In addition, the conference had six special sessions: Computational Neuroscience, History and Philosophy of Computing, Lowness Notions in Computability, Probabilistic Programming and Higher-Order Computation, Smoothed and Probabilistic Analysis of Algorithms, and Transfinite Computations. Speakers in these special sessions were selected by the respective special session organizers and were invited to contribute a paper to this volume.

Computational Neuroscience.
Organizers: Noura Al Moubayed and Jason Connolly
Speakers:
Ulrik Beierholm (Durham University)
Netta Cohen (Leeds University)
Evelyne Sernagor (Newcastle University)
V Anne Smith (University of St. Andrews)

History and Philosophy of Computing.
Organizers: Council of the HaPoC Commission
Speakers:
Tony Hoare (University of Oxford)
Michael Jackson (Open University)
Raymond Turner (University of Essex)

Lowness Notions in Computability.
Organizers: Johanna Franklin and Joseph S. Miller
Speakers:
Kenshi Miyabe (Meiji University)
Benoît Monin (LACL, Créteil University)
Keng Meng Selwyn Ng (Nanyang Technological University)
Don Stull (LORIA)

Probabilistic Programming and Higher-Order Computation
Organizer: Christine Tasson
Speakers:
Thomas Ehrhard (IRIF, Paris Diderot University)
Cameron Freer (MIT)
Joost-Pieter Katoen (RWTH Aachen)
Sam Staton (University of Oxford)

Smoothed and Probabilistic Analysis of Algorithms
Organizer: Bodo Manthey
Speakers:
Sophie Huiberts (CWI, Amsterdam)
Stefan Klootwijk (University of Twente)
Clemens Rösner (University of Bonn)
Sebastian Wild (University of Waterloo)

Transfinite Computations
Organizer: Sabrina Ouazzani
Speakers:
Merlin Carl (University of Konstanz)
Lorenzo Galeotti (University of Hamburg)
Benjamin Rin (Utrecht University)
Philip Welch (University of Bristol)

The members of the Program Committee of CiE 2019 selected for publication in this volume and for presentation at the conference 20 of the 35 non-invited submitted papers. Each paper received at least three reviews by the Program Committee and their subreviewers. In addition to the accepted contributed papers, this volume contains seven invited papers. The production of the volume would have been impossible without the diligent work of our expert referees, Program Committee members and additional reviewers alike. We would like to thank all of them for their excellent work.

All authors who contributed to this conference were encouraged to submit significantly extended versions of their papers, with additional unpublished research content, to *Computability. The Journal of the Association CiE.*

The Steering Committee of the conference series CiE is concerned about the representation of female researchers in the field of computability. In order to increase female participation, the series started the Women in Computability (WiC) program in 2007. In 2016, after the new constitution of the Association CiE allowed for the possibility of creating special interest groups, a Special Interest Group named Women in Computability was established. Since 2016, the WiC program has been sponsored by ACM's Women in Computing. This program includes a workshop, the annual WiC dinner, the mentorship program and a grant program for young female researchers. The Women in Computability workshop continued in 2019, coordinated by Liesbeth De Mol. In 2019, additional funding for the workshop was received from the L'Oréal foundation.

The organizers of CiE 2019 would like to thank the following entities for their financial support (in alphabetical order): the Association for Symbolic Logic (ASL), the Division of History of Science and Technology (DHST), the Division of Logic, Methodology and Philosophy of Science and Technology (DLMPST), Durham University, the European Association for Theoretical Computer Science (EATCS), the Commission for the History and Philosophy of Computing (HaPoC), the London Mathematical Society (LMS), and Springer. We would also like to acknowledge the support of our non-financial sponsor, the Association Computability in Europe.

We gratefully thank Matthew Johnson for his work in the Organizing Committee toward making the conference a successful event, and Eelkje Eppenga for designing the poster of CiE 2019.

We thank Andrej Voronkov for his EasyChair system, which facilitated the work of the Program Committee and the editors considerably.

May 2019   Florin Manea
                                                              Barnaby Martin
                                                             Daniël Paulusma
                                                           Giuseppe Primiero

Organization

Program Committee

Erika Ábrahám	RWTH Aachen University, Germany
Merlin Carl	University of Konstanz, Germany
Erzsébet Csuhaj-Varjú	Eötvös Loránd University, Hungary
Liesbeth De Mol	University of Lille, France
Valeria de Paiva	Nuance Communications, USA
Walter Dean	The University of Warwick, UK
Gianluca Della Vedova	University of Milano-Bicocca, Italy
Alessandra Di Pierro	University of Verona, Italy
Henning Fernau	University of Trier, Germany
Johanna Franklin	Hofstra University, USA
Paweł Gawrychowski	University of Wrocław, Poland
Pinar Heggernes	University of Bergen, Norway
Mathieu Hoyrup	LORIA, France
Peter Jonsson	Linköping University, Sweden
Dietrich Kuske	TU Ilmenau, Germany
Stéphane Le Roux	ENS Paris-Saclay, France
Benedikt Löwe	University of Amsterdam, The Netherlands
Florin Manea	Christian Albrechts University, Germany
Barnaby Martin	Durham University, UK
Elvira Mayordomo	University of Zaragoza, Spain
Klaus Meer	BTU Cottbus-Senftenberg, Germany
Dirk Nowotka	Christian Albrechts University, Germany
Daniël Paulusma	Durham University, UK
Arno Pauly	Swansea University, UK
Daniela Petrisan	Paris Diderot University, France
Giuseppe Primiero	University of Milan, Italy
Christine Tasson	Paris Diderot University, France
Hélène Touzet	University of Lille, France
Peter Van Emde Boas	University of Amsterdam, The Netherlands
Georg Zetzsche	Max Planck Institute for Software Systems, Germany

Organizing Committee

Matthew Johnson
Barnaby Martin
Daniël Paulusma (Chair)
Giuseppe Primiero

Additional Reviewers

Afshari, Bahareh
Barmpalias, George
Bernardinello, Luca
Bienvenu, Laurent
Bridges, Douglas
Bulín, Jakub
Campeanu, Cezar
Chella, Antonio
Csima, Barbara
Day, Joel
Downey, Rod
Engels, Christian
Fiorino, Guido
Gadouleau, Maximilien
Galeotti, Lorenzo
Gazdag, Zsolt
Giacobazzi, Roberto
Grigoreff, Serge
Hill, Darryl
Janicki, Philip
Kihara, Takayuki
Křivka, Zbyněk
Lemay, Aurélien

Lempp, Steffen
Leporati, Alberto
Maletti, Andreas
Masini, Andrea
Masopust, Tomas
Mélès, Baptiste
Melnikov, Alexander
Miller, Russell
Mousavi, Mohammad
Mummert, Carl
Nies, Andre
Patey, Ludovic
Potapov, Igor
San Mauro, Luca
Secco, Gisele
Solomon, Reed
Stephan, Frank
Tedre, Matti
Thiebaut, Jocelyn
Turetsky, Dan
Vaszil, György
Weiermann, Andreas
Welch, Philip

Contents

Recent Advances in the Computation of the Homology of Semialgebraic Sets

Felipe Cucker[(✉)]

Department of Mathematics, City University of Hong Kong,
Kowloon Tong, Hong Kong
macucker@cityu.edu.hk

Abstract. This article describes recent advances in the computation of the homology groups of semialgebraic sets. It summarizes a series of papers by the author and several coauthors (P. Bürgisser, T. Krick, P. Lairez, M. Shub, and J. Tonelli-Cueto) on which a sequence of ideas and techniques were deployed to tackle the problem at increasing levels of generality. The goal is not to provide a detailed technical picture but rather to throw light on the main features of this technical picture, the complexity results obtained, and how the new algorithms fit into the landscape of existing results.

Keywords: Semialgebraic sets · Homology groups · Weak complexity

1 Semialgebraic Sets

Simply put, a *semialgebraic set* is a subset of $\mathbb{R}^n$ which can be described as a Boolean combination of the solution sets of polynomial equalities and inequalities. But let's be more precise.

An *atomic* set is the solution set of one of the following five expressions

$$f(x) = 0, \quad f(x) \geq 0, \quad f(x) > 0, \quad f(x) \leq 0, \quad f(x) < 0, \tag{1}$$

where $f \in \mathbb{R}[X_1,\ldots,X_n]$. A Boolean combination of subsets of $\mathbb{R}^n$ is a set obtained by sequence of Boolean operations (unions, intersections and complements) starting from these subsets. For instance,

$$((A^c \cup B) \cap C)^c \cup (A \cap B) \tag{2}$$

(here S^c denotes the complement of S) is a Boolean combination of A, B, C.

There is a close relationship between the set-operations context of Boolean combinations and the syntaxis of formulas involving polynomials. Indeed, unions, intersections and complements of the sets defined in (1) correspond, respectively, to disjunctions, conjunctions and negations of the formulas defining these sets.

Partially supported by a GRF grant from the Research Grants Council of the Hong Kong SAR (project number CityU 11202017).

F. Manea et al. (Eds.): CiE 2019, LNCS 11558, pp. 1–12, 2019.
https://doi.org/10.1007/978-3-030-22996-2_1

If, for instance, A, B, C are given by $f(x) > 0$, $g(x) = 0$ and $h(x) \leq 0$ then the set in (2) is given by the formula

$$\neg\big((\neg(f(x) > 0) \vee g(x) = 0) \wedge h(x) \leq 0\big) \vee \big(f(x) > 0 \wedge g(x) = 0\big).$$

A formula is *monotone* if it does not contain negations. Any formula can be rewritten into an equivalent monotone formula. A formula is *purely conjunctive* if it contains neither negations nor disjunctions. It is therefore just a conjunction of atomic formulas. The associated semialgebraic subsets of $\mathbb{R}^n$ are said to be *basic*.

By definition, the class of semialgebraic sets is closed under unions, intersections and complements. It is much less obvious, but also true, that it is closed under projections. That is, if $\pi : \mathbb{R}^n \to \mathbb{R}^m$ is a projection map and $S \subset \mathbb{R}^n$ is semialgebraic then so is $\pi(S)$. It follows that if Ψ is a formula involving polynomials $f_1, \ldots, f_q$ in the variables $X_1, \ldots, X_n$ and $\overline{X}_1, \ldots, \overline{X}_\ell$ are (pairwise disjoint) subsets of these variables then the set of solutions of the formula

$$Q_1 \overline{X}_1 \, Q_2 \overline{X}_2 \ldots Q_\ell \overline{X}_\ell \, \Psi \qquad (3)$$

where $Q_i \in \{\exists, \forall\}$, is semialgebraic as well. Note that this is a subset of $\mathbb{R}^s$ where $s = n - (\overline{n}_1 + \ldots + \overline{n}_\ell)$ with $\overline{n}_i$ being the number of variables in $\overline{X}_i$. A well-known example is the following. Consider the set of points $(a, b, c, z) \in \mathbb{R}^4$ satisfying $az^2 + bz + c = 0$. This is, clearly, a semialgebraic subset of $\mathbb{R}^4$. Its projection onto the first three coordinates is the set

$$\exists z \, (az^2 + bz + c = 0)$$

which is the same as

$$\big(a \neq 0 \wedge (b^2 - 4ac \geq 0)\big) \vee \big(a = 0 \wedge b \neq 0\big) \vee \big(a = 0 \wedge b = 0 \wedge c = 0\big)$$

which is a semialgebraic subset of $\mathbb{R}^3$.

Some books with good expositions of semialgebraic sets are [6–8].

2   Some Computational Problems

Because of the pervasive presence of semialgebraic sets in all sorts of contexts, a variety of computational problems for these sets have been studied in the literature. In these problems, one or more semialgebraic sets are given as the data of the problem. The most common way to specify such a set $S \subseteq \mathbb{R}^n$ is by providing a tuple $f = (f_1, \ldots, f_q)$ of polynomials in $\mathbb{R}[X_1, \ldots, X_n]$ and a formula Ψ on these polynomials. We will denote by $\mathsf{S}(f, \Psi)$ the semialgebraic set defined by the pair (f, Ψ).

A few examples (but this list is by no means exhaustive) are the following.

Membership. Given (f, Ψ) and $x \in \mathbb{R}^n$, decide whether $x \in \mathsf{S}(f, \Psi)$.
Feasibility. Given (f, Ψ) decide whether $\mathsf{S}(f, \Psi) \neq \varnothing$.

Dimension. Given (f, Ψ) compute $\dim S(f, \Psi)$.

Counting. Given (f, Ψ) compute the cardinality $|S(f, \Psi)|$ if $\dim S(f, \Psi) = 0$ (and return ∞ if $\dim S(f, \Psi) > 0$ or 0 if $S(f, \Psi) = \varnothing$).

Connected Components. Given (f, Ψ) compute the number of connected components of $S(f, \Psi)$.

Euler Characteristic. Given (f, Ψ) compute the Euler characteristic $\chi(S(f, \Psi))$.

Homology. Given (f, Ψ) compute the homology groups of $S(f, \Psi)$.

Another example is the following.

Quantifier Elimination. Given (f, Ψ) and a quantifier prefix as in (3) compute polynomials g and a quantifier-free formula Ψ' such that the set of solutions of (3) is $S(g, \Psi')$.

In the particular case when the union of $\overline{X}_1, \ldots, \overline{X}_\ell$ is $\{X_1, \ldots, X_n\}$ (in other words, when (3) has no free variables) the formula (3) evaluates to either `True` or `False`. The problem above has thus a natural subproblem.

Decision of Tarski Algebra. Given (f, Ψ) and a quantifier prefix without free variables decide whether (3) is true.

A number of geometric problems can be stated as particular cases of the Decision of Tarski Algebra. For instance the Feasibility problem is so, as deciding whether $S(f, \Psi) \neq \varnothing$ is equivalent to decide the truth of $\exists x \, (x \in S(f, \Psi))$. It is not difficult to see that deciding whether $S(f, \Psi)$ is closed, or compact, or convex, etc. can all be expressed as particular cases of the Decision of Tarski Algebra.

3   Algorithms and Complexity

3.1   Symbolic Algorithms

In 1939 Tarski proved that the first-order theory of the reals was decidable (the publication [39] of this work was delayed by the war). His result was actually stronger; he gave an algorithm that solved the Quantifier elimination problem. At that time, interest was put on computability only. But two decades later, when focus had mostly shifted to complexity, it was observed that the cost of Tarski's procedure (that is, the number of arithmetic operations and comparisons performed) was enormous: a tower of exponentials. Motivated by this bound, Collins [16] and Wüthrich [43] independently devised an algorithm with a better complexity. Given $f = (f_1, \ldots, f_q)$ their algorithm computes a *Cylindrical Algebraic Decomposition* (CAD) associated to f. Once with this CAD at hand, all the problems mentioned in Sect. 2 can be solved (with a cost polynomial in the size of the CAD). The cost of computing the CAD of f is

$$(qD)^{2^{\mathcal{O}(n)}}$$

where D is the maximum of the degrees of the f_i. Whereas this doubly exponential bound is much better than the tower of exponentials for Tarski's algorithm, it is apparent that this approach may be inefficient. Indeed, one must compute a CAD before solving any of the problems above and one would expect that not all of them require the same computational effort.

In the late 1980s Grigoriev and Vorobjov introduced a breakthrough, the *Critical Points Method.* Using new algorithmic ideas they showed that the feasibility problem could be solved with cost $(qD)^{n^{\mathcal{O}(1)}}$ [24] and, more generally, that the truth of a quantifier-free formula as in (3) could be decided with cost $(qD)^{\mathcal{O}(n^{4\ell})}$ [23]. The new algorithmic ideas were further refined to obtain sharper bounds for both deciding emptiness [4,32] and eliminating quantifiers [4] but were also quickly applied to solve other problems such as, for instance, counting connected components [5,14,15,25,27], computing the dimension [28], the Euler-Poincaré characteristic [2], and the first few Betti numbers [3]. For all of these problems, singly exponential time algorithms were devised. But for the computation of homology groups only partial advances were achieved. The generic doubly exponential behavior of CAD remained the only choice.

All the algorithms mentioned above belong to what are commonly called *symbolic* algorithms. Although there is not an agreed upon definition for this class of algorithms, a characteristic feature of them is the use of exact arithmetic. Most of them consider input data to be arrays of integers (or rational numbers) and the size of these integers, as well as of all intermediate computations, is considered both to measure the input size and to define the cost of arithmetical operations. A formal model of such an algorithm is the Turing machine introduced by Turing in [40].

3.2 Numerical Algorithms

Shortly after the end of the war, Turing began working at the National Physics Laboratory in England. In his Turing Lecture [42] Wilkinson gives an account of this period. Central in this account is the story of how a linear system of 18 equations in 18 variables was solved using a computer and how, to understand the quality of the computed solution, Turing eventually came to write a paper [41] which is now considered the birth certificate of the modern approach to numerical algorithms. We will not enter into the details of this story (the reader is encouraged to read both Wilkinson's and Turing's papers). Suffice to say that the central issue here is that the underlying data (the coefficients of the linear system and the intermediate computations) are finite-precision numbers (such as the floating-point numbers dealt with by many computer languages). This implies that every operation—from the reading of the input data to each arithmetic operation—is affected by a small error. The problem is, the accumulation of these errors may end up on a very poor, or totally meaningless, approximation to the desired result. Turing realized that the magnitude of the final error depended on the quality of the input data, a quality that he measured with a *condition number.* This is a real number usually in the interval $[1, \infty]$. Data whose condition number is ∞ are said to be *ill-posed* and correspond to those inputs for which arbitrary small perturbations may have a qualitatively different behavior with respect to the considered problem. For instance, non-invertible matrices viz linear equation solving (as arbitrary small perturbations may change the system from infeasible (no solutions) to feasible). Condition numbers were (and still are) eventually used both for numerical stability analysis (their original role) and for complexity analysis. A comprehensive view of this condition-based approach to algorithmic analysis can be found in [9].

Although there is not an agreed upon definition for the notion of numerical algorithm, a characteristic feature of them is a design with numerical stability in mind. It is not necessary that such an analysis will be provided with the algorithm but some justification of its expected merits in this regard is common. A cost depending on a condition number is also common. Sometimes it is this feature what distinguishes between a numerical algorithm and a symbolic one described in terms of exact arithmetic with data from $\mathbb{R}$.

Numerical algorithms (in this sense) were first developed in the context of linear algebra. In the 1990s Renegar, in a series of influencial papers [33–35], introduced condition-based analysis in optimization while Shub and Smale were doing so for complex polynomial system solving [36]. Probably the first article to deal with semialgebraic sets was [19], where a condition number for the feasibility problem is defined and a numerical algorithm is exhibited solving the problem. Both the cost and the numerical stability of the algorithm were analyzed in terms of this condition number.

It is worth to describe the general idea of this algorithm. One first homogenizes all polynomials so that solutions are considered on the sphere $\mathbb{S}^n$ instead of on $\mathbb{R}^n$. Then one constructs a grid $\mathcal{G}$ of points on this sphere such that any point in $\mathbb{S}^n$ is at a distance at most $\eta = \frac{1}{2}$ of some point in $\mathcal{G}$. At each point x in the grid one checks two conditions. The first one, involving the computation of $g(x)$ and the derivative $\mathrm{D}g(x)$ (for some polynomial tuples g), if satisfied, shows the existence of a point in $\mathsf{S}(f, \Psi)$. The second, involving only the computation of $g(x)$, if satisfied, shows that there are no points of $\mathsf{S}(f, \Psi)$ at a distance at most η to x. If the first condition is satisfied at some $x \in \mathcal{G}$ the algorithm stops and returns `feasible`. If the second condition holds for all $x \in \mathcal{G}$ the algorithm stops and returns `infeasible`. If none of these two events occurs the algorithm replaces $\eta \leftarrow \frac{\eta}{2}$ and repeats. The number of iterations depends on the condition number of the data f and it is infinite, that is, the algorithm loops forever, when the input is ill-posed. The cost of each iteration is exponential in n (because so is the number of points in the grid) and, hence, the total cost of the algorithm is also exponential in n. From a pure complexity viewpoint, a comparison with the symbolic algorithms for feasibility mentioned in Sect. 3.1 is far from immediate. But the motivating reasoning behind this algorithm was not pure complexity. One observes that the symbolic algorithms have ultimately to do computations with matrices of exponential size. In contrast, the numerical algorithm performs an exponential number of computations with polynomial size matrices. *As these computations are independent from one another one can expect a much better behavior under the presence of finite precision.*

3.3 Probabilistic Analyses

A shortcoming of a cost analysis in terms of a condition number is the fact that, unlike the size of the data, condition numbers are not known a priori. The established way of dealing with this shortcoming is to endow the space of data with a probability measure and to eliminate the condition number from the complexity (or numerical stability) bounds in exchange for a probabilistic reliance. We can no longer bound how long will it take to do a computation. We can only have some expectation with a certain confidence. This approach was pioneered by von Neumann and Goldstine [22] (who introduced condition in [30] along with Turing) and was subsequently advocated by Demmel [20] and Smale [37].

The most extended probabilistic analysis in the analysis of algorithms is the *average-case*. Cost bounds of this form show a bound on the expected value of the cost (as opposed to its worst-case). A more recent form of probabilistic analysis is the *smoothed analysis* of Spielman and Teng. This attempts to extrapolate between worst-case and average-case by considering the worst case over the data a of the expectation of the quantity under study over random perturbations of a (see [38] for a vindication of this analysis). Even more recently a third approach, called *weak analysis*, was proposed in [1] by Amelunxen and Lotz. Its aim is to give a theoretical explanation of the efficiency in practice of numerical algorithms whose average complexity is too high. A paradigm of this situation is the power method to compute the leading eigenpair of a Hermitian matrix: this algorithm is very fast in practice, yet the average number of iterations it performs has been shown to be infinite [29]. Amelunxen and Lotz realized that here, as in many other problems, this disagreement between theory and practice is due to the presence of a vanishingly small (more precisely, having a measure exponentially small with respect to the input size) set of outliers, outside of which the algorithm can be shown to be efficient. They called *weak* any complexity bound holding outside such small set of outliers. One can prove, for instance, that the numerical algorithm for the feasibility problem in [19] has weak single exponential cost.

4   Computing the Homology of Semialgebraic Sets

It is against the background of the preceding sections that we can describe some recent progress in the computation of homology groups of semialgebraic sets. As the algorithms behind this progress are numerical a good starting point for its discussion is understanding ill-posedness and condition for the problem.

4.1 Ill-Posedness and Condition

Recall, our data is a pair (f, Ψ) where $f = (f_1, \ldots, f_q)$ is a tuple of polynomials. Consider, to begin with, the case where $q \leq n$ and Ψ corresponds to $f = 0$. This equality defines a real algebraic subset V of $\mathbb{R}^n$. It is well known that in general, if V is non-singular then sufficiently small perturbation in the coefficients of

f will define an algebraic set with the same topology as V. The one exception to this fact is the case where the projective closure of V intersects the hyperplane at infinity in a non-transversal manner. If V is singular, in contrast, arbitrary small perturbations can change its topology. The simplest instance of this occurs when considering a degree two polynomial in one variable. In this case, its zero set is singular if and only if it consists of a single point (which is a double zero). Arbitrarily small perturbation of the coefficients can change the discriminant of the polynomial to either negative (V becomes empty) or positive (the number of connected components of V becomes two). A similar behavior can be observed when the intersection of V with the hyperplane at infinity is singular. The simplest example is now given by a degree two polynomial in two variables. The intersection above is singular if and only if V is a parabola. Again, arbitrarily small perturbations may change V into a hyperbola (and hence increase the number of connected components) or into an ellipse (which, unlike the parabola, is not simply connected).

We conclude that f is ill-posed when either its zero set in $\mathbb{R}^n$ has a singularity or the zero set of (f^h, X_0) in $\mathbb{P}^n$ has one. Here f^h is the homogenization of f with respect to the variable X_0 and $\mathbb{P}^n$ is n-dimensional projective space. This is equivalent to say that the zero set of either f^h or (f^h, X_0) has a singular point in $\mathbb{P}^n$.

Exporting this understanding from algebraic sets to closed semialgebraic sets is not difficult. One only needs to notice that when a closed semialgebraic set (given by a lax formula Ψ) has a change in topology due to a perturbation, there is a change of topology in its boundary as well. As this boundary is the union of (parts of) zero sets of subsets of f it follows that we can define the *set of ill-posed* tuples f to be

$$\Sigma := \{f \in \mathbb{R}[X_1, \ldots, X_n] \mid \exists J \subset \{0, 1, \ldots, q\}, |J| \leq n+1, \exists \xi \in \mathbb{P}^n$$
$$(f_j^h(\xi) = 0 \text{ for } j \in J \text{ and } \operatorname{rank} D_\xi f_J^h < |J|)\}$$

where $f_0^h := X_0$ and $f_J^h = (f_j^h)_{j \in J}$. One can then show that a tuple f is in Σ if and only if for some lax formula Ψ there exist arbitrarily small perturbations of f changing the topology of $\mathsf{S}(f, \Psi)$.

In [12] a condition number $\kappa_{\mathrm{aff}}(f)$ is defined that has several desirable features. Notably among them, the inequality (see [12, Thm. 7.3])

$$\kappa_{\mathrm{aff}}(f) \leq 4D \frac{\|f\|}{d(f, \Sigma)}$$

where $\| \ \|$ is the Weyl norm and d is its induced distance, and the following result [12, Prop. 2.2].

Proposition 1. *There is an algorithm* κ-ESTIMATE *that, given* $f \in \mathbb{R}[X_1, \ldots, X_n]$, *returns a number* K *such that*

$$0.99\kappa_{\mathrm{aff}}(f) \leq \mathsf{K} \leq \kappa_{\mathrm{aff}}(f)$$

if $\kappa_{\text{aff}}(f) < \infty$ *or loops forever otherwise. The cost of this algorithm is bounded by* $(qnD\kappa_{\text{aff}}(f))^{\mathcal{O}(n)}$. □

The condition number κ_{aff} is the last incarnation in a sequence of condition numbers that were defined for a corresponding sequence of problems: the Turing condition number for linear equation solving, the quantity $\mu_{\text{norm}}(f)$ defined by Shub and Smale [36] for complex polynomial equation solving, the value $\kappa(f)$ defined for the problem of real equation solving [17], and the condition number $\kappa^*(f;g)$ for the problem of computing the homology of *basic* semialgebraic sets.

4.2   Main Result

We can finally describe the main result in this overview. To do so, we first define a few notions we have been vague about till now.

Let $\boldsymbol{d} := (d_1, \dots, d_q)$ be a degree pattern and $\mathcal{P}_{\boldsymbol{d}}[q]$ be the linear space of polynomial tuples $f = (f_1, \dots, f_q)$ with $f_i \in \mathbb{R}[X_1, \dots, X_n]$ of degree at most d_i. We endow $\mathcal{P}_{\boldsymbol{d}}[q]$ with the Weyl inner product (see [9, §16.1] for details) and its induced norm and distance. This norm allows us to further endow $\mathcal{P}_{\boldsymbol{d}}[q]$ with the standard Gaussian measure with density

$$\rho(f) = \frac{1}{(2\pi)^{\frac{N}{2}}} e^{-\frac{\|f\|^2}{2}}.$$

Here $N := \dim \mathcal{P}_{\boldsymbol{d}}[q]$. Note that, as the condition number $\kappa_{\text{aff}}(f)$ is scale invariant, i.e., it satisfies $\kappa_{\text{aff}}(f) = \kappa_{\text{aff}}(\lambda f)$ for all real $\lambda \neq 0$, the tail $\text{Prob}\{\kappa_{\text{aff}}(f) \geq t\}$ has the same value when f is drawn from the Gaussian above and when it is drawn from the uniform distribution on the sphere $\mathbb{S}(\mathcal{P}_{\boldsymbol{d}}[q]) = \mathbb{S}^{N-1}$.

Theorem 1. *We exhibit a numerical algorithm* HOMOLOGY *that, given a tuple* $f \in \mathcal{P}_{\boldsymbol{d}}[q]$ *and a Boolean formula* Φ *over* p, *computes the homology groups of* $\mathsf{S}(f, \Phi)$. *The cost of* Homology *on input* (f, Φ), *denoted* $\mathsf{cost}(f, \Phi)$, *that is, the number of arithmetic operations and comparisons in* $\mathbb{R}$, *satisfies:*

$$\mathsf{cost}(f, \Phi) \leq \mathsf{size}(\Phi) q^{\mathcal{O}(n)} (nD\kappa_{\text{aff}}(p))^{\mathcal{O}(n^2)}.$$

Furthermore, if f *is drawn from the Gaussian distribution on* $\mathcal{P}_{\boldsymbol{d}}[q]$ *(or, equivalently, from the uniform distribution on* $\mathbb{S}^{N-1}$*), then*

$$\mathsf{cost}(f, \Phi) \leq \mathsf{size}(\Phi) q^{\mathcal{O}(n)} (nD)^{\mathcal{O}(n^3)}$$

with probability at least $1 - (nqD)^{-n}$. *The algorithm is numerically stable.* □

The algorithm HOMOLOGY uses the same broad idea of the algorithm in [19] for deciding feasibility: it performs some simple computations on the (exponentially many) points of a grid in the sphere $\mathbb{S}^n$. The mesh of this grid, determining how many points it contains, is a function of $\kappa_{\text{aff}}(f)$.

Summarizing to the extreme, the algorithm proceeds as follows. Firstly, we homogenize the data. That is, we pass from $f = (f_1, \ldots, f_q)$ to $f^h := (f_0^h, f_1^h, f_q^h)$ (recall, $f_0^h = X_0$) and from Ψ to $\Psi^h := \Psi \wedge (f_0^h > 0)$. Let $S_\mathbb{S}(f^h, \Psi^h)$ denote the solutions in $\mathbb{S}^n$ for the pair (f^h, Ψ^h). We then have the isomorphism of homology

$$H_*\big(S(f, \Psi)\big) \simeq H_*\big(S_\mathbb{S}(f^h, \Psi^h)\big).$$

Secondly, we use κ-ESTIMATE (recall Proposition 1) to approximate $\kappa_{\mathrm{aff}}(f)$. With this approximation at hand we construct a sufficiently fine grid $\mathcal{G}$ on the sphere. In the basic case, that is when Ψ is purely conjunctive, a subset $\mathcal{X} \subset \mathcal{G}$ and a real ε are constructed such that we have the following isomorphism of homology

$$H_*\big(S_\mathbb{S}(f^h, \Psi^h)\big) \simeq H_*\big(\cup_{x \in \mathcal{X}} B(x, \varepsilon)\big) \tag{4}$$

where $B(x, \varepsilon)$ is the open ball centered at x with radius ε. The Nerve Theorem [26, Corollary 4G.3] then ensures that

$$H_*\big(\cup_{x \in \mathcal{X}} B(x, \varepsilon)\big) \simeq H_*(\mathfrak{C}_\varepsilon(\mathcal{X})) \tag{5}$$

where $\mathfrak{C}_\varepsilon(\mathcal{X})$ is the Čech complex associated to the pair $(\mathcal{X}, \varepsilon)$. And computing the homology of a Čech complex is a problem with known algorithms. The cost analysis of the resulting procedure in terms of $\kappa_{\mathrm{aff}}(f)$—the first bound in Theorem 1—easily follows. The second bound in this theorem relies on a bound for the tail $\mathrm{Prob}\{\kappa_{\mathrm{aff}}(f) \geq t\}$ which follows from the geometry of Σ (its degree as an algebraic set) and the main result in [11].

This is the roadmap followed in [10]. A key ingredient in this roadmap is a bound of the radius of injectivity (or reach) of an algebraic manifold in terms of its condition number obtained in [18]. This bound allows one to get a constructive handle on a result by Niyogi, Smale and Weinberger [31, Prop. 7.1] leading to (4) and, more generally, it allowed to obtain a version of Theorem 1 in [18] for the case of smooth projective sets.

To obtain a pair $(\mathcal{X}, \varepsilon)$ satisfying (4) when Ψ is not purely conjunctive appears to be difficult. The equivalence in (4) follows from showing that, for a sufficiently small $s > 0$, the nearest-point map from the tube around $S_\mathbb{S}(f^h, \Psi^h)$ of radius s onto $S_\mathbb{S}(f^h, \Psi^h)$ is a retraction. This is no longer true if Ψ is not purely conjunctive.

In the case that Ψ is lax the problem was thus approached differently [12]. For every $\propto \in \{\leq, =, \geq\}$ and every $i = 0, \ldots, q$, let $\mathcal{X}_i^\propto \subset \mathcal{G}$ and ε be as above, so that (4) holds with Ψ^h the atomic formula $f_i^h \propto 0$. To each pair $(\mathcal{X}_i^\propto, \varepsilon)$ we associate the Čech complex $\mathfrak{C}_\varepsilon(\mathcal{X}_i^\propto)$. Then, the set $S_\mathbb{S}(f^h, \Psi^h)$ and the complex

$$\Psi\big(\mathfrak{C}_\varepsilon(\mathcal{X}_0^{\leq}), \mathfrak{C}_\varepsilon(\mathcal{X}_0^{=}), \mathfrak{C}_\varepsilon(\mathcal{X}_0^{\geq}), \ldots, \mathfrak{C}_\varepsilon(\mathcal{X}_q^{\leq}), \mathfrak{C}_\varepsilon(\mathcal{X}_q^{=}), \mathfrak{C}_\varepsilon(\mathcal{X}_q^{\geq}))\big)$$

have the same homology groups. This complex is recursively built from the complexes $\mathfrak{C}_\varepsilon(\mathcal{X}_i^\propto)$ in the same manner (i.e., using the same sequence of Boolean operations) that $S_\mathbb{S}(f^h, \Psi^h)$ is built from the atomic sets $f_i^h(x) \propto 0$. One can then algorithmically proceed as in [10]. The proof of this homological equivalence is far from elementary; it relies on an inclusion-exclusion version of the Mayer-Vietoris Theorem and on the use of Thom's first isotopy lemma on a convenient Whitney stratification of $S_S(f^h, \Psi^h)$.

The extension to arbitrary (i.e., not necessarily lax) formulas has been recently done in [13]. It reduces this case to that of lax formulas through a construction of Gabrielov and Vorobjov [21]. This construction, however, was purely qualitative: the existence of a finite sequence of real numbers satisfying a desirable property was established, but no procedure to compute this sequence was given. The core of [13] consists of a quantitative version of Gabrielov and Vorobjov construction. Maybe not surprisingly, a distinguished role in this result is again played by the condition of the data: indeed, this sequence can be taken to be any sequence provided its largest element is less than $\frac{1}{\sqrt{2}\,\kappa_{\mathrm{aff}}(f)}$.

4.3 Final Remarks

It is worth to conclude with some caveats about what exactly are the merits of Theorem 1. This can only be done by comparing this result with the computation of homology groups via a CAD described in Sect. 3.1, a comparison which is delicate as these algorithms are birds of different feather.

A first remark here is that, in the presence of finite precision, the CAD will behave appallingly whereas HOMOLOGY is likely do behave much better. With this in mind, let us assume that arithmetic is infinite-precision and focus in this case.

The main virtue of HOMOLOGY is the fact that, outside a set of exponentially small measure, it has a single exponential cost. Hence, outside this negligibly small set, it runs exponentially faster than CAD. Inside this set of data, however, it can take longer than CAD and will even loop forever for ill-posed data. But this shortcoming has an easy solution: one can run "in parallel" both HOMOLOGY and CAD and halt whenever one of them halts. This procedure results in a weak singly exponential cost with a doubly exponential worst-case.

Once said that, these considerations are of a theoretical nature. In practice, the quality of the bounds is such that only "small" data (i.e., polynomials in just a few variables) can be considered. In this case the difference between the two cost bounds depends much on the constants hidden in the big $\mathcal{O}$ notation.

References

1. Amelunxen, D., Lotz, M.: Average-case complexity without the black swans. J. Complexity **41**, 82–101 (2017)
2. Basu, S.: On bounding the Betti numbers and computing the Euler characteristic of semi-algebraic sets. In: Proceedings of the Twenty-eighth Annual ACM Symposium on the Theory of Computing (Philadelphia, PA, 1996), pp. 408–417. ACM, New York (1996)
3. Basu, S.: Computing the first few Betti numbers of semi-algebraic sets in single exponential time. J. Symbolic Comput. **41**(10), 1125–1154 (2006)

4. Basu, S., Pollack, R., Roy, M.F.: On the combinatorial and algebraic complexity of quantifier elimination. J. ACM **43**, 1002–1045 (1996)
5. Basu, S., Pollack, R., Roy, M.F.: Computing roadmaps of semi-algebraic sets on a variety. J. AMS **33**, 55–82 (1999)
6. Basu, S., Pollack, R., Roy, M.F.: Algorithms in real algebraic geometry. Algorithms and Computation in Mathematics, vol. 10, 2nd edn. Springer, Berlin (2006)
7. Benedetti, R., Risler, J.J.: Real Algebraic and Semi-algebraic Sets, Hermann (1990)
8. Bochnak, J., Coste, M., Roy, M.F.: Real Algebraic Geometry. Ergebnisse der Mathematik und ihrer Grenzgebiete. 3. Folge/A Series of Modern Surveys in Mathematics, vol. 36. Springer, Heidelberg (1998). https://doi.org/10.1007/978-3-662-03718-8. Translated from the 1987 French original, Revised by the authors
9. Bürgisser, P., Cucker, F.: Condition. Grundlehren der mathematischen Wissenschaften, vol. 349. Springer, Heidelberg (2013). https://doi.org/10.1007/978-3-642-38896-5
10. Bürgisser, P., Cucker, F., Lairez, P.: Computing the homology of basic semialgebraic sets in weakly exponential time. J. ACM **66**(1), 5:1–5:30 (2019)
11. Bürgisser, P., Cucker, F., Lotz, M.: The probability that a slightly perturbed numerical analysis problem is difficult. Math. Comput. **77**, 1559–1583 (2008)
12. Bürgisser, P., Cucker, F., Tonelli-Cueto, J.: Computing the homology of semialgebraic sets. I: Lax formulas. Found. Comput. Math. arXiv:1807.06435 (2018)
13. Bürgisser, P., Cucker, F., Tonelli-Cueto, J.: Computing the homology of semialgebraic sets. II: Arbitrary formulas (2019). preprint
14. Canny, J.: Computing roadmaps of general semi-algebraic sets. Comput. J. **36**(5), 504–514 (1993)
15. Canny, J., Grigorev, D., Vorobjov, N.: Finding connected components of a semialgebraic set in subexponential time. Appl. Algebra Eng. Commun. Comput. **2**(4), 217–238 (1992)
16. Collins, G.E.: Quantifier elimination for real closed fields by cylindrical algebraic decompostion. In: Brakhage, H. (ed.) GI-Fachtagung 1975. LNCS, vol. 33, pp. 134–183. Springer, Heidelberg (1975). https://doi.org/10.1007/3-540-07407-4_17
17. Cucker, F.: Approximate zeros and condition numbers. J. Complexity **15**, 214–226 (1999)
18. Cucker, F., Krick, T., Shub, M.: Computing the homology of real projective sets. Found. Comp. Math. **18**, 929–970 (2018)
19. Cucker, F., Smale, S.: Complexity estimates depending on condition and round-off error. J. ACM **46**, 113–184 (1999)
20. Demmel, J.: The probability that a numerical analysis problem is difficult. Math. Comput. **50**, 449–480 (1988)
21. Gabrielov, A., Vorobjov, N.: Approximation of definable sets by compact families, and upper bounds on homotopy and homology. J. Lond. Math. Soc. **80**(1), 35–54 (2009)
22. Goldstine, H., von Neumann, J.: Numerical inverting matrices of high order, II. Proc. AMS **2**, 188–202 (1951)
23. Grigoriev, D.: Complexity of deciding Tarski algebra. J. Symbolic Comput. **5**, 65–108 (1988)
24. Grigoriev, D., Vorobjov, N.: Solving systems of polynomial inequalities in subexponential time. J. Symbolic Comput. **5**, 37–64 (1988)
25. Grigoriev, D., Vorobjov, N.: Counting connected components of a semialgebraic set in subexponential time. Comput. Complexity **2**, 133–186 (1992)
26. Hatcher, A.: Algebraic Topology. Cambridge University Press, Cambridge (2002)

27. Heintz, J., Roy, M.F., Solerno, P.: Single exponential path finding in semi-algebraic sets II: the general case. In: Bajaj, C. (ed.) Algebraic Geometry and its Applications, pp. 449–465. Springer, New York (1994). https://doi.org/10.1007/978-1-4612-2628-4_28

28. Koiran, P.: The real dimension problem is $NP_{\mathbb{R}}$-complete. J. Complexity **15**, 227–238 (1999)

29. Kostlan, E.: Complexity theory of numerical linear algebra. J. Comput. Appl. Math. **22**, 219–230 (1988)

30. von Neumann, J., Goldstine, H.: Numerical inverting matrices of high order. Bull. AMS **53**, 1021–1099 (1947)

31. Niyogi, P., Smale, S., Weinberger, S.: Finding the homology of submanifolds with high confidence from random samples. Discrete Comput. Geom. **39**, 419–441 (2008)

32. Renegar, J.: On the computational complexity and geometry of the first-order theory of the reals. Part I. J. Symbolic Comput. **13**, 255–299 (1992)

33. Renegar, J.: Some perturbation theory for linear programming. Math. Program. **65**, 73–91 (1994)

34. Renegar, J.: Incorporating condition measures into the complexity theory of linear programming. SIAM J. Optim. **5**, 506–524 (1995)

35. Renegar, J.: Linear programming, complexity theory and elementary functional analysis. Math. Program. **70**, 279–351 (1995)

36. Shub, M., Smale, S.: Complexity of Bézout's theorem I: geometric aspects. J. AMS **6**, 459–501 (1993)

37. Smale, S.: Complexity theory and numerical analysis. In: Iserles, A. (ed.) Acta Numerica, pp. 523–551. Cambridge University Press (1997)

38. Spielman, D., Teng, S.H.: Smoothed analysis: an attempt to explain the behavior of algorithms in practice. Commun. ACM **52**(10), 77–84 (2009)

39. Tarski, A.: A Decision Method for Elementary Algebra and Geometry. University of California Press, Berkeley (1951)

40. Turing, A.: On computable numbers, with an application to the Entscheidungsproblem. Proc. London Math. Soc. **S2–42**, 230–265 (1936)

41. Turing, A.: Rounding-off errors in matrix processes. Quart. J. Mech. Appl. Math. **1**, 287–308 (1948)

42. Wilkinson, J.: Some comments from a numerical analyst. J. ACM **18**, 137–147 (1971)

43. Wüthrich, H.R.: Ein Entscheidungsverfahren für die Theorie der reell-abgeschlossenen Körper. In: Strassen, V., Specker, E. (eds.) Komplexität von Entscheidungsproblemen Ein Seminar. LNCS, vol. 43, pp. 138–162. Springer, Heidelberg (1976). https://doi.org/10.1007/3-540-07805-3_10

Surreal Blum-Shub-Smale Machines

Lorenzo Galeotti[1,2,3]([⊠])

[1] Fachbereich Mathematik, Universität Hamburg,
Bundesstraße 55, 20146 Hamburg, Germany
lorenzo.galeotti@gmail.com
[2] Institute for Logic, Language and Computation, Universiteit van Amsterdam,
Postbus 94242, 1090 GE Amsterdam, The Netherlands
[3] Amsterdam University College,
Postbus 94160, 1090 GD Amsterdam, The Netherlands

Abstract. Blum-Shub-Smale machines are a classical model of computability over the real line. In [9], Koepke and Seyfferth generalised Blum-Shub-Smale machines to a transfinite model of computability by allowing them to run for a transfinite amount of time. The model of Koepke and Seyfferth is asymmetric in the following sense: while their machines can run for a transfinite number of steps, they use real numbers rather than their transfinite analogues. In this paper we will use the surreal numbers in order to define a generalisation of Blum-Shub-Smale machines in which both time and register content are transfinite.

1 Introduction

In 1989 Blum, Shub and Smale introduced a model of computation to study computability over rings; see [1]. Of particular interest for us is the notion of computability that Blum-Shub-Smale machines (BSSM) induce over the real numbers. A BSSM for the real numbers is a register based machine in which each register contains a real number. A program for such a machine is a finite list of commands. Each command can be either a computation or branch command. The execution of a computation command allows the machine to apply a rational function to update the content of the registers. A branch command, on the other hand, leaves the content of the registers unchanged and allows the machine to apply a rational function to some register and execute a jump based on the result of this operation, i.e., to jump to a different point of the code if the result is 0 and to continue the normal execution otherwise.

In [9,11], Koepke and Seyfferth defined the notion of *infinite time Blum-Shub-Smale machine* that is a generalised version of Blum-Shub-Smale machines which can carry out transfinite computations over the real numbers.

Infinite time Blum-Shub-Smale machines work essentially as standard BSSMs at successor times apart from the fact that, contrary to classical BSSMs, they can only apply rational functions with rational coefficients[1]. At limit stages an

---

[1] A stronger version of infinite time Blum-Shub-Smale machines could be obtained by allowing infinite time Blum-Shub-Smale machines to use rational functions with real coefficients, but this was not done in [11].

© Springer Nature Switzerland AG 2019
F. Manea et al. (Eds.): CiE 2019, LNCS 11558, pp. 13–24, 2019.
https://doi.org/10.1007/978-3-030-22996-2_2

infinite time Blum-Shub-Smale machine computes the content of each register by taking the limit over the real line of the values that the register assumed at previous stages (if this exists); and updates the program counter to the inferior limit of its values at previous stages. The theory of infinite time Blum-Shub-Smale machine was further studied in [7].

Infinite time Blum-Shub-Smale machines provide an asymmetric generalisation of BSSMs. In particular, while infinite time Blum-Shub-Smale machines are allowed to run for arbitrary transfinite time, they are using real numbers, a set that can be very small compared to the running times. It is then natural to ask whether a symmetric notion can be defined.

The first problem in doing so is, as usual in this context, that of finding a suitable structure which one can use in place of the real line in the generalised context. As we will see, the surreal numbers, a very general number system which contains both real and ordinal numbers, will provide a natural framework to develop this generalised theory.

In this paper, we will introduce a generalised version of Blum-Shub-Smale machines based on surreal numbers and we will show some preliminary results of the theory of these machines.

2   Surreal Numbers

The surreal numbers were introduced by Conway in order to give a mathematical definition of the abstract notion of "number". In this section we will present basic results on surreal numbers; see [2,5] for a detailed introduction. A *surreal number* is a function from an ordinal α to $\{+, -\}$, i.e., a sequence of pluses and minuses of ordinal length. We denote the class of surreal numbers by No. The *length* of a surreal number x (i.e., its domain) is denoted by $\ell(x)$.

For surreal numbers x and y, we define $x < y$ if there exists α such that $x(\beta) = y(\beta)$ for all $\beta < \alpha$, and $x(\alpha) = -$ and either $\alpha = \ell(y)$ or $y(\alpha) = +$, or $\alpha = \ell(x)$ and $y(\alpha) = +$.

In Conway's original construction, every surreal number is generated by filling some gap in the previously generated numbers. The following theorem connects this intuition to the surreal numbers as we have defined them. First, given sets of surreal numbers X and Y, we write $X < Y$ if for all $x \in X$ and $y \in Y$ we have $x < y$.

Theorem 1 (Simplicity theorem). *If L and R are two sets of surreal numbers such that $L < R$, then there is a unique surreal x of minimal length such that $L < \{x\} < R$, denoted by $[L|R]$. Furthermore, for every $x \in$ No we have $x = [L_x|R_x]$ for $L_x = \{y \in$ No $;\ x > y \wedge y \subset x\}$ and $R_x = \{y \in$ No $;\ x < y \wedge y \subset x\}$.*

Given two subsets L and R of surreal numbers such that $L < R$, we will call the pair (L, R) a *cut*. For any surreal number $x \in$ No we define the *canonical representation* of x as the cut (L_x, R_x).

Using the simplicity theorem Conway defined the field operations $+_s$, $\cdot_s$, $-_s$, and the multiplicative inverse over No and proved that these operations satisfy the axioms of real closed fields; see [2,5]. Moreover, Ehrlich proved that No is the universal class real closed field in the sense that every real closed field is isomorphic to a subfield of No; cf., e.g., [3].

3   Generalising Infinite Time BSSMs

The main goal of this paper is to introduce a notion of register machine which generalises infinite time Blum-Shub-Smale machines in order to allow them to work with arbitrary transfinite space. To do so we want to make our machines able to perform transfinite computations over surreal numbers.

A very natural approach to this problem would be to allow infinite time Blum-Shub-Smale machine registers to store surreal numbers keeping the behaviour of the machine the same. This means that at successor stages the machine will still be allowed to either branch or apply a rational function with surreal coefficients to the registers. At limit stages the machine would have then to use limits[2] over No to compute the contents of the registers. Unfortunately this approach does not work: one can show that if a totally ordered field[3] K has cofinality $\kappa > \omega$, then every non-eventually constant sequence of length $<\kappa$ diverges in K. Similarly for surreal numbers we have the following result:

Lemma 2 (Folklore). *For every ordinal α, every non-eventually constant sequence of length α of surreal numbers diverges.*

The previous result tells us that the classical notion of limit is not the right notion in the context of transfinite computability over a field. Note that the phenomenon of diverging sequences is not a special feature of the surreal numbers, but follows from the basic theory of ordered fields. Thus, any generalisation of the theory of BSSMs to a non-archimedean field would need to deal with this issue.

4   Surreal Blum-Shub-Smale Machines

A surreal Blum-Shub-Smale machine (SBSSM) is a register machine. Since, as we will see, the formal definition of SBSSMs is quite involved, let us start by giving a brief informal explanation of how they work. There are two different types of registers in our machines: *normal registers* and *Dedekind registers*. Normal registers are just registers that contain surreal numbers; as we will see, the machine can write and read normally from these registers. Dedekind registers on the other hand are a new piece of hardware. Each Dedekind register R can be thought of as to have three different components S^{L}, S^{R}, and R. The components

---

[2] By this we mean the notion of limit coming from the order topology over No.

[3] A totally ordered field is a field together with a total order $\leq$ such that for all x, y, and z, we have that if $x \leq y$, then $x + z \leq y + z$, and if x and y are positive, then $x \cdot y$ is positive. The cofinality of an ordered field is the least cardinal λ such that there is a sequence of length λ cofinal in the field.

S^L and S^R called left and right stack of R, respectively, can be thought of as two possibly infinite stacks of surreal numbers. The last component R of the register can be thought as a normal register whose content is automatically updated by the machine to the surreal $[S^L|S^R]$. Note that it could be that $S^L \not< S^R$; in this case we will assume that the machine crashes.

A SBSSM is just a finite set of normal and Dedekind registers. A program for such a machine will be a finite linear sequence of commands. As for BSSMs there are two types of commands:

Computation: the machine can apply a rational function to a normal register or to a Dedekind register and save the result in a register (either Dedekind or normal) or in a stack.

Branch: the machine can check if the content of a normal register or of a Dedekind register is bigger than 0 and perform a jump based on the result.

In each program we should specify two subsets of the set of normal registers; one that will contain the input of the program, and the other that will contain the output of the program.

A surreal Blum-Shub-Smale machine will behave as follows: at successor stages our machine just executes the current command and updates content of stacks, registers, and program counter accordingly. At limit stage α, the program counter is set using $\liminf$ as for infinite time Blum-Shub-Smale machines; the content of each normal register is updated as follows: if the content of the register is eventually constant with value x, then we set the value of the register to x; otherwise we set it to 0. For Dedekind registers we proceed as follows: if from some point on the content of the stacks is constant, we leave the content of the stacks, and therefore the content of the register, unchanged. If the content of the stacks is not eventually constant but from some point $\beta < \alpha$ on there is no computation instruction whose result is saved in the register, then we set the value of each stack to the union of its values from β on, and we set the content of the register accordingly. If none of the previous cases occurs, then we set the content of the register to 0 and empty the stacks.

We are now ready to give a formal definition of surreal Blum-Shub-Smale machine.

Given two polynomials $p, q \in \mathrm{No}[X_0, \ldots, X_n]$, we will call $\frac{p(X_0, \ldots, X_n)}{q(X_0, \ldots, X_n)}$ a *formal polynomial quotient* over No in $n + 1$ variables.

Let $n \in \mathbb{N}$ and $F : \mathrm{No}^{n+1} \to \mathrm{No}$ be a partial class function. Then, we say that F is a *rational map* over No if there are polynomials in $n + 1$ variables $p, q \in \mathrm{No}[X_0, \ldots, X_n]$ such that $F(s_0, \ldots, s_n) = \frac{p(s_0, \ldots, s_n)}{q(s_0, \ldots, s_n)}$ for each $s_0, \ldots, s_n \in \mathrm{No}$. In this case, we will say that $\frac{p(X_0, \ldots, X_n)}{q(X_0, \ldots, X_n)}$ is a formal polynomial quotient *defining* F.

Denote by $\overrightarrow{X}$ the set of finite tuples of variables of any length. Then, we will denote by $\mathrm{No}(\overrightarrow{X})$ the class of formal polynomials quotients over No in any number of variables. Given a subclass K of $\mathrm{No}(\overrightarrow{X})$ and a partial class function $F : \mathrm{No}^{m+1} \to \mathrm{No}$ with $m \in \mathbb{N}$, we will say that F is in the class K, in symbols $F \in K$, if there is a formal polynomial quotient in K defining F. Finally, given a subclass K of No we will denote by $K(\overrightarrow{X})$ the the class of *formal polynomial quotients with coefficients* in K.

Definition 3. *Let $\mathfrak{N}$ and $\mathfrak{D}$ be two disjoint sets of natural numbers, I and O be two disjoint subsets of $\mathfrak{N}$, and K be a subclass of* No. *A $(\mathfrak{N}, \mathfrak{D}, I, O, K)$-SBSSM program P is a finite sequence $(C_0, \ldots, C_n)$ with $n \in \mathbb{N}$ such that for every $0 \le m \le n$ the command C_m is of one of the following types:*

1. *Computation $R_i := f(R_{j_0}, \ldots, R_{j_m})$ were $f : \text{No}^{n+1} \to \text{No}$ is a map in $K(\overrightarrow{X})$ and $i \in (\mathfrak{N} \setminus I) \cup \mathfrak{D}$ and $j_0, \ldots, j_m \in \mathfrak{N} \cup \mathfrak{D}$.*
2. *Stack Computation $\text{Push}_d(R_i, R_j)$ were $i \in \mathfrak{D}$, $j \in \mathfrak{N} \cup \mathfrak{D}$ and $d \in \{L, R\}$.*
3. *Branch if R_i then j were $i \in \mathfrak{N} \cup \mathfrak{D}$ and $j \le n$.*

The sets $\mathfrak{N}$ and $\mathfrak{D}$ are the sets of normal and Dedekind registers of our program, respectively; and, I and O are the sets of input and output registers, respectively. When the registers are irrelevant for the argument we will omit, $\mathfrak{N}$, $\mathfrak{D}$, I, and O and call P a $K(\overrightarrow{X})$-SBSSM program.

Definition 4. *Let $\mathfrak{N}$ and $\mathfrak{D}$ be two disjoint sets of natural numbers, the sets $I = (i_0, \ldots, i_m)$ and $O = (i_0, \ldots, i_{m'})$ be two disjoint subsets of $\mathfrak{N}$, K be a subclass of* No, *and $P = (C_0, \ldots, C_n)$ be a $(\mathfrak{N}, \mathfrak{D}, I, O, K)$-SBSSM program. Given $x \in \text{No}^{m+1}$ the SBSSM computation of P with input x is the transfinite sequence[4]*

$$(R^{\text{N}}(t), S^{\text{L}}(t), S^{\text{R}}(t), \text{PC}(t))_{t \in \theta} \in (\text{No}^{\mathfrak{N}} \times \wp(\text{No})^{\mathfrak{D}} \times \wp(\text{No})^{\mathfrak{D}} \times \omega)^{\theta}$$

where

1. *θ is a successor ordinal or $\theta = \text{On}$;*
2. *$\text{PC}(0) = 0$;*
3. *$R^{\text{N}}(0)(i_j) = x(j)$ if $i_j \in I$ and $R^{\text{N}}(0)(i) = 0$ otherwise;*
4. *for all $i \in \mathfrak{D}$ we have $S^{\text{L}}(0)(i) = S^{\text{R}}(0)(i) = \emptyset$;*
5. *if $\theta = \text{On}$ then for every $t < \theta$ we have $0 \le \text{PC}(t) \le n$. If θ is a successor ordinal $\text{PC}(\theta - 1) > n$ and for every $t < \theta - 1$ we have $0 \le \text{PC}(t) \le n$;*
6. *for all $t < \theta$ for all $j \in \mathfrak{D}$ we have $S^{\text{L}}(t)(j) < S^{\text{R}}(t)(j)$;*
7. *for every $t < \theta$ if $0 \le \text{PC}(t) \le n$ and $C_{\text{PC}(t)} = R_i := f(R_{j_0}, \ldots, R_{j_n})$ then $\text{PC}(t + 1) = \text{PC}(t) + 1$ and: $R^{\text{N}}(t + 1)(i) = f(c(j_0), \ldots, c(j_n))$ if $i \in \mathfrak{N}$, $(S^{\text{L}}(t+1)(i), S^{\text{R}}(t+1)(i))$ is the canonical representation of $f(c(0), \ldots, c(n))$ if $i \in \mathfrak{D}$, where for every $m < n$ $c(m) := [S^{\text{L}}(t)(j_m) | S^{\text{R}}(t)(j_m)]$ if $j_m \in \mathfrak{D}$, and $c(m) := R_{j_m}$ otherwise.*
8. *for every $t < \theta$ if $0 \le \text{PC}(t) \le n$ and $C_{\text{PC}(t)} = \text{Push}_d(R_i, R_j)$ then $\text{PC}(t + 1) = \text{PC}(t) + 1$ and $S^d(t + 1)(i) := R^{\text{N}}(t)(j)$ if $j \in \mathfrak{N}$; $S^d(t + 1)(i) := [S^{\text{L}}(t)(j) | S^{\text{R}}(t)(j)]$ if $j \in \mathfrak{D}$. The rest is left unchanged in $t + 1$;*
9. *for every $t < \theta$ if $0 \le \text{PC}(t) \le n$ and $C_{\text{PC}(t)} = $ if R_i then j then: $\text{PC}(t + 1) := j$ if $i \in \mathfrak{N}$ and $R^{\text{N}}(t)(i) > 0$; $\text{PC}(t + 1) := j$ if $i \in \mathfrak{D}$ and $[S^{\text{L}}(t)(i) | S^{\text{R}}(t)(i)] > 0$; $\text{PC}(t + 1) := \text{PC}(t) + 1$ if $i \in \mathfrak{N}$ and $R^{\text{N}}(t)(i) \le 0$; and, $\text{PC}(t + 1) := \text{PC}(t) + 1$ if $i \in \mathfrak{D}$ and $[S^{\text{L}}(t)(i) | S^{\text{R}}(t)(i)] \le 0$. The rest is left unchanged in $t + 1$;*

---

[4] By abuse of notation we write $\wp(\text{No})$ for the class of subsets of No.

10. *for every* $t < \theta$ *if* t *is a limit ordinal then:* $\mathrm{PC}(t) = \liminf_{s < t} \mathrm{PC}(s)$, *for every* $i \in \mathfrak{N}$ *we let* $R^{\mathrm{N}}(t)(i) := R^{\mathrm{N}}(t')(i)$ *if there is* t' *such that* $\forall t > t'' > t' R^{\mathrm{N}}(t')(i) = R^{\mathrm{N}}(t'')(i)$; $R^{\mathrm{N}}(t)(i) := 0$ *if there is no such a* t'.
    For all $i \in \mathfrak{D}$, *if there are* t'_L *and* t'_R *smaller than* t *such that for every* $t'_L < t''_L < t$ *and* $t'_R < t''_R < t$ *we have* $S^{\mathrm{L}}(t''_L)(i) = S^{\mathrm{L}}(t'_L)(i)$ *and* $S^{\mathrm{R}}(t''_R)(i) = S^{\mathrm{R}}L(t'_R)(i)$ *we have* $S^{\mathrm{L}}(t)(i) = S^{\mathrm{L}}(t'_L)(i)$ *and* $S^{\mathrm{R}}(t)(i) = S^{\mathrm{R}}(t'_R)(i)$. *Otherwise, let* $U_{t,i} := \{t'' < t \mid \forall t'' \le t' < t(C_{\mathrm{PC}(t')} = \mathrm{R}_j := f(\mathrm{R}_{j_0}, \dots, \mathrm{R}_{j_n}) \to i \neq j)\}$. *Then* $S^{\mathrm{L}}(t)(i) = \bigcup_{t' \in U_{t,i}} S^{\mathrm{L}}(t')(i)$ *and* $S^{\mathrm{R}}(t)(i) = \bigcup_{t' \in U_{t,i}} S^{\mathrm{R}}(t')(i)$.

If θ *is a successor ordinal, we say that* P *halts on* x *with output* $y := (R^{\mathrm{N}}(\theta - 1)(i))_{i \in O}$ *and write* $P(x) = y$.

In the previous definition, for each $\alpha \in \theta$ and $i \in \mathfrak{N}$, $R^{\mathrm{N}}(\alpha)(i)$ is the content of the normal register i at the αth step of the computation; similarly, $S^{\mathrm{L}}(\alpha)(i)$ and $S^{\mathrm{R}}(\alpha)(i)$ are the sets representing the left and the right stack of the Dedekind register i; moreover, $\mathrm{PC}(\alpha)$ is the value of the *program counter*. Items 2, 3, and 4 describe the initialisation of the machine. In particular, the program counter is set to 0, each normal register but the input registers are initialised to 0, the input registers are initialised to x, and each stack is emptied. Item 5 ensures that the program counter assumes correct values and that the computation *stops*. Items 7, 8, and 9 describe the semantics of the instructions according to our previous description. Finally, item 10 describes the behaviour of the machine at limit stages according to the description we gave before.

Definition 5. *Let* $n, m \in \mathbb{N}$ *and* $F : \mathrm{No}^n \to \mathrm{No}^m$ *be a (partial) class function over the surreal numbers and* K *a subclass of* No. *Then we say that* F *is* $K(\overrightarrow{X})$-*SBSSM computable iff there are* $\mathfrak{N}, \mathfrak{D}, I, O \subset \mathbb{N}$ *with* $|I| = n$, $|O| = m$ *and there is a* $(\mathfrak{N}, \mathfrak{D}, I, O, K)$-*SBSSM program* P *such that for every* n-*tuple* x *of surreal numbers we have that: if* $F(x) = y$ *then* $P(x) = y$, *and if* $x \notin \mathrm{dom}(F)$ *then* $P(x)$ *does not halt. Moreover, we say that* F *is SBSSM computable if it is* $\mathrm{No}(\overrightarrow{X})$-*SBSSM computable.*

As show in [7], infinite time Blum-Shub-Smale machines can only compute reals in the ω^ωth level $\mathbf{L}_{\omega^\omega}$ of constructible universe. Since $\mathbb{R}$ is a subfield of No, every constant real function is $\mathbb{R}(\overrightarrow{X})$-SBSSM computable. Therefore, our SBSSM machines are stronger than infinite time Blum-Shub-Smale machines.

Note that the hardware of our machines in principle does not allow a direct access to the sign sequence representing a surreal number, e.g., there is no instruction which allows us to read the αth sign of a surreal in the register i.

Lemma 6. *Let* K *be a subclass of* No *such that* $\{-1, 0, 1\} \subseteq K$. *Then, the following functions are* $K(\overrightarrow{X})$-*SBSSM computable:*

1. *The function* Lim *that given an ordinal number* α *returns* 1 *if* α *is a limit ordinal and* 0 *otherwise;*
2. *Gödel's pairing function* $\mathfrak{g} : \mathrm{On} \times \mathrm{On} \to \mathrm{On}$;

3. *The function* sgn : No × On → $\{0, 1, 2\}$ *that for every* $\alpha \in$ On *and* $s \in$ No *returns* 0 *if the* $1 + \alpha th$[5] *sign in the sign expansion of* s *is* −, 1 *if the* $1 + \alpha th$ *sign in the sign expansion of* s *is* + *and* 2 *if the sign expansion of* s *is shorter than* $1 + \alpha$;

4. *the function* seg : No × On → No *that given a surreal* s *and an ordinal* $\alpha \in \text{dom}(s)$ *returns the surreal whose sign sequence is the initial segment of* s *of length* α.

5. *The function* cng : No × On × $\{0, 1\}$ → No *that given a surreal* $s \in$ No, sgn $\in$ $\{0, 1\}$ *and* $\alpha \in$ On *such that* $\alpha < \text{dom}(s)$ *returns a surreal* $s' \in$ No *whose sign expansion is obtained by substituting the* $1 + \alpha th$ *sign in the expansion of* s *with* − *if* sgn = 0 *and with* + *if* sgn = 1;

By interpreting 0 as − and 1 as +, every binary sequence corresponds naturally to a surreal number. Therefore, we can represent the content of a tape of Turing machines, infinite time Turing machines (ITTMs), and ordinal Turing machines (OTMs) as a surreal number. Lemma 6 tells us that we can actually access this representation and modify it.

5   Computational Power of Surreal Blum-Shub-Smale Machines

Now that we introduced a notion of computability over No, we shall compare our new model of computation with classical and transfinite models of computation. In this section, we will assume that the reader is familiar with the basic definitions of classical computability theory, infinite time Turing machines computability theory, and ordinal Turing machines computability theory; see, e.g., [6, 8].

Given a class C and a set X we will denote by $X^{<C}$ the class of functions whose domain is in C and codomain is X. Let α be an ordinal, X be a set, and C be a class. Given a sequence $(w_\beta)_{\beta < \alpha}$ of elements in $X^{<\alpha}$, we define $[w_\beta]_{\beta < \alpha}$ to be the concatenation of the w_βs.

We start by fixing a representation of binary sequences in No. Let $\Delta :$ No → $2^{<\text{On}}$ be such that for all $s \in$ No, $\Delta(s)$ is the binary sequence of length $\text{dom}(s)$ obtained by substituting each + in s by a 1 and each − by a 0.

Definition 7. *Given a partial function* $f : 2^{<\text{On}} \to 2^{<\text{On}}$ *and a class of rational functions* $K(\overrightarrow{X})$ *we say that* f *is* $K(\overrightarrow{X})$-SBSSM *computable if there is a* $K(\overrightarrow{X})$-*SBSSM program which computes the surreal function* F *such that* $f = \Delta \circ F \circ \Delta^{-1}$.

As we will see, if K is a subclass of No containing $\{-1, 0, 1\}$ then $K(\overrightarrow{X})$-SBSSMs are very powerful. In order to show this, we will now begin by proving their capability of simulating all the most important classical models of transfinite computation. Using Lemma 6 it is immediate to see that if K is a subclass

---

[5] In this sentence $1 + \alpha$ should be read as the ordinal addition so that for $\alpha \geq \omega$ we have $1 + \alpha = \alpha$.

of No such that $\{-1, 0, 1\} \subseteq K$, then every function computable by an ordinary Turing machine is $K(\vec{X})$-SBSSM computable; moreover, the classical halting problem is $K(\vec{X})$-SBSSM computable.

The following notion was introduced by Hamkins and Lewis in [6] and further studied by several authors; see, e.g., [12]. An ordinal α is *clockable* if there is an ITTM which runs on empty input for exactly α steps. We will denote by λ the supremum of the clockable ordinals.

Theorem 8. *Let K be a subclass of No such that $\{-1, 0, 1\} \subseteq K$. Then, every ITTM-computable function is $K(\vec{X})$-SBSSM computable. Moreover, if $\lambda \in K$, then the halting problem for ITTMs is $K(\vec{X})$-SBSSM computable.*

Proof. We will assume that our ITTM has only one tape; a similar proof works in the general case. We call a *snapshot* of an execution of an ITTM at time α a tuple $(T(\alpha), I(\alpha), H(\alpha)) \in \{0, 1\}^\omega \times \omega \times \omega$ where $T(\alpha)$ is a function representing the tape content of the ITTM at time α, $I(\alpha)$ is the state of the machine at time α, and $H(\alpha)$ is the position of the head at time α. We know that we can code $T(\alpha)$ as a sign sequence of length ω. Moreover, at the successor stages, by Lemma 6, we can modify this sequence in such a way that the result is a sign sequence in No_ω coding the ITTM tape after the operation is performed. Moreover, we know that there is a bound, λ, to the possible halting times of an ITTM. Therefore, we can code the list of the $T(\alpha)$ in the snapshots of an ITTM as a sequence of pluses and minuses length λ; hence, as a surreal number of the same length. Consider the $K(\vec{X})$-SBSSM program that uses two Dedekind registers T and S, and two normal registers I and H. The first Dedekind register is used to keep track of the tapes in the snapshots, the second Dedekind register is used to keep track of how many ITTM instructions have been executed, the register I is used to keep track of the current state of the ITTM, and the register H to keep track of the current head position.

At each step α, if S is a successor ordinal, the program first copies the last ω-many bits of T into a normal register R; then, executes the instruction I with head position[6] $(\omega \times S) + H$ on the string sequence of T writing the result in R. Then, the program computes the concatenation s_α of T and R; and pushes the canonical representation of s_α into the stacks of T. Since for all $\beta < \alpha$, the sign sequence of s_β is an initial segment of s_α, T will contain $\bigcup_{\beta \in \alpha} s_\alpha$ at limit stages.

Now, if S is a limit, the program first computes the content of R as the pointwise $\liminf$ of the snapshots in T. Note that this is computable. Indeed, suppose that the program needs to compute the $\liminf$ of the bit in position i; then it can just look sequentially at the values of the snapshots at i and if it finds a 0 at i in the αth snapshot it pushes $\alpha - 1$ into the left stack of a Dedekind register R'. Once the program has looked through all the snapshots, it will compute the $\liminf$ of the cell in position i as 0 if $R' = S$ and as 1 otherwise. Then, the program will set H to 0 and I to the special limit state and continue the normal execution. This ends the first part of the proof.

---

[6] Once again the operations in $(\omega \times S) + H$ must be interpreted as ordinal operations.

Now, assume that $\lambda \in K$. Note that the $K(\overrightarrow{X})$-SBSSM program we have just introduced can simulate the ITTM and check after the execution of every ITTM step that $S < \lambda$. If at some point the program simulates λ-many steps of the ITTM, i.e., $S \geq \lambda$, the program will just halt knowing that the ITTM can not halt.

Since, by [11, Lemma 5], ITTMs can decide the halting problem of infinite time Blum-Shub-Smale machines we get:

Corollary 9. *Let K be a subclass of No such that $\{-1, 0, 1\} \subseteq K$. Then, every function computable by an infinite time Blum-Shub-Smale machine is $K(\overrightarrow{X})$-SBSSM computable and the halting problem for infinite time Blum-Shub-Smale machine is $K(\overrightarrow{X})$-SBSSM computable.*

As shown in [8, Lemma 6.2], every OTM computable real is in the constructible universe **L**. Therefore, if $\mathbf{V} \neq \mathbf{L}$, we have that the notions of OTM and $\mathbb{R}(\overrightarrow{X})$-SBSSM computability do not coincide. As usual we will denote by ZFC the axioms of set theory with the Axiom of Choice.

Lemma 10. *If ZFC is consistent, so is ZFC+ "there is a function that is $\mathbb{R}(\overrightarrow{X})$-SBSSM computable but not OTM computable".*

Theorem 11. *Let K be a subclass of No such that $\{-1, 0, 1\} \subseteq K$. Then, every OTM computable partial function $f : 2^{<\mathrm{On}} \to 2^{<\mathrm{On}}$ is $K(\overrightarrow{X})$-SBSSM computable.*

Proof. We will assume that our machine has two tapes, one read-only input tape and an output tape; the general case follows.

Our program will be very similar to the one we used for ITTMs. For this reason, we will mostly focus on the differences.

The main difference is that, while for ITTM we can just save the sequence of tape snapshots, for OTM we cannot simply do that because the tape has class length. The problem can be solved by padding. Given a binary sequence $b := [b_\beta]_{\beta \in \alpha}$ where $b_\beta \in \{-, +\}$ for each $\beta < \alpha$, let b^p be the sequence obtained by concatenating the sequence $[+b_\beta+]_{\beta \in \alpha}$ with the sequence $--$. We call b^p the padding of b. With this operation, we can now save the initial meaningful part of the OTM tape in a register.

The program has four Dedekind registers T, S, H_i, I_i, and two normal registers H and I. As for ITTMs, the Dedekind register T is used to keep track of the tapes in the snapshots; the Dedekind register S is used to keep track of how many OTM instructions have been executed; the register I is used to keep track of the current state of the OTM; and the register H to keep track of the current head position. Note that, since at limit stages the head position and the state of the machine need to be set to the lim inf of their previous contents, we added the Dedekind registers H_i and I_i to keep track of the histories of H and I, respectively.

The registers T, H_i and I_i are really the main difference between this program and the one we used to simulate ITTM. At each stage, T will contain the concatenation of the paddings of the previous configurations of the OTM tape. Note that the sequence $--$ works as a delimiter between one snapshot and the next one. Also, since we cannot save all the OTM tape, each time we will just record the initial segment of the OTM tape of length S, i.e., the maximum portion we could have modified.

If $S := \alpha + 1$, the program first copies the last snapshot in T to a normal register s_α removing the padding. At this point, the program can just simulate one step of OTM and then compute the padding s_α^p of s_α, and push the standard representation of s_α^p in T.

Now, the program will take the content of H_i, and will compute the surreal number h_α whose sign sequence is H_i followed by H minuses and one plus. Then, the program will push the canonical representation of h_α into the stacks of H_i. Similarly for I, the program will take the content of I_i, and will compute the surreal number i_α whose sign sequence is I_i followed by I_i minuses and one plus. Then, the program will push the canonical representation of i_α into the stacks of I_i.

Again, note that, as for ITTMs, at limit stages T, H_i and I_i will contain the concatenation of the padded snapshots of the tape, H and I, respectively.

If S is a limit ordinal, with a bit of overhead due to padding, the program can compute the pointwise lim inf of the tape. It is not hard to see that this operation is a minor modification of the one used for ITTMs. Note that, in this case, not all the bits will be present in every snapshot; if we want to compute the ith bit of the limit snapshot we will have to start computing the lim inf from the ith snapshot in T. The rest is essentially the same as what we did for ITTM case. Then, the program will compute the content of I; and, using H_i and I_i, it can compute the lim inf of H only considering the stages where I was the current state. Then, the program can proceed exactly as in the successor case.

As we have seen so far, if K is a subclass of No such that $\{-1, 0, 1\} \subsetneq K$ then $K(\overrightarrow{X})$-SBSSMs are at least as powerful as OTMs. It turns out that, if $K = \{-1, 0, 1\}$, the two models of computation are actually equivalent; see Theorem 14.

As shown in [4], via representations it is possible to use OTMs to induce a notion of computability over surreal numbers. We will take the same approach here.

Let $\delta_{\mathrm{No}} : 2^{<\mathrm{On}} \to \mathrm{No}$ be the function that maps each surreal number to a binary sequence as follows: $\delta_{\mathrm{No}}(p) = q$ iff p is a binary sequence of length $2 \times \ell(q) + 2$, such that $p = [w_\alpha]_{\alpha \in \ell(q)+1}$ where: $w_\alpha := 00$ if $\alpha \in \mathrm{dom}(q)$ and $q(\alpha) = -$ and, $w_\alpha := 11$ if $\alpha \in \mathrm{dom}(q)$ and $q(\alpha) = +$, and $w_{\ell(q)} := 01$.

To avoid unnecessary complications, in the following we will only deal with unary surreal functions.

Definition 12. *Given a partial function $F : \mathrm{No} \to \mathrm{No}$, we say that F is OTM computable if there is an OTM program that computes the function G such that $F = \delta_{\mathrm{No}} \circ G \circ \delta_{\mathrm{No}}^{-1}$.*

Note that the function δ_{No} is essentially[7] an extension to the class of surreal numbers of the function $\delta_{\mathbb{Q}_\kappa}$ introduced in [4]. From this fact, and from the fact that, as shown in [4, Lemma 9 & 10], OTMs are capable of computing surreal operations and convert back and forth from cut representation to sign sequences, it is easy to see that OTMs and $\{-1,0,1\}(\overrightarrow{X})$-SBSSM have the same computational strength.

Theorem 13. *Let K be a subclass of OTM computable elements of No, i.e., such that for every $s \in K$ the sequence $\delta_{No}^{-1}(s)$ is computable by an OTM with no input. Then, every $K(\overrightarrow{X})$-SBSSM computable function is OTM computable. In particular, every $\{-1,0,1\}(\overrightarrow{X})$-computable function is OTM computable.*

So, $\{-1,0,1\}(\overrightarrow{X})$-SBSSM have the same computational power as OTMs. Note that, if we enlarge the class of rational functions our machine is allowed to use, we obtain progressively stronger models of computations. Moreover, it is easy to see that the class of coefficients allowed in the class of rational functions acts as a set of parameters on the OTMs side.

Theorem 14. *Let K be a subclass of No. Then a partial function $F : \mathrm{No} \to \mathrm{No}$ is $K(\overrightarrow{X})$-SBSSM computable iff it is computable by an OTM with parameters in K.*

Corollary 15. *Every partial function $F : \mathrm{No} \to \mathrm{No}$ which is a set is $\mathrm{No}(\overrightarrow{X})$-SBSSM computable.*

In [10], Ethan Lewis defines a notion of computability based on OTMs which allows for infinite programs. We will call these machines *infinite program machines* (IPMs). In [10], Lewis shows that IPMs are equivalent to OTMs with parameters in 2^{On}. Therefore, Theorem 14 tells us that $\mathrm{No}(\overrightarrow{X})$-SBSSM are a register model for IPMs.

We end this paper by introducing halting sets and universal programs for our new model of computation. Using classical coding techniques, given a class of rational functions $K(\overrightarrow{X})$, every $K(\overrightarrow{X})$-SBSSM program can be coded as one (possibly infinite) binary sequence, i.e., a surreal number.

Given two natural numbers n and m, and a subclass K of surreal numbers we will denote by $\mathfrak{P}_K^{n,m}$ the class of $(\mathfrak{N}, \mathfrak{D}, I, O, K)$-SBSSM programs with $|I| = n$, $|O| = m$.

Let K be a class of the surreal numbers. We define the following class[8]:
$$H_K^{n,m} := \{(p,s) \in \mathrm{No} \mid p \text{ is a } K(\overrightarrow{X})\text{-SBSSM program in } \mathfrak{P}_K^{n,m} \text{ halting on } s\}.$$

As usual, we say that a set of surreal numbers is decidable if its characteristic function is computable.

---

[7] The class function δ_{No} is not literally an extension of $\delta_{\mathbb{Q}_\kappa}$ just because in [4] we assumed $\mathrm{dom}(\delta_{\mathbb{Q}_\kappa}) \subset 2^\kappa$ rather than $\mathrm{dom}(\delta_{\mathbb{Q}_\kappa}) \subset 2^{<\kappa}$. This does not make much of a difference in our algorithms as far as we have a marker for the end of the code of the sign sequence (i.e., the last two bits in the definition of δ_{No}).

[8] Note that, if K is a set $H_K^{n,m}$ is also a set.

If we assume that K contains $\{-1, 0, 1\}$ we can use the code of a program, together with the fact that OTMs can simulate $K(\overrightarrow{X})$-SBSSM and can be simulated by $K(\overrightarrow{X})$-SBSSM, to define a universal SBSSM program.

Let $\mathfrak{N}$ and $\mathfrak{D}$ be two disjoint sets of natural numbers, I and O be two disjoint subsets of $\mathfrak{N}$, and K be a class of surreal numbers. A $(\mathfrak{N}, \mathfrak{D}, I, O, K)$-SBSSM program P with $|I| = n + 1$ and $|O| = m$ is called *universal* if for every code p' of a program in $\mathfrak{P}_K^{n,m}$ and for every $x \in \mathrm{No}^n$ we have $P(p', x) = P(x)$.

A straightforward generalisation of the classical arguments shows the following results:

Theorem 16. *Let K be a subclass of* No *containing* $\{-1, 0, 1\}$. *Moroever, let* $\mathfrak{N}, \mathfrak{D}, I, O \subset \mathbb{N}$ *be such that:* $\mathfrak{N}$ *and* $\mathfrak{D}$ *are disjoint; and,* I *and* O *are disjoint subsets of* $\mathfrak{N}$. *Then, there is a universal* $(\mathfrak{N}, \mathfrak{D}, I, O, K)$-SBSSM *program.*

Corollary 17. *Let K be a subclass of* No *containing* $\{-1, 0, 1\}$. *Then* $H_K^{1,1}$ *is not* $K(\overrightarrow{X})$-SBSSM *computable.*

References

1. Blum, L., Shub, M., Smale, S.: On a theory of computation and complexity over the real numbers: NP-completeness, recursive functions and universal machines. Bull. Am. Math. Soc. **21**, 1–46 (1989)
2. Conway, J.H.: On Numbers and Games. A K Peters & CRC Press, Natick (2000)
3. Ehrlich, P.: An alternative construction of Conway's ordered field No. Algebra Universalis **25**(1), 7–16 (1988)
4. Galeotti, L., Nobrega, H.: Towards computable analysis on the generalised real line. In: Kari, J., Manea, F., Petre, I. (eds.) CiE 2017. LNCS, vol. 10307, pp. 246–257. Springer, Cham (2017). https://doi.org/10.1007/978-3-319-58741-7_24
5. Gonshor, H.: An Introduction to the Theory of Surreal Numbers. London Mathematical Society Lecture Note Series, vol. 110. Cambridge University Press, Cambridge (1986)
6. Hamkins, J.D., Lewis, A.: Infinite time turing machines. J. Symbolic Logic **65**, 567–604 (2000). https://doi.org/10.2307/2586556
7. Koepke, P., Morozov, A.S.: The computational power of infinite time Blum-Shub-Smale machines. Algebra Logic **56**(1), 37–62 (2017)
8. Koepke, P.: Turing computations on ordinals. Bull. Symbolic Logic **11**(3), 377–397 (2005)
9. Koepke, P., Seyfferth, B.: Towards a theory of infinite time Blum-Shub-Smale machines. In: Cooper, S.B., Dawar, A., Löwe, B. (eds.) CiE 2012. LNCS, vol. 7318, pp. 405–415. Springer, Heidelberg (2012). https://doi.org/10.1007/978-3-642-30870-3_41
10. Lewis, E.: Computation with Infinite Programs. Master's thesis, ILLC Master of Logic Thesis Series MoL-2018-14, Universiteit van Amsterdam (2018)
11. Seyfferth, B.: Three Models of Ordinal Computability. Ph.D. thesis, Rheinische Friedrich-Wilhelms-Universität Bonn (2013)
12. Welch, P.D.: The length of infinite time turing machine computations. Bull. London Math. Soc. **32**(2), 129–136 (2000)

Non-Recursive Trade-Offs Are "Almost Everywhere"

Markus Holzer[(✉)] and Martin Kutrib

Institut für Informatik, Universität Giessen, Arndtstr. 2, 35392 Giessen, Germany
{holzer,kutrib}@informatik.uni-giessen.de

Abstract. We briefly summarize some of the findings on non-recursive trade-offs, which were first observed by Meyer and Fischer in their seminal paper on "Economy of Description by Automata, Grammars, and Formal Systems" in 1971. This general phenomenon is about conversion problems between different (computational) description models that cannot be solved efficiently. Indeed, they evade solvability *a forteriori* because the change in description size caused by such a conversion cannot be bounded above by any recursive function. Hence, a result on non-recursive trade-offs can alternatively be interpreted as a compression of the description model with *arbitrary* space gains. Since 1971 there has been a steadily growing list of results where this phenomenon has been observed, and it appears that non-recursive trade-offs are "almost everywhere."

1 Introduction

In computer science the *systematic* analysis and classification of the size of formal systems for specifying mathematical objects as opposed to the computational power of such systems was initiated by Meyer and Fischer [22], and is known as descriptional complexity (of formal systems). Descriptional complexity has historically been a multidisciplinary area of study, with contributions from very different areas of computer science such as, for example, automata and formal language theory, computational complexity, cryptography, information theory, etc. In the classification of automata, grammars, and related (formal) systems it turns out that the gain in economy of description heavily depends on the considered system. For instance, it is well known that nondeterministic finite automata can be converted into equivalent deterministic finite automata of at most exponential size. For deterministic pushdown automata accepting a regular language, we know that they can be converted into an equivalent finite automaton of at most doubly-exponential size [27]. In contrast, if we replace "deterministic pushdown automata" with "nondeterministic pushdown automata" then the maximum size blow-up can no longer be bounded by any recursive function—more precisely [22, Proposition 7]:

"For any recursive function f and for arbitrarily large integers n, there is a [nondeterministic pushdown automaton] of size n describing a regular

F. Manea et al. (Eds.): CiE 2019, LNCS 11558, pp. 25–36, 2019.
https://doi.org/10.1007/978-3-030-22996-2_3

(in fact co-finite) set whose reduced finite automaton has at least $f(n)$ states."

Therefore the achievable benefit in description length is of *arbitrary* size. This result is not a coincidence. In fact, it is an expression of a much wider phenomenon which is nowadays known as a non-recursive trade-off between descriptional systems. Since its original encounter there has been a steadily growing list of results where non-recursive trade-offs have been observed, for example, [6,8,9,12,16,17,19,20,25,26,28]. They usually sprout at the wayside of the crossroads of (un)decidability and in many cases proving such trade-offs apparently requires ingenuity and careful automata constructions. From the broad spectrum of descriptional systems that are covered by the previously mentioned papers, it is fair to say that "non-recursive trade-offs are almost everywhere." For a survey on descriptional complexity results, not limited to non-recursive trade-offs, we refer to, for example, [13].

In this paper we briefly give our view of what constitute nice and interesting findings on non-recursive trade-offs. Our focus is on the relation between non-recursive trade-offs and undecidability from a automata and formal language perspective. We do not prove the presented results but we draw attention to the big picture of some of the main ideas involved. Besides techniques and tools to prove non-recursive trade-offs we present an assorted list of examples involving formal description systems of Chomskyan and Lindenmayerian standard formal language families, as well as from classical automata theory. These formal languages are very appealing since they are well studied from different angles in the literature, in particular inclusions relations, incomparabilities, closure properties, decision problems, etc. It is worth mentioning that besides the qualitative classification of transformations it is also possible to quantitatively classifying non-recursive trade-offs. Due to the overwhelming amount of literature on the subject in question it obviously lacks completeness. Nevertheless, we hope to give a detailed overview on how non-recursive trade-offs are triggered by (un)decidability results on the underlying description systems.

The paper is organized as follows: in the next section we introduce some notations and basic properties of descriptional systems and reasonable size measures. Then in Sect. 3 we present a unified proof scheme for non-recursive trade-offs emerging form the literature with some examples applied. Finally, Sect. 4 is devoted to the categorization of non-recursive trade-offs by bounds on their growth rate. To this end, some preliminaries from computability theory are needed. Then it is shown how to derive upper and lower bounds on non-recursive trade-offs by verifying that descriptors of a certain type generate languages that can be described by another description system. This problem is central and it is called the "S_2-*ness of* S_1 *descriptor problem*," where S_1 and S_2 are descriptional systems satisfying certain properties. Moreover, it is shown how non-recursive trade-offs can be deduced by simple non-closure properties and their computational complexity. In all cases we give meaningful assorted examples from the literature.

2   Preliminaries

Let Σ^* denote the set of all words over the finite alphabet Σ. The *empty word* is denoted by λ, and $\Sigma^+ = \Sigma^* \setminus \{\lambda\}$. For the *length* of a word w we write $|w|$. Set inclusion is denoted by $\subseteq$ and strict set inclusion by $\subset$.

We recall some notation for descriptional complexity. Following [13] we say that a *descriptional system* S is a set of finite descriptors such that each $D \in S$ describes a formal language $L(D)$, and the underlying alphabet $\mathrm{alph}(D)$ over which D represents a language can be obtained from D. The *family of languages represented* (or *described*) by S is $\mathscr{L}(S) = \{L(D) \mid D \in S\}$. For every language L, the set $S(L) = \{D \in S \mid L(D) = L\}$ is the set of its descriptors in S. A *complexity measure* for a descriptional system S is a total recursive mapping c with $c : S \to \mathbb{N}$.

Example 1. Finite automata or (deterministic) linear bounded automata can be encoded over some fixed alphabet such that their input alphabets can be extracted from the encodings. The sets of these encodings are descriptional systems S_1 and S_2, and $\mathscr{L}(S_1)$ is the family of regular languages and $\mathscr{L}(S_2)$ is the family of (deterministic) context-sensitive languages.

Examples for complexity measures for finite automata or linear bounded automata are the total number of symbols, that is, the *length of the encoding* (length), or, in the former case, the *number of states* and, in the latter case, the product of the number of states and the number of tape symbols. ∎

Here we only use complexity measures that are recursively related to length. If there is a total recursive function $g : \mathbb{N} \times \mathbb{N} \to \mathbb{N}$ such that

$$\mathrm{length}(D) \leq g(c(D), |\mathrm{alph}(D)|)$$

for all $D \in S$, then c is said to be an *s-measure*. If, in addition, for any alphabet Σ, the set of descriptors in S describing languages over Σ is recursively enumerable in order of increasing size, then c is said to be an *sn-measure*.

Example 2. The number of states is an sn-measure for finite automata. The product of the number of states and the number of tape symbols is an sn-measure for linear bounded automata. ∎

We say that a descriptional system S is *recursive* (*recursively enumerable*), if for each descriptor $D \in S$ the language $L(D)$ is recursive (recursively enumerable). Moreover, if there exists an effective procedure to convert D into a Turing machine that decides (semi-decides) $L(D)$, then the descriptional system is said to be *effectively recursive* (*effectively recursively enumerable*).

Whenever we consider the relative succinctness of two descriptional systems S_1 and S_2, we assume that the intersection $\mathscr{L}(S_1) \cap \mathscr{L}(S_2)$ is non-empty. Let S_1 and S_2 be descriptional systems with complexity measures c_1 and c_2, respectively. A total function $f : \mathbb{N} \to \mathbb{N}$ is an *upper bound* for the increase in complexity when changing from a descriptor in S_1 to an equivalent descriptor in S_2, if for all $D_1 \in S_1$ with $L(D_1) \in \mathscr{L}(S_2)$, there exists a $D_2 \in S_2(L(D_1))$ such that $c_2(D_2) \leq f(c_1(D_1))$.

If there is no recursive upper bound, then the *trade-off* for changing from a description in S_1 to an equivalent description in S_2 *is said to be non-recursive*. In other words, there are no recursive functions serving as upper bounds, that is, whenever the trade-off from one descriptional system to another is non-recursive, one can choose an arbitrarily large recursive function f but the gain in economy of description exceeds f when changing from the former system to the latter. In fact, the non-recursive trade-offs are independent of particular sn-measures. Any two complexity measures c_1 and c_2 for some descriptional system S are related by a function $h(n) = \max\{\, c_2(D) \mid D \in S \text{ such that } c_1(D) = n \,\}$. By the properties of sn-measures, function h is recursive. So, a non-recursive trade-off exceeds any difference caused by applying two sn-measures. Moreover, as long as non-recursive trade-offs are studied one may think safely as the length of the encoding strings to be the complexity measure.

3   Non-recursive Trade-Offs

Most of the non-recursive trade-off proofs appearing in the literature are basically relying on two general results of Hartmanis [8,9]. The following theorem is a slightly generalized and unified form of these proof techniques [13].

Theorem 1. *Let S_1 and S_2 be two descriptional systems for recursive languages such that any descriptor D in S_1 and S_2 can effectively be converted into a Turing machine that decides $L(D)$, and let c_1 be a measure for S_1 and c_2 be an sn-measure for S_2. If there exists a descriptional system S_3 and a property P that is not semi-decidable for descriptors from S_3, such that, given an arbitrary $D_3 \in S_3$, (i) there exists an effective procedure to construct a descriptor D_1 in S_1, and (ii) D_1 has an equivalent descriptor in S_2 if and only if D_3 does not have property P, then the trade-off between S_1 and S_2 is non-recursive.*

So, to some extend, the undecidability of the problem whether some descriptor in S_3 does not have property P is reduced to the computability of the trade-off.

Are non-recursive trade-offs a rare phenomenon? No, they are appearing almost everywhere. For example, between each two separated Turing machine space classes there is always a non-recursive trade-off (see also [9]).

Example 3. Denote the languages accepted by deterministic Turing machines obeying a space bound s by $\mathsf{DSPACE}(s)$. If $\mathsf{DSPACE}(s_1) \supset \mathsf{DSPACE}(s_2)$ for a constructable bound s_1, then the trade-off between s_1-space bounded Turing machines and s_2-space bounded Turing machines is non-recursive.

In order to obtain the trade-offs, we apply Theorem 1 as follows. There exists an s_1-space bounded Turing machine M that has no equivalent s_2-space bounded Turing machine. Let S_3 be the set of deterministic one-tape Turing machines. For any $D_3 \in S_3$ an s_1-space bounded Turing machine D_1 is constructed that works as follows. On input w, machine D_1 generates successively all strings v whose lengths are bounded by $s_1(|w|)$ (s_1 is constructable). For each v it is tested whether it represents a valid (halting) computation of D_3 on empty input. If D_1 does not find this string, it starts simulating M on its input w. If D_1 finds the string, it rejects its input w. So, we obtain $L(D_1) = L(M)$ if D_3 does not halt on empty input. Otherwise, $L(D_1)$ is empty or a finite subset of $L(M)$.

This implies D_1 has an equivalent s_2-space bounded Turing machine if and only if M halts on empty input. So, property P is to run forever on empty input, which is not semi-decidable for Turing machines.   ∎

The example can be modified to work for several other Turing machine classes which are separated by bounding some resource. For example, P $\neq$ NP if and only if the trade-off between NP and P is non-recursive—see [9–11] for further relations between descriptional and computational complexity.

Let us come to more restricted computational models than Turing machines. The first non-recursive trade-off at all was observed by Meyer and Fischer [22] between context-free grammars and finite automata. Also for restricted models the method to derive non-recursive trade-offs presented in Theorem 1 can serve as a powerful tool. Some non-recursive trade-offs follow immediately by known undecidability results.

Let $\mathcal{S}_1$ and $\mathcal{S}_2$ be two descriptional systems that meet the prerequisites of Theorem 1 such that the set

$$R = \{\, D \mid D \in \mathcal{S}_1 \text{ and } D \text{ has no equivalent descriptor in } \mathcal{S}_2 \,\}$$

is not recursively enumerable. Then we apply Theorem 1 as follows: descriptional system $\mathcal{S}_3$ is set to be $\mathcal{S}_1$ and the property P is to have no descriptor in $\mathcal{S}_2$. So, the trade-off between $\mathcal{S}_1$ and $\mathcal{S}_2$ is non-recursive.

Example 4. Deciding the regularity of linear context-free grammars is Σ_2-hard [2] which implies that neither the problem nor its negation is semi-decidable. So, we may apply Theorem 1 for any pairs of descriptional systems whose first component effectively represents the linear context-free and whose second component effectively represents the regular languages. That is, there are non-recursive trade-offs between linear context-free grammars and deterministic finite automata, between one-turn pushdown automata and nondeterministic finite automata, etc.   ∎

Exemplarily, we mention some further non-recursive trade-offs. For more results, references, and discussions about this phenomenon see [13,18].

Theorem 2. *The trade-offs between context-free grammars and unambiguous context-free grammars, and between unambiguous context-free grammars and deterministic pushdown automata are non-recursive.*

So, even if the nondeterminism is restricted to unambiguous computations, the descriptional power of such pushdown automata is much stronger than that of deterministic ones. On the other hand, the impact of the restriction is significant, since the descriptional power of nondeterministic pushdown automata is much stronger than that of unambiguous ones.

Recently, in [14] a new interpretation of nondeterminism has been introduced. The restriction introduced there is that of *one-time* nondeterminism, which means that at the outset the computation is nondeterministic, but whenever it performs a guess, this guess is fixed for the rest of the computation. This is a clear change on the semantics of nondeterminism. The new concept is studied for finite automata (OTNFA) and pushdown automata. Although one-time nondeterminism does not increase the accepting power of ordinary finite state devices, their conciseness is even greater than that of ordinary nondeterministic finite automata compared to deterministic ones. In particular, for the special case of an n-state OTNFA with a sole nondeterministic state, that is nondeterministic only for one input symbol and has nondeterministic degree n, then $(n + 1)^n$ states are a lower bound for any equivalent deterministic finite automaton.

For one-time nondeterministic pushdown automata the results on the descriptional complexity are even more dramatic.

Theorem 3. *The trade-offs between nondeterministic pushdown automata and one-time nondeterministic pushdown automata, and between one-time nondeterministic pushdown automata and deterministic pushdown automata are non-recursive.*

Also between the levels of infinite hierarchies of separated language classes, where intuitively the classes are closer together, there are non-recursive trade-offs. In [21] the trade-offs between $(k + 1)$-turn and k-turn pushdown automata are investigated.

Theorem 4. *Let $k \geq 1$ be an integer. Then the following trade-offs are non-recursive:*

1. *between nondeterministic 1-turn pushdown automata and finite automata,*
2. *between nondeterministic $(k + 1)$-turn pushdown automata and nondeterministic k-turn pushdown automata,*
3. *between nondeterministic pushdown automata and nondeterministic finite-turn pushdown automata, and*
4. *between nondeterministic and deterministic k-turn pushdown automata.*

So, there are infinite hierarchies such that between each two levels there are non-recursive trade-offs. Other results of such flavor have been obtained in [18] where deterministic and nondeterministic one-way k-head finite automata (k-DFAs and k-NFAs, respectively) are considered.

Theorem 5. *Let $k \geq 1$ be an integer. Then the following trade-offs are non-recursive:*

1. *between $(k + 1)$-DFAs and k-DFAs,*

2. *between $(k + 1)$-NFAs and k-NFAs,*
3. *between $(k + 1)$-DFAs and k-NFAs,*
4. *between 2-NFAs and $(k + 1)$-DFAs, and*
5. *between $(k + 1)$-DFAs and nondeterministic pushdown automata.*

In [16] the problem whether there are non-recursive trade-offs between the levels of the hierarchies defined by two-way k-head finite automata (cf. also [17]) has been answered in the affirmative.

Furthermore, Hartmanis [8] raised the question whether the trade-off between two descriptional systems is caused by the fact that in one system it can be proved what is accepted, but that no such proofs are possible in the other system. For example, consider descriptional systems for the deterministic context-free languages. It is easy to verify whether a given pushdown automaton is deterministic, but there is no uniform way to verify that a nondeterministic pushdown automaton accepts a deterministic context-free language. Sticking with this example, one may ask whether the trade-off is affected if so-called *verified nondeterministic pushdown automata* are considered which come with an attached proof that they accept deterministic languages. The following theorem summarizes results from [8].

Theorem 6. *The following trade-offs are non-recursive:*

1. *between verified and deterministic pushdown automata,*
2. *between pushdown automata and verified pushdown automata, and*
3. *between verified ambiguous and unambiguous context-free grammars.*

4 Levels of Non-Recursive Trade-Offs

Non-recursive trade-offs between descriptional systems are investigated in an abstract and more axiomatic fashion in [7]. The aim is to categorize non-recursive trade-offs by bounds on their growth rate, and to show how to deduce such bounds in general. Also criteria are identified which, in the spirit of abstract language theory, allow to deduce non-recursive trade-offs from effective closure properties of language families on the one hand, and differences in the decidability status of basic decision problems on the other.

So, we are now interested in classifying non-recursive trade-offs qualitatively. As it will turn out, the $\mathcal{S}_2$-*ness of* $\mathcal{S}_1$ *descriptors*, that is, the problem given a descriptor $D_1 \in \mathcal{S}_1$ does the language $L(D_1)$ belong to $\mathscr{L}(\mathcal{S}_2)$, plays a central role in this task. We assume the reader to be familiar with the basics of recursively enumerable sets as contained in [23]. In particular we consider the *arithmetic hierarchy*, which is defined as follows:

$$\Sigma_1 = \{\, L \mid L \text{ is recursively enumerable} \,\},$$
$$\Sigma_{n+1} = \{\, L \mid L \text{ is recursively enumerable in some } P \in \Sigma_n \,\},$$

for $n \geq 1$. Here, a language L is said to be recursively enumerable in some P if there is a Turing machine with oracle P that semi-decides L. Let Π_n be the complement of Σ_n, that is, $\Pi_n = \{ L \mid \overline{L}$ is in $\Sigma_n \}$. Moreover, let $\Delta_n = \Sigma_n \cap \Pi_n$, for $n \geq 1$. Observe that $\Delta_1 = \Sigma_1 \cap \Pi_1$ is the class of all recursive sets. Completeness and hardness are always meant with respect to many-one reducibilities $\leq_m$, if not otherwise stated. Let K denote the *halting set*, that is, the set of all encodings of Turing machines that accept their own encoding. For any set A define $A' = K^A$ to be the *jump* or *completion* of A, where K^A is the *A-relativized halting set*, which is the set of all encodings of Turing machines with oracle A that accept their own encoding, and define $A^{(0)} = A$ and $A^{(n+1)} = (A^{(n)})'$, for $n \geq 0$. By Post's Theorem we have that $\emptyset^{(n)}$ is Σ_n-complete ($\overline{\emptyset^{(n)}}$ is Π_n-complete) with respect to many-one reducibility, for $n \geq 1$, where $\emptyset^{(n)}$ is the *n-th jump* of $\emptyset$. Moreover, note that (1) $L \in \Sigma_{n+1}$ if and only if L is recursively enumerable in $\emptyset^{(n)}$ and (2) $L \in \Delta_{n+1}$ if and only if L is recursive in, or equivalently *Turing reducible* to, the jump $\emptyset^{(n)}$. In this case we simply write $L \leq_T \emptyset^{(n)}$, where $\leq_T$ refers to Turing reducibility. In the following we also use the above introduced framework on Turing machines and reductions in order to compute (partial) functions.

4.1   Bounds for Non-Recursive Trade-Offs

In this subsection we generalize the observations before Example 4 the other way around. It will turn out, that whenever a non-recursive trade-off between descriptional systems S_1 and S_2 exists, its (upper) bound is induced by the property of verifying the S_2-ness of an S_1 descriptor, that is, the problem of determining, whether for a given descriptor $D \in S_1$ the language $L(D)$ belongs to $\mathscr{L}(S_2)$. In order to make this more precise we need the following theorem [7].

Theorem 7. *Let S_1 and S_2 be two descriptional systems. The problem of determining for a given descriptor $D_1 \in S_1$ whether there exists an equivalent descriptor in S_2, that is, the S_2-ness of S_1 descriptors, can be solved in Σ_3, if both S_1 and S_2 are effectively recursively enumerable. When both systems are effectively recursive, the problem can be solved in Σ_2.*

A closer look reveals that equivalence between descriptors from S_1 and S_2 can be solved in Π_1 if *both* descriptional systems are effectively recursive. This equivalence problem belongs to the class Π_2 if both systems are constructively recursively enumerable. Thus, the upper bound on the equivalence problem is one less in the level of unsolvability than the S_2-ness of S_1 descriptors.

   The previous theorem can be utilized to prove an upper bound when changing from one system to another one.

Theorem 8. *Let S_1 and S_2 be two descriptional systems, c_1 be an sn-measure for S_1 and c_2 be a measure for S_2. If both S_1 and S_2 are constructively recursively enumerable, then there is a total function $f : \mathbb{N} \to \mathbb{N}$ that serves as an upper bound for the increase in complexity when changing from a descriptor in S_1 to an equivalent descriptor in S_2, satisfying $f \leq_T \emptyset'''$. When both systems are constructively recursive, the function f can be chosen to satisfy $f \leq_T \emptyset''$.*

What about lower bounds on the trade-off function f? In fact, we show that there is a relation between the function f and the equivalence problem between S_1 and S_2 descriptors, in the sense that, whenever the former problem becomes easy, the latter is easy too.

Theorem 9. *Let S_1 and S_2 be two descriptional systems, c_1 be a measure for S_1, c_2 be an sn-measure for S_2, and $f : \mathbb{N} \to \mathbb{N}$ be a total function that serves as an upper bound for the increase in complexity when changing from a descriptor in S_1 to an equivalent descriptor in S_2. Then we have:*

1. *If both S_1 and S_2 are constructively recursively enumerable and $f \leq_T \emptyset''$ then the S_2-ness of S_1 descriptors is recursive in $\emptyset''$.*
2. *If both descriptional systems are constructively recursive and $f \leq_T \emptyset'$ then the S_2-ness of S_1 descriptors is recursive in $\emptyset'$.*

Thus, we can show that only *two* types of non-recursive trade-offs within the recursively enumerable languages exist! First consider the context-free grammars and the right-linear context-free grammars (or equivalently finite automata) as descriptional systems. Thus, we want to consider the trade-off between context-free languages and regular languages. In [22] it was shown that this trade-off is non-recursive. By Theorem 8, one can choose the upper bound function f such that $f \leq_T \emptyset''$. On the other hand, if $f \leq_T \emptyset'$, then by Theorem 9 we deduce that checking regularity for context-free grammars is recursive in $\emptyset'$ and hence belongs to Δ_2. This is a contradiction, because in [4] this problem is classified to be Σ_2-complete. So, we obtain a non-recursive trade-off somewhere in between $\emptyset''$ and $\emptyset'$, that is, $f \leq_T \emptyset''$ but $f \not\leq_T \emptyset'$.

Furthermore, based on the previous observations, in [7] another proof scheme for non-recursive trade-offs is developed. The statement reads as follows.

Theorem 10. *Let S_1 and S_2 be two descriptional systems, c_1 be a measure for S_1, and c_2 be an sn-measure for S_2. Then the trade-off between S_1 and S_2 is non-recursive, if one of the following two cases applies:*

1. *If both S_1 and S_2 are effectively recursively enumerable and the S_2-ness of S_1 descriptors is at least Σ_3-hard, or*
2. *if both descriptional systems are effectively recursive and the S_2-ness of S_1 descriptors is at least Σ_2-hard.*

Here hardness is meant with respect to many-one reducibility.

4.2 Deriving Non-recursive Trade-Offs from Closure Properties

Next we present two rather abstract methods for proving non-recursive trade-offs. In contrast to previous schemes, here we only use properties that are known from the literature for many descriptional systems: these concern the decidability of basic decision problems on the one hand, and closure properties familiar from the study of abstract families of languages on the other hand.

To this end, we define effective closure of descriptional systems under language operations. We illustrate the definition by example of language union. Let S be a descriptional system. We say that S is *effectively closed under union*, if there is an effective construction that, given some pair of descriptors D_1 and D_2 from S, yields a descriptor from S for $L(D_1) \cup L(D_2)$. Effective closure under other language operations is defined in a similar vein.

A descriptional system is called an *effective trio* [15], if it is effectively closed under λ-free homomorphism, inverse homomorphism, and intersection with regular languages. If it is also effectively closed under general homomorphism, we speak of an *effective full trio*. Every trio is also effectively closed under concatenation with regular sets.

Theorem 11. *Let S_1 and S_2 be two descriptional systems that are effective full trios, S_1 be effectively recursively enumerable, c_1 be a measure for S_1, and c_2 be an sn-measure for S_2. If*

1. *the infiniteness problem for S_1 is not semi-decidable and*
2. *the infiniteness problem for S_2 is decidable,*

then the trade-off between S_1 and S_2 is non-recursive.

The advantage of this quite abstract method for proving non-recursive trade-offs lies in the fact that in automata and formal language theory language families and their closure properties are very well investigated. We give an assorted list of examples:

Example 5. For any applicable s-measures, the following trade-offs are non-recursive:

1. between Turing machines and finite automata,
2. between Turing machines and (linear) context-free grammars,
3. between Turing machines and ET0L systems, and
4. between Turing machines and (linear) context-free indexed grammars,

where indexed grammars were introduced in [1], and ET0L systems were studied, for example, in [24]. ∎

Yet another method is given next, where the undecidable problem the non-recursive trade-off relies on is the emptiness problem and not the infiniteness problems as in the previous theorem.

Theorem 12. *Let S_1 and S_2 be two descriptional systems that are effective trios, S_1 be effectively recursively enumerable, c_1 be a measure for S_1, and c_2 be an sn-measure for S_2. If*

1. S_1 *has a decidable word problem but an undecidable emptiness problem, and*
2. S_2 *has a decidable emptiness problem,*

then the trade-off between S_1 and S_2 is non-recursive.

Since the emptiness problem is more likely to be studied for a formal language family, it is somehow easier to apply the previous theorem. Again, we give an assorted list of non-recursive trade-offs between different formal language generating mechanisms.

Example 6. For any applicable s-measures, the following trade-offs are non-recursive:

1. between growing context-sensitive grammars and finite automata,
2. between growing context-sensitive grammars and (linear) context-free grammars,
3. between growing context-sensitive grammars and ET0L systems,
4. between growing context-sensitive grammars and indexed grammars,
5. between context-sensitive grammars and finite automata,
6. between context-sensitive grammars and ET0L systems,
7. between context-sensitive grammars and (linear) context-free grammars, and
8. between context-sensitive grammars and indexed grammars,

where growing context-sensitive grammars are studied, for example, in [3,5]. Observe that context-sensitive grammars form an effective trio. ∎

5   Conclusions

We have given a brief overview on non-recursive trade-offs. What we have not talked about are recursive trade-offs. The best known recursive trade-off is between nondeterministic and deterministic finite automata, but this is only one example of many others. Recursive trade-offs are as present as non-recursive ones. In fact, one can show that there exist two descriptional systems S_1 and S_2 and two sn-measures c_1 and c_2 such that $\Omega(f)$ may serve as a lower bound in the order of magnitude when changing from a descriptor in S_1 to an equivalent descriptor in S_2, where $f \colon \mathbb{N} \to \mathbb{N}$ is any total recursive function [7].

References

1. Aho, A.V.: Indexed grammars - an extension of context-free grammars. J. ACM **15**, 647–671 (1968)
2. Bordihn, H., Holzer, M., Kutrib, M.: Unsolvability levels of operation problems for subclasses of context-free languages. Int. J. Found. Comput. Sci. **16**, 423–440 (2005)
3. Buntrock, G., Loryś, K.: On growing context-sensitive languages. In: Kuich, W. (ed.) ICALP 1992. LNCS, vol. 623, pp. 77–88. Springer, Heidelberg (1992). https://doi.org/10.1007/3-540-55719-9_65
4. Cudia, D.F.: The degree hierarchy of undecidable problems of formal grammars. In: Symposium on Theory of Computing (STOC 1970), pp. 10–21. ACM Press (1970)
5. Dahlhaus, E., Warmuth, M.K.: Membership for growing context-sensitive grammars is polynomial. J. Comput. Syst. Sci. **33**, 456–472 (1986)

6. Goldstine, J., Kappes, M., Kintala, C.M.R., Leung, H., Malcher, A., Wotschke, D.: Descriptional complexity of machines with limited resources. J. UCS **8**, 193–234 (2002)
7. Gruber, H., Holzer, M., Kutrib, M.: On measuring non-recursive trade-offs. J. Autom. Lang. Comb. **15**, 107–120 (2010)
8. Hartmanis, J.: On the succinctness of different representations of languages. SIAM J. Comput. **9**, 114–120 (1980)
9. Hartmanis, J.: On Gödel speed-up and succinctness of language representations. Theoret. Comput. Sci. **26**, 335–342 (1983)
10. Hartmanis, J., Baker, T.P.: Relative succinctness of representations of languages and separation of complexity classes. In: Bečvář, J. (ed.) MFCS 1979. LNCS, vol. 74, pp. 70–88. Springer, Heidelberg (1979). https://doi.org/10.1007/3-540-09526-8_6
11. Hartmanis, J., Baker, T.P.: Succinctness, verifiability and determinism in representations of polynomial-time languages. In: IEEE Symposium on Foundations and Computer Science, pp. 392–396 (1979)
12. Herzog, C.: Pushdown automata with bounded nondeterminism and bounded ambiguity. Theoret. Comput. Sci. **181**, 141–157 (1997)
13. Holzer, M., Kutrib, M.: Descriptional complexity - an introductory survey. In: Scientific Applications of Language Methods, pp. 1–58. Imperial College Press (2010)
14. Holzer, M., Kutrib, M.: One-time nondeterministic computations. Int. J. Found. Comput. Sci. (to appear)
15. Hopcroft, J.E., Ullman, J.D.: Introduction to Automata Theory, Languages, and Computation. Addison-Wesley, Boston (1979)
16. Kapoutsis, C.A.: From $k + 1$ to k heads the descriptive trade-off is non-recursive. In: Descriptional Complexity of Formal Systems (DCFS 2004), pp. 213–224 (2004)
17. Kutrib, M.: On the descriptional power of heads, counters, and pebbles. Theoret. Comput. Sci. **330**, 311–324 (2005)
18. Kutrib, M.: The phenomenon of non-recursive trade-offs. Int. J. Found. Comput. Sci. **16**, 957–973 (2005)
19. Malcher, A.: Descriptional complexity of cellular automata and decidability questions. J. Autom. Lang. Comb. **7**, 549–560 (2002)
20. Malcher, A.: On the descriptional complexity of iterative arrays. IEICE Trans. Inf. Syst. **E87-D**, 721–725 (2004)
21. Malcher, A.: On recursive and non-recursive trade-offs between finite-turn pushdown automata. J. Autom. Lang. Comb. **12**, 265–277 (2007)
22. Meyer, A.R., Fischer, M.J.: Economy of description by automata, grammars, and formal systems. In: Symposium on Switching and Automata Theory (SWAT 1971), pp. 188–191. IEEE (1971)
23. Rogers, H.: Theory of Recursive Functions and Effective Computability. McGraw-Hill, New York (1967)
24. Rozenberg, G., Salomaa, A.: The Mathematical Theory of L Systems. Academic Press, New York (1980)
25. Schmidt, E.M., Szymanski, T.G.: Succinctness of descriptions of unambiguous context-free languages. SIAM J. Comput. **6**, 547–553 (1977)
26. Sunckel, B.: On the descriptional complexity of metalinear CD grammar systems. In: Descriptional Complexity of Formal Systems (DCFS 2004), pp. 260–273 (2004)
27. Valiant, L.G.: Regularity and related problems for deterministic pushdown automata. J. ACM **22**, 1–10 (1975)
28. Valiant, L.G.: A note on the succinctness of descriptions of deterministic languages. Inform. Control **32**, 139–145 (1976)

Probabilistic Analysis of Facility Location on Random Shortest Path Metrics

Stefan Klootwijk$^{(\boxtimes)}$ and Bodo Manthey

Faculty of Electrical Engineering, Mathematics and Computer Science,
University of Twente, P.O. Box 217, 7500 AE Enschede, The Netherlands
{s.klootwijk,b.manthey}@utwente.nl

Abstract. The facility location problem is an $\mathcal{NP}$-hard optimization problem. Therefore, approximation algorithms are often used to solve large instances. Such algorithms often perform much better than worst-case analysis suggests. Therefore, probabilistic analysis is a widely used tool to analyze such algorithms. Most research on probabilistic analysis of $\mathcal{NP}$-hard optimization problems involving metric spaces, such as the facility location problem, has been focused on Euclidean instances, and also instances with independent (random) edge lengths, which are non-metric, have been researched. We would like to extend this knowledge to other, more general, metrics.

We investigate the facility location problem using random shortest path metrics. We analyze some probabilistic properties for a simple greedy heuristic which gives a solution to the facility location problem: opening the κ cheapest facilities (with κ only depending on the facility opening costs). If the facility opening costs are such that κ is not too large, then we show that this heuristic is asymptotically optimal. On the other hand, for large values of κ, the analysis becomes more difficult, and we provide a closed-form expression as upper bound for the expected approximation ratio. In the special case where all facility opening costs are equal this closed-form expression reduces to $O(\sqrt[4]{\ln(n)})$ or $O(1)$ or even $1 + o(1)$ if the opening costs are sufficiently small.

1 Introduction

Large-scale combinatorial optimization problems, such as the facility location problem, show up in many applications. These problems become computationally intractable as the instances grow. This issue is often tackled by (successfully) using approximation algorithms or ad-hoc heuristics to solve these optimization problems. In practical situations these, often simple, heuristics have a remarkable performance, even though theoretical results about them are way more pessimistic.

Over the last decades, probabilistic analysis has become an important tool to explain this difference. One of the main challenges here is to come up with

A full version containing all proofs is available at https://arxiv.org/abs/1903.11980.

F. Manea et al. (Eds.): CiE 2019, LNCS 11558, pp. 37–49, 2019.
https://doi.org/10.1007/978-3-030-22996-2_4

a good probabilistic model for generating instances: this model should reflect realistic instances, but it should also be sufficiently simple in order to make the probabilistic analysis possible.

Until recently, in almost all cases either instances with independent edge lengths, or instances with Euclidean distances have been used for this purpose (e.g. [1,7]). These models are indeed sufficiently simple, but they have shortcomings with respect to reflecting realistic instances: realistic instances are often metric, although not Euclidean, and the independent edge lengths do not even yield a metric space.

In order to overcome this, Bringmann et al. [3] used the following model for generating random metric spaces, which had been proposed by Karp and Steele [12]. Given an undirected complete graph, start by drawing random edge weights for each edge independently and then define the distance between any two vertices as the total weight of the shortest path between them, measured with respect to the random weights. Bringmann et al. called this model *random shortest path metrics*. This model is also known as *first-passage percolation*, introduced by Hammersley and Welsh as a model for fluid flow through a (random) porous medium [8,10].

1.1   Related Work

Although a lot of studies have been conducted on random shortest path metrics, or first-passage percolation (e.g. [5,9,11]), systematic research of the behavior of (simple) heuristics and approximation algorithms for optimization problems on random shortest path metrics was initiated only recently [3]. They provide some structural properties of random shortest path metrics, including the existence of a good clustering. These properties are then used for a probabilistic analysis of simple algorithms for several optimization problems, including the minimum-weight perfect matching problem and the k-median problem.

For the facility location problem, several sophisticated polynomial-time approximation algorithms exist, the best one currently having a worst-case approximation ratio of 1.488 [13]. Flaxman et al. conducted a probabilistic analysis for the facility location problem using Euclidean distances [6]. They expected to show that some polynomial-time approximation algorithms would be asymptotically optimal under these circumstances, but found out that this is not the case. On the other hand, they described a trivial heuristic which is asymptotically optimal in the Euclidean model.

1.2   Our Results

This paper aims at extending our knowledge about the probabilistic behavior of (simple) heuristics and approximation algorithms for optimization problems using random shortest path metrics. We will do so by investigating the probabilistic properties of a rather simple heuristic for the facility location problem, which opens the κ cheapest facilities (breaking ties arbitrarily) where κ only depends on the facility opening costs. Due to the simple structure of this heuristic, our results are more structural than algorithmic in nature.

We show that this heuristic yields a $1 + o(1)$ approximation ratio in expectation if the facility opening costs are such that $\kappa \in o(n)$. For $\kappa \in \Theta(n)$ the analysis becomes more difficult, and we provide a closed-form expression as upper bound for the expected approximation ratio. We will also show that this closed-form expression is $O(\sqrt[4]{\ln(n)})$ if all facility opening costs are equal. This can be improved to $O(1)$ or even $1 + o(1)$ when the facility opening costs are sufficiently small. Note that we will focus on the expected approximation ratio and not on the ratio of expectations, since a disadvantage of the latter is that it does not directly compare the performance of the heuristic on specific instances.

We start by giving a mathematical description of random shortest path metrics and the facility location problem (Sect. 2). After that, we introduce our simple heuristic properly and have a brief look at its behavior (Sect. 3). Then we present some general technical results (Sect. 4) and two different bounds for the optimal solution (Sect. 5) that we use to prove our main results in Sect. 6. We conclude with some final remarks (Sect. 7).

2 Notation and Model

In this paper, we use $X \sim P$ to denote that a random variable X is distributed using a probability distribution P. $\mathrm{Exp}(\lambda)$ is being used to denote the exponential distribution with parameter λ. In particular, we use $X \sim \sum_{i=1}^{n} \mathrm{Exp}(\lambda_i)$ to denote that X is the sum of n independent exponentially distributed random variables with parameters $\lambda_1, \ldots, \lambda_n$. If $\lambda_1 = \ldots = \lambda_n = \lambda$, then X is a Gamma distributed random variable with parameters n and λ, denoted by $X \sim \Gamma(n, \lambda)$.

For $n \in \mathbb{N}$, we use $[n]$ as shorthand notation for $\{1, \ldots, n\}$. If $X_1, \ldots, X_m$ are m random variables, then $X_{(1)}, \ldots, X_{(m)}$ are the order statistics corresponding to $X_1, \ldots, X_m$ if $X_{(i)}$ is the ith smallest value among $X_1, \ldots, X_m$ for all $i \in [m]$. Furthermore we use H_n as shorthand notation for the nth harmonic number, i.e., $H_n = \sum_{i=1}^{n} 1/i$. Finally, if a random variable X is stochastically dominated by a random variable Y, i.e., we have $F_X(x) \geq F_Y(x)$ for all x (where $X \sim F_X$ and $Y \sim F_Y$), we denote this by $X \precsim Y$.

Random Shortest Path Metrics. Given an undirected complete graph $G = (V, E)$ on n vertices, we construct the corresponding random shortest path metric as follows. First, for each edge $e \in E$, we draw a random edge weight $w(e)$ independently from an exponential distribution[1] with parameter 1. Given these random edge weights $w(e)$, the distance $d(u, v)$ between each pair of vertices $u, v \in V$ is defined as the minimum total weight of a u, v-path in G. Note

---

[1] Exponential distributions are technically easiest to handle due to their memorylessness property. A (continuous, non-negative) probability distribution of a random variable X is said to be memoryless if and only if $\mathbb{P}(X > s + t \mid X > t) = \mathbb{P}(X > s)$ for all $s, t \geq 0$. [17, p. 294].

that this definition yields the following properties: $d(v, v) = 0$ for all $v \in V$, $d(u, v) = d(v, u)$ for all $u, v \in V$, and $d(u, v) \leq d(u, s) + d(s, v)$ for all $u, s, v \in V$. We call the complete graph with distances d obtained from this process a random shortest path metric.

Facility Location Problem. We consider the (uncapacitated) facility location problem, in which we are given a complete undirected graph $G = (V, E)$ on n vertices, distances $d : V \times V \to \mathbb{R}_{\geq 0}$ between each pair of vertices, and opening costs $f : V \to \mathbb{R}_{>0}$. In this paper, the distances are randomly generated, according to the random shortest path metric described above. Moreover, w.l.o.g. we assume that the vertices are numbered in such a way that the opening costs satisfy $f_1 \leq f_2 \leq \ldots \leq f_n$ and we assume that these costs are predetermined, independent of the random edge weights. We will use F_k as a shorthand notation for $\sum_{i=1}^{k} f_i$. Additionally, we assume that the ratios between the opening costs are polynomially bounded, i.e., we assume $f_n/f_1 \leq n^q$ for some constant q as $n \to \infty$.

The goal of the facility location problem is to find a nonempty subset $U \subseteq V$ such that the total cost $c(U) := f(U) + \sum_{v \in V} \min_{u \in U} d(u, v)$ is minimal, where $f(U)$ denotes the total opening cost of all facilities in U. This problem is $\mathcal{NP}$-hard [4]. We use OPT to denote the total cost of an optimal solution, i.e.,

$$\mathsf{OPT} = \min_{\varnothing \neq U \subseteq V} c(U).$$

One of the tools we use in our proofs in Sect. 6 involves fixing the number of facilities that has to be opened. We use OPT_k to denote the total cost of the best solution to the facility location problem with the additional constraint that exactly k facilities need to be opened, i.e.,

$$\mathsf{OPT}_k = \min_{\substack{\varnothing \neq U \subseteq V \\ |U|=k}} c(U).$$

Note that $\mathsf{OPT} = \min_{k \in [n]} \mathsf{OPT}_k$ by these definitions.

3   A Simple Heuristic and Some of Its Properties

In this paper we are interested in a rather simple heuristic that only takes the facility opening costs f_i into account while determining which facilities to open and which not, independently of the metric space. Define $\kappa := \kappa(n; f_1, \ldots, f_n) = \max\{i \in [n] : f_i < 1/(i - 1)\}$. Then our heuristic opens the κ cheapest facilities (breaking ties arbitrarily). Note that in the special case where all opening costs are the same, i.e. $f_1 = \ldots = f_n = f$, this corresponds to $\kappa = \min\{\lceil 1/f \rceil, n\}$.

This rather particular value of κ originates from the following intuitive argument. Based on the results of Bringmann et al. [3, Lemma 5.1] (see below) we know that the expected cost of the solution that opens the k cheapest facilities is given by $g(k) := F_k + H_{n-1} - H_{k-1}$. This convex function decreases as long as k satisfies $f_k < 1/(k-1)$. Therefore, at least intuitively, the value of κ that we use is likely to provide a relatively 'good' solution.

We will show that this is indeed the case. Our main result will be split into two parts, based on the actual value of κ. If $\kappa \in o(n)$ (i.e. if there are 'many' relatively expensive facilities), then we will show that our simple heuristic is asymptotically optimal for any polynomially bounded opening costs (that satisfy $\kappa \in o(n)$). On the other hand, if $\kappa \in \Theta(n)$, then the analysis becomes more difficult, and we will only provide a closed-form expression that can be used to determine an upper bound for the expected approximation ratio. We will show that this expression yields an $O(\sqrt[4]{\ln(n)})$ approximation ratio in the special case with $f_1 = \ldots = f_n = f$, and $O(1)$ or even $1 + o(1)$ if f is sufficiently small.

Throughout the remainder of this paper we will use ALG to denote the value of the solution provided by this heuristic.

Probability Distribution of ALG. In this section we derive the probability distribution of the value of the solution provided by our simple greedy heuristic, ALG, and derive its expectation.

If $\kappa = n$, then ALG denotes the cost of the solution which opens a facility at every vertex $v \in V$. So, we have $\mathsf{ALG} = F_n$, and, in particular, $\mathbb{P}(\mathsf{ALG} = F_n) = 1$.

If $1 \le \kappa < n$, then the distribution of ALG is less trivial. In this case, the total opening costs are given by F_κ, whereas, the distribution of the connection costs is known and given by $\sum_{i=\kappa}^{n-1} \mathrm{Exp}(i)$ [3, Sect. 5]. This results in $\mathsf{ALG} - F_\kappa \sim \sum_{i=\kappa}^{n-1} \mathrm{Exp}(i)$.

Using this probability distribution, we can derive the expected value of ALG. If $\kappa = n$, then it follows trivially that $\mathbb{E}[\mathsf{ALG}] = F_n$. If $1 \le \kappa < n$, then we have

$$\mathbb{E}[\mathsf{ALG}] = F_\kappa + \sum_{i=\kappa}^{n-1} \frac{1}{i} = F_\kappa + H_{n-1} - H_{\kappa-1} = F_\kappa + \ln(n/\kappa) + \Theta(1).$$

4   Technical Observations

In this section we present some technical lemmas that are being used for the proofs of our theorems in Sect. 6. These lemmas do not provide new structural insights, but are nonetheless very helpful for our proofs.

First of all, we will use the Cauchy-Schwarz inequality to bound the expected approximation ratio of our simple greedy heuristic. For general random variables X, Y, this inequality states that $|\mathbb{E}[XY]| \le \sqrt{\mathbb{E}[X^2]\mathbb{E}[Y^2]}$.

Secondly, we will bound a sum of exponential distributions by a Gamma distribution. The following Lemma enables us to do so.

Lemma 1 ([18, **Ex. 1.A.24**]). *Let $X_i \sim \mathrm{Exp}(\lambda_i)$ independently, $i = 1, \ldots, m$. Moreover, let $Y_i \sim \mathrm{Exp}(\eta)$ independently, $i = 1, \ldots, m$. Then we have*

$$\sum_{i=1}^{m} X_i \succsim \sum_{i=1}^{m} Y_i \qquad \text{if and only if} \qquad \prod_{i=1}^{m} \lambda_i \le \eta^m.$$

We will use the following upper bound for the expectation of the maximum of a number of (dependent) random variables.

Lemma 2 ([2, Thm. 2.1]). *Let $X_1, \ldots, X_n$ be a sequence of random variables, each with finite mean and variance. Then it follows that*

$$\mathbb{E}\left[\max_i X_i\right] \leq \max_i \mathbb{E}\left[X_i\right] + \sqrt{\frac{n-1}{n} \cdot \sum_{i=1}^{n} \mathrm{Var}(X_i)}.$$

We will also make use of Rényi's representation [15,16] in order to be able to link sums and order statistics of exponentially distributed random variables. It states the following.

Lemma 3. *Let $X_i \sim \mathrm{Exp}(\lambda)$ independently, $i = 1, \ldots, m$, and let $X_{(1)}, \ldots, X_{(m)}$ be the order statistics corresponding to $X_1, \ldots, X_m$. Then, for any $i \in [m]$,*

$$X_{(i)} = \frac{1}{\lambda} \sum_{j=1}^{i} \frac{Z_j}{m - j + 1},$$

where $Z_j \sim \mathrm{Exp}(1)$ independently, and where "$=$" means equal distribution.

A special case of Rényi's representation is given by the following corollary.

Corollary 4. *Let $Y_i \sim \mathrm{Exp}(1)$ independently, $i = 1, \ldots, n-1$, and let $Y_{(1)}, \ldots, Y_{(n-1)}$ be the order statistics corresponding to $Y_1, \ldots, Y_{n-1}$. Then, for any $i \in [n-1]$,*

$$Y_{(n-i)} \sim \sum_{k=i}^{n-1} \mathrm{Exp}(k).$$

Moreover, we use the following bound for the expected value of the ratio X/Y for two dependent nonnegative variables X and Y, conditioned on the event that Y is relatively small.

Lemma 5. *Let X and Y be two arbitrary nonnegative random variables and assume that $\mathbb{P}(Y \leq \delta) = 0$ for some $\delta > 0$. Then, for any y that satisfies $\mathbb{P}(Y < y) > 0$, we have*

$$\mathbb{P}(Y < y) \cdot \mathbb{E}\left[\frac{X}{Y} \,\middle|\, Y < y\right] \leq \frac{1}{\delta^2} \cdot \mathbb{P}(Y < y) + \int_{1/\delta^2}^{\infty} \mathbb{P}(X \geq \sqrt{x}) \, \mathrm{d}x.$$

5 Bounds for the Optimal Solution

Not much is known about the distribution of the value of the optimal solution, OPT, and about the distributions of OPT_k. Therefore, in this section we derive two bounds for these optimal solutions which we can use in Sect. 6.

We start with an upper bound for the cumulative distribution function of OPT that works good for relative small values of OPT (i.e. values close to F_1).

Lemma 6. *Let $z \in [F_1, F_n]$ and define $\zeta := \max\{k : z \geq F_k\}$. Then, for any given opening costs f_i, we have*

$$\mathbb{P}(\mathsf{OPT} < z) \leq \sum_{i=1}^{\zeta} \binom{n}{i} \binom{n-1}{i-1} \left(1 - e^{-(z-F_i)}\right)^{n-i}.$$

Using the result of Lemma 1 we can also derive a stochastic lower bound for OPT_{n-k}.

Lemma 7. *Let $Z_k \sim \Gamma(k, e\binom{n}{2}/k)$. Then we have $\mathsf{OPT}_{n-k} \succsim F_{n-k} + Z_k$.*

6   Main Results

In this section we present our main results. We show that our simple heuristic is asymptotically optimal if $\kappa \in o(n)$, and we provide a closed-form expression as an upper bound for the expected approximation ratio if $\kappa \in \Theta(n)$. Finally we will evaluate this expression for the special case where $f_1 = \ldots = f_n = f$.

Theorem 8. *Define $\kappa := \kappa(n; f_1, \ldots, f_n) = \max\{i \in [n] : f_i < 1/(i-1)\}$ and assume that $\kappa \in o(n)$. Let ALG denote the total cost of the solution which opens, independently of the metric space, the κ cheapest facilities (breaking ties arbitrarily), i.e., the facilities with opening costs $f_1, \ldots, f_\kappa$. Then, it follows that*

$$\mathbb{E}\left[\frac{\mathsf{ALG}}{\mathsf{OPT}}\right] = 1 + o(1).$$

In order to prove this theorem, we consider the following three cases for the opening cost f_1 of the cheapest facility:

1. $f_1 \leq 1/\ln^2(n)$ as $n \to \infty$;
2. $f_1 \in O(\ln(n))$ and $f_1 > 1/\ln^2(n)$ as $n \to \infty$;
3. $f_1 \in \omega(\ln(n))$.

The proofs for Case 1 and 3 follow below, after we have stated two lemmas that are needed for Case 1.

Lemma 9. *Let $f_1 \leq 1/\ln^2(n)$ as $n \to \infty$. For sufficiently large n we have*

$$\int_{1/f_1^2}^{\infty} \mathbb{P}\left(\mathsf{ALG} \geq \sqrt{x}\right) dx \leq O\left(\frac{1}{n}\right).$$

Lemma 10. *Let q be a constant such that $f_n/f_1 \leq n^q$, let $\beta(n) = \ln(n/\kappa)(1 + 1/n)^{-1}$, take $\zeta(n) := \max\{i : \beta(n) \geq F_i\}$ and assume that $\kappa < n$. For sufficiently large n, and for any integer i with $1 \leq i \leq \zeta(n)$, we have*

$$\binom{n}{i}\binom{n-1}{i-1}\left(1 - e^{-(\beta(n) - F_i)}\right)^{n-i} \leq \frac{1}{n^{2q+4}}.$$

Proof (Theorem 8 (Case 1)). Let n be sufficiently large. By definition of κ, it follows that $f_\kappa < 1/(\kappa-1)$ and thus $F_\kappa < \kappa/(\kappa-1) \leq 2$ whenever $\kappa \geq 2$. If $\kappa = 1$, then we have $F_\kappa = f_1 < 1$ as $n \to \infty$. So, in any case we have $F_\kappa = O(1)$. Now, by our observations in Sect. 3 we know that $\mathbb{E}[\mathsf{ALG}] = F_\kappa + \ln(n/\kappa) + \Theta(1) = \ln(n/\kappa) + \Theta(1)$. Set $\beta(n) := \ln(n/\kappa)(1 + 1/n)^{-1}$ and observe that $\beta(n) \in \omega(1)$. Conditioning on the events $\mathsf{OPT} \geq \beta(n)$ and $\mathsf{OPT} < \beta(n)$ yields

$$\mathbb{E}\left[\frac{\mathsf{ALG}}{\mathsf{OPT}}\right] \leq \mathbb{E}\left[\frac{\mathsf{ALG}}{\beta(n)}\right] + \mathbb{P}(\mathsf{OPT} < \beta(n)) \cdot \mathbb{E}\left[\frac{\mathsf{ALG}}{\mathsf{OPT}} \,\middle|\, \mathsf{OPT} < \beta(n)\right].$$

We start by bounding the second part. Applying Lemma 5, with $X = \mathsf{ALG}$, $Y = \mathsf{OPT}$, $y = \beta(n)$ and $\delta = f_1$, we get

$$\mathbb{P}(\mathsf{OPT} < \beta(n)) \, \mathbb{E}\left[\frac{\mathsf{ALG}}{\mathsf{OPT}} \,\middle|\, \mathsf{OPT} < \beta(n)\right] \leq \frac{\mathbb{P}(\mathsf{OPT} < \beta(n))}{f_1^2} + \int\limits_{1/f_1^2}^{\infty} \mathbb{P}\left(\mathsf{ALG} \geq \sqrt{x}\right) \mathrm{d}x.$$

Note that we may use Lemma 5 since OPT can only take any value in $[f_1, \infty)$, and we have $\beta(n) > f_1$, which implies $\mathbb{P}(\mathsf{OPT} < \beta(n)) > 0$. The probability containing OPT can be bounded using Lemma 6, whereas the integral can be bounded by Lemma 9. Together, this yields

$$\mathbb{E}\left[\frac{\mathsf{ALG}}{\mathsf{OPT}}\right] \leq \mathbb{E}\left[\frac{\mathsf{ALG}}{\beta(n)}\right] + \frac{1}{f_1^2} \cdot \sum_{i=1}^{\zeta(n)} \binom{n}{i}\binom{n-1}{i-1}\left(1 - e^{-(\beta(n)-F_i)}\right)^{n-i} + O\left(\frac{1}{n}\right),$$

where $\zeta(n) := \max\{i : \beta(n) \geq F_i\}$. The terms of the summation can be bounded by Lemma 10. Using this lemma, we obtain that

$$\mathbb{E}\left[\frac{\mathsf{ALG}}{\mathsf{OPT}}\right] \leq \mathbb{E}\left[\frac{\mathsf{ALG}}{\beta(n)}\right] + \frac{1}{f_1^2} \cdot \sum_{i=1}^{\zeta(n)} \frac{1}{n^{2q+4}} + O\left(\frac{1}{n}\right) \leq \mathbb{E}\left[\frac{\mathsf{ALG}}{\beta(n)}\right] + \frac{1/f_1^2}{n^{2q+3}} + O\left(\frac{1}{n}\right),$$

since $\zeta(n) \leq n$ by definition. Moreover, since $\kappa \in o(n)$ implies $f_n > 1/n$ as $n \to \infty$, we also have $f_1 \geq f_n/n^q > 1/n^{q+1}$ as $n \to \infty$ for some constant q. This results in

$$\mathbb{E}\left[\frac{\mathsf{ALG}}{\mathsf{OPT}}\right] \leq \mathbb{E}\left[\frac{\mathsf{ALG}}{\beta(n)}\right] + n^{2q+2} \cdot \frac{1}{n^{2q+3}} + O\left(\frac{1}{n}\right) = \mathbb{E}\left[\frac{\mathsf{ALG}}{\beta(n)}\right] + O\left(\frac{1}{n}\right).$$

Since we started with $\beta(n) = \ln(n/\kappa)(1 + 1/n)^{-1}$ and $n/\kappa \in \omega(1)$ (since $\kappa \in o(n)$), it follows that

$$\mathbb{E}\left[\frac{\mathsf{ALG}}{\mathsf{OPT}}\right] \leq \frac{\mathbb{E}[\mathsf{ALG}]}{\beta(n)} + O\left(\frac{1}{n}\right) \leq \frac{\ln(n/\kappa) + \Theta(1)}{\ln(n/\kappa)}\left(1 + \frac{1}{n}\right) + O\left(\frac{1}{n}\right) = 1 + o(1),$$

which finishes the proof of this case. $\square$

Proof (Theorem 8 (Case 3)). For sufficiently large n, we have $f_1 > 1$, and thus $\kappa = 1$ since $f_2 \geq f_1 > 1 = 1/(2-1)$. Therefore, using our observations in Sect. 3, we can derive that $\mathbb{E}[\mathsf{ALG}] = F_1 + \ln(n) + \Theta(1)$ for sufficiently large n. Moreover, we know that $\mathsf{OPT} \geq F_1$. Using this observation, it follows that

$$\mathbb{E}\left[\frac{\mathsf{ALG}}{\mathsf{OPT}}\right] \leq \mathbb{E}\left[\frac{\mathsf{ALG}}{F_1}\right] = \frac{F_1 + \ln(n) + \Theta(1)}{F_1} = 1 + \frac{\ln(n) + \Theta(1)}{\omega(\ln(n))} = 1 + o(1),$$

which finishes the proof of this case. $\qquad\square$

Theorem 11. *Define $\kappa := \kappa(n; f_1, \ldots, f_n) = \max\{i \in [n] : f_i < 1/(i-1)\}$ and assume that $\kappa \in \Theta(n)$. Let ALG denote the total cost of the solution which opens, independently of the metric space, the κ cheapest facilities (breaking ties arbitrarily), i.e., the facilities with opening costs $f_1, \ldots, f_\kappa$. Then we can bound the expected approximation ratio by*

$$\mathbb{E}\left[\frac{\mathsf{ALG}}{\mathsf{OPT}}\right] \leq \sqrt{\max\left\{\frac{1}{F_n^2}, \max_{k \in [n-1]} O\left(\frac{n^4 F_{n-k} + kn^2}{n^4 F_{n-k}^3 + k^4 F_{n-k}}\right)\right\}}$$

$$+ \sqrt[4]{O\left(\frac{n^6 F_{n-1}^2 + n^2}{n^8 F_{n-1}^7 + F_{n-1}^3} + \frac{n^{10} F_{n-2}^3 + n^4}{n^{14} F_{n-2}^9 + F_{n-2}^2} + \sum_{k=3}^{n-1} \frac{k^3 n^{12} F_{n-k}^3 + k^9 n^6}{n^{16} F_{n-k}^9 + k^{16} F_{n-k}}\right)}.$$

Moreover, if $\kappa = n$, then the expected approximation ratio can be bounded by

$$\mathbb{E}\left[\frac{\mathsf{ALG}}{\mathsf{OPT}}\right] \leq F_n \cdot \sqrt{\max\left\{\frac{1}{F_n^2}, \max_{k \in [n-1]} O\left(\frac{n^4 F_{n-k} + kn^2}{n^4 F_{n-k}^3 + k^4 F_{n-k}}\right)\right\}}$$

$$+ F_n \cdot \sqrt[4]{O\left(\frac{n^6 F_{n-1}^2 + n^2}{n^8 F_{n-1}^7 + F_{n-1}^3} + \frac{n^{10} F_{n-2}^3 + n^4}{n^{14} F_{n-2}^9 + F_{n-2}^2} + \sum_{k=3}^{n-1} \frac{k^3 n^{12} F_{n-k}^3 + k^9 n^6}{n^{16} F_{n-k}^9 + k^{16} F_{n-k}}\right)}.$$

The proof of this theorem requires some tedious computations which we used to bound exponential integrals by Padé approximants [14]. The results of these computations are stated in the following lemmas.

Lemma 12. *Let $X_k = 1/(F_{n-k} + Z_k)^2$ where $Z_k \sim \Gamma(k, e\binom{n}{2}/k)$. Then, for any $k \in \{1, \ldots, n-1\}$ it follows that*

$$\mathbb{E}[X_k] \leq O\left(\frac{n^4 F_{n-k} + kn^2}{n^4 F_{n-k}^3 + k^4 F_{n-k}}\right).$$

Lemma 13. *Let $X_k = 1/(F_{n-k} + Z_k)^2$ where $Z_k \sim \Gamma(k, e\binom{n}{2}/k)$. Then, for any $k \in \{3, \ldots, n-1\}$ it follows that*

$$\mathbb{E}[X_k^2] - (\mathbb{E}[X_k])^2 \leq O\left(\frac{k^3 n^{12} F_{n-k}^3 + k^9 n^6}{n^{16} F_{n-k}^9 + k^{16} F_{n-k}}\right),$$

whereas for $k = 1$ and $k = 2$ we have

$$\mathbb{E}\left[X_1^2\right] - \left(\mathbb{E}\left[X_1\right]\right)^2 \leq O\left(\frac{n^6 F_{n-1}^2 + n^2}{n^8 F_{n-1}^7 + F_{n-1}^3}\right),$$

$$\mathbb{E}\left[X_2^2\right] - \left(\mathbb{E}\left[X_2\right]\right)^2 \leq O\left(\frac{n^{10} F_{n-2}^3 + n^4}{n^{14} F_{n-2}^9 + F_{n-2}^2}\right).$$

Proof (Theorem 11). Using the Cauchy-Schwarz inequality for random variables (see Sect. 4), we obtain

$$\mathbb{E}\left[\frac{\mathsf{ALG}}{\mathsf{OPT}}\right] \leq \sqrt{\mathbb{E}\left[\mathsf{ALG}^2\right]} \cdot \sqrt{\mathbb{E}\left[\frac{1}{\mathsf{OPT}^2}\right]}.$$

Recall from Sect. 3 that we know the distribution of ALG. We can use this to compute and bound $\mathbb{E}[\mathsf{ALG}^2]$. If $\kappa < n$, then we obtain

$$\mathbb{E}\left[\mathsf{ALG}^2\right] = (F_\kappa + H_{n-1} - H_{\kappa-1})^2 + \sum_{i=\kappa}^{n-1} \frac{1}{i^2} = (F_\kappa + \ln(n/\kappa) + \Theta(1))^2 + \sum_{i=\kappa}^{n-1} \frac{1}{i^2}$$

which is $O(1)$ since $\kappa \in \Theta(n)$ and $F_\kappa \leq \kappa f_\kappa < \kappa/(\kappa-1) \leq 2$ for such κ. If $\kappa = n$, then we have $\mathbb{E}[\mathsf{ALG}^2] = F_n^2$.

It remains to bound $\mathbb{E}[1/\mathsf{OPT}^2]$. We start by using our final notion from Sect. 2, and subsequently using the result of Lemma 7. This yields

$$\mathbb{E}\left[\frac{1}{\mathsf{OPT}^2}\right] = \mathbb{E}\left[\max_k \frac{1}{\mathsf{OPT}_{n-k}^2}\right] \leq \mathbb{E}\left[\max_k \frac{1}{(F_{n-k} + Z_k)^2}\right],$$

where $Z_k \sim \Gamma(k, e\binom{n}{2}/k)$ and where we take the maximum over $k \in \{0, \ldots, n-1\}$. Next we use the result of Lemma 2 to get the maximum operator out of the expectation. This yields

$$\mathbb{E}\left[\frac{1}{\mathsf{OPT}^2}\right] \leq \max_k \mathbb{E}\left[\frac{1}{(F_{n-k} + Z_k)^2}\right] + \sqrt{\frac{n-1}{n} \cdot \sum_{k=0}^{n-1} \mathrm{Var}\left(\frac{1}{(F_{n-k} + Z_k)^2}\right)}$$

$$\leq \max_k \mathbb{E}\left[X_k\right] + \sqrt{\sum_{k=0}^{n-1} \left(\mathbb{E}\left[X_k^2\right] - (\mathbb{E}\left[X_k\right])^2\right)},$$

where we also used $X_k := 1/(F_{n-k} + Z_k)^2$ to shorten notation, applied the difference formula for the variance, and used the inequality $(n-1)/n \leq 1$.

Since we know the distribution of Z_k, we can compute and subsequently bound the expectations of X_k that occur in this last expression. For $k = 0$ we have $Z_0 = 0$, and thus $\mathbb{E}[X_0] = 1/F_n^2$ and $\mathbb{E}[X_0^2] - (\mathbb{E}[X_0])^2 = 0$. For $k \in [n-1]$, Lemmas 12 and 13 yield the bounds that we need to obtain the desired result.□

Finally, we evaluate the just proven bound for the approximation ratio for the special case where all facility opening costs are equal, i.e., $f_1 = \ldots = f_n = f$.

Corollary 14. *Assume that $f_1 = \ldots = f_n = f$. Define $\kappa := \kappa(n; f_1, \ldots, f_n) = \max\{i \in [n] : f_i < 1/(i-1)\} = \min\{\lceil 1/f \rceil, n\}$ and assume that $\kappa \in \Theta(n)$. Let ALG denote the total cost of the solution which opens, independently of the metric space, κ arbitrarily chosen facilities, e.g., the facilities $\{1, \ldots, \kappa\}$. Then, it follows that*

$$\mathbb{E}\left[\frac{\mathsf{ALG}}{\mathsf{OPT}}\right] = O(1) + O\left(\sqrt[4]{\ln(n)n^3 f^3}\right),$$

which for $f \in O(1/n\sqrt[3]{\ln(n)})$ is equal to $O(1)$. Moreover, if $f \in o(1/n^3)$, then this approximation ratio becomes $1 + o(1)$.

Before we can prove this corollary, we need two more lemmas.

Lemma 15. *Suppose that $f \in O(1/n)$. Then, for any $k \in [n-1]$ we have*

$$O\left(\frac{n^4(n-k)f + kn^2}{n^4(n-k)^3 f^3 + k^4(n-k)f}\right) \leq O\left(\frac{1}{n^2 f^2}\right).$$

Moreover, if $f \in o(1/n^3)$, then for any $k \in [n-1]$ we have

$$O\left(\frac{n^4(n-k)f + kn^2}{n^4(n-k)^3 f^3 + k^4(n-k)f}\right) \leq o\left(\frac{1}{n^2 f^2}\right).$$

Lemma 16. *Suppose that $f \in O(1/n)$. Then we have*

$$O\left(\frac{n^8 f^2 + n^2}{n^{15} f^7 + n^3 f^3} + \frac{n^{13} f^3 + n^4}{n^{25} f^9 + n^2 f^2} + \sum_{k=3}^{n-1} \frac{k^3 n^{12}(n-k)^3 f^3 + k^9 n^6}{n^{16}(n-k)^9 f^9 + k^{16}(n-k)f}\right)$$

$$\leq O\left(\frac{\ln(n)}{nf} + \frac{1}{n^4 f^4}\right).$$

Moreover, if $f \in o(1/n^3)$ then this result can be improved to $O(1/nf^3)$.

Proof (Corollary 14). Observe that $\kappa \in \Theta(n)$ and $\kappa = \min\{\lceil 1/f \rceil, n\}$ implies that $f \in O(1/n)$. We start by bounding the maximum in the first term. Lemma 15 shows that in our special case this maximum is asymptotically bounded by the first element. Using this result, we can now bound the maximum in the first term by $O(1/n^2 f^2)$. Moreover, if $f \in o(1/n^3)$, then for sufficiently large n it follows that the maximum is given by its first element, i.e., it is equal to $1/n^2 f^2$.

Next, we evaluate the sum of the variances. Lemma 16 provides the corresponding result. If $f \in \Theta(1/n)$, then it follows that

$$\mathbb{E}\left[\frac{\mathsf{ALG}}{\mathsf{OPT}}\right] \leq \sqrt{O\left(\frac{1}{n^2 f^2}\right)} + \sqrt[4]{O\left(\frac{\ln(n)}{nf} + \frac{1}{n^4 f^4}\right)} = O\left(\sqrt[4]{\ln(n)}\right).$$

If $f \in o(1/n)$ then we have for sufficiently large n that $\kappa = n$ and therefore

$$\mathbb{E}\left[\frac{\mathsf{ALG}}{\mathsf{OPT}}\right] \leq \sqrt{O\left(\frac{n^2 f^2}{n^2 f^2}\right)} + \sqrt[4]{O\left(\frac{\ln(n)n^4 f^4}{nf} + \frac{n^4 f^4}{n^4 f^4}\right)} = O\left(1 + \sqrt[4]{\ln(n)n^3 f^3}\right),$$

where the last term in general is bounded by $O(\sqrt[4]{\ln(n)})$ (since $f \in O(1/n)$) and more specifically by $O(1)$ if $f \in O(1/n\sqrt[3]{\ln(n)})$. If $f \in o(1/n^3)$, then we obtain

$$\mathbb{E}\left[\frac{\mathsf{ALG}}{\mathsf{OPT}}\right] \leq nf\left(\sqrt{\frac{1}{n^2f^2}} + \sqrt[4]{O\left(\frac{1}{nf^3}\right)}\right) = 1 + O\left(\sqrt[4]{n^3f}\right) = 1 + o(1),$$

which finishes this proof. □

7   Concluding Remarks

We have analyzed a rather simple heuristic for the (uncapacitated) facility location problem on random shortest path metrics. We have shown that in many cases this heuristic produces a solution which is surprisingly close to the optimal solution as the size of the instances grows. A logical next step would be to look at heuristics that are (slightly) more sophisticated, and see whether their performance on random shortest path metrics is better than our simple heuristic.

On the other hand there are many other $\mathcal{NP}$-hard (combinatorial) optimization problems for which it would be interesting to know how they behave on random short path metrics.

References

1. Ahn, S., Cooper, C., Cornuéjols, G., Frieze, A.: Probabilistic analysis of a relaxation for the k-median problem. Math. Oper. Res. **13**(1), 1–31 (1988). https://doi.org/10.1287/moor.13.1.1
2. Aven, T.: Upper (lower) bounds on the mean of the maximum (minimum) of a number of random variables. J. Appl. Probab. **22**(3), 723–728 (1985). https://doi.org/10.2307/3213876
3. Bringmann, K., Engels, C., Manthey, B., Rao, B.V.R.: Random shortest paths: non-euclidean instances for metric optimization problems. Algorithmica **73**(1), 42–62 (2015). https://doi.org/10.1007/s00453-014-9901-9
4. Cornuejols, G., Nemhauser, G.L., Wolsey, L.A.: The uncapacitated facility location problem. In: Mirchandani, P.B., Francis, R.L. (eds.) Discrete Location Theory, pp. 119–171. Wiley, New York (1990). chap. 3
5. Davis, R., Prieditis, A.: The expected length of a shortest path. Inf. Process. Lett. **46**(3), 135–141 (1993). https://doi.org/10.1016/0020-0190(93)90059-I
6. Flaxman, A.D., Frieze, A.M., Vera, J.C.: On the average case performance of some greedy approximation algorithms for the uncapacitated facility location problem. Comb. Probab. Comput. **16**(5), 713–732 (2007). https://doi.org/10.1017/S096354830600798X
7. Frieze, A.M., Yukich, J.E.: Probabilistic analysis of the TSP. In: Gutin, G., Punnen, A.P. (eds.) The Traveling Salesman Problem and Its Variations. Combinatorial Optimization, vol. 12, pp. 257–307. Springer, Boston (2007). https://doi.org/10.1007/0-306-48213-4_7
8. Hammersley, J.M., Welsh, D.J.A.: First-passage percolation, subadditive processes, stochastic networks, and generalized renewal theory. In: Neyman, J., Le Cam, L.M. (eds.) Bernoulli 1713 Bayes 1763 Laplace 1813, pp. 61–110. Springer, Heidelberg (1965). https://doi.org/10.1007/978-3-642-99884-3_7

9. Hassin, R., Zemel, E.: On shortest paths in graphs with random weights. Math. Oper. Res. **10**(4), 557–564 (1985). https://doi.org/10.1287/moor.10.4.557
10. Howard, C.D.: Models of first-passage percolation. In: Kesten, H. (ed.) Probability on Discrete Structures. Encyclopaedia of Mathematical Sciences (Probability Theory), pp. 125–173. Springer, Heidelberg (2004). https://doi.org/10.1007/978-3-662-09444-0_3
11. Janson, S.: One, two and three times log n/n for paths in a complete graph with random weights. Comb. Probab. Comput. **8**(4), 347–361 (1999). https://doi.org/10.1017/S0963548399003892
12. Karp, R.M., Steele, J.M.: Probabilistic analysis of heuristics. In: Lawler, E.L., Lenstra, J.K., Rinnooy Kan, A.H.G., Shmoys, D.B. (eds.) The Traveling Salesman Problem: A Guided Tour of Combinatorial Optimization, pp. 181–205. Wiley (1985)
13. Li, S.: A 1.488 approximation algorithm for the uncapacitated facility location problem. Inf. Comput. **222**, 45–58 (2013). https://doi.org/10.1016/j.ic.2012.01.007
14. Luke, Y.L.: Chapter XIV polynomial and rational approximations for the incomplete gamma function. In: The Special Functions and their Approximations, Mathematics in Science and Engineering, vol. 53, part 2, pp. 186–213. Elsevier (1969). https://doi.org/10.1016/S0076-5392(09)60074-6
15. Nagaraja, H.N.: Order statistics from independent exponential random variables and the sum of the top order statistics. In: Balakrishnan, N., Sarabia, J.M., Castillo, E. (eds.) Advances in Distribution Theory, Order Statistics, and Inference. Statistics for Industry and Technology, pp. 173–185. Birkhäuser, Boston (2006). https://doi.org/10.1007/0-8176-4487-3_11
16. Rényi, A.: On the theory of order statistics. Acta Math. Acad. Scientiarum Hung. **4**(3–4), 191–231 (1953). https://doi.org/10.1007/BF02127580
17. Ross, S.M.: Introduction to Probability Models, 10th edn. Academic Press, Burlington (2010)
18. Shaked, M., Shanthikumar, J.G.: Stochastic Orders. Springer, New York (2007). https://doi.org/10.1007/978-0-387-34675-5

Uniform Relativization

Kenshi Miyabe[⊠]

Department of Mathematics, Meiji University, Kawasaki, Kanagawa, Japan
research@kenshi.miyabe.name

Abstract. This paper is a tutorial on uniform relativization. The usual relativization considers computation using an oracle, and the computation may not work for other oracles, which is similar to Turing reduction. The uniform relativization also considers computation using oracles, however, the computation should work for all oracles, which is similar to truth-table reduction. The distinction between these relativizations is important when we relativize randomness notions in algorithmic randomness, especially Schnorr randomness. For Martin-Löf randomness, its usual relativization and uniform relativization are the same so we do not need to care about this uniform relativization.

We focus on two specific examples of uniform relativization: van Lambalgen's theorem and lowness. Van Lambalgen's theorem holds for Schnorr randomness with the uniform relativization, but not with the usual relativization. Schnorr triviality is equivalent to lowness for Schnorr randomness with the uniform relativization, but not with the usual relativization. We also discuss some related known results.

1 Introduction

1.1 Relativization

In computability theory, many notions are relativized via oracle Turing machines. As an example, a set $A \subseteq \mathbb{N}$ is called *computable* if it is computable by a Turing machine. An oracle Turing machine is a Turing machine with an oracle tape, which is a one-way infinite tape. If one uses $B \subseteq \mathbb{N}$ as an oracle, the oracle Turing machine can ask whether $k \in B$ during the computation. If A is computable by a Turing machine with an oracle B, then we say that A is *Turing reducible to B* or that A is *computable relative to B*. Similarly, many notions, results, and proofs can be relativized.

There are some other reducibilities. One of them is truth-table reducibility or abbreviated by tt-reducibility. If A is Turing reducible to B, then there exists an oracle Turing machine such that it computes A with the oracle B, but this machine may be undefined for an oracle other than B. If the reduction is total and defines a set for every oracle, then the reduction is called tt-reduction, and we say that A is *tt-reducible to B*. Some researchers say that A is *tt-computable relative to B*.

Uniform relativization is, roughly speaking, this tt-version of relativization. To distinguish them, we sometimes call Turing-reducibility version of relativization Turing relativization. Even if a notion is a tt-version of relativization of a

© Springer Nature Switzerland AG 2019
F. Manea et al. (Eds.): CiE 2019, LNCS 11558, pp. 50–61, 2019.
https://doi.org/10.1007/978-3-030-22996-2_5

known notion, the notion can be described using the terminology of tt-reduction in many cases. However, when relativizing a randomness notion, it is not appropriate to describe it via the reductions. We admitted that there are some types of relativization and named it the tt-version uniform relativization.

Uniform relativization of computable sets with an oracle B is nothing but the sets tt-reducible to B. In a more general setting, uniform relativization of a notion with an oracle B is defined by a total operator from oracles to the sets describing the notions and it is no longer a tt-reduction. We require uniformity for the operator, and that is the reason we call it uniform relativization.

Uniform relativization of Schnorr randomness behaves more naturally than Turing relativization of Schnorr randomness. This is where we found this relativization.

1.2   Algorithmic Randomness

Randomness is a central notion in natural science. The theory of algorithmic randomness defines many randomness notions and studies their properties. For simplicity, from now on, the underlying space is the Cantor space 2^ω with the uniform measure μ on it.

Martin-Löf randomness (or ML-randomness) is the most studied notion and a subclass of 2^ω. An interesting result was shown for ML-randomness by van Lambalgen [16]: $X \oplus Y$ is ML-random if and only if X is ML-random and Y is ML-random Turing relative to X. Here, X, Y are infinite binary sequences and $X \oplus Y$ is the sequence alternating between X and Y. Intuitively, if a sequence is random, then the odd-numbered parts should be random, and the even-numbered parts should be random relative to the odd parts, and vice versa. This property should hold for every natural randomness notion and its suitable relativization.

For a randomness notion $R \subseteq 2^\omega$, consider a relativized version $R^A \subseteq 2^\omega$ with an oracle A. If van Lambalgen's theorem holds for this notion, then

$$X \oplus Y \in R \iff X \in R \text{ and } Y \in R^X.$$

Fix R, then the suitable relativization R^X is automatically determined for every $X \in R$. Hence, van Lambalgen's theorem can be used as a criterion of a natural relativization for a natural randomness notion.

Schnorr randomness is another natural randomness notion. This notion comes up naturally in computable measure theory. It turns out that van Lambalgen's theorem holds for Schnorr randomness with the uniform relativization, but not for Schnorr randomness with Turing relativization. This fact has many applications, and uniform relativization is a powerful tool in the study of Schnorr randomness.

Notice that Turing relativization is used in van Lambalgen's theorem for ML-randomness. For ML-randomness, its Turing relativization and uniform relativization are the same. Hence, uniform relativization of ML-randomness is not a new notion.

Another result relating to relativized randomness is *lowness*. A central result on this topic is the equivalence between lowness for ML-randomness and K-triviality. When giving a Schnorr-randomness version, we need uniform relativization of Schnorr randomness. This fact is another evidence that uniformly relativized Schnorr randomness is a fundamental notion.

In Sect. 2 we review some basic definitions and results. In Sect. 3 we introduce uniform relativization and give some related results. In Sect. 4 we gather some results relating to uniform lowness.

2 Preliminaries

2.1 Reduction

We follow standard notations in computability theory. For details, see e.g. [4,30].

We identify a set $A \subseteq \mathbb{N}$ of natural numbers with a binary sequence $A \in 2^\omega = \{0,1\}^\omega$ by $n \in A \iff A(n) = 1$ for all n. Let $(\Phi_e)_{e\in\mathbb{N}}$ be a computable enumeration of all oracle Turing machines. The machine Φ can be seen as an operator with a partial domain from 2^ω to 2^ω as follows: For sets $A, B \in \mathbb{N}$, $B = \Phi^A$ is defined by $\Phi^A(n) = B(n)$ for every n. This Φ is called a *Turing reduction*. If this operator is total, then Φ is called a *tt-reduction*.

For sets $A, B \in \mathbb{N}$, A is *Turing reducible to* B, denoted by $A \leq_T B$, if there is a Turing reduction Φ such that $A = \Phi^B$. The set A is *tt-reducible to* B, denoted by $A \leq_{tt} B$, if there is a tt-reduction Φ such that $A = \Phi^B$.

For a computable set B, we have $A \leq_T B$ if and only if $A \leq_{tt} B$ if and only if A is computable. If $A \leq_{tt} B$, then clearly $A \leq_T B$. The converse does not hold.[1]

2.2 Randomness Notions

We also follow standard notations in the theory of algorithmic randomness. For details, see e.g. [9,25].

Cantor space 2^ω is the set of all infinite binary sequences equipped with the topology generated by the cylinder sets $[\sigma] = \{X \in 2^\omega : \sigma \prec X\}$ where $\sigma \in 2^{<\omega}$ is a finite binary sequence, and $\prec$ is the prefix relation. Let μ be the uniform measure on 2^ω defined by $\mu([\sigma]) = 2^{-|\sigma|}$ for every $\sigma \in 2^{<\omega}$.

A real $x \in \mathbb{R}$ is *computable* if there exists a computable sequence $(q_n)_n$ of rationals such that $|x - q_n| \leq 2^{-n}$ for all n. A real $x \in \mathbb{R}$ is *lower semicomputable* if there exists an increasing computable sequence $(q_n)_n$ of rationals such that $x = \lim_{n\to\infty} q_n$. Every computable real is lower semicomputable, but there is a lower semicomputable real that is not computable.

An open set $U \subseteq 2^\omega$ is *c.e.* if there exists a computable sequence S of finite binary strings such that $U = \bigcup_{\sigma \in S}[\sigma]$. Notice that the measure of a c.e. open set is lower semicomputable, but not computable in general.

---

[1] Every noncomputable c.e. Turing degree contains a hypersimple set [26, Proposition III.3.13] while a hypersimple set is not tt-complete [26, Theorem III.3.10].

A *ML-test* is a uniform sequence $(U_n)_n$ of c.e. open sets such that $\mu(U_n) \leq 2^{-n}$ for all n. A set $X \in 2^\omega$ is *ML-random* if it passes each ML-test, that is, $X \notin \bigcap_n U_n$ for every ML-test $(U_n)_n$. A *Schnorr test* is a ML-test $(U_n)_n$ such that $\mu(U_n)$ is uniformly computable. A set X is *Schnorr random* if it passes each Schnorr test.

2.3   Computable Analysis

To formalize uniform relativization of randomness notions, we use the terminology of computable analysis. For more details, see [2, 32].

Let X be a set. A *representation* of X is a surjective function $\delta :\subseteq \omega^\omega \to X$. For the real line, we usually consider the *Cauchy representation* $\rho_C :\subseteq \omega^\omega \to \mathbb{R}$ defined by

$$\rho_C(p_1, p_2, \cdots) = x \iff \lim_{n \to \infty} \nu_\mathbb{Q}(p_n) = x \text{ and } (\forall i < j)|\nu_\mathbb{Q}(p_i) - \nu_\mathbb{Q}(p_j)| \leq 2^{-i}$$

where $\nu_\mathbb{Q} :\subseteq \omega \to \mathbb{Q}$ is a computable notation of $\mathbb{Q}$. For the class $\mathcal{O}$ of all open sets on 2^ω, we usually consider the *inner representation* $\theta :\subseteq \omega^\omega \to \mathcal{O}$ defined by

$$\theta(p_1, p_2, \cdots) = \bigcup_n \nu(p_n)$$

where ν is a computable notation of the cylinder sets. For 2^ω, we use the identity $\mathrm{Id} :\subseteq \omega^\omega \to 2^\omega$ as a representation.

Let X be a set with a representation δ. If $\delta(p) = x \in X$ for $p \in \omega^\omega$, then p is called a δ-*name* of x. An element $x \in X$ is δ-*computable* if it has a computable δ-name. Then, $x \in \mathbb{R}$ is computable if and only if it is ρ_C-computable. An open set U is c.e. if and only if it is θ-computable.

For sets X_1, X_2 with representations δ_1, δ_2, a function $f :\subseteq X_1 \to X_2$ is (δ_1, δ_2)-*computable* if there is a computable function $g :\subseteq \omega^\omega \to \omega^\omega$ such that $f \circ \delta_1(p) = \delta_2 \circ g(p)$ for every $p \in \mathrm{dom}(\delta_1)$. Roughly speaking, given any δ_1-name p of $x \in X_1$, the function g computes a δ_2-name q of $f(x)$.

3   Uniform Relativization

The goal of this section is to define uniform relativization of Schnorr randomness. As a warm-up, let us begin by defining uniform relativization of more basic objects.

3.1   Uniform Relativization of c.e. Sets

We do not try to define uniform relativization itself. Instead, we define uniform relativization of some notions.

A set $A \in 2^\omega$ is *computable uniformly relative to* B if there exists a tt-reduction Φ such that $A = \Phi^B$ or equivalently $A \leq_{tt} B$. The reduction can use B as an oracle, but should be total. This roughly means that the reduction cannot use a special property of B because the reduction should work for all oracles.

A set $A \subseteq \mathbb{N}$ is *c.e.* if there exists a Turing machine $\Phi :\subseteq \omega \to \omega$ such that $A = \mathrm{dom}(\Phi)$. The machine Φ can be seen as an operator. A set $A \subseteq \mathbb{N}$ is *c.e. relative to B* if there exists an oracle Turing machine $\Phi :\subseteq \omega \to \omega$ such that $A = \mathrm{dom}(\Phi^B)$. While the notion of tt-reduction requires the oracle Turing machine to be total, it is not the case in the relativization of c.e. sets: every oracle Turing machine sends each oracle Y to a set that is c.e. relative to Y, so every oracle Turing machine is defined everywhere in that sense. Thus, Turing relativization of c.e. sets is the same as uniform relativization of c.e. sets.

Recall that the use of tt-reduction has a computable bound. If $A \leq_{tt} B$ via Φ, then there exists a computable function $f : \omega \to \omega$ such that the oracle use of $\Phi^X(n)$ is bounded by $f(n)$. In the computation of $\mathrm{dom}(\Phi^B)$ we do not have such a bound. Uniform relativization is similar to tt-reduction, but it is not appropriate to identify them.

3.2   Uniform Relativization of c.e Open Sets

Let us turn to c.e. open sets. An open set $U \subseteq 2^\omega$ is *c.e.* if it is θ-computable. An open set U is *c.e. relative to $B \in 2^\omega$* if there is a (Id, θ)-computable function $f :\subseteq 2^\omega \to \mathcal{O}$ such that $f(B) = U$. The function f can be partial but we require $B \in \mathrm{dom}(f)$. This is the usual Turing relativization. Notice that the function $f(X)$ produces a c.e. set for every input $X \in 2^\omega$, so again uniform relativization of c.e. openness is the same as its Turing relativization.

We consider the notion of c.e. openness with the measure $\leq 2^{-n}$ for $n \in \mathbb{N}$. This strange notion comes up in the relativization of Martin-Löf randomness. An open set $U \subseteq 2^\omega$ is c.e. and has measure $\leq 2^{-n}$ relative to $B \in 2^\omega$ if there is a (Id, θ)-computable function $f :\subseteq 2^\omega \to \mathcal{O}$ such that $f(B) = U$ and $\mu(U) \leq 2^{-n}$. In this case, the measure of $\mu(f(X))$ may be larger than 2^{-n} for some oracle $X \in 2^\omega$. However, we can modify f by restricting the enumeration of the cylinder sets as long as its measure is $\leq 2^{-n}$. This modified function $\hat{f}$ is also computable, the measure of $\hat{f}(X)$ is $\leq 2^{-n}$ for each $X \in 2^\omega$, and $\hat{f}(B) = U$. Hence, again its uniform relativization is the same as its Turing relativization.

Finally, we consider the notion of c.e. openness with a computable measure. This notion corresponds to the relativization of Schnorr randomness. An open set $U \subseteq 2^\omega$ is c.e. and has a computable measure Turing relative to $B \in 2^\omega$ if there are a (Id, θ)-computable function $f :\subseteq 2^\omega \to \mathcal{O}$ and a (Id, ρ_C)-computable function $g :\subseteq 2^\omega \to \mathbb{R}$ such that $f(B) = U$ and $g(B) = \mu(U)$. Notice that f, g can be partial but we should have $B \in \mathrm{dom}(f) \cap \mathrm{dom}(g)$.

In this case, we can not extend f, g to be total by any computable modification. Let us give a counterexample. Let $A, B \subseteq \mathbb{N}$ be sets such that $A \leq_T B$ but $A \not\leq_{tt} B$. Define $U = \bigcup_{k \in A}[0^k 1]$. Since $A \leq_T B$, the open set U is c.e. Turing relative to B and its measure is computable Turing relative to B.

Suppose that there exist a total (Id, θ)-computable function $f :\subseteq 2^\omega \to \mathcal{O}$ and a total (Id, ρ_C)-computable function $g :\subseteq 2^\omega \to \mathbb{R}$ such that $\mu(f(Y)) = g(Y)$ for every $Y \in 2^\omega$ and $f(B) = U$. Consider the following reduction $\Phi :\subseteq 2^\omega \to 2^\omega$

with an input $Y \in 2^\omega$. For each $n \in \mathbb{N}$, enumerate the inner cylinders of $f(Y)$ until the measure larger than $g(Y) - 2^{-n-1}$. If the intersection between the finite approximation and $[0^n 1]$ is not empty, then $\Phi^Y(n)$ outputs 1. Otherwise, $\Phi^Y(n)$ outputs 0.

Since $\mu(f(Y)) = g(Y)$ for every $Y \in 2^\omega$, the reduction can find the finite approximation for every n and Φ is total. Suppose $Y = B$ and $n \in A$. Then, the finite approximation U_n of $f(B)$ should intersect with $[0^n 1]$, otherwise $\mu(U_n) > \mu(g(Y)) - 2^{-n-1} = \mu(U) - 2^{-n-1}$ and $\mu(U_n \cup [0^n 1]) = \mu(U_n) + 2^{-n-1} > \mu(U)$, which contradicts with $U_n \cup [0^n 1] \subseteq U$. Hence, $\Phi^B(n) = 1 = A(n)$. Suppose $Y = B$ and $n \notin A$. Then, the finite approximation U_n of $f(B)$ can not intersect with $[0^n 1]$ because U_n is the inner approximation of $U = \bigcup_{k \in A} [0^k 1]$. Hence, $\Phi^B(n) = 0 = A(n)$. This contradicts with $A \not\leq_{tt} B$.

Now we know that uniform relativization of c.e. openness with a computable measure is different from its Turing relativization.

3.3 Uniform Relativization of Schnorr Randomness

We are now ready to define uniform relativization of Schnorr randomness, or abbreviated by uniform Schnorr randomness. The definition is complicated, but the idea is the same as the basic notions defined in the above.

Definition 1 ([23]). *A* uniform Schnorr test *is a pair of computable functions f, g satisfying the follows:*

1. *$f : 2^\omega \times \omega \to \mathcal{O}$ is $(\mathrm{Id}, \mathrm{Id}_\omega, \theta)$-computable with $\mu(f(X, n)) \leq 2^{-n}$ for all $X \in 2^\omega$ and $n \in \omega$.*
2. *$g : 2^\omega \times \omega \to \mathbb{R}$ is $(\mathrm{Id}, \mathrm{Id}_\omega, \rho_C)$-computable such that $g(X, n) = \mu(f(X, n))$ for all $X \in 2^\omega$ and $n \in \omega$.*

Here, $\mathrm{Id}_\omega : \omega \to \omega$ is the identity function on ω. A set $A \in 2^\omega$ is Schnorr random uniformly relative to B if $A \notin \bigcap_n f(B, n)$ for each uniform Schnorr test $\langle f, g \rangle$.

Each Schnorr test uniformly relative to B is a Schnorr test Turing relative to B. Thus, each Schnorr random Turing relative to B is Schnorr random uniformly relative to B. Hence, uniform relativized randomness is weaker than Turing relativized randomness.

We need this relativization for van Lambalgen's theorem for Schnorr randomness to hold. For sets $A, B \subseteq \mathbb{N}$, let $A \oplus B = \{2n : n \in A\} \cup \{2n + 1 : n \in B\}$.

Theorem 1 ([23]). *The set $A \oplus B$ is Schnorr random if and only if A is Schnorr random and B is Schnorr random uniformly relative to A.*

Theorem 2 ([18,33] **and** [25, **Remark 3.5.22**]). *Van Lambalgen's theorem fails for Schnorr randomness with Turing relativization.*

In particular, uniform relativization of Schnorr randomness is different from its Turing relativization.

Notice that, if A is Schnorr random and B is Schnorr random Turing relative to A, then $A \oplus B$ is Schnorr random. This is because uniformly relativized Schnorr randomness is weaker than Turing relativized Schnorr randomness. For the other direction, assume that A is Schnorr random and B is covered by a Schnorr test relative to A. One needs uniformity or totality of the test to construct a Schnorr test covering $A \oplus B$. Uniform relativization naturally comes up when looking at the proofs of van Lambalgen's theorem.

3.4   Other Characterizations

In the above, we defined uniform Schnorr randomness via tests. Schnorr randomness has characterizations by martingales, computable measure machines, and integral tests. We can also characterize uniform Schnorr randomness by them. The proofs are straightforward, but we need to check that everything works uniformly in oracles. We give the definitions to look at how to uniformly relativize these notions.

A *martingale* is a function $d : 2^{<\omega} \to \mathbb{R}^+$ such that $2d(\sigma) = d(\sigma 0) + d(\sigma 1)$ for every $\sigma \in 2^{<\omega}$ where $\mathbb{R}^+$ is the set of all nonnegative reals. A set $X \in 2^\omega$ is ML-random if and only if $\sup_n d(X \upharpoonright n) < \infty$ for all left-c.e. martingales d. A set $X \in 2^\omega$ is Schnorr random if and only if $d(X \upharpoonright n) < f(n)$ for at most finitely many n for every computable martingale d and every computable order f. These are classical results by Schnorr [28, 29]. Here, an order is an unbounded nondecreasing function from ω to ω. Franklin and Stephan [10] observed that X is not Schnorr random if and only if there is a computable martingale d and a computable function f such that $(\exists^\infty n)d(X \upharpoonright f(n)) \geq n$.

A *uniformly computable martingale* is a computable map $d : 2^\omega \times 2^{<\omega} \to \mathbb{R}^+$ such that $d^Z := d(Z, \cdot)$ is a martingale for every $Z \in 2^\omega$. A set X is Schnorr random uniformly relative to A if and only if $d^A(X \upharpoonright n) < f(n)$ for almost all n for each uniformly computable martingale d and a computable order f if and only if $d^A(X \upharpoonright h(n)) < n$ for almost all n for each uniformly computable martingale d and a strictly increasing computable function h [19]. We can replace the computable order f above with $\hat{f}^A$ such that $\hat{f} : 2^\omega \times \omega \to \omega$ is a computable function and $\hat{f}^Z$ is an order for each $Z \in 2^\omega$. This is because, for such $\hat{f}$, we can find a computable order f such that $\hat{f}^Z(n) \geq f(n)$ for each $n \in \mathbb{N}$ and each $Z \in 2^\omega$ by compactness of 2^ω.

Franklin and Stephan [10] defined tt-Schnorr random set X relative to A as a set such that there are no martingale $d \leq_{tt} A$ and no function $g \leq_{tt} A$ such that $(\exists n)d(X \upharpoonright h(n)) \geq n$. We can replace $h \leq_{tt} A$ with a computable function h [10, Remark 2.4]. This notion is equivalent to Schnorr randomness uniformly relative to A [23, Proposition 6.1]. However, there are some subtle points to note in the tt-relativization. See Sect. 6 in [23] for details.

Let $M :\subseteq 2^{<\omega} \to 2^{<\omega}$ be a Turing machine. The *measure* (or halting probability) of M is $\sum_\sigma \{2^{-|\sigma|} : M(\sigma) \downarrow\}$. The measure of a prefix-free Turing machine is less than or equal to 1 by Kraft's inequality. The measure of a universal prefix-free Turing machine U is called Chaitin's omega, denoted by Ω_U, which is ML-random [3], hence not computable. A prefix-free Turing machine with a computable measure is called a *computable measure machine*. A set X is Schnorr random if and only if $K_M(X \restriction n) > n - O(1)$ for every computable measure machine M [7].

An oracle prefix-free Turing machine $M :\subseteq 2^\omega \times 2^{<\omega} \to 2^{<\omega}$ is a *uniformly computable measure machine* if the maps $X \mapsto \sum_\sigma \{2^{-|\sigma|} : M(X, \sigma) \downarrow\}$ is a total computable function. A set X is Schnorr random uniformly relative to A if and only if $K_{M^A}(X \restriction n) > n - O(1)$ for every uniformly computable measure machine M (essentially due to [19]).

An *integral test* is an integrable nonnegative lower semicomputable function $f : 2^\omega \to \mathbb{R}^+$. A set $X \in 2^\omega$ is ML-random if and only if $f(X) < \infty$ for each integral test, which is by Levin: see e.g. [17, Subsection 4.5.6, 4.7]. A set $X \in 2^\omega$ is Schnorr random if and only if $f(X) < \infty$ for each nonnegative lower semicomputable function $f : 2^\omega \to \mathbb{R}^+$ such that $\int f \, d\mu$ is a computable real [20]. Such a function is called a *Schnorr integral test*.

A *uniform Schnorr integral test* is a lower semicomputable function $f : 2^\omega \times 2^\omega \to \mathbb{R}^+$ such that $X \mapsto \int f(X, Z)\mu(dZ)$ is a computable function from 2^ω to $\mathbb{R}$. The first component is for oracles and the second for the tested sets A set Y is Schnorr random uniformly relative to X if and only if $f(X, Y) < \infty$ for each uniform Schnorr integral test f [23, Proposition 4.1].

3.5   Related Work

Van Lambalgen's theorem for uniform Schnorr randomness was further generalized to noncomputable measures [27], and was used in the study of Schnorr reducibility and total-machine reducibility [22]. Van Lambalgen's theorem for uniform relativization of computable randomness holds in a weaker form [23, Theorem 5.1]. Van Lambalgen's theorem for uniform Kurtz randomness was studied in [13]. Van Lambalgen's theorem for Demuth randomness was studied in [5], where they used "partial relativization."

4   Uniform Lowness

Another topic relating to relativized randomness is lowness. First, we recall some results on lowness for ML-randomness. Then, we see that uniform lowness was needed to give a Schnorr-randomness version.

4.1   Characterization of Triviality via Lowness

Many randomness notions have characterizations via complexity. The Levin-Schnorr theorem says that $A \in 2^\omega$ is ML-random if and only if $K(A \restriction n) > n - O(1)$ where K is the prefix-free Kolmogorov complexity. Roughly speaking, a set is random if the complexities of its initial segments are high. Thus, the complexity is a measure of randomness. A set $A \in 2^\omega$ is K-reducible to $B \in 2^\omega$ if $K(A \restriction n) < K(B \restriction n) + O(1)$. This is one formalization of saying that A is not more random than B. The class of K-trivial sets is the bottom degree of this reducibility. A set $A \in 2^\omega$ is K-trivial if $K(A \restriction n) \leq K(n) + O(1)$. Obviously, every computable set is K-trivial. In contrast, there is a noncomputable K-trivial set.

Interestingly, K-triviality can be characterized by lowness. A set $A \in 2^\omega$ is *low for ML-randomness* if every ML-random set Turing relative to A is already (unrelativized) ML-random. This means that the set A can not derandomize any ML-random set. These notions coincide, that is, a set $A \in 2^\omega$ is K-trivial if and only if A is low for ML-randomness. For details on this topic, see e.g. [24].

We have Schnorr-randomness counterparts of these notions as follows.

Definition 2 ([7]). *A set $A \in 2^\omega$ is Schnorr reducible to $B \in 2^\omega$ denoted by $A \leq_{Sch} B$ if, for every computable measure machine M, there exists a computable measure machine N such that $K_N(A \restriction n) \leq K_M(B \restriction n) + O(1)$. A set $A \in 2^\omega$ is called* Schnorr trivial *if $A \leq_{Sch} \emptyset$.*

This notion can be characterized by uniform lowness for Schnorr randomness. A set $A \in 2^\omega$ is called *uniformly low for Schnorr randomness* if every Schnorr random set uniformly relative to A is already Schnorr random.

Theorem 3 (essentially due to [10]). *A set $A \in 2^\omega$ is Schnorr trivial if and only if A is uniformly low for Schnorr randomness.*

Since there is a Turing-complete Schnorr trivial set [8], some Schnorr trivial sets are not Turing low for Schnorr randomness. Hence, uniform lowness and Turing lowness for Schnorr randomness are different.

4.2   Other Characterizations

The class of K-trivial sets has many characterizations, and so does the class of Schnorr trivial sets.

The first one is by traceability. A *trace* is a sequence $\{T_n\}$ of sets. A trace for a function f is a trace $\{T_n\}$ with $f(n) \in T_n$ for all n. For a function h, a trace $\{T_n\}$ is h-bounded if $|T_n| \leq h(n)$ for all n. A set A is *computable tt-traceable* if there is a computable order h such that all functions $f \leq_{tt} A$ are traced by an h-bounded computable trace. Roughly speaking, the values computable from traceable sets have limited possibilities. Many variants were studied in Hölzl and Merkle [12].

Franklin and Stephan [10] showed that uniform lowness of Schnorr randomness is equivalent to computable tt-traceability. There is no counterpart for ML-randomness. This result is a modification of the one that Turing lowness of Schnorr randomness is equivalent to computable (Turing) traceability [15,31].

The next one is by lowness for machines. A set A is called *low for K* if $K(n) \leq K^A(n) + O(1)$. This means that the set A can not compress n more than without it. In fact, a set is K-trivial if and only if it is low for K.

A Schnorr-randomness version is as follows. We say that a set A is *uniformly low for* computable measure machines if, for every uniformly computable measure machine M, there exists a computable measure machine N such that

$K_N(n) \leq K_{M^A}(n) + O(1)$. Then, uniform lowness for computable measure machines is equivalent to computable tt-traceability [19], hence to Schnorr triviality. The proof was given by straightforward modification of the fact that Turing lowness for computable measure machines is equivalent to computable traceability [6].

The class of K-trivial sets also has a base-type characterization. A set A is a *base for* ML-randomness if there exists a ML-random set X relative to A such that $A \leq_T X$. Notice that each computable set is a base for ML-randomness. If A has much information, the class of ML-random sets relative to A is so small that we can not find such a set in the Turing degrees above A.

Its Schnorr-randomness version is not straightforward. See the discussion in [10, Section 6]. We say that a set A is a *base for Schnorr randomness* if there is no $X \geq_T A$ such that X is Schnorr random Turing relative to A. Franklin, Stephan, and Yu [11] showed that this is equivalent to saying that the set A does not compute the halting problem.

One adaptation is as follows. A set A is a *tt-base for uniformly computable martingales* if, for each uniformly computable martingale d, there exists a set $X \geq_{tt} A$ such that $\sup_n d^A(X \upharpoonright n) < \infty$. The last condition roughly means that X is computably random uniformly relative to A only for this d. It turns out that Schnorr triviality is equivalent to being a tt-base for uniformly computable martingales [21, Theorem 6.4].

4.3   Related Work

Decidable prefix-free machines also characterize ML-randomness and Schnorr randomness [1]. Schnorr reducibility can be characterized by complexity for prefix-free decidable machines by adding a computable order. We write $A \leq_{wdm} B$ if, for each decidable prefix-free machine M and a computable order g, there exists a decidable prefix-free machine N such that $K_N(A \upharpoonright n) \leq K_M(B \upharpoonright n) + g(n) + O(1)$. In fact, $A \leq_{wdm} B$ if and only if $A \leq_{Sch} B$ [21, Theorem 3.5]. In particular, Schnorr triviality has a characterization by decidable prefix-free machines.

Schnorr triviality is also equivalent to not totally i.o. complex [12], which is a characterization by total machines.

The equivalence between lowness for ML-randomness and lowness for K was strengthened to the equivalence between $\leq_{LR}$ and $\leq_{LK}$ [14]. Its uniform Schnorr-randomness version was proved in [21, Theorem 5.1] and Turing relativized Schnorr-randomness version in [22].

Computable traceability was characterized by order-lowness for prefix-free decidable machines [1, Theorem 24]. Recall that computable traceability is equivalent to Turing lowness for Schnorr randomness.

Acknowledgements. The author thanks the anonymous reviewers for their careful reading of the manuscript and their suggestions.

References

1. Bienvenu, L., Merkle, W.: Reconciling data compression and kolmogorov complexity. In: Arge, L., Cachin, C., Jurdziński, T., Tarlecki, A. (eds.) ICALP 2007. LNCS, vol. 4596, pp. 643–654. Springer, Heidelberg (2007). https://doi.org/10.1007/978-3-540-73420-8_56
2. Brattka, V., Hertling, P., Weihrauch, K.: A tutorial on computable analysis. In: Cooper, S.B., Löwe, B., Sorbi, A. (eds.) New Computational Paradigms, pp. 425–491. Springer, New York (2008). https://doi.org/10.1007/978-0-387-68546-5_18
3. Chaitin, G.J.: A theory of program size formally identical to information theory. J. Assoc. Comput. Mach. **22**, 329–340 (1975)
4. Cooper, S.B.: Computability Theory. CRC Press, New York (2004)
5. Diamondstone, D., Greenberg, N., Turetsky, D.: A van Lambalgen theorem for Demuth randomness. In: Downey, R., Brendle, J., Goldblatt, R., Kim, B. (eds.) Proceedings of the 12th Asian Logic Conference, pp. 115–124 (2013)
6. Downey, R., Greenberg, N., Mihailovic, N., Nies, A.: Lowness for computable machines. In: Chong, C.T., Feng, Q., Slaman, T.A., Woodin, W.H., Yang, Y. (eds.) Computational Prospects of Infinity: Part II. Lecture Notes Series, pp. 79–86. World Scientific Publishing Company, Singapore (2008)
7. Downey, R., Griffiths, E.: Schnorr randomness. J. Symbolic Logic **69**(2), 533–554 (2004)
8. Downey, R., Griffiths, E., LaForte, G.: On Schnorr and computable randomness, martingales, and machines. Math. Logic Q. **50**(6), 613–627 (2004)
9. Downey, R.G., Hirschfeldt, D.R.: Algorithmic Randomness and Complexity. Theory and Applications of Computability. Springer, New York (2010). https://doi.org/10.1007/978-0-387-68441-3
10. Franklin, J.N.Y., Stephan, F.: Schnorr trivial sets and truth-table reducibility. J. Symbolic Logic **75**(2), 501–521 (2010)
11. Franklin, J.N.Y., Stephan, F., Yu, L.: Relativizations of randomness and genericity notions. Bull. London Math. Soc. **43**(4), 721–733 (2011)
12. Hölzl, R., Merkle, W.: Traceable sets. In: Calude, C.S., Sassone, V. (eds.) TCS 2010. IAICT, vol. 323, pp. 301–315. Springer, Heidelberg (2010). https://doi.org/10.1007/978-3-642-15240-5_22
13. Kihara, T., Miyabe, K.: Uniform Kurtz randomness. J. Logic Comput. **24**(4), 863–882 (2014)
14. Kjos-Hanssen, B., Miller, J.S., Solomon, D.R.: Lowness notions, measure, and domination. J. London Math. Soc. **85**(3), 869–888 (2012)
15. Kjos-Hanssen, B., Nies, A., Stephan, F.: Lowness for the class of Schnorr random reals. SIAM J. Comput. **35**(3), 647–657 (2005)
16. van Lambalgen, M.: Random sequences. Ph.D. thesis, University of Amsterdam (1987)
17. Li, M., Vitányi, P.: An Introduction to Kolmogorov Complexity and Its Applications. TCS. Springer, New York (2008). https://doi.org/10.1007/978-0-387-49820-1
18. Merkle, W., Miller, J., Nies, A., Reimann, J., Stephan, F.: Kolmogorov-Loveland randomness and stochasticity. Ann. Pure Appl. Logic **138**(1–3), 183–210 (2006)
19. Miyabe, K.: Truth-table Schnorr randomness and truth-table reducible randomness. Math. Logic Q. **57**(3), 323–338 (2011)
20. Miyabe, K.: L^1-computability, layerwise computability and Solovay reducibility. Computability **2**, 15–29 (2013)

21. Miyabe, K.: Schnorr triviality and its equivalent notions. Theory Comput. Syst. **56**(3), 465–486 (2015)
22. Miyabe, K.: Reducibilities relating to Schnorr randomness. Theory Comput. Syst. **58**(3), 441–462 (2016)
23. Miyabe, K., Rute, J.: Van Lambalgen's Theorem for uniformly relative Schnorr and computable randomness. In: Downey, R., Brendle, J., Goldblatt, R., Kim, B. (eds.) Proceedings of the 12th Asian Logic Conference, pp. 251–270, July 2013
24. Nies, A.: Eliminating concepts. In: Computational prospects of infinity II. IMS Lecture Notes Series, vol. 15, pp. 225–248 (2008)
25. Nies, A.: Computability and Randomness, Oxford Logic Guides, vol. 51. Oxford University Press, Oxford (2009). https://doi.org/10.1093/acprof:oso/9780199230761.001.0001
26. Odifreddi, P.: Classical Recursion Theory, vol. 1. North-Holland (1990)
27. Rute, J.: Schnorr randomness for noncomputable measures. Inf. Comput. **258**, 50–78 (2018)
28. Schnorr, C.P.: Zufälligkeit und Wahrscheinlichkeit. LNM, vol. 218. Springer, Heidelberg (1971). https://doi.org/10.1007/BFb0112458
29. Schnorr, C.: A unified approach to the definition of a random sequence. Math. Syst. Theory **5**, 246–258 (1971)
30. Soare, R.I.: Turing Computability. Theory and Applications of Computability. Springer, Heidelberg (2016). https://doi.org/10.1007/978-3-642-31933-4
31. Terwijn, S.A., Zambella, D.: Computational randomness and lowness. J. Symbolic Logic **66**(3), 1199–1205 (2001)
32. Weihrauch, K.: Computable Analysis: An Introduction. Springer, Berlin (2000). https://doi.org/10.1007/978-3-642-56999-9
33. Yu, L.: When van Lambalgen's theorem fails. Proc. Am. Math. Soc. **135**(3), 861–864 (2007)

Correctness, Explanation and Intention

Raymond Turner[✉]

School of Computer Science and Electronic Engineering,
University of Essex, Colchester, UK
turnr@essex.ac.uk

Abstract. There appear to be two fundamentally different notions of program correctness that emanate from two different notions of program: the mathematical correctness of abstract programs and the empirical correctness of their implemented physical manifestations [2,16,17]. In the abstract case, a program is taken to be correct when it meets its specification. This is a mathematical affair with all the precision and clarity that follows. But physical correctness raises some concerns and puzzles that have their origins in Putnam's notion of physical computation [15]. Moreover, these concerns would appear to effect the mathematical case. Comparing the two cases will draw out some underling philosophical issues in the traditional approaches to correctness. In particular, we examine the different concepts of explanation that accompany the different notions of correctness, and expose the underlying role of agency in both.

1 Physical Correctness

The nature of program correctness has been a central concern in the philosophy of computer science [16,17] for some time (e.g. [4,5]). But much of this debate has been aimed at the correctness of abstract programs relative to their specification. However, on the face of it, physical correctness can be characterised in the same way as its mathematical analogue: the physical artefact has to meet its functional specification. But matters are not quite so straightforward.

To see why, suppose that the physical artefact is a machine of some kind with states and operations that move the machine between those states. Physical programs are sequences of such operations. We assume that the specification is given via an abstract device. For correctness of the physical device relative to the abstract one, we would expect the physical machine to behave in harmony with the abstract one, i.e., the physical state transitions of the machine should be in harmony with the abstract ones. One proposal that reflects these intuitions is due to Hillary Putnam [15] who expressed matters in terms of Turing machines. The following is a generalisation: the so-called *Simple Mapping Account* of physical computation (SMA) [13,14].

A physical system P correctly performs a computation A if and only if the following holds: there is a mapping from the states of the physical system P to the states of an abstract system A, such that the following is true.

© Springer Nature Switzerland AG 2019
F. Manea et al. (Eds.): CiE 2019, LNCS 11558, pp. 62–71, 2019.
https://doi.org/10.1007/978-3-030-22996-2_6

The state transitions between the physical states mirror the state transitions between the computational states: for any abstract state transition $s_1, \Rightarrow s_2$, if the system is in the physical state that maps onto s_1, it then goes into the physical state that maps onto s_2.

On the face of it, this appears to be exactly the right conditions for the physical system to be a correct implementation of the abstract one. However, there is a caveat here.

The original proposal was aimed at a characterisation of physical computation. The following is how the latter notion is introduced in [13].

> In our ordinary discourse, we distinguish between physical systems that perform computations, such as computers and calculators, and physical systems that don't, such as rocks. Among computing devices, we distinguish between more and less powerful ones. These distinctions affect our behaviour: if a device is computationally more powerful than another, we pay more money for it. What grounds these distinctions? What is the principled difference, if there is one, between a rock and a calculator, or between a calculator and a computer? Answering these questions is more difficult than it may seem.

What is the principled difference, if there is one, between a rock and a computer? If we are concerned with characterising which physical devices compute and which ones do not, then our goal must be to provide an abstract model of physical computation. In contrast, our concern is with the correctness of an artefact. Consequently, we start with the abstract specification and then demand that the constructed artefact meets the abstract specification. We shall have more about this difference later.

Despite these different intentions, the simple mapping account is a natural place to start for both enterprises: we shall explore the consequences of taking the SMA as a characterisation of physical correctness.

Unfortunately, puzzles come thick and fast: if SMA is taken as the central definition of physical correctness it trivialises matters. To see why, consider a concrete case involving digital circuits. Suppose that the abstract specification of the device is given by the truth table definition of conjunction. The input states are the usual truth values and their conjunction forms the output state. The intended artefact might then be taken to be an electronic device that is given by an NMOS AND gate [12]. According to SMS, if correctly constructed, such a gate will satisfy the specification, but it is not the only thing that does. According to SMA, the state-to-state behaviour of the abstract and physical systems must be in harmony. But what does this amount to? To answer this notice that the contained conditional

if the system is in the physical state that maps onto s_1, it then goes into the physical state that maps onto s_2

would normally be taken to be the standard material conditional. But this exposes the demands of the specification, and highlights the fact that they are very easy to meet: because the material conditional is employed, the requirement is purely extensional in the sense that the specification only demands a one-to-one correspondence between the states of the abstract and physical machines. This leads to a situation where almost anything implements a given abstract specification. In other words, extensionality only demands that we have some physical device whose possible state transitions are in one-to-one correspondence with the abstract ones. Physical correctness is trivialised: any arrangement of physical objects that is isomorphic to the truth table for conjunction will satisfy the demands of SMA. This leads to the following simple characterisation.

A physical system P is correct relative to an abstract system A if (for each operation of the machine) there is a one-to-one correspondence between the state-to-state table of P and the state-to-state table of A.

In other words, the two tables/relations are in one-to-one correspondence. Cleary, there is one obvious objection to such an account: in principle almost any physical table of the right size will provide a correct implementation of a given abstract operation. So the proposed account of physical correctness seems too liberal. In fact, any solution that consists of an enumeration of the results of the abstract machine calculation will be a correct implementation. However, even when we are dealing with abstract machines such as finite state machines with a finite number of states, extensional solutions are impractical since the results of the computation must be computed ahead of time. In practice, this is a crucial part of the objection to such solutions admitted by the simple mapping account: enumerated solutions are at best impractical, and at worse impossible.

2   Mathematical Correctness

Moreover, this particular concern is not restricted to the physical situation [17]. To see why things might go awry in the mathematical case, consider the following specification [7] of the square root for natural numbers.

$$SQRT(x : Num, y : Real)$$

$$\longleftrightarrow$$

$$y * y = x \wedge y > 0$$

$SQRT$ defines a relation between natural numbers and real numbers where Num and $Real$ are the data types for the specification language. Logically, this unpacks to the following demand.

A program P operating over these data types is taken to satisfy this specification if the following holds.

$$\forall x : Num. \forall y : Real \cdot SQRT(x, y) \leftrightarrow P(x) = y.$$

Observe that in implemented programming languages we will always be dealing with finite sets and types and finite operations on them. So we restrict our

Input	1	2	3	4	5
Output	1	1.41421356237	1.73205080757	2	2.2360679775

discussion accordingly. Indeed, for pedagogical reasons we assume that numbers $1, 2, 3, 4, 5$ constitute the data type *Num* of both languages. Now consider an abstract table A that associates the square root with the numbers $1, 2, 3, 4, 5$.

For this table, we then have

$$\forall x : Num. \forall y : Real \cdot SQRT(x, y) \longleftrightarrow A(x) = y.$$

In other words, the two relations are in extensional agreement. In this sense the abstract table satisfies the specification. We can put such a table in a more familiar programming style using case statements as follows.

```
Case  x=1  then  y=1
Case  x=2  then  y=1.41421356237
Case  x=3  then  y=1.73205080757
Case  x=4  then  y=2
Case  x=5  then  y=2.2360679775
```

Abstract mathematical correctness is reduced to extensional agreement. So what is common to both the physical and mathematical accounts is that they admit solutions based upon enumerating the results.

3   Explanation and Correctness

In general, such *extensional* solutions are not highly valued. This applies to the simple mapping account based upon the material conditional in the physical case and with enumerative solutions in the mathematical one. In so far as they satisfy the functional requirements, they are solutions. However, they are *bad* solutions: any solution that requires all the computations to be done ahead of time is impractical. Moreover, there are underlying conceptual reasons for this badness. There is something that is common to both the physical and the mathematical cases: *good* solutions appeal to some notion of explanation, even if rather different ones.

In the physical case, we expect some causality to be part of the characterisation [1,3]. After all, the electronic circuit is a causal device. Consequently, one natural proposal is to use a conditional such as the counterfactual conditional that, through its appeal to causal laws, embodies causation in its truth conditions. This offers some causal account of why the physical machine, the electronic circuit, moves from state to state. Such causal solutions are explanatory.

Kroes provides a similar account of why technical artefacts satisfy their functional requirements. This is given in terms of physical laws, the physical makeup and configuration of the artefact, and the dynamic behaviours and causal interactions of its components. Of course, as Kroes [9,10] points out, this explains what the artefact actually does, not necessarily what it was intended to do. This is given by its functional specification. He argues that the relation between abstract structure and physical devices can be conceived in terms of pragmatic rules of action that are grounded in such causal considerations. Consider again the AND gate.

– If an electronic AND gate is used properly in appropriate circumstances, by appeal to an appropriate causal account, it will compute the Boolean operation of conjunction.

We may then infer the following rule of action:

– In order to compute the Boolean *AND*, use an AND gate.

Explanation effects the mathematical case as well [11]. Mathematicians seek new proofs of known theorems because the new proofs throw more light on matters - they provide better explanations of the result.

In the case of program correctness, the extensional solutions must give way to more explanatory algorithms and programs. And their explanatory value depends upon their proofs of correctness. Such proofs provide an explanation of why the algorithm/program satisfies its specification. For example, consider a solution for the square root specification based upon Newton's iterative method.

$$x_{k+1} = \frac{1}{2}\left(x_k + \frac{n}{x_k}\right)$$

At each stage the new value is computed from the old one, and the process continues until the old and new values (x_k and x_{k+1}) converge. The Newtonian solution is explanatory: the algorithm returns the square root of n since, when the two values of x converge, we obtain a quadratic equation for x whose solution is the square root of n.

$$x^2 = n$$

This is obvious in the case of the algorithm, but how does the proof of correctness for a corresponding program explain how it works? Newton's method can be programmed in Python using a simple iteration.

```
1  def newtonSqrt(n):
2  approx = 0.5 * n
3  better = 0.5 * (approx + n/approx)
4  while better != approx:
5      approx = better
6      better = 0.5 * (approx + n/approx)
7  return approx
```

Without spelling out all the details of a Hoare style correctness proof [7], we indicate why they supply explanations. In the Hoare calculus, theorem statements take the following form.

$$\{\phi\}P\{\psi\}.$$

This asserts that if the predicate calculus assertion ϕ is true before the program P runs, then ψ will be true afterwards. The calculus consists of rules for all constructs in the language of this form. For example, the rule for sequencing is given as follows.

$$\frac{\{\phi\}P\{\psi\} \qquad \{\psi\}Q\{\mu\}}{\{\phi\}P;Q\{\mu\}}.$$

For sequencing, the output state for the first program becomes the input state for the second. This can be rephrased in explanatory terms as follows.

– In the state ϕ, $P;Q$ will end in state μ because, in state ϕ, the program P will end in state ψ, and in state ψ the program Q will end in state μ

This is a deductive explanation of why the program for sequencing meets its specification in transforming state ϕ into state μ.

In the case of iteration, the rule takes the following form.

$$\frac{\{B\bigwedge\varphi\}P\{\varphi\}}{\{\varphi\}whileBdoP\{\neg B\bigwedge\varphi\}}$$

The premise tells us that the invariant of the loop is ϕ where B is the Boolean condition. The conclusion informs us that after the iteration, the invariant still holds but the Boolean condition is false. This can be rephrased in explanatory terms as follows.

– In the state ϕ, *while B do P* (if it terminates) will end in state $\neg B\bigwedge\varphi$ because, in state $B \wedge \phi$, the program P will end in state ϕ.

The invariant of the loop for Newton's method is the following.

$$better = 0.5 * (approx + n/approx)$$

The invariant forms the basis of the quadratic equation whose solution yields the square root up to the supplied level of approximation. While other proof techniques offer similar explanations of why a program meets its specification, for Hoare style proofs we can dig a little deeper.

Lang [11] offers an account in terms of a symmetrical relationship between the statement of the theorem and the structure of the proof. In Hoare style proofs, each step in the formal proof relates directly back to the statement of the main theorem. So, if we follow the whole proof through, each step mirrors the statement of the theorem. The proofs are structured around the symmetry induced throughout the proof by the Hoare-triples. Each step is a miniature form of the theorems statement. There is a uniformity to the proof that centres on the statement of correctness, and the correctness of the whole is compositionally constructed in a way that follows the structure of the program. In this sense, the proof provides a uniform transparent explanation of why the program satisfies its specification.

Generally speaking extensional solutions are impractical and lack explanatory power. The two cases, the abstract and the physical, appeal to different notions of explanation in order to enhance the standard extensional correctness demands[1]. But there is another hidden philosophical concern that governs both.

4 Intention and Correctness

An assumption underlying enumerative solutions is that computation and programming are notions that can be fully characterised independently of any intentions. Wittgenstein implicitly criticised this long before the simple mapping account appeared on the scene.

> There might be a caveman who produced regular sequences of marks for himself. He amused himself, e.g., by drawing on the wall of the cave. But he is not following the general expression of a rule. And when we say that he acts in a regular way that is not because we can form such an expression. That is, the fact that we could construct a rule to describe the regularity of his behaviour does not entail that he was following that rule [18] (6.41). Imagine a caveman accidentally scratches the following table in the sand with a stick:

1	2	3	4	5
1	1.41421356237	1.73205080757	2	2.2360679775

Do we say he has correctly programmed some fragment of the square root? Obviously not: he had no intention of constructing a program for the square root. Nor can he explain why it is a solution. Just getting the answer right, by accident, involves no intention of any kind. The caveman is merely scribbling in the sand. Programming and computing are intentional activities, and establishing and explaining why a program meets its specification is part of the activity of programming. Hiving off the extensional aspect of this activity, and considering it in isolation from its intentional context, is at the core of the problem raised by the simple mapping account. In terms of technical artefacts, removing agency and intention is to treat the program as a thing with no function.

The significance of agency can be clearly seen in the logical difference between a specification and a theory or model. Suppose instead that the specification is taken as a *model* of the program. We would still demand the same expression of the relationship.

$$\forall x : Num.\forall y : Real \cdot SQRT(x, y) \longleftrightarrow A(x) = y.$$

But being a *model* of the program has consequences. If the model is wrong, does not correctly predict the outcome of the program, we change the model.

---

[1] [6] addresses the relationship between the physical and the abstract in terms of levels of abstraction.

This is the opposite of what we would do if SQRT is taken as a specification: if things go wrong we change the program not the specification. The intentional stance is quite different. If the relationship between SQRT and the program is merely characterised extensionally, as it is in the simple mapping account, this distinction is ignored. The distinction between specification and model cannot be catered for extensionally but must involve this intentional aspect.

We can illustrate the intentional stance in a more familiar setting. The standard semantics for the language of first-order logic provides a definition of truth for the language. Consequently, we may employ it to guide the construction of a proof system: the semantics provides the correctness conditions for the construction of any such system. Success is traditionally taken as the soundness and completeness of the rules: soundness establishes the legitimacy of the rules, and completeness demonstrates that we have not missed any. If there is disagreement between the proof theory and the semantics, we blame the proof theory: we change the rules to gain soundness and completeness.

Conversely, if we take the proof theory to have normative priority, our aim might be to construct a semantic theory that is sound and complete with respect to it. In justifying the semantics, we would still be engaged in a mathematical activity, and we would still attempt to prove soundness and completeness. However, soundness now establishes that the semantics is a correct reflection of the rules, and completeness demonstrates that the semantics does not sanction illegitimate ones. The proof theory now has governance. In particular, if there is disagreement of either kind, we now blame the semantics and change it.

Notice that we would prove the same mathematical results whether we took the semantics or the proof theory to have priority. The correctness criteria are identical. However, the intentional stance is different: we may be engaged in the same proofs but the underling intentions are different. Indeed, what I take to have definitional priority determines the interpretation I give to soundness and completeness.

The lambda calculus and its semantics provides a more convincing example where the proof theory dominates. Any set-theoretic interpretation of the calculus would only be taken to be a *model* if the rules were sound. In this sense, the proof theory of the calculus has normative priority over the semantics.

This intentional aspect of correctness goes beyond any purely extensional requirements. *Programming*, *proving* and *explaining* are intentional activities.

5   Physical Computation and Physical Correctness

Intentional stance is important in regard to the distinction with which we began: there is a difference between characterising physical computation and systems and characterising the correctness of a physical artefact relative to its specification. They may have the same conditions of correctness, but they are intentionally different. One contemporary characterisation of physical computing systems is due to Piccinini [14].

A physical system is a computing system just in case it is a mechanism one of whose functions is to manipulate vehicles based solely on differences between different portions of the vehicles according to a rule defined over the vehicles.

Judgements about success or failure of such characterisations partly depend upon our ordinary ability to distinguish between physical systems that perform computations and those that do not. Of course, philosophical considerations drive much of the investigation. Deciding which devices are taken to be computers and which are not often involves philosophical analysis. Nevertheless the endeavour is partly an empirical one where these intuitive judgements determine the success or failure of the proposed model or characterisation. Ordinarily, we take it that computers and calculators compute, but rocks do not. Any characterisation that gets this wrong will probably be rejected. When the model/characterisation does not agree with such judgments we reject it.

There is an even more empirical approach to the characterisation of physical computation that is based upon the central notion of representation between models and physical systems used in physics [8].

The key to the interaction between abstract and physical entities in physics is via the representation relation. This is the method by which physical systems are given abstract descriptions: an atom is represented as a wave function, a billiard ball as a point in phase space, a black hole as a metric tensor and so on. That this relation is possible is a prerequisite for physics: without a way of describing objects abstractly, we cannot do science. ....We argue that a 'computer' is a physical system about which we have a set of physical theories from which we derive both the full representation relation and the dynamics.

This enterprises has a much stronger empirical feel to it than the philosophical investigation of [14]. However, our general point does not depend upon such fine distinctions.

With the characterisation of correctness, it is not the physical device that determines whether the specification is true. Instead, the aim is to show that an artefact satisfies its abstract specification. If the artefact does not satisfy the specification we reject the artefact not the specification. The extensional relationship between the abstract device and the physical one might well be the same for both. The intuitions that underlie the simple mapping account would seem to apply to both. However, the intentional stance is different. When things go wrong we blame and change different things. Of course, what is a specification at one level of abstraction is a device at another. See [17] for a detailed discussion of this issue.

References

1. Chalmers, D.: Does a rock implement every finite-state automaton? Synthese **108**, 309–333 (1996)
2. Colburn, T.: Philosophy and Computer Science. M.E. Sharp Publishers, New York and London (2000)
3. Copeland, J.: What is computation? Synthese **108**(3), 335–359 (1996)
4. De Millo, R.A., Lipton, R.J., Perlis, A.J.: Social processes and proofs of theorems and programs. Commun. ACM CACM **22**(5), 271–280 (1979)
5. Fetzer, J.H.: Program verification: the very idea. Commun. ACM **31**(9), 1048–1063 (1988)
6. Primiero, G.: Information in the philosophy of computer science. In: Floridi, L. (ed.) The Routledge Encyclopedia on the Philosophy of Information, pp. 90–106 (2016)
7. Hoare, C.A.: An axiomatic basis for computer programming. Commun. ACM **12**(10), 576–580 (1969)
8. Horsman, C., Stepney, S., Wagner, R., Kendon, V.: When does a physical system compute? Proc. R. Soc. A Math. Phys. Eng. Sci. **470**(2169), 20140182 (2014)
9. Kroes, P.: Technological explanations: the relation between structure and function of technological objects. Techné Res. Philos. Technol. **3**(3), 124–134 (1988)
10. Kroes, P.: Technical Artefacts: Creations of Mind and Matter: A Philosophy of Engineering Design. Springer, Dordrecht (2012). https://doi.org/10.1007/978-94-007-3940-6
11. Lang, M.: Aspects of mathematical explanation: symmetry, unity and salience. Philos. Rev. **123**(4), 485–531 (2014)
12. https://en.wikipedia.org/wiki/NMOS_logic
13. Piccinini, G.: Computation in Physical Systems. The Stanford Encyclopedia of Philosophy (Summer 2015 Edition)
14. Piccinini, G.: Physical Computation: A Mechanistic Account. Oxford University Press, Oxford (2015)
15. Putnam, H.: Minds and machines. In: Hook, S. (ed.) Dimensions of Mind: A Symposium, pp. 138–164. Collier, New York (1960)
16. Turner, R., Angius, N.: The Philosophy of Computer Science. The Stanford Encyclopaedia of Philosophy (Spring 2017 Edition)
17. Turner, R.: Towards a philosophy of computer science. Computational Artifacts, pp. 13–19. Springer, Heidelberg (2018). https://doi.org/10.1007/978-3-662-55565-1_2
18. Wittgenstein, L.: Remarks on the foundations of mathematics. von Wright, G.H., Rhees, R., Anscombe, G.E.M. (eds.). Oxford (1978)

Higher Type Recursion for Transfinite Machine Theory

Philip Welch[(⊠)]

School of Mathematics, University of Bristol, Bristol BS8 1TW, England
p.welch@bristol.ac.uk
http://people.maths.bris.ac.uk/~mapdw/

Abstract. We look at some preliminary work in the theory of transfinite Turing machines generalised in the manner of Kleene to higher type recursion theory. The underlying philosophy is that ordinary Turing computability and inductive definability is replaced by the example here of Infinite Time Turing Machine computability and quasi-inductive definability.

1 Introduction

The purpose of this paper is to give a purely descriptive account of how notions of 'recursion' obtained from transfinite computational machines could be harnessed to yield a theory of *higher type of recursion* using those machines. (To make it clear from the outset: type 0 objects are of the form: $n \in \omega$; type 1 are of the form $x : \omega \to \omega$, and type 2 are of the form $F : (\mathbb{N}^{\mathbb{N}}) \to \omega$ *etc.* We shall not deal with objects here of type higher than 2.)

We restrict ourself here to ideas and definitions. We summarise some results that characterise the semi-decidable sets for such notions, but all proofs must be omitted. The point is to indicate how analogies with Kleene's theory of Higher Type recursion from the late '50's and early '60's can be used to develop these ideas in the transfinite context.

Our transfinite machine will be the ω length tape Infinite Time Turing Machine ("ittm") model of Hamkins and Kidder [8] with which we shall assume the reader is familiar. Much of what we say generalises to machines with longer tapes.

We shall give analogies to Kleene's type-2 recursion and the objects that naturally arise there, but formulated for type-2 recursion using ittm's. We don't claim to give the final form of this: there are a number of decisions and choices along the way, that could have been made differently. Kleene's theory can be cast in that of *monotone inductive definitions* which we first recall. The concept corresponding to this for ittm-theory is that of a *quasi-inductive definition*. In Sect. 2 we give first a sketch of Kleene's theory applied to wellfounded trees of Turing machines ("tm" will always denote a regular Turing machine) and the type-2 objects that naturally occur here. The theory of hyperarithmetic sets and the fact that 'semi-decidable' in this context corresponds to Π_1^1 are of great

© Springer Nature Switzerland AG 2019
F. Manea et al. (Eds.): CiE 2019, LNCS 11558, pp. 72–83, 2019.
https://doi.org/10.1007/978-3-030-22996-2_7

weight in what follows. In Sect. 3 we give a description of the ittm version of this, according to a choice of type-2 oracles. As much for motivation, or additional justification for our structure, as for anything else, in Sect. 4 we state some applications results in low levels of determinacy.

2   Inductive Operators

Let $\Phi : \mathcal{P}(\mathbb{N}) \to \mathcal{P}(\mathbb{N})$ be any arithmetic Γ operator (that is '$n \in \Gamma(X)$' is an arithmetic relation of X and n.)

Definition 1. *(i)* Φ *is* monotone *iff* $\forall X \subseteq Y \subseteq \mathbb{N} \quad \longrightarrow \quad \Phi(X) \subseteq \Phi(Y);$
(ii) Φ *is* progressive *iff* $\forall X \subseteq \mathbb{N} \quad (X \subseteq \Phi(X)).$

In either case we set: $\Phi_0(X) = X$ and then: $\Phi_\alpha(X) = \Phi(\bigcup_{\beta<\alpha}(\Phi_\beta(X)))$. We call Φ *inductive* if it is monotone or progressive. Clearly Φ inductive implies there will be *fixed points* the least of which will be: $\Phi_\infty(X) =_{df} \Phi_\alpha(X)$ where α is least with $\Phi_\alpha(X) = \Phi_{\alpha+1}(X)$; clearly α will be countable.

The theory of inductive operators was heavily investigated in the 1960's and early 70's by Spector, Gandy, Hinman, Richter, Aczel, Moschovakis, Aanderaa, Cenzer and others. From this work developed Moschovakis's theory of *generalised definability* and *inductive definitions* over *abstract structures* [15]. This tied in with previous work in *admissibility theory* "The next admissible set" (Barwise-Gandy-Moschovakis Theorem, [1]), and the Spector-Gandy Theorem that: "$\Pi^1_1 = \Sigma_1(L_{\omega_1^{ck}})$" - $L_{\omega_1^{ck}}$ being the least admissible set over $\mathbb{N}$.

Definition 2 (Quasi-inductive operators). *Let* Φ *be any operator. Define iterates* Φ *as before except for limits* $\lambda \leq On$:

$$\Phi_\lambda(X) = \liminf_{\alpha \to \lambda} \Phi_\alpha(X) = \bigcup_{\alpha<\lambda} \bigcap_{\lambda>\beta>\alpha} \Phi_\beta(X).$$

For arithmetic operators this is, in effect, due to Burgess [3], but which has its roots in the notion of *revision theoretic definability* of Gupta and Belnap [6].

Lemma 1. *Any such operator has a least countable* $\zeta = \zeta(\Phi, X)$ *with* $\Phi_\zeta(X) = \Phi_{On}(X)$. *Moreover there is a cub class of ordinals, closed and unbounded beneath any uncountable cardinal, of ordinals* ξ, *with* $\Phi_\zeta(X) = \Phi_{On}(X)$.

There are not a huge number of examples of quasi-inductive operators in the literature, but an important one is that of an *infinite time turing machine* (ittm) where we regard the ω-length tape(s) as a sequence of cells whose contents are revised according to the transition table of the program. This results in a *recursive* operator Φ which moreover only updates at most one cell, so one integer of X, at each stage. All the active new work takes place at the limit stages with the lim inf rule.

Kleene in [10] developed an equational calculus, itself evolving out of his analysis of the Gödel-Herbrand General Recursive Functions (on integers) from the 1930's, but now enlarged for dealing with recursion in objects of finite type. A particular type-2 functional was that derived from the *ordinary jump* oJ, where

$$\mathsf{oJ}(e, \boldsymbol{m}, x) = \begin{cases} 1 & \text{if } [e](\boldsymbol{m}, x)\downarrow \text{ (meaning has a } \textit{defined value} \text{ or } \textit{converges}) \\ 0 & \text{otherwise.} \end{cases}$$

(We shall also use the notation "$[e]^{\mathsf{l}}(p)$" rather than "$\{e\}^{\mathsf{l}}(p)$" to indicate that we are using Kleene recursion using tm's, and reserve the latter more usual notation for ittm recursion.) Here $\boldsymbol{m}$ is a string of integers, and x a function $x :$ $\omega \to \omega$ (thus an object of type 1) and e the index number of an ordinarily Turing recursive functional of type-1 objects. (A vector of such functions will be denoted in bold.) The reader should note the use of the downarrow in $[e](\boldsymbol{m}, x)\downarrow$ to mean just what it says: the expression is *defined*, and for which we use *convergence* as a synonym. Similarly $[e](\boldsymbol{m}, \boldsymbol{x})\uparrow$ will mean the expression is *undefined* with synonym of *divergence*. Functions of type greater than 1 are conventionally called 'functionals', but we may occasionally let this slip.

The functional oJ can be considered as a functional just on type-2 objects (absorbing objects of lower type by their type-2 counterparts). Using coding of vectors of functions we ultimately think of this as oJ having domain $\omega \times {}^{k}\omega \times {}^{l}({}^{\omega}\omega)$ for any $k, l \in \omega$.

Kleene then developed (see Hinman [9] Ch. VI) a theory of generalised recursion in type-2 (and higher) functionals; in this theory a designation such as '$[e]^{\mathsf{l}}$' refers to the e'th function *recursive in the type-2 functional* l. (Warning: this is not just the simple use of the oracle l in a linear computation as the notation might suggest, but refers to a tree of computation with calls to the oracle.) During a computation of, say, $[e]^{\mathsf{l}}(\boldsymbol{n}, \boldsymbol{y})$ oracle steps are allowed whereby the result of a query $(f, \boldsymbol{m}, \boldsymbol{x})$ is directly asked of l, and an integer result, $\mathsf{l}(f, \boldsymbol{m}, \boldsymbol{x})$, is returned. (Of course even to make the query the values of each of the infinitely many values of the functions $\boldsymbol{x}$ have to already have been calculated; calculating each of these values can in turn require asking the oracle l for further values *etc.*; thus such a recursion can be represented by a tree, which if convergent is well founded, but is potentially infinitely branching at any node, with each branch calculating some $x(k)$ say.) In this formalism the index set H_{oJ} defined by:

$$H_{\mathsf{oJ}}(e) \leftrightarrow [e]^{\mathsf{oJ}}(e)\downarrow$$

is a complete semi-recursive (in oJ) set of integers, and Kleene showed that this is in turn a complete Π^1_1 set of integers. Further he showed that the oJ-*recursive sets of integers, i.e.* those sets R for which

$$R(n) \leftrightarrow [e]^{\mathsf{oJ}}(n)\downarrow 1 \ \wedge \ \neg R(n) \leftrightarrow [e]^{\mathsf{oJ}}(n)\downarrow 0$$

for some index e, are precisely the hyperarithmetic ones. (See Hinman [9] Ch. VII.1 for a discussion of this.)

Kleene gave his account of recursion in objects of finite type which we have alluded to above in [10,13]. In order to give further weight to his definition he then showed it was extensionally equivalent to an alternative given by a Turing machine model enhanced with oracle calls to a higher type functional, see [11, 12]; this was just as for the case of ordinary Turing computation. Many different concepts of computation on numbers turned out to be equivalent. By showing that the equational model had the same functions as a Turing machine model he was emulating the same conceptual move Turing had made. In the first paper [11] he showed how any Turing machine computation of finite type could be achieved on the generalized recursive equational approach. The second paper [12] showed the reverse. In both directions a convergent, (so defined) computation could be represented, not as a finite tree of computations as for ordinary recursion, but now as a well-founded but in general infinitely branching tree of computations of function values - which in general required calculating infinite objects (as we indicated above), such as all values $x(n)$ for a function $x : \mathbb{N} \longrightarrow \mathbb{N}$, at some level in the tree before submitting that completely calculated function itself as an argument to a function of higher type at the level above. The wellfounded tree of either functional calculations, or of Turing machine computational calls, depending on the representation, witnessed a successfully defined or convergent computation. The tree occurs dynamically as part of the computational process.

Our account here is motivated in spirit by that latter approach. Instead of using an equational calculus we shall couch our model not just in terms of the Kleenean Turing machine, but in terms of ittm's and their computations, viewed as quasi-inductive operators now recursive in a certain operator iJ in place of Kleene's oJ.

Viewed as a class of quasi-inductive operators, the output tape (or every third cell say of a single tape model) of an ittm represents an element of Cantor space at any stage; that output tape may or may not converge to a fixed value. If it does then the the real there is to be regarded as the output of the computation. Notice this is a more generalised notion than that of the machine *halting* and hence with a fixed output tape for that reason. Halting is really just a special case of the basic phenomenon of 'fixed output'. The idea that a tape is *eventually settled* is broader in the infinite time context: a calculation can continue indefinitely, without any changes to the output tape section. It must have seemed natural to consider only the halting computations when first thinking about ittm behaviour, but as [17] showed, even to characterise those halting calculations required stepping back and analysing the whole class of eventually settled computations: the latter we regard as more fundamental, and as characteristic of the ittm process. To analyse ittm behaviour is to analyse the eventually settled outputs (which we shall call 'fixed outputs' below), and to find out what they are capable of computing requires analysing those fixed outputs, not just the more specialised halting outputs. The Spector class naturally associated to this form of definability by itttm's is precisely that using this fixed output rather than the proper subclass using nominally halting output. And of course this is in accord with the quasi-inductive scheme above.

Given a set $A \subseteq \omega \cup {}^{\omega}2$, this can be used as an oracle during a computation on an ittm in a familiar way: ? *Is the integer on (or is the whole of) the current output tape contents an element of A?* and receive a 1/0 answer for "Yes"/"No".

We identify elements of ω as coded up in $^{\omega}2$ in some fixed way, and so may consider such A as always subsets of $^{\omega}2$. But further: since having A respond with one 0/1 at a time can be repeated, by using an ω-sequence of queries/responses, we could equally well allow A to return an element $f \in {}^{\omega}2$ as a response (we have no shortage of time). We could then allow as functionals also $\mathsf{I} : {}^{\omega}2 \longrightarrow {}^{\omega}2$. However for this paper we shall only consider functionals into ω. Some examples follow. As is usual we let $\{e\}$ represent the partial function computed by the e'th ittm programmed machine P_e.

Definition 3. *(The infinite time jump* iJ*)*
(i) We write $\{e\}(\boldsymbol{m}, \boldsymbol{x})\!\downarrow$ *if the* e*'th ittm-computable function with input* $\boldsymbol{m}, \boldsymbol{x}$ *has a fixed output* $c \in 2^{\mathbb{N}}$, *in which case we write* $\{e\}(\boldsymbol{m}, \boldsymbol{x}) = c$.
(ii) We then define iJ *by:*

$$\mathsf{iJ}(e, \boldsymbol{m}, \boldsymbol{x}) = \begin{cases} 1 & \text{if } \{e\}(\boldsymbol{m}, \boldsymbol{x})\!\downarrow; \\ 0 & \text{otherwise (for which we write } \{e\}(\boldsymbol{m}, \boldsymbol{x})\!\uparrow). \end{cases}$$

The functional iJ then is the counterpart of the standard tm operator oJ.

Definition 4. *For x a real, the complete (ordinary) ittm-semirecursive-in-x set, denoted by* $\tilde{x}$ *is the set of integers* $\{e \mid \{e\}(e, x)\!\downarrow\}$.

One consequence of the (relativized to a real x) λ-ζ-Σ-Theorem (*cf.* [16] Thm 2.6) is that $\tilde{x}$ is recursively isomorphic to the complete Σ_2-Theory of $L_{\zeta^x}[x]$.

3   Higher Type Recursion

In the Kleenean recursion in type-2 functionals, in [11,12] a successful computation (meaning one with output) could be effected by imagining tm's placed at nodes on a wellfounded tree, with computations proceeding at nodes that make computation calls to a lower node, seeking the value of some $x(k)$ say. The computation time at each node, regarding each call to a lower node as being just one step in the computation of the calling node, is then finite. (For otherwise the computation at the node is never completed and the whole overall computation will fail.) An overall computation may fail by instituting a series of calls to subcomputations that form an infinite descending path in the tree. In such cases the machines on the path all hang after finitely many steps, all waiting for data to be passed up from the immediate subcomputation it has called.

In the ittm case we may again conceive of an overall or master ittm computation taking place at the top level; such a computation may take infinitely many steps in time, and will be considered as successful if the output tape is fixed from some point in time onwards. The master computation may make queries of a type-2 functional I in which the computation is considered recursive. It may call subcomputations of exactly the same type: ittm's with the capability to make oracle queries of I.

We give a more detailed description of this as a representation in terms of underlying ittm's. $\{e\}^{\mathsf{I}}(\boldsymbol{m}, \boldsymbol{x})$ will represent the e'th program in the usual format, say transition tables, but designed with appeal to oracle calls possible. We are thus considering computations of a partial function $\{e\}^{\mathsf{l}} : {}^k\omega \times {}^{\mathsf{l}}({}^\omega 2) \to {}^\omega 2$. Such a computation has potentially computation time, or stages, unbounded in the ordinals.

The computation of $P^{\mathsf{l}}_e(\boldsymbol{m}, \boldsymbol{x})$ proceeds in the usual ittm-fashion, working as a tm at successor ordinals and taking lim inf's of cell values *etc.* at limit ordinals. (We take lim inf's rather than lim sup's as this accords more with the notion characteristic functions of quasi-inductive operators. This makes no difference to the computational possibilities of ittms's or here at higher types.) At a time α an oracle query may be initiated. We may conventionally fix that the real number subject to query is that infinite string on the even numbered cells of the scratch type. If this string is $(f, m, y_0, y_1 \ldots,)$ then setting $y = y_0, y_1 \ldots$, the *query* or *oracle call* which we shall denote $Q(\mathsf{l}, f, m, y)$ is the question: ?*What is* $\mathsf{l}(z)$ *where* $P^{\mathsf{l}}_f(m, y)\downarrow z$? and at stage $\alpha + 1$ receives the value $\mathsf{l}(z)$. If it is not the case that $P^{\mathsf{l}}_f(m, y)\downarrow z$ for any z, *i.e.*, it fails to have a fixed output, then there is no z to which l can be applied, and the overall computation fails. (We could try to stay closer to the Kleenean setting, where a tree branches infinitely often downwards, to compute for some $z \in {}^\omega\omega$, $z(0), z(1), \ldots$ in turn, and then can ask for $\mathsf{l}(z)$. There, if any of the computations $z(k)$ failed, then the query to l did not take place, and the overall computation failed. But one thing we have with ittm computation is plenty of time, so we can amalgamate the individual computations $z(k)$ as simply one computation of all of z.)

Space prohibits a formal definition of the representation above, but we can determine its effect as follows *via* an inductive operator I. Just as the Kleene equational calculus can be seen to build up in an inductive fashion a set of indices $\Omega[\mathsf{l}]$ for successful computations recursive in l (see Hinman [9], pp. 259–261), so we can define the fixed point of a monotone operator $I = I^{\mathsf{l}}$ on $(\omega \times \omega^{<\omega} \times (\omega^\omega)^{<\omega}) \times \omega^\omega$ which will give us the successful ittm-computations recursive in l.

Definition 5. *We set $I(X) =:$*

$$\{\langle\langle e, \boldsymbol{m}, \boldsymbol{x}\rangle, z\rangle \mid P^X_e(\boldsymbol{m}, \boldsymbol{x})\downarrow z \text{ is an ittm-computation making only oracle calls}$$
$$Q(X, e', \boldsymbol{m}', \boldsymbol{x}') \text{ and receiving back } \mathsf{l}(z') \text{ where } X(\langle e', \boldsymbol{m}', \boldsymbol{x}'\rangle) = z' \}.$$

As this is monotone, we may let
$I^0 = \varnothing;\ I^{<\alpha} = \bigcup_{\beta<\alpha} I^\beta$ & $I^\alpha = I(I^{<\alpha})$ *in the usual way, and reach a least fixed point I^∞.*

Then:

Theorem 1 (The $\{e\}$'th function generalised recursive in l). *Using I^∞:*
$\{e\}^{\mathsf{l}}(\boldsymbol{m}, \boldsymbol{x})$ *is defined, or convergent, with output z iff $I^\infty(\langle e, \boldsymbol{m}, \boldsymbol{x}\rangle) = z$.*
In which case we set $\{e\}^{\mathsf{l}}(\boldsymbol{m}, \boldsymbol{x}) = z$. Otherwise it is undefined or divergent.

Overall we have a *computation tree* - also called a *tree of subcomputations*, with subcomputation calls performed at branching nodes below the top level. However, although the computation is most easily represented by a tree, we may think of the computation as a linear sequential process as we visit each node of the tree in turn.

We therefore make the following conventions. During the calculation of $\{e\}^{\mathsf{l}}(\boldsymbol{m}, \boldsymbol{x})$ the initial calculation takes place at the topmost node ν_0 which we declare to be *at Level* 0 in our computation tree $\mathfrak{T} = \mathfrak{T}^{\mathsf{l}}(\boldsymbol{e}, \boldsymbol{m}, \boldsymbol{x})$. Let us suppose the first oracle query concerning $\{f_0\}^{\mathsf{l}}(n_0, y_0)$ is made at some stage. The tree $\mathfrak{T}$ will then have a node ν_1 below ν_0, labelled with $\langle f_0, n_0, y_0 \rangle$ and we declare the computation $\{f_0\}^{\mathsf{l}}(n_{0,} y_0)$ to be performed at this Level 1. Thus 'control' of the overall process is at the level of the node. Further, we may define the *overall length function* $H = H(\mathsf{l}, e, \boldsymbol{m}, \boldsymbol{x})$ as the length of the computation that occurs at the nodes of the wellfounded part of $\mathfrak{T}$. Sequentially H totals up the ordinal number of stages of operation at each of the nodes where control currently resides.

Definition 6. *(i) The* level *of the computation* $\{e\}^{\mathsf{l}}(\boldsymbol{m}, \boldsymbol{x})$ *at time* α *(as given by H), denoted* $\Lambda(e, \mathsf{l}, (\boldsymbol{m}, \boldsymbol{x}), \alpha)$, *is the level of the node* ν_ι *at which the overall computation is being performed at time* α, *where:*
(ii) the level *of a node* ν_ι *is the length of the path in the tree from* ν_0 *to* ν_ι.
(iii) By Level n *we accordingly mean the set of nodes in the tree with level n.*

Thus for a convergent computation, at any time the level is a finite number ('depth' would have been an equally good choice of word). A *divergent computation* is one in which either (i) an oracle call resulting in a calculation at some node fails to produce an output z (and so no value $\mathsf{l}(z)$ can be returned to the level above) or (ii) $\mathfrak{T}$ is illfounded (with a rightmost path of order type then ω).

Recall that a 'snapshot' at time γ in a computation by an ittm is the ω-sequence of bits of information consisting of the current read/write head position, transition state number, and the sequence of cell values. The snapshots up to the stage in a calculation $P_e^{\mathsf{l}}(\boldsymbol{m}, \boldsymbol{x})$ where it ends its first loop (if this occurs) will have all the relevant information then in the calculation: everything thereafter is mere repetition. (This would be undefined if the computation tree is illfounded). We say that a computation '*exhibits final looping behaviour*' ('at stage σ', or 'by stage τ'), if there are stages or times $\xi < \sigma \, (\leq \tau)$ with at the top level (a) identical snapshots at ξ and σ, and moreover (b) no cell that had a stable value at time ξ changes that value in the interval (ξ, σ).

ITTM Recursion in $^2\mathsf{E}$. We shall draw to a close the discussion of generalised recursion in functions l as this will take us too far from our goal, and shall leave this for future work. For us, as for Kleene, recursion in $^2\mathsf{E}$ is fundamental. Recall, for $y \in \mathbb{N}^{\mathbb{N}}$, $^2\mathsf{E}(y) = 0$ if $\exists n \, y(n) = 0$ and $^2\mathsf{E}(y) = 1$ otherwise. Many of the theorems of type-2 recursion about functionals F have to be prefixed with the requirement that $^2\mathsf{E}$ is recursive in F. (Such F are called *normal*.)

Definition 7. *We say* F *is* (generalised) ittm-partial recursive in G *if there is an index e so that* $\mathsf{F} = \{e\}^{\mathsf{G}}$. F *is* ittm-recursive in G *if it is partial recursive in G and total. A relation R is* ittm-recursive in l *if its characteristic function is. R is* ittm semi-recursive in l *if it is the domain of a functional ittm-partial recursive in l.*

Kleene showed that the functionals $^2\mathsf{E}$ and oJ are mutually (Kleene) recursive in each other (*cf.* [9] VI.1.4.). We shall have this too:

Theorem 2. *The functionals* iJ *and* $^2\mathsf{E}$ *are mutually ittm-recursive.*

We wish to apply this theory to the particular case of iJ - the *infinite jump*.

A number of elementary facts concerning computation trees $\mathfrak{T}$ living in transitive admissible sets M may be proven.

Lemma 2. *Suppose* $\{e\}^{\mathsf{iJ}}(\boldsymbol{m},\boldsymbol{x})$ *has a computation tree* $\mathfrak{T} \in M$, *and with* $\boldsymbol{x} \in M$, *where* M *is a transitive admissible set, closed under the function* $y \longmapsto \tilde{y}$. *Then* $(\{e\}^{\mathsf{iJ}}(\boldsymbol{m},\boldsymbol{x})$ *is convergent* $)^M \longleftrightarrow \{e\}^{\mathsf{iJ}}(\boldsymbol{m},\boldsymbol{x})$ *is convergent.*

It was an essential feature of ordinary ittm-theory that if a computation $P_e(m)$ produced an output it would always have done this by stage ζ where ζ is least so that for some $\Sigma > \zeta$ we had $L_\zeta \prec_{\Sigma_2} L_\Sigma$; this was shown in the "λ-ζ-Σ Theorem" (see [16] 2.1 and 2.3). The Σ_2 liminf nature of the limit rule underlay this, and the same is true here.

Definition 8. *A pair of ordinals* (μ,ν) *is a* Σ_2-*extendible pair if* $L_\mu \prec_{\Sigma_2} L_\nu$ *and moreover* ν *is the least such with this property with respect to* μ. *We say* μ *is* Σ_2-*extendible if there exists* ν *with* (μ,ν) *a* Σ_2-*extendible pair. By relativisation, a pair of ordinals* (μ,ν) *is an* $(\boldsymbol{x},\mathsf{I})$-$\Sigma_2$-*extendible pair, and* μ *is* $(\boldsymbol{x},\mathsf{I})$-$\Sigma_2$-*extendible, if* $L_\mu[\boldsymbol{x},\mathsf{I}] \prec_{\Sigma_2} L_\nu[\boldsymbol{x},\mathsf{I}]$.

Then of importance for our purposes are:

Lemma 3. *The computation* $\{e\}^{\mathsf{I}}(\boldsymbol{m},\boldsymbol{x})$ *exhibits final looping behaviour if and only if there exists some* $(\boldsymbol{x},\mathsf{I})$-$\Sigma_2$-*extendible pair* (ζ,Σ) *so that* $\Lambda(e,\mathsf{I},\boldsymbol{x},\zeta) = 0$.

The dependence on I in the above is natural. With iJ it can be dropped:

Lemma 4. *The computation* $\{e\}^{\mathsf{iJ}}(\boldsymbol{m},\boldsymbol{x})$ *exhibits final looping behaviour if and only if there exists some* $\boldsymbol{x}$-Σ_2-*extendible pair* (ζ,Σ) *so that* $\Lambda(e,\mathsf{iJ},\boldsymbol{x},\zeta) = 0$.

Usual methods prove an S^n_m-theorem and:

Theorem 3 (The I-Recursion theorem). *If* $\mathsf{F}(e,\boldsymbol{m},\boldsymbol{x})$ *is ittm-recursive in* I, *there is* $e_0 \in \omega$ *so that*

$$\{e_0\}^{\mathsf{I}}(\boldsymbol{m},\boldsymbol{x}) = \mathsf{F}(e_0,\boldsymbol{m},\boldsymbol{x}).$$

Another Example: Lubarsky's Feedback-ittm Recursions
We are indebted to Lubarsky's work in [14] and grateful for discussions with him on his earlier FITTM's (= Feedback ITTM's). His notion of 'feedback' uses the concept of properly halting where the basic outcome occurs when an ITTM *halts* rather than having, as here, a fixed output. (We have indicated above why we consider 'fixed output' to mean a fixed output tape.) He describes wellfounded

computation trees, not as arising from a Kleene style recursion, but as ITTM's (with extra tapes) that have the additional state of "an oracle query *does the Feedback ITTM with program the content of the first additional tape on input the content of the second [tape] converge?*" which will receive a Yes/No answer. (As intimated, convergence is halting.) He then describes the semantics of such computations as wellfounded trees, where subcalls are again queries of the same type (*"Does* $\{e\}_{FITTM}(x)$ *halt?"*). An FITTM computation *freezes* if the tree becomes illfounded. He asks a number of questions, such as to the ranks for the wellfounded trees occurring, what are the reals output or appearing on tapes of such machines. We briefly state answers to these below.

We may describe an induction building up directly the class of successful FITTM-computations as a fixed point of a monotone operator, in this spirit, just as in Definition 5 above. However we construe this fixed point as that arising from a type-2 operator, let us call it here hJ, from an ittm-recursion defined as in this paper. Recursion in hJ then also becomes an example of ittm-recursion in ^2E with Theorem 2 applying again: hJ and ^2E are mutually ittm-recursive in each other. Thus for us Feedback-ITTM computations become a particular example of this higher type ittm recursion, and the class of $x \in \mathbb{N}^{\mathbb{N}}$ FITTM-computable coincide with those ittm-recursive in ^2E.

Computation Times. The Lemmata 3 and 4 above give sufficient conditions for a computation $\{e_0\}^{iJ}(m, x)$ to converge. We need to find out exactly how long computations in iJ take in order to characterise the iJ-recursive and semi-recursive sets. The clue is that ordinary ittm-computations can compute the theories and codes for the levels of the L-hierarchy up to the end of the first Σ_2-extendible pair interval (ζ, Σ). (This is shown in [17], and in [4] a programme is explicitly given that shows how codes and theories can be simultaneously produced by an ordinary ittm recursion on α for $\alpha < \Sigma$; after stage Σ it drops into a repeating loop of reproducing the results on its output tape of the $\alpha \in (\zeta, \Sigma)$.) A machine that writes out codes for L_α's must in some sense be, at least akin to, a universal machine, since by absoluteness, any ittm computation on integer input can be run in L. Here we have, in effect, ittm's that can whilst within such a Σ_2-extendible interval, call other ittm's as part of a subroutine. It might not be inconceivable that such behaviour is overall fashioned, when they try to write codes for levels L_α's, by their reaching levels of the L-hierarchy, where the Σ_2-extendible pairs become *nested*.

Definition 9. *For $m \geq 1$ an m-depth Σ_2-nesting of an ordinal α is a sequence* $(\zeta_n, \sigma_n)_{n<m}$ *so that*
(i) if $m = 1$ then $\zeta_0 < \alpha < \sigma_0$;
(ii) if $0 < n + 1 < m$ then $\zeta_n \leq \zeta_{n+1} < \alpha < \sigma_{n+1} < \sigma_n$;
(iii) if $k < m$ then $L_{\zeta_k} \prec_{\Sigma_2} L_{\sigma_k}$.

We may show that there are processes generalised ittm-recursive in $iJmf$ that compute levels of L up to the points where any finite depth of nesting occurs, where each additional depth of nesting corresponds to computing up to repeating snapshots at one depth lower in $\mathfrak{T}$. It then seems inevitable that illfounded trees must ultimately occur by some ordinal corresponding to *infinite depth nesting*. But *prima facie* there is no such ordinal, since there can be no infinite descending chain $\sigma_{n+1} < \sigma_n$ in the above definition.

We thus shall want to consider *non-standard admissible models* (M, E) of KP together with some other properties. We let WFP(M) be the wellfounded part of the model. By the so-called 'Truncation Lemma' it is well known (v. [2]) that this wellfounded part must also be an admissible set. Usually for us the model will also be a countable one of "$V = L$". Let M be such and let $\alpha = \text{On} \cap \text{WFP}(M)$. By the above α is thus an admissible ordinal, *i.e.* L_α will also be a KP model. As remarked, an 'ω-depth' nesting cannot exist by the wellfoundedness of the ordinals. However an illfounded model M when viewed from the outside may have infinite descending chains of M-ordinals in its illfounded part. These considerations motivate the following definition.

Definition 10. *An* infinite depth Σ_2-nesting of α based on M *is a sequence* $(\zeta_n, s_n)_{n < \omega}$ *with:*
$$(i)\ \zeta_n \leq \zeta_{n+1} < \alpha \subset s_{n+1} \subset s_n;\quad (ii)\ s_n \in \text{On}^M;\quad (iii)\ (L_{\zeta_n} \prec_{\Sigma_2} L_{s_n})^M.$$

Thus the s_n form an infinite descending E-chain (where, as above, E is the membership relation of the illfounded model) through the illfounded part of the model M.

Whilst any countable transitive admissible set can be extended to have an illfounded part, (again v. [2]) and, for example, there are illfounded end-extensions of $L_{\omega_1^{ck}}$, that does not mean that this latter model can be extended to an illfounded model M which supports an infinite depth Σ_2-nesting: a relatively large countable admissible β is needed for that:

Definition 11. *Let* β_0 *be the least ordinal* β *so that* L_β *forms the wellfounded part of an admissible end-extension* (M, E) *based on which there exists an infinite depth* Σ_2-nesting of β.

It turns out that L_{β_0} is a model of Σ_1-Separation. Hence it has a proper, and so least, Σ_1-elementary submodel: $L_{\alpha_0} \prec_{\Sigma_1} L_{\beta_0}$. These ordinals feature in what follows.

4   Conclusions

Theorem 4. *(i) If a recursion* $\{e\}^{iJ}(\boldsymbol{m})$ *converges, then it does so by time* α_0, *and the latter ordinal is the supremum (over e and $\boldsymbol{m}$) of convergence times of such computations. (ii) There is a recursion* $\{h\}^{iJ}(\boldsymbol{m})$ *that only diverges at* β_0, *and all such divergent computations diverge before or at this time.*

Lemma 5. *(i) The* iJ-*recursive sets of integers are precisely those of* L_{α_0}; *(ii) the* iJ-*semi-recursive sets are those* $\Sigma_1(L_{\alpha_0})$.   Q.E.D.

The following answers two questions of Lubarsky:

Corollary 1. *The reals appearing on the tapes of freezing fittm-computations of* [14] *are precisely those of* L_{β_0}*; similarly the supremum of the ranks of the wellfounded parts of freezing fittm-computation trees is also* β_0.

Lemma 6. *The complete ittm-semidecidable-in-*iJ *set of integers*

$$K = \{(e, m) \in \omega \times \omega \mid \{e\}^{\mathsf{iJ}}(e)(e, m) = 1\}$$

as well as

$$H^{\mathsf{iJ}}(e) \leftrightarrow \{e\}^{\mathsf{iJ}}(e)\downarrow$$

are recursively isomorphic to the complete Σ_1*-Theory of* $\langle L_{\alpha_0}, \in \rangle$.

These last two lemmata can be compared with a result of Kleene *et al.*:

Theorem 5. *The complete (Kleene)-semidecidable in* oJ *set of integers is recursively isomorphic to the complete* Σ_1*-Theory of* $\langle L_{\omega_1^{ck}}, \in \rangle$. *The* oJ*-recursive sets of integers are precisely those of* $L_{\omega_1^{ck}}$, *that is, the hyperarithmetic sets.*

A Postlude. In earlier work we had located in the L-hierarchy winning strategies for Σ_3^0 two person perfect information games. The games in [18] connected to nested Σ_2-extendability. The presumed connection with ittm's becomes a intriguing question, and most of this work was motivated by trying to understand this. The summary above indeed ties in with these results, which we mention here without explaining the connection. See [19].

Theorem 6. *Let* η *be least so that for any* Σ_3^0*-game there is a winning strategy for one of the players definable over* L_η. *Then* $\eta = \beta_0$.

Subsequently S. Hachtman ([7]) found another remarkable characterisation of β_0:

- *Let* γ *be least so that, as a model of a fragment of second order arithmetic,* $\mathbb{R} \cap L_\gamma$ *is a model of* Π_2^1*-monotone induction. Then* $\gamma = \beta_0$.

Open Questions. As can perhaps be seen from this sketch there are more open questions than known facts. A closer analysis of ittm recursions in general type 2 functionals needs to be done:

Q1 Formulate a Stage Comparison Theorem *for ittm-recursion. (See* [9].VI*)*
Much as there are several approaches to the hyperarithmetic sets uses Kleene recursion, there are notation systems associated with ittm-theory. One can use the theory of $\partial\Sigma_3^0$-monotone operators to obtain *norms*, thus prewellorderings, on ittm semi-decidable sets. Presumably many features of Kleene recursion have some analogue for ittm recursion.

Q2 What is the correct definition and properties of the superjump *(due to Gandy for Kleene recursion) for ittm higher type recursions? (See* [9].VI*)*
We have only considered type-2 recursions to date.

Q3 Is there a suitable notion of ittm-recursion in this spirit at types-3 and above?
In another direction one can enlarge the notion of computation by taking on the *hypermachines* of [5]. Such machines may have loops at Σ_n-extendible ordinals by analogy with the ittm's.

Q4 Develop a theory of higher type hypermachine recursion.

References

1. Barwise, J., Gandy, R., Moschovakis, Y.: The next admissible set. J. Symb. Log. **36**(1), 108–120 (1971)
2. Barwise, K.: Admissible Sets and Structures. Perspectives in Mathematical Logic, vol. 2. Springer, Heidelberg (1975)
3. Burgess, J.: The truth is never simple. J. Symb. Log. **51**(3), 663–681 (1986)
4. Friedman, S.D., Welch, P.: Two observations regarding infinite time turing machines. In: Dimitriou, I. (ed.) Bonn International Workshop on Ordinal Computability, pp. 44–48. Hausdorff Centre for Mathematics, University of Bonn, Bonn (2008)
5. Friedman, S.D., Welch, P.: Hypermachines. J. Symb. Log. **76**(2), 620–636 (2011)
6. Gupta, A., Belnap, N.: The Revision Theory of Truth. MIT Press, Cambridge (1993)
7. Hachtman, S.: Determinacy and monotone inductive definitions. Isr. J. Math. (to appear)
8. Hamkins, J., Lewis, A.: Infinite time Turing machines. J. Symb. Log. **65**(2), 567–604 (2000)
9. Hinman, P.: Recursion-Theoretic Hierarchies. Ω Series in Mathematical Logic. Springer, Berlin (1978)
10. Kleene, S.: Recursive quantifiers and functionals of finite type I. Trans. Am. Math. Soc. **91**, 1–52 (1959)
11. Kleene, S.: Turing-machine computable functionals of finite type I. In: Proceedings 1960 Conference on Logic, Methodology and Philosophy of Science, pp. 38–45. Stanford University Press, Stanford (1962)
12. Kleene, S.: Turing-machine computable functionals of finite type II. Proc. Lond. Math. Soc. **12**, 245–258 (1962)
13. Kleene, S.: Recursive quantifiers and functionals of finite type II. Trans. Am. Math. Soc. **108**, 106–142 (1963)
14. Lubarsky, R.: ITTM's with feedback. In: Schindler, R.D. (ed.) Ways of Proof Theory. Ontos (2010)
15. Moschovakis, Y.: Elementary Induction on Abstract structures. Studies in Logic series, vol. 77. North-Holland, Amsterdam (1974)
16. Welch, P.D.: Eventually infinite time Turing degrees: infinite time decidable reals. J. Symb. Log. **65**(3), 1193–1203 (2000)
17. Welch, P.D.: The length of infinite time Turing machine computations. Bull. Lond. Math. Soc. **32**, 129–136 (2000)
18. Welch, P.D.: Weak systems of determinacy and arithmetical quasi-inductive definitions. J. Symb. Log. **76**, 418–436 (2011)
19. Welch, P.D.: $G_{\delta\sigma}$-games. Isaac Newton Preprint Series. No. NI12050-SAS (2012)

Effective Embeddings
for Pairs of Structures

Nikolay Bazhenov[1,2], Hristo Ganchev[3], and Stefan Vatev[3(✉)]

[1] Sobolev Institute of Mathematics, Novosibirsk, Russia
bazhenov@math.nsc.ru
[2] Novosibirsk State University, Novosibirsk, Russia
[3] Faculty of Mathematics and Informatics, Sofia University,
5 James Bourchier blvd., 1164 Sofia, Bulgaria
{ganchev,stefanv}@fmi.uni-sofia.bg

Abstract. We study computable embeddings for pairs of structures, i.e. for classes containing precisely two non-isomorphic structures. We show that computable embeddings induce a non-trivial degree structure for two-element classes consisting of computable structures, in particular the pair of linear orders $\{\omega, \omega^\star\}$, which are the order types of the positive integers and the negative integers, respectively.

1 Introduction

We study computability-theoretic complexity for classes of countable structures. A widely used approach to investigating algorithmic complexity involves comparing different classes of structures by using a particular notion of *reduction* between classes. Examples of such reductions include computable embeddings [4,5], Turing computable embeddings [7,13], Σ-reducibility [6,16], computable functors [10,15], enumerable functors [18], etc. If a class $\mathcal{K}_0$ is reducible to a class $\mathcal{K}_1$ and there is no reduction from $\mathcal{K}_1$ into $\mathcal{K}_0$, then one can say that $\mathcal{K}_1$ is computationally "harder" than $\mathcal{K}_0$. In a standard computability-theoretic way, a particular reduction gives rise to the corresponding degree structure on classes. Nevertheless, note that there are other ways to compare computability-theoretic complexity of two classes of structures, see, e.g., [11,14].

Friedman and Stanley [8] introduced the notion of *Borel embedding* to compare complexity of the classification problems for classes of countable structures. Calvert, Cummins, Knight, and Miller [4] (see also [13]) developed two notions, *computable embeddings* and *Turing computable embeddings*, as effective counterparts of Borel embeddings. The formal definitions of these embeddings are given in Sect. 2. Note that if there is a computable embedding from $\mathcal{K}_0$ into $\mathcal{K}_1$, then there is also a Turing computable embedding from $\mathcal{K}_0$ into $\mathcal{K}_1$. The converse is not true, see Sect. 2 for details.

The first author was partially supported by NSF Grant DMS #1600625, which allowed him to visit Sofia. The second and third authors were partially supported by BNSF Bilateral Grant DNTS/Russia 01/8 from 23.06.2017.

F. Manea et al. (Eds.): CiE 2019, LNCS 11558, pp. 84–95, 2019.
https://doi.org/10.1007/978-3-030-22996-2_8

In this paper, we follow the approach of [4] and study computable embeddings for pairs of structures, i.e. for classes $\mathcal{K}$ containing precisely two non-isomorphic structures. Our motivation for investigating pairs of structures is two-fold. First, these pairs play an important role in computable structure theory. The technique of pairs of computable structures, which was developed by Ash and Knight [1,2], found many applications in studying various computability-theoretic properties of structures (in particular, their degree spectra and effective categoricity, see, e.g., [2,3,9]).

Second, pairs of computable structures constitute the simplest case, which is significantly different from the case of one-element classes: It is not hard to show that for any computable structures $\mathcal{A}$ and $\mathcal{B}$, the one-element classes $\{\mathcal{A}\}$ and $\{\mathcal{B}\}$ are equivalent with respect to computable embeddings. On the other hand, our results will show that computable embeddings induce a non-trivial degree structure for two-element classes consisting of computable structures.

In this paper, we concentrate on the pair of linear orders ω and $\omega^\star$. By $\deg_{tc}(\{\omega, \omega^\star\})$ we denote the degree of the class $\{\omega, \omega^\star\}$ under Turing computable embeddings. Quite unexpectedly, it turned out that a seemingly simple problem of studying computable embeddings for classes from $\deg_{tc}(\{\omega, \omega^\star\})$ requires developing new techniques.

The outline of the paper is as follows. Section 2 contains the necessary preliminaries. In Sect. 3, we give a necessary and sufficient condition for a pair of structures $\{\mathcal{A}, \mathcal{B}\}$ to belong to $\deg_{tc}(\{\omega, \omega^\star\})$. In Sect. 4, we show that the pair $\{1 + \eta, \eta + 1\}$ is the greatest element inside $\deg_{tc}(\{\omega, \omega^\star\})$, with respect to computable embeddings. Section 5 proves the following result: inside $\deg_{tc}(\{\omega, \omega^\star\})$, there is an infinite chain of degrees induced by computable embeddings. Section 6 contains further discussion.

2 Preliminaries

We consider only computable languages, and structures with domain contained in ω. We assume that any considered class of structures $\mathcal{K}$ is closed under isomorphism, modulo the restriction on domains. In addition, we assume that all the structures from $\mathcal{K}$ have the same language. For a structure $\mathcal{S}$, $D(\mathcal{S})$ denotes the atomic diagram of $\mathcal{S}$. We will often identify a structure and its atomic diagram.

Let $\mathcal{K}_0$ be a class of L_0-structures, and $\mathcal{K}_1$ be a class of L_1-structures. In the definition below, we use the following convention: An *enumeration operator* Γ is treated as a c.e. set of pairs (α, φ), where α is a finite set of basic $(L_0 \cup \omega)$-sentences, and φ is a basic $(L_1 \cup \omega)$-sentence.

Definition 1 ([4,13]). *An enumeration operator Γ is a computable embedding of $\mathcal{K}_0$ into $\mathcal{K}_1$, denoted by $\Gamma\colon \mathcal{K}_0 \leq_c \mathcal{K}_1$, if Γ satisfies the following:*

1. *For any $\mathcal{A} \in \mathcal{K}_0$, $\Gamma(\mathcal{A})$ is the atomic diagram of a structure from $\mathcal{K}_1$.*
2. *For any $\mathcal{A}, \mathcal{B} \in \mathcal{K}_0$, we have $\mathcal{A} \cong \mathcal{B}$ if and only if $\Gamma(\mathcal{A}) \cong \Gamma(\mathcal{B})$.*

Any computable embedding has an important property of *monotonicity*: If $\Gamma\colon \mathcal{K}_0 \leq_c \mathcal{K}_1$ and $\mathcal{A} \subseteq \mathcal{B}$ are structures from $\mathcal{K}_0$, then we have $\Gamma(\mathcal{A}) \subseteq \Gamma(\mathcal{B})$ [4, Proposition 1.1].

Definition 2 ([4,13]). *A Turing operator* $\Phi = \varphi_e$ *is a* Turing computable embedding *of* $\mathcal{K}_0$ *into* $\mathcal{K}_1$, *denoted by* $\Phi \colon \mathcal{K}_0 \leq_{tc} \mathcal{K}_1$, *if* Φ *satisfies the following:*

1. *For any* $\mathcal{A} \in \mathcal{K}_0$, *the function* $\varphi_e^{D(\mathcal{A})}$ *is the characteristic function of the atomic diagram of a structure from* $\mathcal{K}_1$. *This structure is denoted by* $\Phi(\mathcal{A})$.
2. *For any* $\mathcal{A}, \mathcal{B} \in \mathcal{K}_0$, *we have* $\mathcal{A} \cong \mathcal{B}$ *if and only if* $\Phi(\mathcal{A}) \cong \Phi(\mathcal{B})$.

Proposition (Greenberg and, independently, Kalimullin; see [12,13])
If $\mathcal{K}_0 \leq_c \mathcal{K}_1$, *then* $\mathcal{K}_0 \leq_{tc} \mathcal{K}_1$. *The converse is not true.*

Both relations $\leq_c$ and $\leq_{tc}$ are preorders. If $\mathcal{K}_0 \leq_{tc} \mathcal{K}_1$ and $\mathcal{K}_1 \leq_{tc} \mathcal{K}_0$, then we say that $\mathcal{K}_0$ and $\mathcal{K}_1$ are *tc-equivalent*, denoted by $\mathcal{K}_0 \equiv_{tc} \mathcal{K}_1$. For a class $\mathcal{K}$, by $\deg_{tc}(\mathcal{K})$ we denote the family of all classes which are *tc*-equivalent to $\mathcal{K}$. Similar notations can be introduced for the *c*-reducibility.

For L-structures $\mathcal{A}$ and $\mathcal{B}$, we say that $\mathcal{A} \equiv_1 \mathcal{B}$ if $\mathcal{A}$ and $\mathcal{B}$ satisfy the same $\exists$-sentences. Let α be a computable ordinal. The formal definition of a *computable Σ_α formula* (or a Σ_α^c formula, for short) can be found in [2, Chap. 7]. By Σ_α^c-$Th(\mathcal{A})$, we denote the set of all Σ_α^c sentences which are true in $\mathcal{A}$. Note that in this paper we will use Σ_α^c formulas only for $\alpha \leq 2$.

Pullback Theorem (Knight, Miller, and Vanden Boom [13])
Suppose that $\mathcal{K}_1 \leq_{tc} \mathcal{K}_2$ *via a Turing operator* Φ. *Then for any computable infinitary sentence* ψ *in the language of* $\mathcal{K}_2$, *one can effectively find a computable infinitary sentence* $\psi^\star$ *in the language of* $\mathcal{K}_1$ *such that for all* $\mathcal{A} \in \mathcal{K}_1$, *we have* $\mathcal{A} \models \psi^\star$ *if and only if* $\Phi(\mathcal{A}) \models \psi$. *Moreover, for a non-zero* $\alpha < \omega_1^{CK}$, *if* ψ *is a* Σ_α^c *formula, then so is* $\psi^\star$.

When we work with pairs of structures, we use the following convention: Suppose that Γ is a computable embedding from a class $\{\mathcal{A}, \mathcal{B}\}$ into a class $\{\mathcal{C}, \mathcal{D}\}$. Then for convenience, we always assume that $\mathcal{A}$ and $\mathcal{B}$ are not isomorphic, $\Gamma(\mathcal{A}) \cong \mathcal{C}$, and $\Gamma(\mathcal{B}) \cong \mathcal{D}$.

Notice that here we abuse the notations: Formally speaking, we identify the two-element class $\{\mathcal{A}, \mathcal{B}\}$ with the family containing all isomorphic copies of $\mathcal{A}$ and all isomorphic copies of $\mathcal{B}$.

3   The *tc*-degree of $\{\omega, \omega^\star\}$

In this section, we give a characterization of the *tc*-degree for the class $\{\omega, \omega^\star\}$:

Theorem 1. *Let* $\mathcal{A}$ *and* $\mathcal{B}$ *be infinite L-structures such that* $\mathcal{A} \not\cong \mathcal{B}$. *Then the following conditions are equivalent:*

(i) $\{\mathcal{A}, \mathcal{B}\} \equiv_{tc} \{\omega, \omega^\star\}$.
(ii) *Both* $\mathcal{A}$ *and* $\mathcal{B}$ *are computably presentable,* $\mathcal{A} \equiv_1 \mathcal{B}$,

$$\Sigma_2^c\text{-}Th(\mathcal{A}) \smallsetminus \Sigma_2^c\text{-}Th(\mathcal{B}) \neq \emptyset, \ \text{and} \ \Sigma_2^c\text{-}Th(\mathcal{B}) \smallsetminus \Sigma_2^c\text{-}Th(\mathcal{A}) \neq \emptyset. \tag{1}$$

In order to prove the theorem, first, we establish the following useful fact:

Proposition 1. *Let $\mathcal{A}$ and $\mathcal{B}$ be infinite computable L-structures such that $\mathcal{A} \not\cong \mathcal{B}$. If $\mathcal{A} \equiv_1 \mathcal{B}$, then $\{\omega, \omega^\star\} \leq_{tc} \{\mathcal{A}, \mathcal{B}\}$.*

Proof (of Theorem 1). $(i) \Rightarrow (ii)$. Since $\{\omega, \omega^\star\} \leq_{tc} \{\mathcal{A}, \mathcal{B}\}$, both $\mathcal{A}$ and $\mathcal{B}$ have computable copies.

Suppose that $\mathcal{A} \not\equiv_1 \mathcal{B}$. W.l.o.g., one may assume that there is an $\exists$-sentence ψ which is true in $\mathcal{A}$, but not true in $\mathcal{B}$. By applying Pullback Theorem to the reduction $\Phi \colon \{\omega, \omega^\star\} \leq_{tc} \{\mathcal{A}, \mathcal{B}\}$, we obtain a Σ_1^c sentence $\psi^\star$ such that $\psi^\star$ is true in ω and false in $\omega^\star$. This gives a contradiction, since $\omega \equiv_1 \omega^\star$. Hence, $\mathcal{A} \equiv_1 \mathcal{B}$.

Consider $\exists\forall$-sentences ξ_0 and ξ_1 in the language of linear orders which say the following: "there is a least element" and "there is a greatest element," respectively. Since $\{\mathcal{A}, \mathcal{B}\} \leq_{tc} \{\omega, \omega^\star\}$, one can apply Pullback Theorem to ξ_0 and ξ_1, and obtain condition (1).

$(ii) \Rightarrow (i)$. Proposition 1 implies that $\{\omega, \omega^\star\} \leq_{tc} \{\mathcal{A}, \mathcal{B}\}$. Now we need to build a Turing operator $\Phi \colon \{\mathcal{A}, \mathcal{B}\} \leq_{tc} \{\omega, \omega^\star\}$.

Fix Σ_2^c sentences φ and ψ such that $\mathcal{A} \models \varphi \, \& \, \neg\psi$ and $\mathcal{B} \models \neg\varphi \, \& \, \psi$. W.l.o.g., one may assume that

$$\varphi = \exists\bar{x} \bigwedge_{i \in \omega} \forall \bar{y}_i \varphi_i(\bar{x}, \bar{y}_i), \quad \psi = \exists\bar{u} \bigwedge_{j \in \omega} \forall \bar{v}_j \psi_j(\bar{u}, \bar{v}_j).$$

Let $\mathcal{S}$ be a copy of one of the structures $\mathcal{A}$ or $\mathcal{B}$. We give an informal description of how to build the structure $\Phi(\mathcal{S})$. Formal details can be recovered from the proof of Proposition 1, mutatis mutandis.

Suppose that $dom(\mathcal{S}) = \{d_0 <_\mathbb{N} d_1 <_\mathbb{N} d_2 <_\mathbb{N} \dots \}$, where $\leq_\mathbb{N}$ is the standard order of natural numbers. By $dom(\mathcal{S})[s]$ we denote the set $\{d_0, d_1, \dots, d_s\}$. We say that a tuple $\bar{d}$ from $dom(\mathcal{S})[s]$ is a $\varphi[s]$-*witness* if for any $i \leq s$ and any tuple $\bar{y}_i$ from $dom(\mathcal{S})[s]$, we have $\mathcal{S} \models \varphi_i(\bar{d}, \bar{y}_i)$. The notion of a $\psi[s]$-*witness* is defined in a similar way. The order of witnesses is induced by their Gödel numbers.

At a stage $s + 1$, we consider the following four cases.

Case 1. There are no $\varphi[s + 1]$-witnesses and no $\psi[s + 1]$-witnesses. Then extend $\Phi(\mathcal{S})[s]$ to $\Phi(\mathcal{S})[s + 1]$ by copying (a finite part of) ω. In particular, put into $\Phi(\mathcal{S})[s + 1]$ an element which is $\leq_{\Phi(\mathcal{S})}$-greater than every element from $\Phi(\mathcal{S})[s]$.

Case 2. There is a $\varphi[s + 1]$-witness, and there are no $\psi[s + 1]$-witnesses. Proceed as in the previous case.

Case 3. There is a $\psi[s + 1]$-witness, and there are no $\varphi[s + 1]$-witnesses. Extend $\Phi(\mathcal{S})[s]$ to $\Phi(\mathcal{S})[s + 1]$ by copying $\omega^\star$.

Case 4. There are both $\varphi[s + 1]$-witnesses and $\psi[s + 1]$-witnesses. If the least $\varphi[s+1]$-witness is $\leq_\mathbb{N}$-less than the least $\psi[s+1]$-witness, then copy ω. Otherwise, copy $\omega^\star$.

It is not difficult to show that the construction gives a Turing operator Φ with the following properties. If $\mathcal{S}$ is a copy of $\mathcal{A}$, then $\Phi(\mathcal{S}) \cong \omega$. If $\mathcal{S} \cong \mathcal{B}$, then $\Phi(\mathcal{S}) \cong \omega^\star$. Theorem 1 is proved. $\square$

4   The Top Pair

Our goal in this section is to prove that among all pairs of linear orders, which are tc-equivalent to $\{\omega, \omega^*\}$, there is a greatest pair under computable embeddings, namely $\{1 + \eta, \eta + 1\}$.

Let us denote by $\mathcal{E}$ the equivalence structure with infinitely many equivalence classes of infinite size. By $\mathcal{E}_k$ we shall denote the equivalence structure with infinitely many equivalence classes of infinite size and *exactly* one equivalence class of size k, and $\hat{\mathcal{E}}_k$ denotes the equivalence structure with infinitely many equivalence classes of infinite size and *infinitely many* equivalence classes of size k. It is straightforward to see that $\{\mathcal{E}_1, \mathcal{E}_2\} \leq_c \{\hat{\mathcal{E}}_1, \hat{\mathcal{E}}_2\}$.

Proposition 2. *Let $\mathcal{L}_1$ and $\mathcal{L}_2$ be two linear orders such that $\mathcal{L}_1$ has a least element and no greatest element, and $\mathcal{L}_2$ has a greatest element and no least element. Then $\{\mathcal{L}_1, \mathcal{L}_2\} \leq_c \{\mathcal{E}_1, \mathcal{E}_2\}$.*

By transitivity of $\leq_c$, we immediately get the following corollary.

Corollary 1. *Let $\mathcal{L}_1$ and $\mathcal{L}_2$ be two linear orders such that $\mathcal{L}_1$ has a least element and no greatest element, and $\mathcal{L}_2$ has a greatest element and no least element. Then $\{\mathcal{L}_1, \mathcal{L}_2\} \leq_c \{\hat{\mathcal{E}}_1, \hat{\mathcal{E}}_2\}$.*

In the end of this section, we will need the following special cases.

Corollary 2. $\{1 + \eta, \eta + 1\} \leq_c \{\mathcal{E}_1, \mathcal{E}_2\}$ *and* $\{1 + \eta, \eta + 1\} \leq_c \{\hat{\mathcal{E}}_1, \hat{\mathcal{E}}_2\}$.

The next result is not so useful in itself, because isomorphic copies of the input structure produce non-isomorphic copies of the output structure, but when we replace every element of the output structure by a copy of η, we will get the same structure.

Proposition 3. *Let $\mathcal{K}_1$ be the class of linear orders, which have a least element and no greatest element, and let $\mathcal{K}_2$ be the class of linear orders, which have no least element and no greatest element. Then $\{\hat{\mathcal{E}}_1, \hat{\mathcal{E}}_2\} \leq_c \{\mathcal{K}_1, \mathcal{K}_2\}$, which means that there is an enumeration operator Γ such that for any copy $\hat{\mathcal{S}}_i$ of $\hat{\mathcal{E}}_i$, $\Gamma(\hat{\mathcal{S}}_i) \in \mathcal{K}_i$, for $i = 0, 1$.*

Proof. Given as input a structure $\mathcal{S}$ in the language of equivalence structures, the enumeration operator Γ will output a linear order with domain D consisting of tuples of elements from $\mathcal{S}$, where

$$D = \{(x_0, \ldots, x_n) \mid \bigwedge_{i < n} (x_i <_{\mathbb{N}} x_{i+1} \ \& \ |[x_i]_{\sim}| \geq 2)\},$$

and for two such tuples $\bar{x} = (x_0, \ldots, x_n)$ and $\bar{y} = (y_0, \ldots, y_k)$, we will say that $\bar{x} \prec \bar{y}$ if

- $\bar{x}$ is a proper extension of $\bar{y}$;
- otherwise, if the first index where $\bar{x}$ and $\bar{y}$ differ is i, then $x_i < y_i$.

Now we consider the linear orders $\mathcal{L}_1 = \Gamma(\hat{\mathcal{E}}_1)$ and $\mathcal{L}_2 = \Gamma(\hat{\mathcal{E}}_2)$. First we will show that $\mathcal{L}_1 \in \mathcal{K}_1$. Given the structure $\hat{\mathcal{E}}_1$, let $\hat{x}$ be the least element in the domain of $\hat{\mathcal{E}}_1$, in the order of natural numbers, describing an equivalence class with exactly one element in $\hat{\mathcal{E}}_1$ and let $\bar{x} = (x_0, \ldots, x_n)$ be all elements of $\hat{\mathcal{E}}_1$, ordered as natural numbers, less that $\hat{x}$ such that

$$x_0 <_{\mathbb{N}} x_1 <_{\mathbb{N}} \cdots <_{\mathbb{N}} x_{n-1} <_{\mathbb{N}} x_n = \hat{x}.$$

It is easy to see that $\bar{x}$ cannot have proper extensions in the domain of $\mathcal{L}_1$ because the equivalence class of x_n has size one. Then if $(y_0, \ldots, y_k) \prec (x_0, \ldots, x_n)$, we have $y_i < x_i$, for some i, but this is impossible since $\bar{x}$ contains all elements less that $\hat{x}$ ordered in a strictly increasing order. We conclude that $\bar{x}$ is the least element of $\mathcal{L}_1$. It is easy to see that $\mathcal{L}_1$ does not have a greatest element. It follows that $\mathcal{L}_1 \in \mathcal{K}_1$.

Similarly, it is clear that $\mathcal{L}_2 = \Gamma(\hat{\mathcal{E}}_2)$ does not have a greatest element. We will show that $\mathcal{L}_2$ does not have a least element. Consider an arbitrary $\bar{x} = (x_0, \ldots, x_n)$ in the domain of $\mathcal{L}_2$. Define $\bar{x}' = (x_0, \ldots, x_n, x')$, for some x' such that $x_n <_{\mathbb{N}} x'$. Then $\bar{x}' \prec \bar{x}$, because $\bar{x}'$ is a proper extension of $\bar{x}$.   □

Corollary 3. $\{\hat{\mathcal{E}}_1, \hat{\mathcal{E}}_2\} \leq_c \{1 + \eta, \eta\}$.

Proof (Sketch). Given the enumeration operator Γ from the proof of Proposition 3, we produce a new enumeration operator, where every element of $\Gamma(\hat{\mathcal{E}}_i)$ is replaced by an interval of rational numbers of the form $[p, q)$.   □

Proposition 4. *Let* $\mathcal{K}_1$ *be the class of infinite linear orders, which have a least element and no greatest element, and let* $\mathcal{K}_2$ *be the class of infinite linear orders, which have no least element and no greatest element. Then* $\{\hat{\mathcal{E}}_1, \hat{\mathcal{E}}_2\} \leq_c \{\mathcal{K}_2, \mathcal{K}_1\}$.

Proof (Sketch). Given as input a structure $\mathcal{S}$ in the language of equivalence structures, the enumeration operator Γ will output a linear order with domain D consisting of tuples of elements from $\mathcal{S}$, where

$$D = \{(x_0, \ldots, x_n) \mid \bigwedge_{i<n} (x_i <_{\mathbb{N}} x_{i+1} \ \& \ |[x_i]_\sim| \geq 3) \ \& \ |[x_n]_\sim| \geq 2\}.$$

Now we essentially repeat the proof of Proposition 3.   □

Corollary 4. $\{\hat{\mathcal{E}}_1, \hat{\mathcal{E}}_2\} \leq_c \{\eta, \eta + 1\}$.

Proof (Sketch). First we reverse the relation in the construction of Γ from Proposition 4 to produce a linear order with a greatest element. Then we produce a new enumeration operator, where every element of the linear order is replaced by an interval of rational numbers of the form $(p, q]$.   □

Corollary 5. $\{\hat{\mathcal{E}}_1, \hat{\mathcal{E}}_2\} \leq_c \{1 + \eta, \eta + 1\}$.

Proof (Sketch). We concatenate the results of the enumeration operators from Corollaries 3 and 4 observing that $1 + \eta + \eta = 1 + \eta$ and $\eta + \eta + 1 = \eta + 1$.   □

Theorem 2. $\{1 + \eta, \eta + 1\} \equiv_c \{\hat{\mathcal{E}}_1, \hat{\mathcal{E}}_2\} \equiv_c \{\mathcal{E}_1, \mathcal{E}_2\}$.

Proof. By Corollary 5, we have that $\{\mathcal{E}_1, \mathcal{E}_2\} \leq_c \{\hat{\mathcal{E}}_1, \hat{\mathcal{E}}_2\} \leq_c \{1 + \eta, \eta + 1\}$ and by Proposition 2 we have that $\{1 + \eta, \eta + 1\} \leq_c \{\mathcal{E}_1, \mathcal{E}_2\}$. □

Proposition 5. *Suppose that $\mathcal{A}$ and $\mathcal{B}$ are structures in the same language, for which there exists a Σ_2^c sentence ϕ such that $\mathcal{A} \models \phi$ and $\mathcal{B} \models \neg\phi$. Then $\{\mathcal{A}, \mathcal{B}\} \leq_c \{\hat{\mathcal{E}}_1, \mathcal{E}\}$.*

Corollary 6. *Suppose that $\mathcal{A}$ and $\mathcal{B}$ are structures for which there exist Σ_2^c sentences ϕ and ψ such that $\mathcal{A} \models \phi$ & $\neg\psi$ and $\mathcal{B} \models \neg\phi$ & ψ. Then*

$$\{\mathcal{A}, \mathcal{B}\} \leq_c \{\hat{\mathcal{E}}_1, \hat{\mathcal{E}}_2\}.$$

Proof (Sketch). For the formula ϕ, we apply Proposition 5 and produce an enumeration operator Γ_1 such that $\{\mathcal{A}, \mathcal{B}\} \leq_c \{\hat{\mathcal{E}}_1, \mathcal{E}\}$. It is trivial to modify the proof of Proposition 5 and apply it for the formula ψ to obtain an enumeration operator Γ_2 such that $\{\mathcal{A}, \mathcal{B}\} \leq_c \{\mathcal{E}, \hat{\mathcal{E}}_2\}$. Then we combine the two operators into one by simply taking a disjoint union of their outputs. □

Theorem 3. *The pair $\{1 + \eta, \eta + 1\}$ is the greatest one under computable embedding in the tc-equivalence class of the pair $\{\omega, \omega^\star\}$.*

Proof. Consider $\{\mathcal{A}, \mathcal{B}\} \equiv_{tc} \{\omega, \omega^\star\}$. By Theorem 1, there exist Σ_2^c sentences ϕ and ψ such that $\mathcal{A} \models \phi$ & $\neg\psi$ and $\mathcal{B} \models \neg\phi$ & ψ. By combining Corollary 6 with Theorem 2, we conclude that $\{\mathcal{A}, \mathcal{B}\} \leq_c \{1 + \eta, \eta + 1\}$. □

5   An Infinite Chain of Pairs

In this section we will show that there is an infinite chain of pairs of linear orderings under computable embeddings inside the *tc*-degree of $\{\omega, \omega^\star\}$. More concretely, we will show that have the following picture:

$$\{\omega, \omega^\star\} <_c \{\omega \cdot 2, \omega^\star \cdot 2\} <_c \cdots <_c \{\omega \cdot 2^k, \omega^\star \cdot 2^k\} <_c \cdots <_c \{1 + \eta, \eta + 1\}.$$

Here we work only with structures in the language of linear orders. We denote by α, β, γ finite linear orders. For an enumeration operator Γ, and a finite linear order α, we define

$$\alpha \Vdash_\Gamma x < y \overset{\text{def}}{\iff} \neg(\exists \beta \supseteq \alpha)[\, \Gamma(\beta) \models y \leq x \,].$$

Moreover, we shall say that α *decides* x and y if $\Gamma(\alpha) \models x < y$ or $\Gamma(\alpha) \models y \leq x$. For two finite linear orders α and β with disjoint domains, we shall write $\alpha + \beta$ for the finite linear order obtained by merging α and β so that the greatest element of α is less than the least element of β. Following Rosenstein [17], we will use the notation $\sum_{i \in \omega} \alpha_i$ for the linear order $\alpha_0 + \alpha_1 + \cdots + \alpha_n + \cdots$, and the notation $\sum_{i \in \omega^\star} \alpha_i$ for the linear order $\cdots + \alpha_n + \cdots + \alpha_1 + \alpha_0$.

Proposition 6. *Suppose* $x, y \in \Gamma(\alpha)$ *and* $x \neq y$. *Then*

$$\alpha \Vdash_\Gamma x < y \ or \ \alpha \Vdash_\Gamma y < x.$$

Proof. Towards a contradiction, assume that

$$\alpha \nVdash_\Gamma x < y \ and \ \alpha \nVdash_\Gamma y < x.$$

This means that there is some $\alpha' \supset \alpha$ such that $\Gamma(\alpha') \models y < x$ and some $\alpha'' \supset \alpha$ such that $\Gamma(\alpha'') \models x < y$. Consider some new extension β of α, such that $\alpha' \cap \beta = \alpha'' \cap \beta = \alpha$, which decides x and y. Without loss of generality, suppose that $\Gamma(\beta) \models x < y$. Then let $\gamma = \beta \cup \alpha'$. By monotonicity of Γ, since $\alpha' \subset \gamma$, we have $\Gamma(\gamma) \models y < x$ and since $\beta \subset \gamma$, we have $\Gamma(\gamma) \models x < y$. We reach a contradiction. □

Corollary 7. *Let* $x_0, \ldots, x_n$ *be distinct elements and* $x_0, \ldots, x_n \in \Gamma(\alpha)$. *Then there is some permutation* π *of* $(0, \ldots, n)$ *such that*

$$\alpha \Vdash_\Gamma x_{\pi(0)} < x_{\pi(1)} < \cdots < x_{\pi(n)}.$$

Proposition 7. *Suppose* $x, y \in \Gamma(\alpha)$ *and* $x \neq y$. *Then*

$$\alpha \Vdash_\Gamma x < y \iff \alpha \nVdash_\Gamma y < x.$$

Proof. (1) $\rightarrow$ (2). Since $x, y \in \Gamma(\alpha)$, there is some $\beta \supseteq \alpha$ which decides x and y. Since $\alpha \Vdash_\Gamma x < y$, it follows that we must have $\Gamma(\beta) \models x < y$ and hence $\alpha \nVdash_\Gamma y < x$.

(2) $\rightarrow$ (1). Suppose that $\alpha \nVdash_\Gamma y < x$. By Proposition 6, we have $\alpha \Vdash_\Gamma x < y$. □

Corollary 8. *Let* $x_0, \ldots, x_n$ *be distinct elements and* $x_0, \ldots, x_n \in \Gamma(\alpha)$. *Then there exists* exactly one *permutation* π *of* $(0, \ldots, n)$ *for which*

$$\alpha \Vdash_\Gamma x_{\pi(0)} < x_{\pi(1)} < \cdots < x_{\pi(n)}.$$

Proposition 8. *Suppose that* $\{\mathcal{A}, \mathcal{B}\} \leq_c \{\mathcal{C}, \mathcal{D}\}$ *via* Γ, *where* $\mathcal{C}$ *has no infinite descending chains, and* $\mathcal{D}$ *has no infinite ascending chains. Then* $\Gamma(\alpha)$ *is finite for any finite linear order* α.

Proof. Towards a contradiction, assume the opposite. We can choose distinct elements $x_i \in \Gamma(\alpha)$, for $i < \omega$, for which, by Corollary 8, we have either

$$(\forall i < \omega)[\ \alpha \Vdash_\Gamma x_i < x_{i+1}\ ] \ or \ (\forall i < \omega)[\ \alpha \Vdash_\Gamma x_{i+1} < x_i\ ].$$

In the first case, we extend α to a copy $\hat{\mathcal{B}}$ of $\mathcal{B}$. Then $\Gamma(\hat{\mathcal{B}}) \models \bigwedge_{i<\omega} x_i < x_{i+1}$, which is a contradiction. In the second case, we extend α to a copy $\hat{\mathcal{A}}$ of $\mathcal{A}$ and again reach a contradiction. □

By Proposition 8, there are infinitely many finite linear orders α such that $\Gamma(\alpha) \neq \emptyset$. In what follows, we will always suppose that we consider only such finite linear orders α.

Proposition 9. *Suppose that α and β are finite linear orders with disjoint domains, and $x, y \in \Gamma(\alpha) \cap \Gamma(\beta)$, $x \neq y$. Then $\alpha \Vdash_\Gamma x < y$ iff $\beta \Vdash_\Gamma x < y$.*

Proof. Let $\alpha \Vdash_\Gamma x < y$ and, by Proposition 6, assume that $\beta \Vdash_\Gamma y < x$. Since α and β are disjoint, form the finite linear orders γ by merging α and β in some way. It follows that $\gamma \Vdash_\Gamma x < y$ and $\gamma \Vdash_\Gamma y < x$, which is a contradiction by Proposition 7. □

Proposition 10. *Suppose that $\{\mathcal{A}, \mathcal{B}\} \leq_c \{\mathcal{C}, \mathcal{D}\}$ via Γ, where $\mathcal{A}$, $\mathcal{C}$ have no infinite descending chains, and $\mathcal{B}$, $\mathcal{D}$ have no infinite ascending chains. There are at most finitely many elements x with the property that there exist α and β with disjoint domains and $x \in \Gamma(\alpha) \cap \Gamma(\beta)$.*

Proof. Towards a contradiction, assume the opposite. By Proposition 8, there is an infinite sequence of mutually disjoint finite linear orders α_i and β_i, $i < \omega$, and distinct elements $x_i \in \Gamma(\alpha_i) \cap \Gamma(\beta_i)$. Consider a copy $\hat{\mathcal{A}}$ of $\mathcal{A}$ extending $\sum_{i \in \omega} \alpha_i$. Clearly $x_i \in \Gamma(\hat{\mathcal{A}})$ and hence

$$\Gamma(\hat{\mathcal{A}}) \models \bigwedge_{i < \omega} x_{\pi(i)} < x_{\pi(i+1)},$$

for some permutation π of ω. For simplicity, suppose that π is the identity function. Then we have that $\alpha_i + \alpha_{i+1} \Vdash_\Gamma x_i < x_{i+1}$. Now, since $x_i \in \Gamma(\beta_i)$ and $x_{i+1} \in \Gamma(\beta_{i+1})$, by Proposition 9, we have $\beta_{i+1} + \beta_i \Vdash_\Gamma x_i < x_{i+1}$. In this way we can build a copy $\hat{\mathcal{B}}$ of $\mathcal{B}$ extending $\sum_{i \in \omega^\star} \beta_i$, and obtain

$$\Gamma(\hat{\mathcal{B}}) \models \bigwedge_{i < \omega} x_i < x_{i+1},$$

which is a contradiction, because $\Gamma(\hat{\mathcal{B}})$ is a copy of $\mathcal{D}$, which has no infinite ascending chains. □

Let us call (x, α) a Γ-pair if $x \in \Gamma(\alpha)$. In view of Proposition 10, for any sequence of $(\alpha_i)_{i<\omega}$ such that $\Gamma(\alpha) \neq \emptyset$, there is an infinite sequence of Γ-pairs $(x_i, \alpha_{k_i})_{i<\omega}$, where all x_i are distinct elements.

Proposition 11. *For any two sequences of Γ-pairs $(x_i, \alpha_i)_{i \in \omega}$ and $(y_i, \beta_i)_{i \in \omega}$, the following are equivalent:*

(i) $\Gamma(\sum_{i \in \omega} \alpha_i + \sum_{i \in \omega} \beta_i) \models \bigwedge_{i < \omega} x_i < y_i < x_{i+1};$
(ii) $\Gamma(\sum_{i \in \omega^\star} \alpha_i + \sum_{i \in \omega^\star} \beta_i) \models \bigwedge_{i < \omega} x_i < y_i < x_{i+1}.$

Proof (Sketch). The two directions are symmetrical. Without loss of generality, suppose that

$$\Gamma(\sum_{i \in \omega} \alpha_i + \sum_{i \in \omega} \beta_i) \models \bigwedge_{i < \omega} x_i < y_i < x_{i+1}. \tag{2}$$

It is enough to show that for an arbitrary i,

$$\alpha_{i+1} + \alpha_i + \beta_{i+1} + \beta_i \Vdash_\Gamma x_i < y_i < x_{i+1} < y_{i+1}.$$

Since $x_i \in \Gamma(\alpha_i)$ and $y_i \in \Gamma(\beta_i)$, by the monotonicity of Γ and (2), we have $\alpha_i + \beta_i \Vdash_\Gamma x_i < y_i$. Similarly, we have

$$\alpha_{i+1} + \beta_i \Vdash_\Gamma y_i < x_{i+1} \text{ and } \alpha_{i+1} + \beta_{i+1} \Vdash_\Gamma x_{i+1} < y_{i+1}.$$

Since all these four finite linear orders are disjoint, we can place α_{i+1} before α_i and β_{i+1} before β_i to obtain $\alpha_{i+1} + \alpha_i + \beta_{i+1} + \beta_i \Vdash_\Gamma x_i < y_i < x_{i+1} < y_{i+1}$. $\square$

Proposition 12. *Suppose* $\{\omega \cdot 2, \omega^* \cdot 2\} \leq_c \{\mathcal{D}_0, \mathcal{D}_1\}$ *via* Γ, *where* $\mathcal{D}_0$ *is a linear order without infinite descending chains and* $\mathcal{D}_1$ *is an infinite order without infinite ascending chains. For any two sequences of* Γ-*pairs* $(x_i, \alpha_i)_{i \in \omega}$ *and* $(y_i, \beta_i)_{i \in \omega}$, *there is a number* q *such that either*

$$\Gamma(\sum_{i \in \omega} \alpha_i + \sum_{i \in \omega} \beta_i) \models \bigwedge_{i,j>q} x_i < y_j \text{ or } \Gamma(\sum_{i \in \omega} \alpha_i + \sum_{i \in \omega} \beta_i) \models \bigwedge_{i,j>q} y_j < x_i.$$

Proof. Assume that for some two sequences of Γ-pairs $(x_i, \alpha_i)_{i<\omega}$ and $(y_i, \beta_i)_{i<\omega}$, we have the opposite. We will show that we can build two infinite subsequences $(x_{s_i}, \alpha_{s_i})_{i<\omega}$ and $(y_{t_i}, \beta_{t_i})_{i<\omega}$, such that

$$\Gamma(\sum_{i \in \omega} \alpha_{s_i} + \sum_{i \in \omega} \beta_{t_i}) \models \bigwedge_i x_{s_i} < y_{t_i} < x_{s_{i+1}}.$$

Then we will apply Proposition 11 to reach a contradiction. Suppose we have built the sequences up to index ℓ, i.e. we have the finite sequences of Γ-pairs $(x_{s_i}, \alpha_{s_i})_{i \leq \ell}$ and $(y_{t_i}, \beta_{t_i})_{i < \ell}$, such that

$$\alpha_{s_0} + \alpha_{s_1} + \cdots + \alpha_{s_\ell} + \beta_{t_0} + \beta_{t_1} + \cdots + \beta_{t_{\ell-1}} \Vdash_\Gamma \bigwedge_{i<\ell} x_{s_i} < y_{t_i} < x_{s_{i+1}}.$$

Start with some indices i and j such that $s_\ell < i$, $t_{\ell-1} < j$, and

$$\Gamma(\sum_{i \in \omega} \alpha_i + \sum_{i \in \omega} \beta_i) \models x_{s_\ell} < x_i < y_j.$$

Now we find some indices j' and i' such that $i < i'$, $j < j'$, and

$$\Gamma(\sum_{i \in \omega} \alpha_i + \sum_{i \in \omega} \beta_i) \models y_j < y_{j'} < x_{i'}.$$

Since $\Gamma(\sum_{i \in \omega} \alpha_i + \sum_{i \in \omega} \beta_i)$ does not contain an infinite descending chain, and by our assumption, we know that we can find such indices. We let $s_{\ell+1} = i'$ and $t_\ell = j'$. By the properties of $\Vdash_\Gamma$, it is clear that

$$\alpha_{s_0} + \alpha_{s_1} + \cdots + \alpha_{s_{\ell+1}} + \beta_{t_0} + \beta_{t_1} + \cdots + \beta_{t_\ell} \Vdash_\Gamma \bigwedge_{i<\ell+1} x_{s_i} < y_{t_i} < x_{s_{i+1}}.$$

Now by Proposition 11, we have the following:

$$\Gamma(\sum_{i \in \omega^*} \alpha_i + \sum_{i \in \omega^*} \beta_i) \models \bigwedge_{i \in \omega} x_{s_i} < y_{t_i} < x_{s_{i+1}},$$

which is a contradiction with the fact that $\Gamma(\sum_{i \in \omega^*} \alpha_i + \sum_{i \in \omega^*} \beta_i)$ does not contain an infinite ascending chain.

Theorem 4. *Fix some $k \geq 2$ and suppose $\{\omega \cdot k, \omega^\star \cdot k\} \leq_c \{\mathcal{D}_0, \mathcal{D}_1\}$ via Γ, where $\mathcal{D}_0$ is a linear order without infinite descending chains and $\mathcal{D}_1$ is an infinite order without infinite ascending chains. Then $\mathcal{D}_0$ includes $\omega \cdot k$ and $\mathcal{D}_1$ includes $\omega^\star \cdot k$.*

Proof (Sketch). For simplicity, let us consider the case of $k = 3$, the general case being a straightforward generalization. Let $\mathcal{A}$, $\mathcal{B}$, and $\mathcal{C}$ be copies of ω, with disjoint domains, partitioned in the following way:

$$\mathcal{A} = \sum_{i \in \omega} \alpha_i, \; \mathcal{B} = \sum_{i \in \omega} \beta_i, \; \text{and } \mathcal{C} = \sum_{i \in \omega} \gamma_i.$$

We use Proposition 12 at most three times. Start with two arbitrary sequences of Γ-pairs $(x_i, \alpha_{k_i})_{i<\omega}$ and $(y_i, \beta_{m_i})_{i<\omega}$ and, without loss of generality, suppose that there is a number ℓ_1 such that

$$\Gamma\Big(\sum_{i \in \omega} \alpha_{k_i} + \sum_{i \in \omega} \beta_{m_i}\Big) \models \bigwedge_{i,j > \ell_1} x_i < y_j.$$

Now we take a sequence of Γ-pairs $(z_i, \gamma_{n_i})_{i<\omega}$ and we may suppose that there is a number ℓ_2 such that

$$\Gamma\Big(\sum_{i \in \omega} \alpha_{k_i} + \sum_{i \in \omega} \gamma_{n_i}\Big) \models \bigwedge_{i,j > \ell_2} x_i < z_j.$$

We must apply Proposition 12 one more time to the two sequences of Γ-pairs $(y_i, \beta_{k_i})_{i<\omega}$ and $(z_i, \gamma_{n_i})_{i<\omega}$. We may suppose that there is a number ℓ_3 such that

$$\Gamma\Big(\sum_{i \in \omega} \beta_{m_i} + \sum_{i \in \omega} \gamma_{n_i}\Big) \models \bigwedge_{i,j > \ell_3} y_i < z_j.$$

By monotonicity of Γ, it follows that for $\ell_0 = \max\{\ell_1, \ell_2, \ell_3\}$, we have

$$\Gamma\Big(\sum_{i \in \omega} \alpha_{k_i} + \sum_{i \in \omega} \beta_{m_i} + \sum_{i \in \omega} \gamma_{n_i}\Big) \models \bigwedge_{i,j,k > \ell_0} x_i < y_j < z_k.$$

Again by monotonicity of Γ, we have $\Gamma(\mathcal{A} + \mathcal{B} + \mathcal{C}) \models \bigwedge_{i,j,k > \ell_0} x_i < y_j < z_k$. We conclude that $\Gamma(\mathcal{A} + \mathcal{B} + \mathcal{C})$ includes a copy of $\omega \cdot 3$. □

Corollary 9. *For any $k < \omega$, $\{\omega \cdot 2^k, \omega^\star \cdot 2^k\} <_c \{\omega \cdot 2^{k+1}, \omega^\star \cdot 2^{k+1}\}$.*

It follows that we have the following chain above $\{\omega, \omega^\star\}$:

$$\{\omega, \omega^\star\} <_c \{\omega \cdot 2, \omega^\star \cdot 2\} <_c \cdots <_c \{\omega \cdot 2^k, \omega^\star \cdot 2^k\} <_c \cdots <_c \{1 + \eta, \eta + 1\}.$$

6   Future Work

Apart from extending our results to more general classes of pairs of structures, we believe that by exploiting the ideas introduced in Sect. 5, we can prove the following conjecture.

Conjecture 1. For any two natural numbers n and k,

$$n \mid k \iff \{\omega \cdot n, \omega^\star \cdot n\} \leq_c \{\omega \cdot k, \omega^\star \cdot k\}.$$

References

1. Ash, C.J., Knight, J.F.: Pairs of recursive structures. Ann. Pure Appl. Logic **46**(3), 211–234 (1990). https://doi.org/10.1016/0168-0072(90)90004-L
2. Ash, C.J., Knight, J.F.: Computable Structures and the Hyperarithmetical Hierarchy. Studies in Logic and the Foundations of Mathematics, vol. 44. Elsevier Science B.V., Amsterdam (2000)
3. Bazhenov, N.: Autostability spectra for decidable structures. Math. Struct. Comput. Sci. **28**(3), 392–411 (2018). https://doi.org/10.1017/S096012951600030X
4. Calvert, W., Cummins, D., Knight, J.F., Miller, S.: Comparing classes of finite structures. Algebra Logic **43**(6), 374–392 (2004). https://doi.org/10.1023/B:ALLO.0000048827.30718.2c
5. Chisholm, J., Knight, J.F., Miller, S.: Computable embeddings and strongly minimal theories. J. Symb. Log. **72**(3), 1031–1040 (2007). https://doi.org/10.2178/jsl/1191333854
6. Ershov, Y.L., Puzarenko, V.G., Stukachev, A.I.: *HF*-Computability. In: Cooper, S.B., Sorbi, A. (eds.) Computability in Context, pp. 169–242. Imperial College Press, London (2011). https://doi.org/10.1142/9781848162778_0006
7. Fokina, E., Knight, J.F., Melnikov, A., Quinn, S.M., Safranski, C.: Classes of Ulm type and coding rank-homogeneous trees in other structures. J. Symb. Log. **76**(3), 846–869 (2011). https://doi.org/10.2178/jsl/1309952523
8. Friedman, H., Stanley, L.: A Borel reducibility theory for classes of countable structures. J. Symb. Log. **54**(3), 894–914 (1989). https://doi.org/10.2307/2274750
9. Goncharov, S., Harizanov, V., Knight, J., McCoy, C., Miller, R., Solomon, R.: Enumerations in computable structure theory. Ann. Pure Appl. Logic **136**(3), 219–246 (2005). https://doi.org/10.1016/j.apal.2005.02.001
10. Harrison-Trainor, M., Melnikov, A., Miller, R., Montálban, A.: Computable functors and effective interpretability. J. Symb. Log. **82**(1), 77–97 (2017). https://doi.org/10.1017/jsl.2016.12
11. Hirschfeldt, D.R., Khoussainov, B., Shore, R.A., Slinko, A.M.: Degree spectra and computable dimensions in algebraic structures. Ann. Pure Appl. Logic **115**(1–3), 71–113 (2002). https://doi.org/10.1016/S0168-0072(01)00087-2
12. Kalimullin, I.S.: Computable embeddings of classes of structures under enumeration and Turing operators. Lobachevskii J. Math. **39**(1), 84–88 (2018). https://doi.org/10.1134/S1995080218010146
13. Knight, J.F., Miller, S., Vanden Boom, M.: Turing computable embeddings. J. Symb. Log. **72**(3), 901–918 (2007). https://doi.org/10.2178/jsl/1191333847
14. Miller, R.: Isomorphism and classification for countable structures. Computability (2018). https://doi.org/10.3233/COM-180095, published online
15. Miller, R., Poonen, B., Schoutens, H., Shlapentokh, A.: A computable functor from graphs to fields. J. Symb. Log. **83**(1), 326–348 (2018). https://doi.org/10.1017/jsl.2017.50
16. Puzarenko, V.G.: A certain reducibility on admissible sets. Sib. Math. J. **50**(2), 330–340 (2009). https://doi.org/10.1007/s11202-009-0038-z
17. Rosenstein, J.G.: Linear Orderings. Pure and Applied Mathematics, vol. 98. Academic Press, New York (1982)
18. Rossegger, D.: On functors enumerating structures. Sib. Elektron. Mat. Izv. **14**, 690–702 (2017). https://doi.org/10.17377/semi.2017.14.059

Bounded Reducibility for Computable Numberings

Nikolay Bazhenov[1,2] (ID), Manat Mustafa[3(✉)] (ID), and Sergei Ospichev[1,2] (ID)

[1] Sobolev Institute of Mathematics,
4 Acad. Koptyug Avenue, Novosibirsk 630090, Russia
{bazhenov,ospichev}@math.nsc.ru
[2] Novosibirsk State University, 2 Pirogova Street, Novosibirsk 630090, Russia
[3] Department of Mathematics, School of Science and Technology,
Nazarbayev University, 53 Qabanbaybatyr Avenue, Astana 010000, Kazakhstan
manat.mustafa@nu.edu.kz

Abstract. The theory of numberings gives a fruitful approach to studying uniform computations for various families of mathematical objects. The algorithmic complexity of numberings is usually classified via the reducibility $\leq$ between numberings. This reducibility gives rise to an upper semilattice of degrees, which is often called the Rogers semilattice. For a computable family S of c.e. sets, its Rogers semilattice $R(S)$ contains the $\leq$-degrees of computable numberings of S. Khutoretskii proved that $R(S)$ is always either one-element, or infinite. Selivanov proved that an infinite $R(S)$ cannot be a lattice.

We introduce a bounded version of reducibility between numberings, denoted by $\leq_{bm}$. We show that Rogers semilattices $R_{bm}(S)$, induced by $\leq_{bm}$, exhibit a striking difference from the classical case. We prove that the results of Khutoretskii and Selivanov cannot be extended to our setting: For any natural number $n \geq 2$, there is a finite family S of c.e. sets such that its semilattice $R_{bm}(S)$ has precisely $2^n - 1$ elements. Furthermore, there is a computable family T of c.e. sets such that $R_{bm}(T)$ is an infinite lattice.

1 Introduction

Uniform computations for families of mathematical objects constitute a classical line of research in computability theory. Formal methods for studying such computations are provided by the theory of numberings. The theory goes back to the seminal article of Gödel [17], where an effective numbering of first-order formulae was used in the proof of the incompleteness theorems. One of the first results, which gave rise to the systematic study of numberings, was obtained by

The work was supported by Nazarbayev University Faculty Development Competitive Research Grants N090118FD5342. The first author was partially supported by the grant of the President of the Russian Federation (No. MK-1214.2019.1). The third author was partially supported by the program of fundamental scientific researches of the SB RAS No. I.1.1, project No. 0314-2019-0002.

F. Manea et al. (Eds.): CiE 2019, LNCS 11558, pp. 96–107, 2019.
https://doi.org/10.1007/978-3-030-22996-2_9

Kleene [25]: he gave a construction of a universal partial computable function. After that, the foundations of the modern theory of numberings were developed by Kolmogorov and Uspenskii [26,36] and, independently, by Rogers [34].

Let S be a countable set. A *numbering* of S is a surjective map ν from ω onto S. A standard tool for measuring the algorithmic complexity of numberings is provided by the notion of *reducibility* between numberings: A numbering ν is *reducible* to another numbering μ (denoted by $\nu \leq \mu$) if there is total computable function $f(x)$ such that $\nu(x) = \mu(f(x))$ for all $x \in \omega$. In other words, there is an effective procedure which, given a ν-index of an object from S, computes a μ-index for the same object. In general, however, the goal is for f to be a readily understandable function, so that we can actually obtain some information from the reduction.

In this paper, we consider only families S containing subsets of ω, i.e., we always assume that $S \subset P(\omega)$ and S is countable.

Let Γ be a complexity class (e.g., Σ_1^0, d-Σ_1^0, Σ_n^0, or Π_n^1). A numbering ν of a family S is Γ-*computable* if the set $\{\langle x, n \rangle : x \in \nu(n)\}$ belongs to the class Γ. We say that a family S is Γ-*computable* if it has a Γ-computable numbering.

Following the literature, the term *computable numbering* will be used as a synonym of a Σ_1^0-computable numbering. In particular, a *computable family* is a family with a Σ_1^0-computable numbering.

In a standard recursion-theoretical way, the notion of reducibility between numberings give rise to the *Rogers upper semilattice* (or *Rogers semilattice* for short) of a family S: For a given complexity class Γ, this semilattice contains the degrees of all Γ-computable numberings of S. Here two numberings have the same degree if they are reducible to each other, see Sect. 2 for the formal details.

There is a large body of literature on Rogers semilattices of computable families. To name only a few, computable numberings were studied by Badaev [4,5], Ershov [11,12], Friedberg [14], Goncharov [18,19], Lachlan [27,28], Mal'tsev [29], Pour-El [33], and many other researchers. Note that computable numberings are closely connected to algorithmic learning theory (see, e.g., the recent papers [1,9,23]). For a survey of results and bibliographical references on computable numberings, the reader is referred to the seminal monograph [12] and the articles [3,6,13].

Goncharov and Sorbi [21] started developing the theory of generalized computable numberings: In particular, this area includes investigations of Γ-computable numberings. The approach of [21] proved to be fruitful for classifying Rogers semilattices in hyperarithmetical hierarchy [3,8,32] and the Ershov hierarchy [7,20,22,31].

In the paper, we introduce the following *bounded version* of the reducibility between numberings:

Definition 1.1. *Let ν and μ be numberings. We say that ν is bm-reducible to μ if there is a total computable function $f(x)$ with the following properties:*

(a) for every $x \in \omega$, we have $\nu(x) = \mu(f(x))$;
(b) for every $y \in \omega$, the preimage $f^{-1}(y)$ is a finite set.

We write $f: \nu \leq_{bm} \mu$ if a function f bm-reduces ν to μ.

The notation $\leq_{bm}$ is a tribute to the little-known paper of Maslova [30]. She introduced a bounded version of m-reducibility on sets: Suppose that $f(x)$ is a computable function, A and B are subsets of ω. Then $f: A \leq_{bm} B$ iff $f: A \leq_m B$ and f satisfies the condition (b) above [30, Definition 1].

Nowadays various types of reductions are commonly used to study properties of mathematical structures (e.g., in Borel reducibility theory [15,16] or in the theory of ceers [2]). Following this line of research, we are introducing the reducibility $\leq_{bm}$, and we aim to investigate the complexity of the corresponding Rogers semilattices and their structural properties.

One would expect that investigating Rogers semilattices under bm-reducibility makes little or no difference for most of the known results on numberings. Quite strikingly, this is *not the case*. In the paper, we illustrate this by considering two algebraic properties of Rogers semilattices.

Historically, the first two major problems on Rogers semilattices were raised by Ershov [10] (see also [6] for a detailed discussion): Let $\mathcal{R}$ be the Rogers semilattice of a computable family $\mathcal{S}$.

Problem A. What is a possible cardinality of $\mathcal{R}$?

Problem B. Can $\mathcal{R}$ be a lattice?

In the classical case (i.e. for the reducibility $\leq$), the problems were solved in 1970s:

A. Khutoretskii [24] proved that $\mathcal{R}$ either has only one element, or is countably infinite.
B. Selivanov [35] proved that an infinite $\mathcal{R}$ cannot be a lattice.

Unexpectedly, the theorems of Khutoretskii and Selivanov *cannot* be extended to the case of bm-reducibility. We obtain the following results:

A′. For every natural number $n \geq 2$, there is a finite family of c.e. sets such that its Rogers semilattice under bm-reducibility has cardinality $2^n - 1$. A similar result is proved for infinite computable families.
B′. There is a computable family of c.e. sets such that its Rogers semilattice under bm-reducibility is an infinite lattice.

These results witness that the bm-reducibility of numberings is an interesting object of study in itself.

The outline of the paper is as follows. Section 2 contains the necessary preliminaries and some general observations about bm-reducibility. In Sect. 3, we obtain an infinite lattice under bm-reducibility (Result B′). Section 4 deals with the possible cardinalities of semilattices under bm-reducibility (Result A′). Section 5 contains further discussion.

2   Preliminaries and General Facts

In all the sections of the paper, except the last one, we consider only computable numberings.

Suppose that ν is a numbering of a family $\mathcal{S}_0$, and μ is a numbering of a family $\mathcal{S}_1$. Note that the condition $\nu \leq \mu$ always implies that $\mathcal{S}_0 \subseteq \mathcal{S}_1$. Clearly, if $\nu \leq_{bm} \mu$, then $\nu \leq \mu$.

Numberings ν and μ are *equivalent* (denoted by $\nu \equiv \mu$) if $\nu \leq \mu$ and $\mu \leq \nu$. The *bm-equivalence* $\equiv_{bm}$ is defined in a similar way. The numbering $\nu \oplus \mu$ of the family $\mathcal{S}_0 \cup \mathcal{S}_1$ is defined as follows:

$$(\nu \oplus \mu)(2x) = \nu(x), \quad (\nu \oplus \mu)(2x+1) = \mu(x).$$

The following fact is well-known (see, e.g., Proposition 3 in [12, p. 36]): If $\trianglelefteq \in \{\leq, \leq_{bm}\}$ and ξ is a numbering of a family $\mathcal{T}$, then

$$(\nu \trianglelefteq \xi \,\&\, \mu \trianglelefteq \xi) \;\Leftrightarrow\; (\nu \oplus \mu \trianglelefteq \xi).$$

Let $\mathcal{S}$ be a computable family of c.e. sets. By $Com_1^0(\mathcal{S})$ we denote the set of all computable numberings of $\mathcal{S}$. Suppose that $\sim$ is the equivalence relation induced by a preorder $\trianglelefteq \in \{\leq, \leq_{bm}\}$. Since the relation $\sim$ is a congruence on the structure $(Com_1^0(\mathcal{S}); \trianglelefteq, \oplus)$, we use the same symbols $\trianglelefteq$ and $\oplus$ on numberings of $\mathcal{S}$ and on $\sim$-equivalence classes of these numberings.

The quotient structure $Q_\sim(\mathcal{S}) := (Com_1^0(\mathcal{S})/\sim; \trianglelefteq, \oplus)$ is an upper semilattice. We say that $Q_\sim(\mathcal{S})$ is the *Rogers semilattice* of the family $\mathcal{S}$ under the reducibility $\trianglelefteq$. For the sake of convenience, we use the following notations:

$$\mathcal{R}_m(\mathcal{S}) := Q_\equiv(\mathcal{S}); \quad \mathcal{R}_{bm}(\mathcal{S}) := Q_{\equiv_{bm}}(\mathcal{S}).$$

Note that $card(\mathcal{R}_m(\mathcal{S})) \leq card(\mathcal{R}_{bm}(\mathcal{S}))$.

Numberings ν and μ are *computably isomorphic* if $\nu = \mu \circ f$, where f is a computable permutation of ω. If ν is a numbering, then by η_ν we denote the corresponding equivalence relation on ω:

$$m \, \eta_\nu \, n \;\Leftrightarrow\; \nu(m) = \nu(n).$$

A numbering ν is *decidable* if the relation η_ν is computable. Numbering ν is *Friedberg* if η_ν is the identity relation.

In our proofs, we will often refer to the following simple fact about *bm*-reducibility:

Lemma 2.1. *Suppose that ν and μ are numberings, and $\nu(x) = \mu(y)$. If the class $[x]_{\eta_\nu}$ is infinite and $[y]_{\eta_\mu}$ is finite, then $\nu \not\leq_{bm} \mu$.*

Proof. Assume that $f \colon \nu \leq_{bm} \mu$. Since $\nu(x) = \mu(y)$, we have $f^{-1}([y]_{\eta_\mu}) = [x]_{\eta_\nu}$. By the pigeonhole principle, there is an element $z \in [y]_{\eta_\mu}$ such that $f^{-1}(z)$ is infinite, which contradicts the definition of *bm*-reducibility.   □

It is well-known that any decidable numbering ν of a family $\mathcal{S}$ induces a *minimal* element in the semilattice $\mathcal{R}_m(\mathcal{S})$. It is easy to show that a similar result fails for the structure $\mathcal{R}_{bm}(\mathcal{S})$:

Corollary 2.1. *Suppose that $\mathcal{S}$ is a computable infinite family, and ν is a decidable, computable numbering of $\mathcal{S}$. Then ν is minimal in $\mathcal{R}_{bm}(\mathcal{S})$ if and only if for every $x \in \omega$, the class $[x]_{\eta_\nu}$ is finite.*

For reasons of space, the proof of Corollary 2.1 is omitted.

A countable family $\mathcal{S}$ of sets is *discrete* if there is a family of finite sets $\mathcal{F}$ with the following properties:

- for any $X \in \mathcal{F}$, there is at most one $W \in \mathcal{S}$ with $X \subseteq W$;
- for every $W \in \mathcal{S}$, there is at least one $X \in \mathcal{F}$ such that $X \subseteq W$.

A family $\mathcal{S}$ is *effectively discrete* if it is discrete, and for the witnessing family $\mathcal{F}$, there is a strongly computable sequence of finite sets $(F_i)_{i \in \omega}$ such that $\mathcal{F} = \{F_i : i \in \omega\}$.

3   Lattices

Let $\mathcal{S}$ be a computable family of c.e. sets. Here we show that the Rogers semilattice $\mathcal{R}_{bm}(\mathcal{S})$ can be an infinite lattice.

Theorem 3.1. *Consider a family $\mathcal{S} := \{\{k\} : k \in \omega\}$. Then the structure $\mathcal{R}_{bm}(\mathcal{S})$ is isomorphic to the lattice of all Π_2^0 sets (under inclusion).*

Proof. Let ν be a computable numbering of the family $\mathcal{S}$. We define a set

$$\mathrm{Inf}(\nu) := \{k \in \omega : \text{the set } \{k\} \text{ has infinitely many } \nu\text{-numbers}\}.$$

It is not hard to show that $\mathrm{Inf}(\nu)$ is a Π_2^0 set.

Lemma 3.1. *Let X be an arbitrary Π_2^0 set. Then there is a computable numbering μ of the family $\mathcal{S}$ such that $\mathrm{Inf}(\mu) = X$.*

Proof. Choose a computable predicate $R(e, y)$ with the following property: for any $e \in \omega$,

$$e \in X \Leftrightarrow \exists^\infty y R(e, y).$$

W.l.o.g., one may assume that $R(e, 0)$ is true for every e. Fix a computable injective function $f \colon \omega \to \omega^2$ such that $range(f) = R$.

For $n \in \omega$, we define $\mu(n) := \{e_n\}$, where $f(n) = (e_n, y_n)$. It is not hard to show that μ is a computable numbering of the family $\mathcal{S}$. Furthermore, the following holds:

(a) If $e \in X$, then there are infinitely many n with $R(e, y_n)$ true. For each such n, we have $\mu(n) = \{e\}$.
(b) If $e \notin X$, then there are only finitely many numbers n with $\mu(n) = \{e\}$.

Therefore, we deduce that $\mathrm{Inf}(\mu) = X$. $\square$

Lemma 3.1 implies that in order to prove the theorem, it is sufficient to establish the following fact:

Lemma 3.2. *Let ν and μ be computable numberings of the family S. Then $\nu \leq_{bm} \mu$ if and only if $\mathrm{Inf}(\nu) \subseteq \mathrm{Inf}(\mu)$.*

Proof. Assume that $\nu \leq_{bm} \mu$. Then Lemma 2.1 shows the following: If the set $\{k\}$ has infinitely many ν-numbers, then $\{k\}$ also has infinitely many μ-numbers. Hence, $\mathrm{Inf}(\nu) \subseteq \mathrm{Inf}(\mu)$.

Now suppose that $\mathrm{Inf}(\nu) \subseteq \mathrm{Inf}(\mu)$. We build a bm-reduction $f \colon \nu \leq_{bm} \mu$.

For a number $e \in \omega$, we choose effective enumerations $\{a_i\}_{i \in I}$ and $\{b_j\}_{j \in J}$ (without repetitions), which enumerate the set of all ν-numbers of $\{e\}$ and the set of all μ-numbers of $\{e\}$, respectively. Here we assume that $I = \bigcup_{s \in \omega} I[s]$ and $J = \bigcup_{s \in \omega} J[s]$, where all $I[s]$ and $J[s]$ are finite initial segments of ω, $I[0] = J[0] = \{0\}$, $I[s] \subseteq I[s+1]$, $J[s] \subseteq J[s+1]$, and $card(I[s+1] \setminus I[s]) \leq 1$. Moreover, $\{I[s]\}_{s \in \omega}$ and $\{J[s]\}_{s \in \omega}$ are strongly computable sequences of finite sets.

The desired function f is built in stages.

Stage 0. Set $f(a_0) = b_0$.

Stage $s+1$. If $I[s+1] = I[s]$, then proceed to the next stage. Otherwise, find n such that $I[s+1] \setminus I[s] = \{n\}$. Let k be the greatest number with $b_k \in f(I[s])$. Consider the following two cases:

1. If $k + 1 \in J[s]$, then define $f(a_n) := b_{k+1}$.
2. If $k + 1 \notin J[s]$, then set $f(a_n) := b_k$.

Note that the described procedure is effective, uniformly in $e \in \omega$. Thus, it is easy to see that f is a total computable function such that $f \colon \nu \leq \mu$.

Assume that there is a number y such that the set $f^{-1}(y)$ is infinite. Suppose that $\mu(y) = \{e\}$ and $y = b_k$. The description of the construction implies that there is a number n such that for all $m \geq n$, we have $m \in I$ and $f(a_m) = b_k$. Thus, $k + 1 \notin J[s]$ for every s. We deduce that $e \in \mathrm{Inf}(\nu) \setminus \mathrm{Inf}(\mu)$, which contradicts our original assumption. Therefore, the function f provides a bm-reduction from ν onto μ. Lemma 3.2 is proved.

This concludes the proof of Theorem 3.1. □

The proof of Theorem 3.1 can be easily modified to obtain the following:

Corollary 3.1. *Let $S = \{A_i : i \in \omega\}$ be a computable family of c.e. sets. Suppose that there is a strongly computable sequence of finite sets $(F_i)_{i \in \omega}$ such that $F_i \subseteq A_i$ and $F_i \not\subseteq A_j$, for all $i \neq j$. Then the structure $\mathcal{R}_{bm}(S)$ is isomorphic to the lattice of all Π_2^0 sets.*

4   Cardinalities of Rogers Semilattices

Here we attack Problem A from the introduction. For finite families S, we obtain a complete description of possible cardinalities of $\mathcal{R}_{bm}(S)$ (Subsect. 4.1). In Subsect. 4.2, we show that the cardinalities from the previous subsection can also be realized via infinite families S. In order to prove this, we give a computable infinite family $\mathcal{T}$ such that all its computable numberings are computably isomorphic (Theorem 4.2). We also provide two sufficient conditions for $\mathcal{R}_{bm}(S)$ being infinite.

First, we recall the following classical result:

Lemma 4.1 (folklore). *Let S be a computable family of c.e. sets. If S contains sets A and B such that $A \subsetneq B$, then the semilattice $\mathcal{R}_m(S)$ is infinite. In particular, this implies that $\mathcal{R}_{bm}(S)$ is also infinite.*

4.1   Finite Families

Theorem 4.1. *Suppose that S is a finite family of c.e. sets. If S contains precisely n sets, then $card(\mathcal{R}_{bm}(S))$ is either equal to $2^n - 1$ or countably infinite. Furthermore, for $n \geq 2$, both these cases can be realized.*

Proof. Let $S = \{A_1, A_2, \ldots, A_n\}$ be a family of c.e. sets. If there are numbers $i \neq j$ with $A_i \subsetneq A_j$, then by Lemma 4.1, the semilattice $\mathcal{R}_{bm}(S)$ is infinite.

Assume that $A_i \not\subseteq A_j$ for all $i \neq j$. Now it is sufficient to show that the structure $\mathcal{R}_{bm}(S)$ contains precisely $2^n - 1$ elements.

Note that the family S is effectively discrete. Indeed, for every $i \neq j$, choose an element $a_{i,j} \in A_i \setminus A_j$, and define the set $F_i := \{a_{i,j} : j \neq i\}$. It is easy to see that for any i and k, the condition $F_k \subseteq A_i$ holds iff $k = i$. Therefore, if ν is an arbitrary computable numbering of S, then for all $i \leq n$ and $x \in \omega$, the following conditions are equivalent:

$$\nu(x) = A_i \iff F_i \subseteq \nu(x) \iff \text{for every } j \neq i, \ F_j \not\subseteq \nu(x).$$

This implies that the numbering ν is decidable.

Let D be a non-empty subset of $\{1, 2, \ldots, n\}$, and d be the least number from D. We define a decidable numbering μ_D of the family S as follows:

$$\mu_D(x) = A_{x+1}, \quad \text{for } x < n;$$

$$\mu_D(\langle i, j\rangle) = \begin{cases} A_i, & \text{if } i \in D, \\ A_d, & \text{otherwise,} \end{cases} \quad \text{where we assume that } \langle i, j\rangle \geq n.$$

It is not hard to establish the following properties:

(a) Lemma 2.1 implies that for finite sets $D \neq E$, we have $\mu_D \not\equiv_{bm} \mu_E$.
(b) Consider an arbitrary computable numbering ν of S. We define a non-empty set $D_\nu := \{i : A_i \text{ has infinitely many } \nu\text{-numbers}\}$. Using the decidability of ν, one can obtain that $\nu \equiv_{bm} \mu_{D_\nu}$.

These properties show that the cardinality of $\mathcal{R}_{bm}(S)$ is equal to the number of non-empty subsets of the set $\{1, 2, \ldots, n\}$. Thus, $card(\mathcal{R}_{bm}(S)) = 2^n - 1$. $\square$

4.2   Infinite Families

First, we build infinite computable families $\mathcal{S}$ with finite semilattices $\mathcal{R}_{bm}(\mathcal{S})$. Recall that numberings ν and μ are *computably isomorphic* if there is a computable permutation g of ω such that $\nu = \mu \circ g$. We establish the following fact:

Theorem 4.2. *There is an infinite computable family $\mathcal{T}$ such that any two computable numberings of $\mathcal{T}$ are computably isomorphic. In particular, the semilattice $\mathcal{R}_{bm}(\mathcal{T})$ contains only one element.*

Proof. In the proof of Theorem 3.3 from [2], Andrews and Sorbi built a uniform sequence $(E_i)_{i \in \omega}$ of computably enumerable equivalence relations on ω (or *ceers* for short) with the following properties: Each E_i has infinitely many equivalence classes, and if a c.e. set W intersects infinitely many E_i-classes, then it intersects *every* E_i-class.

Fix such a ceer $E := E_0$. Note that every E-class is non-computable: Indeed, if a class $[x]_E$ is computable, then the c.e. set $\omega \setminus [x]_E$ intersects all but one E-classes.

We define the desired family $\mathcal{T}$ by arranging its computable numbering: For $x \in \omega$, set $\theta(x) := [x]_E$.

Lemma 4.2. *1. The family $\mathcal{T}$ is effectively discrete.*
2. Let $\mathcal{S} \subsetneq \mathcal{T}$. If $\mathcal{S}$ is infinite, then it does not have computable numberings.
3. If $\nu \in Com_1^0(\mathcal{S})$, then every set $A \in \mathcal{S}$ has infinitely many ν-numbers.

Proof. (1) If $A \neq B$ are sets from $\mathcal{T}$, then $A \cap B = \emptyset$. Thus, the sequence of finite sets $(\{k\})_{k \in \omega}$ witnesses the effective discreteness of the family $\mathcal{T}$.

(2) Assume that ν is a computable numbering of an infinite family $\mathcal{S} \subsetneq \mathcal{T}$. Then the c.e. set $W := \bigcup_{n \in \omega} \nu(n)$ intersects infinitely many E-classes, but it does not intersect all E-classes. This contradicts the choice of the ceer E.

(3) Suppose that A has only finitely many ν-numbers. W.l.o.g., one may assume that there is a natural number n_0 such that $\nu(x) = A$ iff $x \leq n_0$. Then a numbering $\mu(x) := \nu(x + n_0 + 1)$ is a computable numbering of the family $\mathcal{T} \setminus \{A\}$, which contradicts the previous item of the lemma.

$\square$

We say that a numbering ν is 1-*reducible* to a numbering μ (denoted by $\nu \leq_1 \mu$) if there is an injective, total computable function $f(x)$ such that $\nu = \mu \circ f$. The following analogue of Myhill Isomorphism Theorem is known (see, e.g., Corollary 2 in [12, p. 208]): If $\nu \leq_1 \mu$ and $\mu \leq_1 \nu$, then ν is computably isomorphic to μ.

Therefore, it is sufficient to show that for any $\nu, \mu \in Com_1^0(\mathcal{T})$, we have $\nu \leq_1 \mu$. A desired 1-reducibility $f \colon \nu \leq_1 \mu$ can be built in stages. At a stage s, we find an element k enumerated into the c.e. set $\nu(s)$. After that, we search for a number m such that $m \notin range(f[s])$ and $k \in \mu(m)$. Such a number m exists by the third item of Lemma 4.2. Moreover, it is easy to see that $\mu(m) = \nu(s)$. Thus, we set $f(s) := m$ and proceed to the next stage. Theorem 4.2 is proved. $\square$

Corollary 4.1. *For any natural number $n \geq 1$, there is a computable infinite family S such that $card(\mathcal{R}_{bm}(S)) = 2^n - 1$.*

Proof. Consider the family $\mathcal{T}$ from the theorem above. If $n = 1$, then one can just choose $S := \mathcal{T}$.

Suppose that $n \geq 2$. Choose a finite family $\mathcal{V}$ from Theorem 4.1 such that $card(\mathcal{R}_{bm}(\mathcal{V})) = 2^n - 1$. Then the desired family S contains the following sets: For any $A \in \mathcal{T}$, we add the set $\{2x : x \in A\}$ into S. For every $B \in \mathcal{V}$, we put the set $\{2y + 1 : y \in B\}$. It is not difficult to show that for this S, the Rogers bm-semilattice contains precisely $2^n - 1$ elements. □

The next two propositions give sufficient conditions for a semilattice $\mathcal{R}_{bm}(S)$ being infinite.

Proposition 4.1. *Let S be a computable infinite family. Suppose that there is a computable numbering ν of S with the following property: there are infinitely many sets $A \in S$ such that the set $\nu^{-1}[A] = \{x \in \omega : \nu(x) = A\}$ is computable. Then the semilattice $\mathcal{R}_{bm}(S)$ is infinite.*

Proof (sketch). Suppose that $A_0, A_1, \ldots, A_n$ are distinct sets from S such that $\nu^{-1}[A_i]$, $i \leq n$, are computable. For $i \leq n$, fix the least number m_i such that $\nu(m_i) = A_i$. W.l.o.g., we may assume that $m_i > 0$. We define computable numberings

$$\mu(x) := \begin{cases} \nu(x), & \text{if } \nu(x) \notin \{A_0, A_1, \ldots, A_n\}, \\ A_i, & \text{if } x = m_i, \\ \nu(0), & \text{otherwise}; \end{cases}$$

$$\theta_i(2x) := \mu(x), \quad \theta_i(2x + 1) := A_i, \ i \leq n.$$

Lemma 2.1 implies that the numberings θ_i, $i \leq n$, are pairwise incomparable under bm-reducibility. Therefore, the semilattice $\mathcal{R}_{bm}(S)$ is infinite. □

Corollary 4.2. *If an infinite family S has a decidable, computable numbering, then the semilattice $\mathcal{R}_{bm}(S)$ is infinite.*

Recall that an infinite set $X \subset \omega$ is *immune* if there is no infinite c.e. set W with $W \subseteq X$. A set $Y \subseteq \omega$ is *co-immune* if its complement is immune.

Proposition 4.2. *Let S be a computable infinite family. Suppose that there is a computable numbering ν of S with the following property: there are infinitely many sets A from S such that $\nu^{-1}[A]$ is co-immune. Then the semilattice $\mathcal{R}_{bm}(S)$ contains an infinite antichain.*

Proof (sketch). Given a set A from S, we define a computable numbering

$$\mu_A(2x) := \nu(x), \quad \mu_A(2x + 1) := A.$$

Suppose that A and B are distinct sets from S such that both $\nu^{-1}[A]$ and $\nu^{-1}[B]$ are co-immune. Assume that $f : \mu_A \leq_{bm} \mu_B$. Then the set $V := \{f(2x + 1)/2 : x \in \omega\}$ is an infinite c.e. subset of $\omega \setminus \nu^{-1}[B]$, which contradicts the co-immunity of $\nu^{-1}[B]$. Thus, μ_A and μ_B are incomparable under bm-reducibility. Therefore, the semilattice $\mathcal{R}_{bm}(S)$ contains an infinite antichain. □

5 Further Discussion

First, we briefly discuss related results on hyperarithmetical numberings.

Let α be a computable ordinal such that $\alpha \geq 2$. Consider a family of Σ_α^0-sets $\mathcal{S}$, which has a Σ_α^0-computable numbering. Following the lines of Sect. 2, one can introduce the Rogers semilattices $\mathcal{R}_{\alpha;m}^0(\mathcal{S})$ and $\mathcal{R}_{\alpha;bm}^0(\mathcal{S})$, which are induced by the degrees of Σ_α^0-computable numberings of $\mathcal{S}$, under the reducibilities $\leq$ and $\leq_{bm}$, respectively.

Proposition 5.1. *Let $\alpha \geq 2$ be a computable successor ordinal. Suppose that $\mathcal{S}$ is a Σ_α^0-computable family such that $\mathcal{S}$ contains at least two elements. Then the Rogers semilattice $\mathcal{R}_{\alpha;bm}^0(\mathcal{S})$ is infinite, and it is not a lattice.*

Proof. Recall that $card(\mathcal{R}_{\alpha;m}^0(\mathcal{S})) \leq card(\mathcal{R}_{\alpha;bm}^0(\mathcal{S}))$. Goncharov and Sorbi [21, Theorem 2.1] proved that the semilattice $\mathcal{R}_{\alpha;m}^0(\mathcal{S})$ is infinite.

Furthermore, in [21, Proposition 2.8] the following result was obtained. If $\mathcal{S}$ is infinite, then one can build a uniform sequence $(\nu_i)_{i\in\omega}$ of Σ_α^0-computable numberings of $\mathcal{S}$ with the following property: If $i \neq j$, then there is no Σ_α^0-computable numbering μ of $\mathcal{S}$ such that $\mu \leq \nu_i$ and $\mu \leq \nu_j$.

This implies that for an infinite $\mathcal{S}$, both structures $\mathcal{R}_{\alpha;m}^0(\mathcal{S})$ and $\mathcal{R}_{\alpha;bm}^0(\mathcal{S})$ are not lower semilattices. Note that the results of Goncharov and Sorbi are formulated and proved only for finite ordinals α. Nevertheless, essentially the same proofs also work for infinite successor α.

Now assume that a Σ_α^0-computable family $\mathcal{S}$ is equal to $\{A_0, A_1, \ldots, A_n\}$, and consider the following Σ_α^0-computable numberings of $\mathcal{S}$:

$$\nu_i(x) := \begin{cases} A_x, & \text{if } x \leq n, \\ A_i, & \text{otherwise.} \end{cases}$$

Lemma 2.1 shows that the numberings ν_i, $i \leq n$, are pairwise bm-incomparable. Moreover, it is not hard to show that for $i \neq j$, there is no numbering μ of the family $\mathcal{S}$ such that $\mu \leq \nu_i$ and $\mu \leq \nu_j$. Hence, $\mathcal{R}_{\alpha;bm}^0(\mathcal{S})$ is not a lattice.   □

We note that the methods of [22] can be used to transfer the obtained existence results (such as Theorem 3.1 and Corollary 4.1) into non-limit levels of the Ershov hierarchy, see Theorems 2 and 17 in [22] for the details.

In conclusion, we formulate two problems that are left open.

Question 5.1. Let $\mathcal{S}$ be a computable infinite family of c.e. sets. Describe all possible cardinalities of the Rogers semilattice $\mathcal{R}_{bm}(\mathcal{S})$.

Note that all our examples of computable families $\mathcal{S}$ possess the following property: If $\mathcal{R}_{bm}(\mathcal{S})$ is an infinite lattice, then the structure $\mathcal{R}_m(\mathcal{S})$ has only one element.

Question 5.2. Is there a computable family $\mathcal{S}$ such that $\mathcal{R}_m(\mathcal{S})$ is infinite and $\mathcal{R}_{bm}(\mathcal{S})$ is a lattice?

Acknowledgements. Part of the research contained in this paper was carried out while the first and the last authors were visiting the Department of Mathematics of Nazarbayev University, Astana. The authors wish to thank Nazarbayev University for its hospitality. The authors also thank the anonymous reviewers for their helpful suggestions.

References

1. Ambos-Spies, K., Badaev, S., Goncharov, S.: Inductive inference and computable numberings. Theor. Comput. Sci. **412**(18), 1652–1668 (2011). https://doi.org/10.1016/j.tcs.2010.12.041
2. Andrews, U., Sorbi, A.: Joins and meets in the structure of ceers. Computability (2018). https://doi.org/10.3233/COM-180098, published online
3. Badaev, S., Goncharov, S.: Computability and numberings. In: Cooper, S.B., Löwe, B., Sorbi, A. (eds.) New Computational Paradigms, pp. 19–34. Springer, New York (2008). https://doi.org/10.1007/978-0-387-68546-5_2
4. Badaev, S.A.: Computable enumerations of families of general recursive functions. Algebra Logic **16**(2), 83–98 (1977). https://doi.org/10.1007/BF01668593
5. Badaev, S.A.: Minimal numerations of positively computable families. Algebra Logic **33**(3), 131–141 (1994). https://doi.org/10.1007/BF00750228
6. Badaev, S.A., Goncharov, S.S.: Theory of numberings: open problems. In: Cholak, P., Lempp, S., Lerman, M., Shore, R. (eds.) Computability Theory and Its Applications. Contemporary Mathematics, vol. 257, pp. 23–38. American Mathematical Society, Providence (2000). https://doi.org/10.1090/conm/257/04025
7. Badaev, S.A., Lempp, S.: A decomposition of the Rogers semilattice of a family of D.C.E. sets. J. Symb. Logic **74**(2), 618–640 (2009). https://doi.org/10.2178/jsl/1243948330
8. Bazhenov, N., Mustafa, M., Yamaleev, M.: Elementary theories and hereditary undecidability for semilattices of numberings. Arch. Math. Logic (2018). https://doi.org/10.1007/s00153-018-0647-y, published online
9. Case, J., Jain, S., Stephan, F.: Effectivity questions for Kleene's recursion theorem. Theor. Comput. Sci. **733**, 55–70 (2018). https://doi.org/10.1016/j.tcs.2018.04.036
10. Ershov, Y.L.: Enumeration of families of general recursive functions. Sib. Math. J. **8**(5), 771–778 (1967). https://doi.org/10.1007/BF01040653
11. Ershov, Y.L.: On computable enumerations. Algebra Logic **7**(5), 330–346 (1968). https://doi.org/10.1007/BF02219286
12. Ershov, Y.L.: Theory of Numberings. Nauka, Moscow (1977). (in Russian)
13. Ershov, Y.L.: Theory of numberings. In: Griffor, E.R. (ed.) Handbook of Computability Theory. Studies in Logic and the Foundations of Mathematics, vol. 140, pp. 473–503. North-Holland, Amsterdam (1999). https://doi.org/10.1016/S0049-237X(99)80030-5
14. Friedberg, R.M.: Three theorems on recursive enumeration. I. Decomposition. II. Maximal set. III. Enumeration without duplication. J. Symb. Logic **23**(3), 309–316 (1958). https://doi.org/10.2307/2964290
15. Friedman, H., Stanley, L.: A Borel reducibility theory for classes of countable structures. J. Symb. Log. **54**(3), 894–914 (1989). https://doi.org/10.2307/2274750
16. Gao, S.: Invariant Descriptive Set Theory. CRC Press, Boca Raton (2009)
17. Gödel, K.: Über formal unentscheidbare Sätze der Principia Mathematica und verwandter Systeme. I. Monatsh. Math. Phys. **38**(1), 173–198 (1931). https://doi.org/10.1007/BF01700692

18. Goncharov, S.S.: Computable single-valued numerations. Algebra Logic **19**(5), 325–356 (1980). https://doi.org/10.1007/BF01669607
19. Goncharov, S.S.: Positive numerations of families with one-valued numerations. Algebra Logic **22**(5), 345–350 (1983). https://doi.org/10.1007/BF01982111
20. Goncharov, S.S., Lempp, S., Solomon, D.R.: Friedberg numberings of families of n-computably enumerable sets. Algebra Logic **41**(2), 81–86 (2002). https://doi.org/10.1023/A:1015352513117
21. Goncharov, S.S., Sorbi, A.: Generalized computable numerations and nontrivial Rogers semilattices. Algebra Logic **36**(6), 359–369 (1997). https://doi.org/10.1007/BF02671553
22. Herbert, I., Jain, S., Lempp, S., Mustafa, M., Stephan, F.: Reductions between types of numberings (2017, preprint)
23. Jain, S., Stephan, F.: Numberings optimal for learning. J. Comput. Syst. Sci. **76**(3–4), 233–250 (2010). https://doi.org/10.1016/j.jcss.2009.08.001
24. Khutoretskii, A.B.: On the cardinality of the upper semilattice of computable enumerations. Algebra Logic **10**(5), 348–352 (1971). https://doi.org/10.1007/BF02219842
25. Kleene, S.C.: Introduction to Metamathematics. Van Nostrand, New York (1952)
26. Kolmogorov, A.N., Uspenskii, V.A.: On the definition of an algorithm. Uspehi Mat. Nauk **13**(4), 3–28 (1958). (in Russian)
27. Lachlan, A.H.: Standard classes of recursively enumerable sets. Z. Math. Logik Grundlagen Math. **10**(2–3), 23–42 (1964). https://doi.org/10.1002/malq.19640100203
28. Lachlan, A.H.: On recursive enumeration without repetition. Z. Math. Logik Grundlagen Math. **11**(3), 209–220 (1965). https://doi.org/10.1002/malq.19650110305
29. Mal'cev, A.I.: Positive and negative numerations. Sov. Math. Dokl. **6**, 75–77 (1965)
30. Maslova, T.M.: Bounded m-reducibilities. In: Golunkov, Y.V. (ed.) Veroyatnostnye Metody i Kibernetika, vol. XV, pp. 51–60. Kazan University, Kazan (1979). (in Russian), MR0577649
31. Ospichev, S.S.: Friedberg numberings in the Ershov hierarchy. Algebra Logic **54**(4), 283–295 (2015). https://doi.org/10.1007/s10469-015-9349-2
32. Podzorov, S.Y.: Arithmetical D-degrees. Sib. Math. J. **49**(6), 1109–1123 (2008). https://doi.org/10.1007/s11202-008-0107-8
33. Pour-El, M.B.: Gödel numberings versus Friedberg numberings. Proc. Am. Math. Soc. **15**(2), 252–256 (1964). https://doi.org/10.2307/2034045
34. Rogers, H.: Gödel numberings of partial recursive functions. J. Symb. Logic **23**(3), 331–341 (1958). https://doi.org/10.2307/2964292
35. Selivanov, V.L.: Two theorems on computable numberings. Algebra Logic **15**(4), 297–306 (1976). https://doi.org/10.1007/BF01875946
36. Uspenskii, V.A.: Systems of denumerable sets and their enumeration. Dokl. Akad. Nauk SSSR **105**, 1155–1158 (1958). (in Russian)

Complexity of Conjunctive Regular Path Query Homomorphisms

Laurent Beaudou[1,2], Florent Foucaud[1], Florent Madelaine[3(✉)],
Lhouari Nourine[1], and Gaétan Richard[4]

[1] LIMOS, Université Clermont Auvergne, Aubière, France
{laurent.beaudou,lhouari.nourine}@uca.fr,
florent.foucaud@gmail.com
[2] Higher School of Economics, National Research University,
3 Kochnovsky Proezd, Moscow, Russia
[3] LACL, Université Paris-Est Créteil, Créteil, France
lorent.madelaine@u-pec.fr
[4] GREYC, Université Caen Normandie, Caen, France
gaetan.richard@unicaen.fr

Abstract. A graph database is a digraph whose arcs are labelled with symbols from a fixed alphabet. A regular graph pattern (RGP) is a digraph whose edges are labelled with regular expressions over the alphabet. RGPs model navigational queries for graph databases, more precisely, conjunctive regular path queries. A match of a navigational RGP query in the database is witnessed by a special navigational homomorphism of the RGP to the database. We study the complexity of deciding the existence of a homomorphism between two RGPs. Such homomorphisms model a strong type of containment between two navigational RGP queries. We show that this problem can be solved by an EXP-TIME algorithm (while general query containment in this context is EXPSPACE-complete). We also study the problem for restricted RGPs over a unary alphabet, that arise from some applications like XPath, and prove that certain interesting cases are polynomial-time solvable.

1 Introduction

Graphs are a fundamental way to store and organize data. Most prominently, graph database systems have been developed for three decades and are widely used; recently, such systems have seen an increased interest both in academic research and in industry [3]. A graph database can be seen as a directed graph with arc-labels (possibly also vertex-labels). Various methods are used to retrieve data in such systems, see for example the very recently developed graph query

The authors acknowledge the support received from the Agence Nationale de la Recherche of the French government throught the program "Investissements d'Avenir" (16-IDEX-0001 CAP 20-25), the IFCAM project "Applications of graph homomorphisms" (MA/IFCAM/18/39), and by the ANR project HOSIGRA (ANR-17-CE40-0022).

© Springer Nature Switzerland AG 2019
F. Manea et al. (Eds.): CiE 2019, LNCS 11558, pp. 108–119, 2019.
https://doi.org/10.1007/978-3-030-22996-2_10

language G-CORE [2] for graph databases. Classically, matching queries in graph databases can be modeled as graph homomorphisms [16]. In this setting, a query is itself a graph, and a match is modeled by a homomorphism of the query to the database, that is, a vertex-mapping that preserves the graph adjacencies and labels. Graph databases can be very large, thus it is important to study the algorithmic complexity of such queries. In modern applications, classic homomorphisms are often not powerful enough to model realistic graph data queries. In recent years, *navigational queries* have been developed [3]. Such queries are more powerful than classical queries, since they allow for non-local pattern matching, by means of arbitrary paths or walks instead of arcs. Such queries can also be modeled as a more general kind of homomorphism, called *navigational homomorphism* (*n-homomorphism* for short). The most studied type of navigational queries is the one of *regular path queries*, that is based on regular expressions [3,10,17]. A *conjunctive regular path query* is modeled by a *regular graph pattern* (RGP for short), that is, a digraph whose arcs are labelled by regular expressions, each representing a regular path query. The associated notion of navigational homomorphism is called *RGP homomorphism*. The study of the algorithmic complexity of RGP homomorphisms has been recently initiated in [19], where the authors focused on homomorphisms of RGPs to graph databases. In the present paper, we continue this study by focusing on conjunctive regular path query containment, as modeled by the existence of a RGP homomorphism between two queries.

We delay to the next section for formal definitions. We consider the following decision problem

RGPHOM
Input: Two RGPs P and Q.
Question: Does P admit an n-homomorphism to Q?

and its non-uniform version, defined as follows for a fixed RGP Q.

RGPHOM(Q)
Input: A RGP P.
Question: Does P admit an n-homomorphism to Q?

The latter was introduced in [19] for the restricted case where Q is a graph database (*i.e.* labels are letters rather than more complex regular expressions) – which amounts to *RGP evaluation* – and showed that this class of problems follows a dichotomy between Ptime and NP-complete. Indeed they showed it to be equivalent to the (classical) homomorphism problems a.k.a. the constraint satisfaction problems [9], whose complexity delineation follows a dichotomy based on specific algebraic properties of the template Q, as shown independently by Bulatov [5] and Zhuk [22].

In this paper, we initiate the study of these problems in full generality, that is when Q is not a graph database but any RGP. We cannot expect a Ptime/NP-complete dichotomy in the style of the result of [19], since RGPHOM(Q) is in fact PSPACE-hard already for very simple cases, as it can model the problem of deciding the inclusion between regular languages. We detail these lower bounds in Sect. 3.

We show in Sect. 4 that RGPHOM is decidable by an EXPTIME algorithm. This shows that n-homomorphism-based query containment is less expensive than general query containment, which, in the case of RGP queries, is known to be EXPSPACE-complete [7,10].

In Sect. 5, we address the simpler case of a unary alphabet $\Sigma = \{a\}$, when all arcs are labelled either by "a" or "a^+". This includes not only all classic homomorphism problems and CSPs, but also queries over hierarchical data reminiscent of SPARQL and XPath [8,15,18]. In particular, we give a simple Ptime/NP-complete complexity dichotomy for the case of undirected (or symmetric) RGPs in the style of Hell and Nešetřil's dichotomy for H-colouring [11]. Furthermore, we show that even for arbitrary (directed) RGPs the problem follows a dichotomy by relating it to (classical) homomorphism problems. Finally, we focus on certain queries of interest. We relate the case of path templates that have only "a" labels to an interesting (Ptime) scheduling problem. We extend this result and show also that for all directed path RGP templates Q with arc labels "a" or "a^+", RGPHOM(Q) is in Ptime.

2   Preliminaries

Let Σ be a fixed countable alphabet. A *graph database* $B = (D_B, E_B)$ over Σ is an arc-labelled digraph, where D_B is a finite digraph with vertex set $V(D_B)$ and arc set $A(D_B)$, and $E_B : A(D_B) \to \Sigma$ is an arc-label function. We may adapt the notion of graph homomorphism to graph databases following [16] and view the existence of a homomorphism from a graph database $Q = (D_Q, E_Q)$ – which models a *query* – to $B = (D_B, E_B)$ as the fact that *the database B matches the query Q*. A homomorphism is a mapping f of $V(D_Q)$ to $V(D_B)$ such that for every arc (x, y) in D_Q, there is an arc $(f(x), f(y))$ in D_B with $E_Q(x, y) = E_B(f(x), f(y))$. If such a homomorphism exists, we note $Q \to B$. Every homomorphic image $f(Q)$ of Q to B is a match of the query Q in B. In this classic setting, queries and graph databases coincide and the existence of a homomorphism between queries models *query containment*. Thus, the *evaluation problem* (deciding whether a query Q has a match in a database B) and the *containment problem* (deciding whether for two queries Q_1 and Q_2, for any database B, if B matches Q_1 then B matches Q_2) both amount to the following decision problem.

HOM

Input: Two arc-labelled digraphs G and H.

Question: Does G admit a homomorphism to H?

HOM is generally NP-complete, even when H is a small fixed graph (for example a symmetric triangle, in which case HOM is equivalent to the graph 3-colourability problem). To better understand the complexity of HOM, the following version, called the *non-uniform* homomorphism problem, has been studied extensively. Here, H is a fixed arc-labelled digraph called the *template*.

> HOM(H)
> Input: An arc-labelled digraph G.
> Question: Does G admit a homomorphism to H?

In a digraph D, a *directed walk* is a sequence of arcs of the digraph, such that the head of each arc is the same vertex as the tail of the next arc. A *directed path* is a directed walk where each vertex occurs in at most two arcs in the sequence.

Standard homomorphisms are not powerful enough to model all queries used in modern graph database systems. In particular, a homomorphism of a query Q to a database B can only match a subgraph of B that is no larger than the query Q itself. To the contrary, *navigational queries* are types of queries where we may allow arbitrarily large subgraphs of the database to match the query. In this setting, we still model the query Q (for database B) as an arc-labelled digraph, but the arcs are labelled with *sets of words* over the alphabet Σ, rather than letters. Now, a match of Q in B is a vertex-mapping f from $V(D_Q)$ to $V(D_B)$ such that for an arc (x, y) of Q labelled with a set $E(x, y)$ of words, there exists a directed walk (or path, depending on applications) W_{xy} in D_B from $f(x)$ to $f(y)$ such that the concatenation of labels of the arcs of W_{xy} is a word of $E(x, y)$.

Perhaps the most popular navigational queries are *conjunctive regular path queries*, studied in many contexts [3,4,8,15,18,19]. These navigational queries are based on regular languages: the labels on query arcs are regular expressions over the alphabet Σ. The advantage of considering such queries is that regular languages are a relatively simple yet powerful way of defining sets of words, that is both well-understood and sufficiently expressive for many applications. A conjunctive regular path query of this type has been called a *regular graph pattern* (RGP).

For a fixed countable alphabet Σ, we denote by $RegExp(\Sigma)$ the set of regular expressions over alphabet Σ, with the symbols $+$ (union), $*$ (Kleene star), and $\cdot$ (concatenation; sometimes this symbol is omitted). Moreover, for a regular expression X, as a notation we let $X^+ := X \cdot X^*$. For any regular expression X in $RegExp(\Sigma)$, we denote by $L(X)$ the regular language defined by X. We will use the following decision problems for regular languages.

> REGULAR LANGUAGE INCLUSION
> Input: Two regular expressions E_1 and E_2 (over the same alphabet).
> Question: Is $L(E_1) \subseteq L(E_2)$?

> REGULAR LANGUAGE UNIVERSALITY
> Input: A regular expression E over alphabet Σ.
> Question: Is $L(E) = \Sigma^*$?

Note that REGULAR LANGUAGE UNIVERSALITY is the special case of REGULAR LANGUAGE INCLUSION where $E_1 = \Sigma^*$ and $E_2 = E$. The following are classic results.

Theorem 1 ([1,21]). REGULAR LANGUAGE UNIVERSALITY *and* REGULAR LANGUAGE INCLUSION *are PSPACE-complete.*

A RGP P over an alphabet Σ is a pair (D_P, E_P), where D_P is a digraph with vertex set $V(D_P)$ and arc set $A(D_P)$ and $E_P : A(D_P) \to RegExp(\Sigma)$ is an arc-label function. Given a directed walk $W = a_{1,2} \ldots a_{k-1,k}$ in a RGP P, the label $E_P(W)$ of W is the regular expression over Σ formed by the concatenation $E_P(a_{1,2}) \ldots E_P(a_{k-1,k})$.

A *sub-RGP* P' of P is induced by a subset $V(D_{P'})$ of $V(D_P)$ where $A(D_{P'})$ is $A(D_P) \cap V(D_{P'}) \times V(D_{P'})$ and the arc label function $E_{P'}$ is the restriction of E_P to $A(D_{P'})$.

Given two RGPs P and Q over alphabet Σ, a *navigational homomorphism* (*n-homomorphism* for short) of P to Q is a mapping f of $V(D_P)$ to $V(D_Q)$ such that for each arc (x,y) in D_P, there is a directed walk W in Q from $f(x)$ to $f(y)$ such that the language $L(E_Q(W))$ is contained in the language $L(E_P(x,y))$. When such an n-homomorphism exists, we write $P \xrightarrow{n} Q$. It is not hard to see that $\xrightarrow{n}$ is transitive.

This definition also applies to graph databases and amounts in fact to *RGP query evaluation* when Q is a graph database and P an RGP.

A digraph D is called a *core* if it has no homomorphism to a proper subdigraph of itself; in other words, every endomorphism is an automorphism. Similarly we define the notion of a *navigational core* (*n-core* for short): a RGP P is an n-core if it has no n-homomorphism to a proper sub-RGP of itself. When studying the problem RGPHOM(Q), we may always assume that Q is an n-core, since RGPHOM(Q) has the same complexity as RGPHOM(C_Q), where C_Q is a sub-RGP of Q that is an n-core. Unfortunately, it is coNP-complete to decide whether a graph is a core [12] (thus deciding whether a RGP is an n-core is coNP-hard, even if it is a graph database).

With respect to classic digraph homomorphisms, any digraph has (up to isomorphism) a unique minimal subgraph to which it admits a homomorphism, called *the* core. This is not the case for n-cores of RGPs and n-homomorphisms. For example, any two RGPs each consisting of a unique directed cycle with all arc labels equal to "a^+" have an n-homomorphism to each other. Thus, if we identify one vertex of two such cycles of different lengths, we obtain a RGP P with two minimal sub-RGPs of P (the two cycles) to which P has an n-homomorphism, thus these two are non-isomorphic n-cores of P.[1]

3   Lower Bounds

The next proposition is proved by a very simple reduction from REGULAR LANGUAGE INCLUSION to RGPHOM for two RGPs each having a single arc.

Proposition 1. RGPHOM *is PSPACE-hard.*

---

[1] There exist more complicated examples where, furthermore, the two non-isomorphic n-cores have the same size.

As witnessed by the simplicity of the reduction given in Proposition 1, the PSPACE-hardness of RGPHOM is inherently caused by the hardness of the underlying regular language problem. This phenomenon arises also in the non-uniform case for a very simple template.

Proposition 2. *Let Σ be a fixed alphabet of size at least 2, and let D_2^Σ be the RGP of order 2 over Σ consisting of a single arc labelled Σ^*. Then, RGPHOM(D_2^Σ) is PSPACE-complete.*

4   An EXPTIME Algorithm for RGPHOM

In certain models where *simple directed paths* rather than directed walks are considered, like in [17], or when the target RGP is acyclic, there is a simple PSPACE algorithm to decide RGPHOM. Indeed, in those cases, the length of a walk in Q is at most $|V(D_Q)|$. Thus, we can iterate over each possible mapping f and for each mapped arc (x, y), we iterate over each possible walk W from $f(x)$ to $f(y)$, and check in polynomial space whether $L(E_Q(W)) \subseteq L(E_P(x, y))$.

However, in general, the walks may be arbitrarily long. As we will see, we can still bound their maximum length. Note that for two RGPs P and Q, if $P \xrightarrow{n} Q$ then the query Q is contained in the query P, but there are examples where the converse does not hold (see Fig. 1). Thus, the problem RGPHOM for two RGPs does not fully capture RGP QUERY CONTAINMENT. Nevertheless, we will show that the former can be solved in EXPTIME, which is better than the (tight) EXPSPACE complexity of RGP QUERY CONTAINMENT shown in [7, 10].

Fig. 1. Two n-core RGPs P and Q over alphabet $\{a, b\}$ which have no n-homomorphism in either direction. From P to Q because one can not map suitably the arc labelled by b, in the other direction because neither b nor a^+ is included in a. However, any database that matches the RGP P would contain a walk of arcs all labelled by a (because of the arc with label a^+ in P). The database would clearly also match Q. So Q is contained in P.

For a regular language L over alphabet Σ and a positive integer n, we denote by $L_{|n}$ the *n-truncation* of L, i.e. the set of words of L with length at most n.

Lemma 1. *Let $A, B_1, \dots B_k$ be a collection of regular expressions over alphabet Σ, and let n_A, n_i be the minimum number of states of an NFA recognizing $L(A)$ and $L(B_i)$, respectively. Then, we have that $L(B_1) \cdots L(B_k) \subseteq L(A)$ if, and only if, $L(B_1)_{|n_A n_1} \cdots L(B_k)_{|n_A n_k} \subseteq L(A)$ (the left hand side denotes the product of languages).*

Proof. Since $L(B_i)_{|n_An_i} \subseteq L(B_i)$ for every i with $1 \le i \le k$, if $L(B_1) \cdots L(B_k) \subseteq L(A)$, then it holds also for truncations and $L(B_1)_{|n_An_1} \cdots L(B_k)_{|n_An_k} \subseteq L(A)$.

For the converse, we assume that $L(B_1)_{|n_An_1} \cdots L(B_k)_{|n_An_k} \subseteq L(A)$. That is, any word $w_1 \cdots w_k$ of $L(B_1) \cdots L(B_k)$ with $|w_i| \le n_An_i$ for every i with $1 \le i \le k$, belongs to $L(A)$. We need to prove that all words of $L(B_1) \cdots L(B_k)$ (without length restriction) belong to $L(A)$.

We proceed by induction on the vectors of subword lengths of words in $L(B_1) \cdots L(B_k)$. For such a word $w_1 \cdots w_k$, this associated vector is $(|w_1|, \ldots, |w_k|)$, and these vectors are ordered lexicographically. The induction hypothesis is that all words of $L(B_1) \cdots L(B_k)$ whose associated vector is at most $(l_1, \ldots, l_k)$ (where for any i with $1 \le i \le k$, l_i is a positive integer), belongs to A. By our assumption, the case where $l_i \le n_An_i$ is true.

Now, consider a word $w = w_1 \cdots w_k$ of $L(B_1) \cdots L(B_k)$, whose associated vector is $(|w_1|, \ldots, |w_k|)$, and where for some $j \in \{1, \ldots, k\}$, $|w_j| = l_j + 1$; whenever $i \ne j$, $|w_i| \le l_i$. Let $\mathcal{A}$ and $\mathcal{A}_j$ be two NFAs recognizing A and B_j with smallest numbers n_A and n_j of states, respectively.

We consider the product automaton $\mathcal{A} \times \mathcal{A}_j$ of $\mathcal{A}$ and $\mathcal{A}_j$, with set of states $S \times S_j$ (where S and S_j are the sets of states of $\mathcal{A}$ and $\mathcal{A}_j$, respectively), and a transition $((s_1, s_2), a, (s_1', s_2'))$ only if we have the transitions (s_1, a, s_1') and (s_2, a, s_2') in $\mathcal{A}$ and $\mathcal{A}_j$, respectively (all other transitions are "dummy transitions" to a "garbage state"). Consider the run of $\mathcal{A} \times \mathcal{A}_j$ for the word w_j. The crucial observation is that, because $|w_j| = l_j + 1 > n_An_j$, this run necessarily visits two states of $\mathcal{A} \times \mathcal{A}_j$ twice, that is, the run contains a directed cycle. Consider the shorter run obtained by pruning this cycle. The two runs start and end at the same two states of $\mathcal{A} \times \mathcal{A}_j$. The shorter run corresponds to a word w_j' of length at most $|w_j| - 1 \le l_j$. Since $w_j \in L(B_j)$, the end state of these runs is a pair containing an accepting state of $\mathcal{A}_j$ (thus w_j' belongs to $L(B_j)$ as well). Thus, the word w' obtained from w by replacing w_j with w_j' belongs to $L(B_1) \cdots L(B_k)$, and w' satisfies the induction hypothesis. Thus, w' belongs to $L(A)$. But now, considering the pruned cycle in $\mathcal{A} \times \mathcal{A}_j$, we can build a valid run for w_j in $\mathcal{A} \times \mathcal{A}_j$ that leads to a valid run for w in $\mathcal{A}$. This proves the inductive step and concludes the proof.

Proposition 3. *Let E be a regular expression over alphabet Σ, and $\mathcal{A}_E$ an NFA with n_E states recognizing $L(E)$. Let $Q = (D_Q, E_Q)$ be an RGP over Σ. For any two vertices u and v in Q, we can compute a walk W from u to v satisfying $L(E_Q(W)) \subseteq L(E)$ (if one exists), in time $2^{O(n_E|Q| \log(|E|+|Q|))}$. Moreover, if such a walk exists, then there exists one of length at most $2^{n_E}|Q|$.*

Proof. Since E and Q are finite, we will assume that $|\Sigma| \le |E| + |Q|$ (if not, we simply remove the unused symbols from Σ.)

By Lemma 1, there is a walk W in Q from u to v such that $L(E_Q(W)) \subseteq L(E)$ if and only if there exists one in the RGP Q' obtained from Q by replacing each arc-label $E_Q(x, y)$ by a regular expression $E_{Q'}(x, y)$ defining the n_En_B-truncation of $L(E_Q(x, y))$ (where n_B is the smallest number of states of an NFA recognizing $L(E(x, y))$). Thus, we first compute Q'. Note that $L(E_{Q'}(x, y))$ contains at most $|\Sigma|^{n_En_B}$ words.

Next, we will construct an auxiliary digraph $G(E, Q, u, v)$. This digraph has vertex set $2^S \times V(Q)$, where S is the set of states of $\mathcal{A}_E$.

Given two states s_1 and s_2 of $\mathcal{A}_E$ and a word w over Σ, we say that w reaches s_2 from s_1 in $\mathcal{A}_E$ if there exists a sequence of transitions of $\mathcal{A}_E$ starting at s_1 and ending at s_2 using the sequence of letters of w.

Now, for two vertices (S_1, x) and (S_2, y) of $G(E, Q, u, v)$, we create the arc $((S_1, x), (S_2, y))$ if, and only if, for each state s of S_1 and each word w of $L(E_{Q'}(x, y))$, w reaches a state of S_2 in $\mathcal{A}_E$.

Deciding whether $((S_1, x), (S_2, y))$ is an arc of $G(E, Q, u, v)$ takes time at most $|S_1| 2^{n_E} |L(E_{Q'}(x, y))|$, which is at most $|\Sigma|^{O(n_E|Q|)}$. Since there are $(2^{n_E}|Q|)^2$ pairs of vertices of $G(E, Q, u, v)$, overall the construction of $G(E, Q, u, v)$ can be done in time $|\Sigma|^{O(n_E|Q|)}$.

Now, we claim that there exists a walk W from u to v with $L(E_Q(W)) \subseteq L(E)$ if and only if there is a directed path in $G(E, Q, u, v)$ from a vertex $(\{s_0\}, u)$ to a vertex (S_f, v), where s_0 is the initial state of $\mathcal{A}_E$, and S_f is a subset of the accepting states of $\mathcal{A}_E$. Indeed, such a path corresponds precisely to a walk W from u to v in Q, such that all the words of $L(W)$ are accepted by $\mathcal{A}_E$.

This check can be done in linear time in the size of $G(E, Q, u, v)$ using a standard BFS search, thus we obtain an additional time complexity of $(2^{n_E}|Q|)^2$, which is also at most $|\Sigma|^{O(n_E|Q|)}$. Since $|\Sigma| \leq |E| + |Q|$ we obtain $2^{O(n_E|Q| \log(|E|+|Q|))}$.

Finally, it is clear that the length of an obtained directed path of $G(E, Q, u, v)$ is at most the number of vertices of $G(E, Q, u, v)$, which is $2^{n_E}|Q|$, as claimed. This completes the proof. $\qquad\square$

Theorem 2. RGPHOM *is in EXPTIME.*

Proof. We proceed as follows. First, we go through all possible vertex-mappings of $V(P)$ to $V(Q)$ (there are $|V(Q)|^{|V(P)|}$ such possible mappings). Consider such a vertex-mapping, f.

For each arc (x, y) in P with label $E_P(x, y)$, we proceed as follows. Let $\mathcal{A}$ be an NFA recognizing $L(E_P(x, y))$ with smallest possible number n_A of states. We apply Proposition 3 to $E = E_P(x, y)$, $\mathcal{A}$ and Q, with $u = f(x)$ and $v = f(y)$: thus we can decide in time $2^{O(n_A|Q| \log(|E_P(x,y)|+|Q|))}$ whether the mapping f satisfies the definition of an n-homomorphism for the arc (x, y). If yes, we proceed to the next arc; otherwise, we abort and try the next possible mapping. If we find a valid mapping, we return YES. Otherwise, we return NO.

Our algorithm has a time complexity of $|V(Q)|^{|V(P)|} \cdot |P| \cdot 2^{O(|P||Q| \log(|P|+|Q|))}$. Let $n = |P| + |Q|$ be the input size. We obtain an overall running time of $2^{O(n^2 \log n)}$, which is an EXPTIME running time. $\qquad\square$

5   RGPs Over a Unary Alphabet: The $\{a, a^+\}$ Case

In this section, we consider a unary alphabet $\Sigma = \{a\}$. For unary regular languages, REGULAR LANGUAGE INCLUSION and REGULAR LANGUAGE UNIVERSALITY are no longer PSPACE-complete but they are coNP-complete (see [13] and [21], respectively) and the lower bounds from Sect. 3 do not apply.

The case where all arc-labels of the considered RGPs are equal to "a" is equivalent to the problem of classic digraph homomorphisms, and is known to capture all CSPs [9]. When each label is either "a" or "a^+"– in which case we speak of $\{a, a^+\}$-RGP – we have two kinds of constraints: arcs labelled "a" must map in a classic, local, way, while arcs labelled "a^+" can be mapped to an arbitrary path in the target RGP. Thus, this setting is useful for example to model descendance relations in hierarchical data such as XML. This setting is for example used in languages like SPARQL or XPath for XML documents, that are tree-structured [8,15,18].

For an $\{a, a^+\}$-RGP Q, let $D(Q)$ be the arc-labelled digraph with labels $\{a, t\}$ and the same vertices as Q obtained from Q as follows. The arcs labelled by a coincide. The set of arcs labelled by t in $D(Q)$ is the transitive closure of the arcs of Q (labelled by either label a or a^+).

Proposition 4. *For any $\{a, a^+\}$-RGP Q, RGPHom(Q) for $\{a, a^+\}$-RGP inputs is Ptime equivalent to* HOM$(D(Q))$. *Thus, the class of $\{a, a^+\}$-RGP-restricted non-uniform* RGPHom *problems follows a dichotomy between Ptime and NP-complete.*

Proof. Let D'_P be the digraph obtained from $P = (D_P, E_P)$ by replacing all arc-labels "a^+" by labels "t". For any pair of $\{a, a^+\}$-RGP P and Q, we have $P \xrightarrow{n} Q$ (n-homomorphism) if and only if $D'_P \to D(Q)$ (classic homomorphism of arc-labelled digraphs). This provides us with a Ptime reduction from RGPHom(Q) to HOM$(D(Q))$. Conversely, given a digraph D'_P with arcs labelled by a and t, let P be the $\{a, a^+\}$-RGP obtained from D'_P by replacing t labels by a^+. This is a Ptime reduction from HOM$(D(Q))$ to RGPHom(Q).

The dichotomy follows from the CSP dichotomy of [5,22].   □

Proposition 5. RGPHom *for $\{a, a^+\}$-RGPs is Ptime reducible to* HOM *for two-arc-labelled digraphs. Thus this uniform problem is in NP.*

The Ptime algorithms based on algebraic methods proposed by Bulatov [5] and Zhuk [22] for tractable CSPs are somewhat contrived and a bit overkill for the class of non-uniform RGPHom problems restricted to $\{a, a^+\}$-RGPs. This motivates us to look for simple direct combinatorial characterisations and simple algorithms for interesting $\{a, a^+\}$-RGPs that model natural queries.

5.1   Undirected $\{a, a^+\}$-RGPs

We now consider *undirected* $\{a, a^+\}$-RGPs, where arcs are pairs of vertices, called *edges* (equivalently, for each arc from x to y, we have its symmetric arc from y to x). In an n-homomorphism between two undirected $\{a, a^+\}$-RGPs P and Q, "a"-edges must be preserved (as in a classic graph homomorphism), while the endpoints of "a^+"-edges need to be mapped to two vertices of Q that are connected by some path. Thus, this variant extends classic graph homomorphisms via additional (binary) connectivity constraints. We provide an analogue of the Hell-Nešetřil dichotomy for HOM(H) [11] in this setting.

Theorem 3. *Let Q be an undirected and connected n-core $\{a, a^+\}$-RGP. If Q has at most one edge, RGPHOM(Q) is in Ptime. Otherwise, RGPHOM(Q) is NP-complete.*

5.2 Directed Path $\{a\}$-RGPs

We first consider RGPHOM(Q) when Q is a directed path whose arc labels are all "a" (Q is called an $\{a\}$-RGP) — arguably the simplest RGP directed graph example — and where inputs are $\{a, a^+\}$-RGPs. This case turns out to have an interesting connection to the following scheduling problem, which enjoys a Ptime algorithm by reduction to a shortest path problem in edge-weighted digraphs [20, Chapter 4.4, p. 666].

> PARALLEL JOB SCHEDULING WITH RELATIVE DEADLINES
> Input: A set J of jobs, a duration function $d : J \to \mathbb{N}$, a relative deadline function $r : J \times J \to \mathbb{Z}$, and a maximum time t_{max}.
> Question: Is there a feasible schedule for the jobs, that is, an assignment $t : J \to \mathbb{N}$ of start times such that every job finishes before time t_{max} and for any two jobs j_1 and j_2, j_1 starts before the time $t(j_2) + r(j_1, j_2)$?

Theorem 4. *For any directed path $\{a\}$-RGP Q, RGPHOM(Q) for $\{a, a^+\}$-RGP inputs is Ptime-reducible to* PARALLEL JOB SCHEDULING WITH RELATIVE DEADLINES. *Thus, RGPHOM(Q) is in Ptime when restricted to such inputs.*

As we will see in the next section, the second part of Theorem 4 can be generalized to all directed path $\{a, a^+\}$-RGPs using a different method.

5.3 Directed Path $\{a, a^+\}$-RGPs

Our next result is more general than Theorem 4, as we use a stronger method. It also extends a result from [18], where the statement is proved for directed tree input RGPs. Here we prove it for all kinds of inputs.

For an arc-labelled digraph D and a positive integer k, the *product digraph* D^k is the digraph with vertices $V(D)^k$ and with an arc labelled ℓ from $(x_1, \ldots, x_k)$ to $(y_1, \ldots, y_k)$ iff all pairs (x_i, y_i) with $1 \leq i \leq k$ are arcs labelled ℓ in D. A homomorphism of D^k to D is called a (k-ary) *polymorphism* of D. For a set S, a function f from S^3 to S is a *majority function* if for all x, y in S, $f(x, x, y) = f(x, y, x) = f(y, x, x) = x$.

Theorem 5. ([14]). *Let D be an arc-labelled digraph that has a ternary polymorphism that is a majority function. Then,* HOM(D) *is in Ptime.*

It is well known that the above applies to directed paths, a result that we can lift to $\{a, a^+\}$-RGPs.

Theorem 6. *Let Q be an $\{a, a^+\}$-RGP whose underlying digraph is a directed path. Then, $D(Q)$ admits a majority polymorphism and thus RGPHOM(Q) is in Ptime.*

We remark that Theorem 6 also applies to RGPs with vertex-labels (where the mapping must preserve the vertex-labels). Indeed, vertex-labels are modeled as unary relations, which trivially satisfy the properties for having a majority polymorphism. Moreover, using the same method, Theorem 6 extends to labels of the form "a^*", "a^k" or "$a^{\leq k}$" for $k \in \mathbb{N}$.

6   Conclusion

We have seen that RGPHOM, which is generally PSPACE-hard (but in NP when the target RGP is a graph database), is in EXPTIME. This favorably compares to the general complexity of RGP query containment, which is EXPSPACE-complete [7,10], and motivates the use of RGP n-homomorphisms to approximate query containment. It remains to close the gap between the PSPACE lower bound and the EXPTIME upper bound.

We have also seen that the case of $\{a, a^+\}$-RGPs (a case that is also in NP, and that corresponds to XPath and SPARQL queries), we have a complete classification of the NP-complete and Ptime cases for undirected RGPs, and all RGPs whose underlying digraph is a directed path are in Ptime. It was proved in [18] that when both the input and target is a directed tree $\{a, a^+\}$-RGP, RGPHOM is in Ptime. Is it true that (for general inputs) RGPHOM(Q) is in Ptime when Q is a directed tree $\{a, a^+\}$-RGP? [2] When all arc-labels are "a", then the only n-core RGPs whose underlying digraphs are directed trees are, in fact, the directed paths. But there are many more directed tree $\{a, a^+\}$-RGP n-cores, see Fig. 2 for a simple example.

Fig. 2. An n-core directed tree $\{a, a^+\}$-RGP. Doubled edges are labelled "a^+", the others are labelled "a".

References

1. Aho, A.V., Hopcroft, J.E., Ullman, J.D.: The Design and Analysis of Computer Algorithms. Addison-Wesley, Reading (1974)
2. Angles, R., et al.: G-CORE: a core for future graph query languages. In: SIGMOD Conference 2018. ACM (2018). https://doi.org/10.1145/3183713.3190654
3. Baeza, P.B.: Querying graph databases. In: PODS 2013 (2013). https://doi.org/10.1145/2463664.2465216

---

[2] This is not true for all acyclic RGPs: there are trees T such that HOM(T) is NP-hard [6]. Thus, for the corresponding $\{a\}$-RGP $Q(T)$, RGPHOM($Q(T)$) is NP-hard.

4. Barceló, P., Romero, M., Vardi, M.Y.: Semantic acyclicity on graph databases. SIAM J. Comput. **45**(4), 1339–1376 (2016). https://doi.org/10.1137/15M1034714
5. Bulatov, A.A.: A dichotomy theorem for nonuniform CSPs. In: FOCS 2017 (2017). https://doi.org/10.1109/FOCS.2017.37
6. Bulin, J.: On the complexity of H-coloring for special oriented trees. Eur. J. Comb. **69**, 54–75 (2018). https://doi.org/10.1016/j.ejc.2017.10.001
7. Calvanese, D., De Giacomo, G., Lenzerini, M., Vardi, M.Y.: Containment of conjunctive regular path queries with inverse. In: KR 2000 (2000)
8. Czerwinski, W., Martens, W., Niewerth, M., Parys, P.: Optimizing tree patterns for querying graph- and tree-structured data. SIGMOD Rec. **46**(1), 15–22 (2017). https://doi.org/10.1145/3093754.3093759
9. Feder, T., Vardi, M.Y.: The computational structure of monotone monadic SNP and constraint satisfaction: a study through datalog and group theory. SIAM J. Comput. **28**(1), 57–104 (1998). https://doi.org/10.1137/S0097539794266766
10. Florescu, D., Levy, A.Y., Suciu, D.: Query containment for conjunctive queries with regular expressions. In: PODS 1998 (1998). https://doi.org/10.1145/275487.275503
11. Hell, P., Nesetril, J.: On the complexity of H-coloring. J. Comb. Theory Ser. B **48**(1), 92–110 (1990). https://doi.org/10.1016/0095-8956(90)90132-J
12. Hell, P., Nesetril, J.: The core of a graph. Discrete Math. **109**(1–3), 117–126 (1992). https://doi.org/10.1016/0012-365X(92)90282-K
13. Hunt, H.B., Rosenkrantz, D.J., Szymanski, T.G.: On the equivalence, containment, and covering problems for the regular and context-free languages. J. Comput. Syst. Sci. **12**, 222–268 (1976)
14. Jeavons, P., Cohen, D.A., Cooper, M.C.: Constraints, consistency and closure. Artif. Intell. **101**(1–2), 251–265 (1998). https://doi.org/10.1016/S0004-3702(98)00022-8
15. Kimelfeld, B., Sagiv, Y.: Revisiting redundancy and minimization in an XPath fragment. In: EDBT 2008 (2008). https://doi.org/10.1145/1353343.1353355
16. Kolaitis, P.G., Vardi, M.Y.: Conjunctive-query containment and constraint satisfaction. J. Comput. Syst. Sci. **61**(2), 302–332 (2000). https://doi.org/10.1006/jcss.2000.1713
17. Mendelzon, A.O., Wood, P.T.: Finding regular simple paths in graph databases. SIAM J. Comput. **24**(6), 1235–1258 (1995). https://doi.org/10.1137/S009753979122370X
18. Miklau, G., Suciu, D.: Containment and equivalence for a fragment of XPath. J. ACM **51**(1), 2–45 (2004). https://doi.org/10.1145/962446.962448
19. Romero, M., Barceló, P., Vardi, M.Y.: The homomorphism problem for regular graph patterns. In: LICS 2017 (2017). https://doi.org/10.1109/LICS.2017.8005106
20. Sedgewick, R., Wayne, K.: Algorithms, 4th edn. Addison-Wesley, Reading (2011)
21. Stockmeyer, L.J., Meyer, A.R.: Word problems requiring exponential time: preliminary report. In: STOC 1973 (1973). https://doi.org/10.1145/800125.804029
22. Zhuk, D.: A proof of CSP dichotomy conjecture. In: FOCS 2017 (2017). https://doi.org/10.1109/FOCS.2017.38

Towards Uniform Online
Spherical Tessellations

Paul C. Bell[1]([✉]) and Igor Potapov[2]

[1] Department of Computer Science, Liverpool John Moores University,
James Parsons Building, Byrom Street, Liverpool L3-3AF, UK
p.c.bell@ljmu.ac.uk
[2] Department of Computer Science, University of Liverpool, Ashton Building,
Ashton Street, Liverpool L69-3BX, UK
potapov@liverpool.ac.uk

Abstract. The problem of uniformly placing N points onto a sphere finds applications in many areas. An online version of this problem was recently studied with respect to the *gap ratio* as a measure of uniformity. The proposed online algorithm of Chen et al. was upper-bounded by 5.99 and then improved to 3.69, which is achieved by considering a circumscribed dodecahedron followed by a recursive decomposition of each face. We analyse a simple tessellation technique based on the regular icosahedron, which decreases the upper-bound for the online version of this problem to around 2.84. Moreover, we show that the lower bound for the gap ratio of placing up to three points is $\frac{1+\sqrt{5}}{2} \approx 1.618$. The uniform distribution of points on a sphere also corresponds to uniform distribution of unit quaternions which represent rotations in 3D space and has numerous applications in many areas.

Keywords: Online algorithms · Discrepancy theory ·
Spherical trigonometry · Uniform point placement

1 Introduction

One of the central problems of classical discrepancy theory is to maximize the uniformity of distributing a set of n points into some metric space [5,11]. For example, this includes questions about arranging points over a unit cube in a d-dimensional space, a polyhedral region, a sphere, a torus or even over a hyperbolic plane, etc. Applications of modern day discrepancy theory include those in number theory (Ramsey theory), problems in numerical integration, financial calculations, computer graphics and computational physics [11].

In order to measure the discrepancy from uniformity, the *gap ratio* metric introduced by Teramoto et al. [16] for analysing dynamic discrepancy has been widely accepted, i.e. the ratio between the maximum and minimal gap, where the maximum gap is the diameter of the largest empty circle on the sphere and the minimal gap is just the minimum pairwise distance. One might consider defining uniformity just by measuring the closest two points, however this does

I. Potapov has been partially supported by EPSRC grant EP/R018472/1.

F. Manea et al. (Eds.): CiE 2019, LNCS 11558, pp. 120–131, 2019.
https://doi.org/10.1007/978-3-030-22996-2_11

not take into account large undesirable gaps that may be present in the point set. Alternatively one may use the standard measure from discrepancy theory; define some fixed geometric shape R and count the number of inserted points that are contained in R, whilst moving it all over some space. This measure has two main disadvantages – that of computational hardness of calculating the discrepancy at each stage and also that we must decide upon a given shape R, each of which may give different results [16].

The more challenging problem of generating a point set on 2-sphere which minimizes criteria such as energy functions, discrepancy, dispersion and mutual distances has been extensively studied in the offline setting [7,10,13–15,19,20]. Some motivations and applications of this problem when restricted to the 2-sphere stretch from the classical *Thompson problem* of determining a configuration of N electrons on the surface of a unit sphere that minimizes the electrostatic potential energy [12,17], to search and rescue/exploration problems as well as problems related to extremal energy, crystallography and computational chemistry [13]. In the original offline version of the problem of distributing points over some space, the number of points is predetermined and the goal is to distribute all points as uniformly as possible at the end of the process.

In contrast, the *online* problem requires that the points should be dynamically inserted one at a time on the surface of a sphere without knowing the set of points in advance and the objective is to distribute the points as uniformly as possible at every instance of inserting a point. Note that in this case once a point has been placed it cannot be later moved. The points on the sphere correspond to unit quaternions and the group of rotations SO(3) [3,8], which have a large number of engineering applications in the online setting (also known as "incremental generation") and plays a crucial role in applications ranging from robotics and aeronautics to computer graphics [20].

The online variant of distributing points in a given space has been already studied, e.g. for inserting integral points on a line [1] or on a grid [21], inserting real points over a unit cube [16] and also recently as a more complex version of inserting real points on a surface of a sphere [6,20,22]. A good strategy for online distribution of points on the plane has been found in [2,16] based on the Voronoi insertion, where the gap ratio is proved to be at most 2. For insertion on a two-dimensional grid, algorithms with a maximal gap ratio $2\sqrt{2} \approx 2.828$ were shown in [21]. The same authors showed that the lower bound for the maximal gap ratio is 2.5 in this context. The other important direction was to solve the problem in a one-dimensional line and an insertion strategy with a uniformity of 2 has been found in [1]. An approach of using *generalised spiral points* was discussed in [12,13], which performs well for minimizing extremal energy, but this approach is strictly offline (number of points N known in advance).

Recently the solution for *online distribution of points on 2-sphere* has been proposed in [6] where a *two phase point insertion algorithm* with an overall upper bound of 5.99 was designed. The first phase uses a circumscribed dodecahedron to place the first twenty vertices, achieving a maximal gap ratio 2.618. After that, each of the twelve pentagonal faces can be recursively divided. This procedure is efficient and leads to a gap ratio of no more than 5.99. With more complex analysis, the bound was recently decreased to at most 3.69 in [22].

Fig. 1. Delauney triangulation applied twice   **Fig. 2.** Recursive triangle dissection (twice)   **Fig. 3.** Deformation of triangles in projection

One may consider whether such two phase algorithms may perform well for this problem by either modifying the initial shape used in the first phase of the algorithm (such as using initial points derived from other Platonic solids) or else whether the recursive procedure used in phase two to tessellate each regular shape may be improved. We may readily identify an advantage to choosing a Platonic solid for which each face is a triangle (the tetrahedron, octahedron and icosahedron), since in this case at least two procedures for tessellating each triangle immediately spring to mind – namely to recursively place a new point at the centre of *each edge*, denoted *triangular dissection* (creating four subtriangles, see Fig. 2), or else the Delaunay tessellation, which is to place a new point at the *centre of each triangle* (creating three new triangles, see Fig. 1).

It can be readily seen that the Delaunay tessellation of each spherical triangle rapidly gives a poor gap ratio, since points start to become dense around the centre of edges of the initial tessellation. The second recursive tessellation strategy (Fig. 2), was conjectured to give a poor ratio in [6]. This intuition seems reasonable, since as we recursively decompose each spherical triangle by this strategy the gap ratio increases as for such a triangle this decomposition deforms with each recursive step as can be seen in Fig. 3. It can also be seen that the gap ratio at each level of the triangular dissection is increasing (see Lemma 5). Nevertheless, we show in this paper that as long as the initial tessellation (stage 1) does not create too 'large' spherical triangles (with high curvature), then the gap ratio of stage 2 has an upper limit, and performs much better than the tessellation of the regular dodecahedron proposed in [6,22].

In this paper, we provide a new algorithm and utilise an circumscribed regular icosahedron and the recursive triangular dissection procedure to reduce the bound of 3.69 derived in [22] to $\frac{\pi}{\arccos\left(1/\sqrt{5}\right)} \approx 2.8376$. Apart from a better upper bound, an advantage of our triangular tessellation procedure is its generalisability and more efficient tessellation as we only need to compute the spherical median between two locally introduced points at every step.

Another natural point insertion algorithm to consider is a greedy algorithm, where points are iteratively added to the centre of the largest empty circle. However, the decomposition of a 2-sphere according to the greedy approach leads to complex non-regular local structures and it soon becomes intractable to determine the next point to place [6,22] (such points can in general even be difficult to describe in a computationally efficient way).

Full details of missing proofs are available in the full version of this paper [4].

2 Notation

2.1 Spherical Trigonometry

Given a set P, we denote by 2^P the power set of P (the set of all subsets). Let $\mathcal{S}$ denote the 3-dimensional unit sphere. We will deal almost exclusively with unit spheres, since for our purposes the gap ratio (introduced formally later) is not affected by the spherical radius. Let $u_1, u_2, u_3 \in \mathbb{R}^3$ be three unit length vectors, then $T = \langle u_1, u_2, u_3 \rangle$ denotes the spherical triangle on $\mathcal{S}$ with vertices u_1, u_2 and u_3. Given some set of points $\{u_1, u_2, u_3\} \cup \{v_j | 1 \le j \le k\}$, a spherical triangle $T = \langle u_1, u_2, u_3 \rangle$ is called *minimal* over that set of points if no v_j for $1 \le j \le k$ lies on the interior or boundary of T. As an example, in Fig. 5, triangle $\langle u_1, u_{113}, u_{112} \rangle$ is minimal, but $\langle u_1, u_{13}, u_{12} \rangle$ is not, since points u_{113} and u_{112} lie on the boundary of that triangle.

The edges of a spherical triangle are arcs of great circles. A great circle is the intersection of $\mathcal{S}$ with a central plane, i.e one which goes through the centre of $\mathcal{S}$. We denote the length of a path connecting two points u_1, u_2 on the unit sphere by $\zeta(u_1, u_2)$ (the spherical length).

Given a non-degenerate spherical triangle (i.e. one with positive area, defined later) with two edges e_1 and e_2 which intersect at a point P, then we say that the angle of P is the angle of P measured when projected to the plane tangent at P. We constrain all spherical triangles to have edge lengths strictly between 0 and π, which avoids issues with *antipodal triangles*. Two points on the unit sphere are called antipodal if the angle between them is π (i.e. they lie on opposite sides of the unit sphere) and an antipodal triangle contains two antipodal points. Several results in spherical trigonometry (and in this paper) are derived by projections of points/edges to planes tangent to a point on the sphere; in all such cases the projection is from the centre of the sphere.

The following results are all standard from spherical trigonometry, see [18] for proofs and further details. The length of an arc belonging to a great circle corresponds with the angle of the arc, see Fig. 4. Furthermore, given an arc between two points u_1 and u_2 on $\mathcal{S}$, the length of the line connecting u_1 and the projection of u_2 to the plane tangent to u_1 is given by $\tan(\zeta(u_1, u_2))$, see Fig. 4.

Lemma 1 (The Spherical Laws of Sines and Cosines). *Given a spherical triangle with sides a, b, c and angles A, B, C opposite to side a, b, c resp., then:* $\cos(c) = \cos(a)\cos(b) + \sin(a)\sin(b)\cos(C)$; $\cos(C) = -\cos(A)\cos(B) + \sin(A)\sin(B)\cos(c)$ *and* $\frac{\sin(a)}{\sin(A)} = \frac{\sin(b)}{\sin(B)}$.

The sum of angles within a spherical triangle is between π (as the volume approaches zero) and 3π (as the triangle fills the whole sphere). The *spherical excess* of a triangle is the sum of its angles minus π radians.

Theorem 1 (Girard's theorem). *The area of a spherical triangle is equal to its spherical excess.*

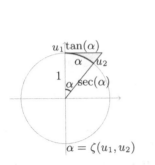

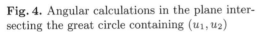

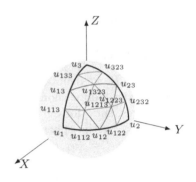

Fig. 4. Angular calculations in the plane intersecting the great circle containing (u_1, u_2)

Fig. 5. σ tessellations

2.2   Online Point Placing on the Unit Sphere

Our aim is to insert a sequence of 'uniformly distributed' points onto $\mathcal{S}$ in an online manner. After placing a point, it cannot be moved in the future. Let p_i be the i'th point thus inserted and let $S_i = \{p_1, p_2, \ldots, p_i\}$ be the configuration after inserting the i'th point. Teramoto et al. introduced the gap ratio [16], which defines a measure of uniformity for point samples and we use this metric (similarly to [6]).

Let $\rho_{\min} : 2^{\mathcal{S}} \to \mathbb{R}$ denote the minimal distance of a set of points, defined by $\rho_{\min}(S_i) = \min_{p,q \in S_i, p \neq q} \zeta(p, q)$. Recall that notation $2^{\mathcal{S}}$ means the set of all points lying on the 2-sphere $\mathcal{S}$. Let $\rho_{\max}^{\mathcal{S}'} : 2^{\mathcal{S}} \to \mathbb{R}$ denote the maximal spherical diameter of the largest empty circle centered at some point of $\mathcal{S}' \subseteq \mathcal{S}$ avoiding a given set of points, defined by $\rho_{\max}^{\mathcal{S}'}(S_i) = \max_{p \in \mathcal{S}'} \min_{q \in S_i} 2 \cdot \zeta(p, q)$. We then define $\rho^{\mathcal{S}'}(S_i) = \frac{\rho_{\max}^{\mathcal{S}'}(S_i)}{\rho_{\min}(S_i)}$ to be the *gap ratio* of S_i over $\mathcal{S}'$. When $\mathcal{S}' = \mathcal{S}$ (i.e. when points can be placed anywhere on the sphere), we define that $\rho(S_i) = \rho^{\mathcal{S}}(S_i)$.

We denote an *equilateral* spherical triangle as one for which each side has the same length. By Lemma 1 (the spherical law of cosines), having three equal length edges implies that an equilateral spherical triangle has the same three angles. By Theorem 1 (Girard's theorem), each such angle is greater than $\frac{\pi}{3}$ (for an equilateral triangle of positive volume). Let $\Delta \subseteq \mathcal{S}$ denote the set of all spherical triangles on the unit sphere.

Consider a spherical triangle $T \in \Delta$. We define a *triangular dissection* function $\sigma : \Delta \to 2^{\Delta}$ in the following way. If $T \in \Delta$ is defined by $T = \langle u_1, u_2, u_3 \rangle$, then $\sigma(T) = \{T_1, T_2, T_3, T_4\} \subset \Delta$, where $T_1 = \langle u_1, u_{12}, u_{13} \rangle$, $T_2 = \langle u_{12}, u_2, u_{23} \rangle$, $T_3 = \langle u_3, u_{13}, u_{23} \rangle$ and $T_4 = \langle u_{12}, u_{13}, u_{23} \rangle$, with u_{ij} being the midpoints (on the unit sphere) of the arc connecting u_i and u_j (see Fig. 5). Define $\sigma_E(T)$ as the set of nine induced edges: $\{(u_1, u_{12}), (u_{12}, u_2), (u_2, u_{23}), (u_{23}, u_3), (u_3, u_{13}),$ $(u_{13}, u_1), (u_{13}, u_{23}), (u_{23}, u_{12}), (u_{12}, u_{13})\}$.

We extend the domain of σ to sets of spherical triangles: $\sigma(\{T_1, T_2, \ldots, T_k\}) = \{\sigma(T_1), \sigma(T_2), \ldots, \sigma(T_k)\}$; thus $\sigma : 2^{\Delta} \to 2^{\Delta}$. Given a spherical triangle $T \in \Delta$, we then define that $\sigma^1(T) = \sigma(T)$ and $\sigma^k(T) = \sigma(\sigma^{k-1}(T))$ for $k > 1$.

For notational convenience, we also define that $\sigma^0(T) = T$ (the identity tessellation). We similarly extend $\sigma_E(T)$ to a set of triangles: $\sigma_E(\{T_1, T_2, \ldots, T_k\}) = \{\sigma_E(T_1), \sigma_E(T_2), \ldots, \sigma_E(T_k)\}$ and let $\sigma_E^k(T) = \sigma_E(\sigma^k(T_1), \sigma^k(T_2), \ldots, \sigma^k(T_k))$. See Fig. 5 for an example showing the tessellation of T to depth 2 (e.g. $\sigma^2(T)$) and the set of edges $\sigma_E^2(T)$.

Let $\mu : \Delta \to 2^S$ be a function which, for an input spherical triangle, returns the (unique) set of three points defining that triangle. For example, given a spherical triangle $T = \langle p_1, p_2, p_3 \rangle$, then $\mu(T) = \{p_1, p_2, p_3\}$. Clearly μ may be extended to sets of triangles by defining that $\mu(\{T_1, \ldots, T_k\}) = \{\mu(T_1), \ldots, \mu(T_k)\}$; thus $\mu : 2^\Delta \to 2^S$. When there is no danger of confusion, by abuse of notation, we sometimes write T rather than $\mu(T)$. This allows us to write $\rho(T)$ (or $\rho(\sigma^k(T))$) for example, as the gap ratio of the three points defining spherical triangle T (resp. the set of points in the k-fold triangular dissection $\sigma^k(T)$).

We will also require an ordering on the set of points generated by a tessellation $\sigma^k(T)$. Essentially, we wish to order the points as those of $\sigma^0(T) = T$ first (in any order), then those of $\sigma^1(T)$ in any order *but omitting the points of $\sigma^0(T) = T$*, then the points of $\sigma^2(T)$, omitting points in triangles of $\sigma^0(T)$ or $\sigma^1(T)$ etc. To capture this notion, we introduce a function $\tau : \Delta \times \mathbb{Z}^+ \to 2^S$ defined thus:

$$\tau(T, k) = \begin{cases} \mu(\sigma^k(T)) - \mu(\sigma^{k-1}(T)) \; ; \; \text{if } k \geq 1 \\ \mu(T) \qquad\qquad\qquad\quad ; \; \text{if } k = 0 \end{cases}$$

As an example, in Fig. 5, $\tau(T, 0) = \{u_1, u_2, u_3\}$, $\tau(T, 1) = \{u_{12}, u_{13}, u_{23}\}$, and $\tau(T, 2) = \{u_{112}, u_{122}, u_{232}, u_{323}, u_{133}, u_{113}, u_{1323}, u_{1213}, u_{1223}\}$. By abuse of notation, we redefine $\sigma^k(T)$ such that $\sigma^k(T) = \tau(T, 0) \cup \tau(T, 1) \cup \cdots \cup \tau(T, k)$ is an ordered set.

3   Overview of Online Vertex Insertion Algorithm

Our algorithm is a two stage strategy. In stage one, we project the 12 vertices of the regular icosahedron onto the unit sphere. The first two points inserted should be opposite each other (antipodal points), but the remaining 10 points can be inserted in any order, giving a stage one gap ratio of $\dfrac{\pi}{\arccos\left(\frac{1}{\sqrt{5}}\right)} \approx 2.8376$.

In the second stage, we treat each of the 20 equilateral spherical triangles of the regular icosahedron in isolation. We show in Lemma 5 that the gap ratio for our tessellation is 'local' and depends only on the local configuration of vertices around a given point. This allows us to consider each triangle separately. During stage two, we use the fact that these twenty spherical triangles are equilateral and apply Lemma 7 to independently tesselate each triangle recursively in order to derive an upper bound of the gap ratio in stage two of $\dfrac{2(3 - \sqrt{5})}{\arcsin\left(\frac{1}{2}\sqrt{2 - \frac{2}{\sqrt{5}}}\right)} \approx 2.760$.

We note here that the radius of the sphere does not affect the gap ratio of the point insertion problem, and thus we assume a unit sphere throughout.

The algorithmic procedure to generate an infinite set of points is shown in Algorithm 1. To generate a set of k points $\{p_1, p_2, \ldots, p_k\}$, we choose the first k points generated by the algorithm.

Stage one: Project 12 vertices of the icosahedron to the unit sphere:
   Place two antipodal points on the unit sphere.
   Place the remaining ten points in any order.
   Arbitrarily label the 20 minimal spherical triangles $T = \{T_1, \ldots, T_{20}\}$.
Stage two: Recursively tessellate minimal triangles
      Let $T' \leftarrow T$
      while TRUE **do**
         for all minimal spherical triangles $R \in T$ **do**
               Let $T' \leftarrow (T' \cup \sigma(R)) - R$
         end for
         Let $T \leftarrow T'$
      end while

Algorithm 1. Placing infinitely many points on the unit sphere using our recursive tessellation procedure on the regular icosahedron.

4   Gap Ratio of Equilateral Spherical Triangles

We will require several lemmata regarding tessellations of spherical triangles. The following lemma is trivial from the spherical sine rule and Girard's theorem.

Lemma 2. *Let $T \in \Delta$ be an equilateral triangle. Then the central triangle in the tessellation $\sigma(T)$ is also equilateral.*

It is worth again noting in Lemma 2 that the other three triangles in the triangular dissection of an equilateral triangle are *not* equilateral, and have a strictly smaller area than the central triangle. This *deformation* of the recursive triangular dissection makes the analysis of the algorithm nontrivial. The following lemma equates the distance from the centroid of an equilateral spherical triangle to a vertex of that triangle.

Lemma 3. *Let $T = \langle u_1, u_2, u_3 \rangle \in \Delta$ be an equilateral triangle with centroid u_c and edge length $\zeta(u_1, u_2) = \alpha$. Then $\zeta(u_1, u_c) = \zeta(u_2, u_c) = \zeta(u_3, u_c) = \arcsin\left(2\sin(\frac{\alpha}{2})/\sqrt{3}\right)$.*

Given an equilateral spherical triangle T, we will also need to determine the maximal and minimal edge lengths in $\sigma_E^k(T)$ for $k \geq 1$, which we now show.

Lemma 4. *Let $T = \langle u_1, u_2, u_3 \rangle \in \Delta$ be an equilateral triangle such that $\alpha = \zeta(u_1, u_2) \in (0, \frac{\pi}{2}]$ and $k \geq 1$. Then the minimal length edge in $\sigma_E^k(T)$ is given by any edge lying on the boundary of T. The maximal length edge of $\sigma_E^k(T)$ is any of the edges of the central equilateral triangle of $\sigma^k(T)$.*

Proof. Consider Fig. 4. The lemma states that in $\sigma^2(T)$ shown, the *shortest* length edge of $\sigma_E^2(T)$ is (u_1, u_{113}), or indeed any such edge on the boundary of triangle $\langle u_1, u_2, u_3 \rangle$. The lemma similarly states that the *longest* edge of $\sigma_E^2(T)$ is edge (u_{1213}, u_{1223}), or indeed any edge of the central equilateral triangle $\langle u_{1213}, u_{1223}, u_{1323} \rangle$.

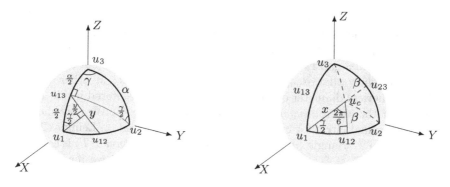

Fig. 6. Max and min lengths of σ tessellations of an equilateral triangle.

Fig. 7. Centroid calculations

Consider now Fig. 6 illustrating $T = \langle u_1, u_2, u_3 \rangle$. Point u_{12} (resp. u_{13}) is at the midpoint of spherical edge (u_1, u_2) (resp. (u_1, u_3)). Let $\alpha = \zeta(u_1, u_2) = \zeta(u_2, u_3) = \zeta(u_1, u_3)$ be the edge length. The intersection of spherical edges (u_2, u_{13}) and (u_1, u_3) forms a spherical right angle. We denote $y = \zeta(u_{13}, u_{12})$, thus y is the edge length of the central equilateral triangle of $\sigma(T)$ (and $\frac{\alpha}{2} = \zeta(u_1, u_{13})$ is the edge length of the minimal length edge of one of the non-central triangles in $\sigma(T)$; note that this is the same for each such triangle).

By the spherical sine rule, $\sin\left(\frac{\gamma}{2}\right) = \frac{\sin\frac{\alpha}{2}}{\sin\alpha}$, which is illustrated by triangle $\langle u_2, u_{13}, u_3 \rangle$. Thus $\sin(\frac{\gamma}{2}) = \frac{1}{2}\sec\left(\frac{\alpha}{2}\right)$. Here we used the identity that $\frac{\sin\left(\frac{\alpha}{2}\right)}{\sin\alpha} = \frac{1}{2}\sec\left(\frac{\alpha}{2}\right)$. Further, one can see by the spherical sine rule that $\sin\left(\frac{y}{2}\right) = \sin\left(\frac{\alpha}{2}\right) \cdot \sin\left(\frac{\gamma}{2}\right) = \frac{1}{2}\frac{\sin\left(\frac{\alpha}{2}\right)}{\cos\left(\frac{\alpha}{2}\right)} = \frac{1}{2}\tan\left(\frac{\alpha}{2}\right)$. This implies $\zeta(u_{13}, u_{12}) = y = 2\arcsin\left(\frac{1}{2}\tan\frac{\alpha}{2}\right)$ which is larger than $\frac{\alpha}{2}$ for $\alpha \in (0, \frac{\pi}{2}]$. To prove this, let $f(\alpha) = 2\arcsin(\frac{1}{2}\tan\frac{\alpha}{2})$ then $\frac{df}{d\alpha} = \frac{2}{\cos^2\left(\frac{\alpha}{2}\right)\sqrt{4-\tan^2\left(\frac{\alpha}{2}\right)}}$ as is not difficult to prove. Noting that if $\alpha \in (0, \frac{\pi}{2}]$, then $\cos^2\left(\frac{\alpha}{2}\right) \in [\frac{1}{2}, 1]$ and $\sqrt{4-\tan^2\left(\frac{\alpha}{2}\right)} \in [\sqrt{3}, 2]$, then $\frac{df}{d\alpha} > \frac{1}{2} = \frac{d\frac{\alpha}{2}}{d\alpha}$ and thus since $f(\alpha) = 0 = \frac{\alpha}{2}$ when $\alpha = 0$, then $y > \frac{\alpha}{2}$ for $\alpha \in (0, \frac{\pi}{2}]$.

For any depth k-tessellation $\sigma^k(T)$, the maximal edge length of $\sigma^k_E(T)$ will thus be given by the length of the edges of the central equilateral triangle and the minimal length edges will be located on the boundary of T as required. $\square$

Given an equilateral spherical triangle $T = \langle u_1, u_2, u_3 \rangle$, we now consider the gap ratio implied by the restriction of points to those of T. The first part of this lemma shows that the gap ratio of a depth-k tessellation is lower than the gap ratio of a depth-$k+1$ tessellation (when restricted to points of T), and the second part shows that in the limit, the upper bound converges.

Lemma 5. *Let $T = \langle u_1, u_2, u_3 \rangle$ be an equilateral spherical triangle with spherical edge length α, then:*

(i) $\rho^T(\mu(\sigma^k(T))) < \rho^T(\mu(\sigma^{k+1}(T)))$;
(ii) $\lim_{k\to\infty} \rho^T(\mu(\sigma^k(T))) = \dfrac{4\sin\left(\frac{\alpha}{2}\right)}{\alpha\sqrt{3-4\sin^2\left(\frac{\alpha}{2}\right)}}$.

Proof Sketch. Consider Fig. 7 and let $\alpha = \zeta(u_1, u_2) = \zeta(u_2, u_3) = \zeta(u_1, u_3)$ be the edge length of the equilateral triangle $T = \langle u_1, u_2, u_3 \rangle$. We begin by calculating $\rho^T(\sigma^0(T)) = \rho^T(T)$[1]. Recall that $\rho^T(T)$ denotes the gap ratio of point set $\mu(T)$ when the maximal gap ratio calculation is restricted to points of T.

We may show that $\rho_{\min}(T) = \alpha$ and $\rho^T_{\max}(T) = 2x$; in other words the maximal spherical diameter of the largest empty circle centered inside T should be placed at the centroid u_c of T. This implies $\rho^T(T) = \frac{2x}{\alpha}$. We may then prove that the maximal circle for $\sigma^k(T)$ is centered at u_c, based on Lemmas 2 and 4.

By Lemma 2, triangle $\langle u_{12}, u_{23}, u_{13} \rangle$ in the decomposition $\sigma(T)$ is equilateral. It is clear that $y > \frac{\alpha}{2}$ in Fig. 6 (proven in Lemma 4) and we may prove that $\beta < x$ in Fig. 7 by the spherical sine rule, since $\gamma > \pi/3$ (by Girard's theorem).

Therefore, $\rho_{\min}(\sigma^1(T)) = \frac{\alpha}{2}$, $\rho^T_{\max}(\sigma^1(T)) = 2\beta$ and thus $\rho^T(\sigma^1(T)) = \frac{4\beta}{\alpha}$. Since the depth 0 tessellation $\sigma^0(T)$ has a gap ratio of $\frac{2x}{\alpha}$ and the depth 1 tessellation has a gap ratio of $\frac{4\beta}{\alpha}$, we may then show that $\frac{2x}{\alpha} < \frac{4\beta}{\alpha}$ which proves the first result of the lemma by induction. The proof of this fact uses a projection of T to the tangent plane of the unit sphere centered at the triangle's centroid u_c and then standard geometry on the projected triangle.

The second statement of the lemma is proven by infinitesimal calculus on this projected triangle, see [4] for details. □

5 Regular Icosahedral Tessellation

As explained in Sect. 3 and Algorithm 1, our algorithm consists of two stages. Using the lemmata of the previous section, we are now ready to show that the stage one gap ratio is no more than $\frac{\pi}{\arccos\left(\frac{1}{\sqrt{5}}\right)} \approx 2.8376$ and the second stage gap ratio is no more than $\frac{2(3-\sqrt{5})}{\arcsin\left(\frac{1}{2}\sqrt{2-\frac{2}{\sqrt{5}}}\right)} \approx 2.760$.

Lemma 6. *The gap ratio of stage one is no more than* $\frac{\pi}{\arccos\left(\frac{1}{\sqrt{5}}\right)} \approx 2.8376$.

Proof. The points of a regular icosahedron can be defined by taking circular permutations of $(0, \pm 1, \pm \phi)$, where $\phi = \frac{1+\sqrt{5}}{2}$ is the golden ratio. Let V' be the set of the twelve such vertices. Normalising each element of V' gives a set V. Note that the area of each spherical triangle is given by $\frac{\pi}{5}$ since we have a unit sphere and twenty identical spherical triangles forming a tessellation. By Girard's theorem (Theorem 1), this implies that $3\gamma - \pi = \frac{\pi}{5}$, where γ is the interior angle of the equilateral triangle and thus $\gamma = \frac{2\pi}{5}$. By the second spherical law of cosines (Lemma 1), this implies that the spherical distance between adjacent vertices, α, is thus given by $\cos(\alpha) = \frac{\cos(\frac{2\pi}{5}) + \cos^2(\frac{2\pi}{5})}{\sin^2(\frac{2\pi}{5})} = \frac{\frac{1}{4}(\sqrt{5}-1) + \frac{3}{8} + (\frac{\sqrt{5}}{8})}{\frac{5}{8} + \frac{\sqrt{5}}{8}} = \frac{\frac{1}{8}(1+\sqrt{5})}{\frac{1}{8}(5+\sqrt{5})} = \frac{1}{\sqrt{5}}$ and therefore $\alpha \approx 1.1071$. The first two points are placed opposite to other, for example $u_2 \approx (0, -0.5257, 0.8507)$ and $u_3 \approx (0, 0.5257, -0.8507)$. At this stage, the gap ratio is 1, since the largest circle may be placed on the equator

---

[1] Note by abuse of notation that we write $\rho^T(T)$ rather than the more formal $\rho^T(\mu(T))$, as explained previously.

(with u_2 and u_3 at the poles) with a diameter of π, whereas the spherical distance between u_2 and u_3 can be calculated as π. The remaining ten vertices of the normalised regular icosahedron are placed in any order. The minimal distance between them is given by α above, and thus the gap ratio during stage one is no more than $\dfrac{\pi}{\arccos\left(\frac{1}{\sqrt{5}}\right)} \approx 2.8376$, as required. $\qquad\square$

As explained in Sect. 3 and Algorithm 1, we start with the twenty equilateral spherical triangles produced in stage one, denoted the depth-0 tessellation of $\mathcal{S}$. We apply σ to each such triangle to generate $20 * 4 = 80$ smaller triangles (note that not all such triangles are equilateral, in fact only eight triangles at each depth tessellation are equilateral). At this stage we have the depth-1 tessellation of $\mathcal{S}$. We recursively apply σ to each spherical triangle at depth-k to generate the depth-$k + 1$ tessellation, which contains $20 * 4^{k+1}$ spherical triangles.

Lemma 7. *The gap ratio of stage two is no more than* $\dfrac{12-4\sqrt{5}}{\arccos\left(\frac{1}{\sqrt{5}}\right)} \approx 2.760$.

Proof. At the start of stage two, we have 20 equilateral spherical triangles, $T_1, \ldots, T_{20}$ which are identical (up to rotation). Note that moving from depth-k tessellation to depth-$k + 1$, each edge of $\sigma_E^k(T_i)$ will be split at its midpoint, therefore applying σ to any spherical triangle will only 'locally' change the gap ratio of at most two adjacent triangles. Thus the order in which σ is applied to each triangle at depth k is irrelevant.

Assume that we have a (complete) depth-k tessellation with $20 * 4^k$ triangles. Lemma 5 tell us that the gap ratio increases from the depth-k to the depth-$k + 1$ tessellations for all $k \geq 0$ and in the limit, the gap ratio of the depth-k tessellation of each T_i is given by:

$$\lim_{k \to \infty} \rho^T(\sigma^k(T_i)) = \frac{4\sin\left(\frac{\alpha}{2}\right)}{\alpha\sqrt{3 - 4\sin^2\left(\frac{\alpha}{2}\right)}} \tag{1}$$

where α is the length of the edges of T_i (i.e. the length of those triangles produced by stage 1 via the icosahedron). When we start to tessellate the depth-k spherical triangles, until we have a complete depth-$k + 1$ tessellation, applying σ to each of the triangles may decrease the minimal gap ratio at most by a factor up to 2 overall (since we split each edge at its midpoint). The maximal gap ratio cannot increase, but decreases upon completing the depth $k + 1$ tessellation. Therefore, we multiply Eq. (1) by 2 to obtain an upper bound of the gap ratio for the entire sequence, not only when some depth-k tessellation is complete. We now solve Eq. (1), after multiplying by 2, by substituting $\alpha = \arccos\left(\frac{1}{\sqrt{5}}\right) \approx 1.1071$. This is laborious, but by noting that $\sin(\frac{\alpha}{2}) = \sqrt{\frac{1}{10}(5 - \sqrt{5})}$ and $\sin^2(\frac{\alpha}{2}) = \frac{1}{10}(5-\sqrt{5})$,

then: $\dfrac{8\sin\left(\frac{\alpha}{2}\right)}{\alpha\sqrt{3-4\sin^2\left(\frac{\alpha}{2}\right)}} = \dfrac{8\sqrt{\frac{1}{10}(5-\sqrt{5})}}{\arccos\left(\frac{1}{\sqrt{5}}\right)\sqrt{1+\frac{2}{\sqrt{5}}}} = \dfrac{12-4\sqrt{5}}{\arccos\left(\frac{1}{\sqrt{5}}\right)} \approx 2.760$. Therefore, the gap ratio during stage two is upper bounded by 2.76. $\qquad\square$

Theorem 2. *The gap ratio of the icosahedral triangular dissection is equal to* $\dfrac{\pi}{\arccos\left(\frac{1}{\sqrt{5}}\right)} \approx 2.8376$.

Table 1. The gap ratio of stage one and two of various regular Platonic solids. Italic elements show which value defines the overall gap ratio in each case.

	Tetrahedron	Octahedron	Dodecahedron	Icosahedron
Stage 1	2.289	2.0	2.618	*2.8376*
Stage 2	*5.921*	*3.601*	*5.995*	2.760

We can now prove the first nontrivial lower bound when we have only 2 or 3 points on the sphere in this online version of the problem.

Theorem 3. *The gap ratio for the problem of placing points on the sphere cannot be less than $\frac{1+\sqrt{5}}{2} \approx 1.6180$.*

Proof. Let us first estimate the ratio with only two points when one point is located at the north pole of the 2-sphere and the other is shifted by a distance x from the south pole. The gap ratio in this case is $\frac{\pi+x}{\pi-x}$, which increases from 1 to ∞ when $x \geq 0$.

Let us now consider the case with three points. If we place the third point on the plane P defined by the center of the sphere and the other two points, then the gap ratio will be $\frac{\pi}{\frac{\pi+x}{2}} = \frac{2\pi}{\pi+x}$ as in this case the maximal diameter of an empty circle is π regardless of the position of the third point on P. Note that the diameter of a largest circle will be on the orthogonal plane to P and the smallest function for the gap ratio in terms of x can be defined by positioning the third point at the largest distance from initial two points which is $\frac{\pi+x}{2}$.

If the third point is not on the plane P then the ratio would be equal to some value $\frac{a}{b}$ that is larger than $\frac{2\pi}{\pi+x}$. This follows from the fact that the value a which is the maximal gap would be greater than π and the minimal gap b would be less than $\frac{\pi+x}{2}$. So the minimal gap ratio that can be achieved for the three points will be represented by the expression $\frac{2\pi}{\pi+x}$.

By solving the equation where the left hand side represents the gap ratio in the case of 3 points (decreasing function) and the right hand side representing the case with 2 points (increasing function), we find a positive value of the one unknown x: $\frac{2\pi}{\pi+x} = \frac{\pi+x}{\pi-x}$. The only positive value x satisfying the above equation has the value $\pi(\sqrt{5} - 2)$ and the gap ratio for this value x is equal to $\frac{1+\sqrt{5}}{2}$. $\square$

6   Conclusion

To determine the most appropriate stage 1 shape, we derived (theoretically and via a computer simulation) the gap ratios of the stage 1 and 2 tessellations of various Platonic solids, shown in Table 1. The results for the dodecahedron are from [6] using a different tessellation (dodecahedra have non triangular faces).

The results match our intuition, that a finer grained initial tessellation performs better in stage 2 than a coarse grained initial tessellation. It would be interesting to consider modifications of the stage 2 procedure which may allow the octahedron to be utilised, given its low stage 1 gap ratio.

References

1. Asano, T.: Online uniformity of integer points on a line. Inf. Process. Lett. **109**(1), 57–60 (2008)
2. Asano, T., Teramoto, S.: On-line uniformity of points. In: 8th Hellenic-European Conference on Computer Mathematics and Its Applications, pp. 21–22 (2007)
3. Bell, P., Potapov, I.: Reachability problems in quaternion matrix and rotation semigroups. Inf. Comput. **206**(11), 1353–1361 (2008)
4. Bell, P.C., Potapov: I.: Towards uniform online spherical tessellations. ArXiV Preprint https://arxiv.org/abs/1810.01786 (2019)
5. Chazelle, B.: The Discrepancy Method: Randomness and Complexity. Cambridge University Press, New York (2000)
6. Chen, C., Lau, F.C.M., Poon, S.-H., Zhang, Y., Zhou, R.: Online inserting points uniformly on the sphere. In: Poon, S.-H., Rahman, M.S., Yen, H.-C. (eds.) WAL-COM 2017. LNCS, vol. 10167, pp. 243–253. Springer, Cham (2017). https://doi.org/10.1007/978-3-319-53925-6_19
7. Hardin, D.P., Saff, E.B.: Discretizing manifolds via minimum energy points. Not. Am. Math. Soc. **51**(10), 1186–1194 (2004)
8. Kirk, D.: Graphics Gems III, pp. 124–132. Academic Press Professional, Inc., San Diego (1992)
9. Lang, R.J.: Twists, Tilings, and Tessellations: Mathematical Methods for Geometric Origami. CRC Press, Boca Raton (2017)
10. Lubotsky, A., Phillips, R., Sarnak, P.: Hecke operators and distributing points on the sphere I. Commun. Pure Appl. Math. **39**, S149–S186 (1986)
11. Matoušek, J.: Geometric Discrepancy: An Illustrated Guide. AC, vol. 18. Springer, Heidelberg (1999). https://doi.org/10.1007/978-3-642-03942-3
12. Rakhmanov, E.A.: Minimal discrete energy on the sphere. Math. Res. Lett. **1**, 647–662 (1994)
13. Saff, E.B., Kuijlaars, A.B.J.: Distributing many points on a sphere. Math. Intell. **19**(1), 5–14 (1997)
14. Sloane, N.J.A., Duff, T.D.S., Hardin, R.H., Conway, J.H.: Minimal-energy clusters of hard spheres. Disc. Comput. Geom. **14**, 237–259 (1995)
15. Sun, X., Chen, Z.: Spherical basis functions and uniform distribution of points on spheres. J. Approx. Theory **151**(2), 186–207 (2008)
16. Teramoto, S., Asano, T., Katoh, N., Doerr, B.: Inserting points uniformly at every instance. IEICE (Institute Electronics, Information and Communication Engineers) Trans. Inf. Syst. **89-D**(8), 2348–2356 (2006)
17. Thompson, J.J.: On the structure of the atom: an investigation of the stability and periods of oscillation of a number of corpuscles arranged at equal intervals around the circumference of a circle; with application of the results to the theory of atomic structure. Philos. Magaz. Ser. 6, **7**(39) 237–265 (1904)
18. Todhunter, I.: Spherical Trigonometry. Macmillan and Co., London (1886)
19. Wagner, G.: On a new method for constructing good point sets on spheres. J. Disc. Comput. Geom. **9**(1), 119–129 (1993)
20. Yershova, A., Jain, S., Lavalle, S.M., Mitchell, J.C.: Generating uniform incremental grids on $SO(3)$ using the Hopf fibration. Int. J. Robot. Res. **29**(7), 801–812 (2010)
21. Zhang, Y., Chang, Z., Chin, F.Y.L., Ting, H.-F., Tsin, Y.H.: Uniformly inserting points on square grid. Inf. Process. Lett. **111**, 773–779 (2011)
22. Zhou, R., Chen, C., Sun, L., Lau, F.C.M., Poon, S.-H., Zhang, Y.: Online uniformly inserting points on the sphere. Algorithms **11**(1), 156 (2018)

Complexity of Maximum Fixed Point Problem in Boolean Networks

Florian Bridoux[1(✉)], Nicolas Durbec[1], Kevin Perrot[1], and Adrien Richard[2,3]

[1] Aix-Marseille Université, Université de Toulon, CNRS, LIS, Marseille, France
{florian.bridoux,nicolas.durbec,ken.perrot}@lis-lab.fr
[2] Laboratoire I3S, CNRS, Université Côte d'Azur, Nice, France
[3] CMM, UMI CNRS 2807, Universidad de Chile, Santiago, Chile
richard@unice.fr

Abstract. A *Boolean network* (BN) with n components is a discrete dynamical system described by the successive iterations of a function $f : \{0,1\}^n \to \{0,1\}^n$. This model finds applications in biology, where fixed points play a central role. For example in genetic regulation they correspond to cell phenotypes. In this context, experiments reveal the existence of positive or negative influences among components: component i has a positive (resp. negative) influence on component j, meaning that j tends to mimic (resp. negate) i. The digraph of influences is called *signed interaction digraph* (SID), and one SID may correspond to multiple BNs. The present work opens a new perspective on the well-established study of fixed points in BNs. Biologists discover the SID of a BN they do not know, and may ask: given that SID, can it correspond to a BN having at least k fixed points? Depending on the input, this problem is in P or complete for NP, NP$^{\#P}$ or NEXPTIME.

1 Introduction

A *Boolean network* (BN) with n components is a discrete dynamical system described by the successive iterations of a function

$$f : \{0,1\}^n \to \{0,1\}^n, \qquad x = (x_1, \ldots, x_n) \mapsto f(x) = (f_1(x), \ldots, f_n(x)).$$

The structure of the network is often described by a signed digraph G, called *signed interaction digraph* (SID) of f, catching effective positive and negative dependencies among components: the vertex set is $[n] := \{1, \ldots, n\}$ and, for all $i, j \in [n]$, there is a positive (resp. negative) arc from i to j if $f_j(x) - f_j(y)$ is positive (resp. negative) for some $x, y \in \{0,1\}^n$ that only differ in $x_i > y_i$. The SID provides a very rough information about f. Hence, given a SID G, the set $F(G)$ of BNs f whose SID is G, is generally huge.

BNs have many applications. In particular, since the seminal papers of Kauffman [14,15] and Thomas [30,31], they are very classical models for the dynamics of gene networks. In this context, the first reliable experimental information often concern the SID of the network, while the actual dynamics are very difficult to

© Springer Nature Switzerland AG 2019
F. Manea et al. (Eds.): CiE 2019, LNCS 11558, pp. 132–143, 2019.
https://doi.org/10.1007/978-3-030-22996-2_12

observe [18, 32]. One is thus faced with the following question: *What can be said about the dynamics described by f according to G only?*

Among the many dynamical properties that can be studied, fixed points are of special interest, since they correspond to stable patterns of gene expression at the basis of particular cellular phenotypes [3, 31]. As such, they are arguably the property which has been the most thoroughly studied. The number of fixed points and its maximization in particular is the subject of a stream of work, *e.g.* in [4–7, 11, 12, 24, 26].

From the complexity point of view, previous works essentially focused on decision problems of the following form: given f and a dynamical property P, what is the complexity of deciding if the dynamics described by f has the property P. For instance, it is well-known that deciding if f has a fixed point is NP-complete in general (see [17] and the references therein), and in P for some families of BNs, such as monotone or non-expansive BNs [10, 13]. However, as mentioned above, in practice, f is often unknown while its SID is well approximated. Hence, a much more natural question is: given a SID G and dynamical property P, what is the complexity of deciding if the dynamics described by some $f \in F(G)$ has the property P. Up to our knowledge, there is, perhaps surprisingly, no work concerning this kind of questions.

In this paper, we study this class of decision problems, focusing on the maximum number of fixed points. More precisely, given a SID G, we denote by $\phi(G)$ the maximum number of fixed points in a BN $f \in F(G)$, and we study the complexity of deciding if $\phi(G) \geq k$.

After the definitions in Sect. 2, we first study the problem when the positive integer k is fixed. We prove in Sect. 3 that, given a SID G, deciding if $\phi(G) \geq k$ is in P if $k = 1$. We also prove in Sect. 4 that the same problem is NP-complete if $k \geq 2$. Furthermore, these results remain true if the maximum in-degree $\Delta(G)$ is bounded by any constant $d \geq 2$. The case $k = 2$ is of particular interest since many works have been devoted to finding necessary conditions for the existence of multiple fixed points, both in the discrete and continuous settings, see [16, 24, 25, 28] and the references therein. Section 5 considers the case where k is part of the input. We prove that, given a SID G and a positive integer k, deciding if $\phi(G) \geq k$ is NEXPTIME-complete, and becomes $NP^{\#P}$-complete if $\Delta(G)$ is bounded by a constant $d \geq 2$. Note that, from these results, we immediately obtain complexity results for the dual decision problem $\phi(G) < k$. A summary is given in Table 1.

In the case where k is fixed, while proving that the problem $\phi(G) \geq k$ belongs to NP, we study a decision problem of independent interest, called *extension* or *consistency* problem [2, 8, 9]. Here, the property P consists of a partial BN, that is, a function $h : X \to \{0, 1\}^n$ where $X \subseteq \{0, 1\}^n$. This partial BN may represent some experimental observations about the dynamics. Given a SID G, we prove that we can check in $\mathcal{O}(|X|^2 n^2)$ time if there is a BN $f \in F(G)$ which is consistent with h, that is, such that $f(x) = h(x)$ for all $x \in X$. Thus, the task consists in extending h to a global BN f under the constraint that the SID of f is G.

Table 1. Complexity results.

Problem	$\Delta(G) \leq d$	$k = 1$	$k \geq 2$	k given in input
$\phi(G) \geq k$	Yes	P	NP-complete	NP$^{\#P}$-complete
	No			NEXPTIME-complete
$\phi(G) < k$	Yes		coNP-complete	coNP$^{\#P}$-complete
	No			coNEXPTIME-complete

2   Definitions and Notations

Let V be a finite set. A *Boolean network* (BN) with component set V is defined as a function $f : \{0,1\}^V \to \{0,1\}^V$. A *configuration* $x \in \{0,1\}^V$ assigns a state $x_i \in \{0,1\}$ to each component $i \in V$. During an application of f, the state of component i evolves according to the *local* function $f_i : \{0,1\}^V \to \{0,1\}$, which is the coordinate i of f, i.e. $f_i(x) = f(x)_i$ for all $x \in \{0,1\}^V$. When $V = [n]$, we write $x = (x_1, \ldots, x_n)$ and $f(x) = (f_1(x), \ldots, f_n(x))$.

Given a configuration $x \in \{0,1\}^V$ and $I \subseteq V$, we denote by x_I the configuration $y \in \{0,1\}^I$ such that $y_i = x_i$ for all $i \in I$. Given $i \in V$, we denote the i-base vector e_i, that is, $(e_i)_i = 1$ and $(e_i)_j = 0$ for all $j \neq i$. If $x, y \in \{0,1\}^V$ then $x \oplus y$ is the configuration $z \in \{0,1\}^V$ such that $z_i = x_i \oplus y_i$ for all $i \in V$, where the addition is computed modulo two. Hence, $x \oplus e_i$ is the configuration obtained from x by flipping component i only.

A *signed digraph* $G = (V, A, \sigma)$ is a digraph (V, A) with an arc-labeling function σ from A to $\{-1, 0, 1\}$, that gives a sign (negative, null or positive) to each arc (i,j), denoted σ_{ij}. We say that G is *simple* if it has no null sign. Given a vertex i and $s \in \{-1, 0, 1\}$, we denote by $N_G^s(i)$ the set of in-neighbors j of i such that $\sigma_{ij} = s$, and we drop G in the notations when it is clear from the context. We call $N^1(i)$ (resp. $N^{-1}(i)$) the set of positive (resp. negative) in-neighbors of i. We also simply denote $N(i)$ the set of all in-neighbors of i. In the following, it is very convenient to set $\tilde{\sigma}_{ij} = 0$ if $\sigma_{ij} \geq 0$ and $\tilde{\sigma}_{ij} = 1$ otherwise.

The *signed interaction digraph* (SID) of a BN f with component set V is the signed digraph $G_f = (V, A, \sigma)$ defined as follows. First, given $i, j \in V$, there is an arc $(i,j) \in A$ if and only if there exists a configuration x such that $f_j(x \oplus e_i) \neq f_j(x)$ (i.e. the state of component i influences the state of component j). Second, the sign σ_{ij} of an arc $(i,j) \in A$ depends on whether the state of j tends to mimic or negate the state of i, and is defined as

$$\sigma_{ij} = \begin{cases} 1 & \text{if } f_j(x \oplus e_i) \geq f_j(x) \text{ for all } x \in \{0,1\}^n \text{ with } x_i = 0, \\ -1 & \text{if } f_j(x \oplus e_i) \leq f_j(x) \text{ for all } x \in \{0,1\}^n \text{ with } x_i = 0, \\ 0 & \text{otherwise.} \end{cases}$$

Given $j \in V$, we say that f_j is the AND (resp. OR) function if it is the ordinary logical and (resp. or) but inputs with a negative sign are flipped, i.e

$$f_j(x) = \bigwedge_{i \in N(j)} x_i \oplus \tilde{\sigma}_{ij} \quad (\text{resp. } f_j(x) = \bigvee_{i \in N(j)} x_i \oplus \tilde{\sigma}_{ij}).$$

Given a signed digraph G, we know that G is a SID (*i.e.* there exists a BN f with $G_f = G$), if and only if there is no vertex i such that $|N(i)| \leq 2$ and $|N^0(i)| = 1$ [23]. In particular, a simple signed digraph is always a SID.

A fundamental remark regarding the present work is that multiple BNs may have the same SID. Given a SID G with vertex set V, we denote by $F(G)$ the set of BNs admitting G as SID:

$$F(G) = \{f : \{0, 1\}^V \to \{0, 1\}^V \mid G_f = G\}.$$

The size of $F(G)$ is generally huge. If a component i has in-degree d in G, then the number of possible local functions f_i is doubly exponential according to d, thus it scales as the number of Boolean functions on d variables, 2^{2^d}. Hence, $|F(G)|$ is at least doubly exponential according to its maximum in-degree, denoted $\Delta(G)$. The precise value of $|F(G)|$ is not trivial, see A006126 on the OEIS [1].

A *fixed point* of f is a configuration x such that $f(x) = x$, which is equivalent to $f_i(x) = x_i$ for all $i \in [n]$. We denote by $\Phi(f)$ the set of fixed points of f and $\phi(f) = |\Phi(f)|$. We are interested in a decision problem related to the maximum number of fixed points of BNs within $F(G)$, denoted

$$\phi(G) = \max \{\phi(f) \mid f \in F(G)\}.$$

More precisely, we will study the complexity of deciding if $\phi(G) \geq k$, where k is a positive integer, fixed or not. This gives the two following decision problems.

---

k-MAXIMUM FIXED POINT PROBLEM (k-MFPP)
Input: a SID G.
Question: $\phi(G) \geq k$?

MAXIMUM FIXED POINT PROBLEM (MFPP)
Input: a SID G and an integer $k \geq 1$.
Question: $\phi(G) \geq k$?

Cycles of interactions (in the SID) are known to play a fundamental role in the dynamical complexity of BN (the cycles we consider are always directed and without repeated vertices). Indeed, if G_f is acyclic then $\phi(f) = 1$ [26]. The *sign* of a cycle or a path in a signed digraph is the product of the signs of its arcs. It is well-known that if all the cycles of G_f are positive (resp. negative) then $\phi(f) \geq 1$ (resp. $\phi(f) \leq 1$), see [4,25]. Hence, if all the cycles of a SID G are negative, then $\phi(G) \leq 1$. The previous notions are illustrated in Fig. 1.

3 k-MAXIMUM FIXED POINT PROBLEM for $k = 1$

A strongly connected component H in a signed digraph G is *trivial* if it has a unique vertex and no arc, and *initial* if G has no arc (i, j) where j is in H but not i. We first have a lemma to concentrate on simple signed digraphs.

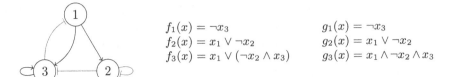

$$f_1(x) = \neg x_3 \qquad\qquad g_1(x) = \neg x_3$$
$$f_2(x) = x_1 \vee \neg x_2 \qquad\qquad g_2(x) = x_1 \vee \neg x_2$$
$$f_3(x) = x_1 \vee (\neg x_2 \wedge x_3) \qquad g_3(x) = x_1 \wedge \neg x_2 \wedge x_3$$

Fig. 1. Example of simple signed digraph G with two BNs $f, g \in F(G)$. BN f has no fixed point, and g has one fixed point (110), which is the maximum for BNs in $F(G)$, that is $\phi(G) = 1$. Note that G has two positive cycles and two negative cycles.

Lemma 1. *For any SID G, there is a simple SID G' such that $\phi(G) \geq 1 \iff \phi(G') \geq 1$, and G' is computable from G in constant parallel time.*

Proof. From G, the construction of G' is made component by component, independently, by removing incoming arcs. For $j \in [n]$,

- If $|N^0(j)| \geq 2$ then we delete all incoming arcs of j. If there exists $f \in F(G)$ and $y \in \Phi(f)$, then we can take $f' \in F(G')$ equal to f, except for $f_j'(x) = y_j$ (a constant). Conversely, if there exists $f' \in F(G')$ and $y \in \Phi(f')$, then we can take $f \in F(G)$ equal to f', except for

$$f_j(x) = \left(b_j \oplus \bigoplus_{i \in N^0(j)} x_i\right) \wedge \bigwedge_{i \in N(j) \setminus N^0(j)} (x_i \oplus \tilde{\sigma}_{ij})$$

with $b_j = \bigoplus_{i \in N^0(j)} y_i$, in the case $y_j = 0$ (the case $y_j = 1$ is symmetric, with OR instead of AND function). We have $f_j'(y) = f_j(y) = y_j$ hence $y \in \Phi(f)$.
- If $|N^0(j)| = 1$, then we delete this arc. One can check that, if $y \in \Phi(f)$ with $f \in F(G)$ (resp. $y \in \Phi(f')$ with $f' \in F(G')$), then there exists $i \in N(j) \setminus N^0(j)$ such that $y_i \oplus \tilde{\sigma}_{ij} = y_j$. Consequently, if there exists $f \in F(G)$ and $y \in \Phi(f)$ then we can take $f' \in F(G')$ equal to f, except that f_j' is the AND function if $y_j = 0$ and the OR function otherwise. Conversely, suppose there exists $f' \in F(G')$ and $y \in \Phi(f')$, and let $\{k\} = N^0(j)$. In the case $y_j = 0$, we can construct a function $f \in F(G)$ equal to f', except for

$$f_j(x) = \left((x_i \oplus \tilde{\sigma}_{ij}) \vee (x_k \oplus y_k)\right) \wedge \bigwedge_{\ell \in N(j) \setminus \{i,k\}} \left((x_\ell \oplus \tilde{\sigma}_{\ell j}) \vee (x_k \oplus \neg y_k)\right).$$

We have $f_j(y) = 0 = y_j$ because the left hand side of the conjunction is false, thus $y \in \Phi(f)$ (the case $y_j = 1$ is symmetric by switching OR and AND functions, and replacing y_k with $\neg y_k$). $\qquad\square$

Lemma 2. *Let G be a simple SID. Then $\phi(G) \geq 1$ if and only if each non-trivial initial strongly connected component of G contains a positive cycle.*

Proof. The left to right implication has been proved by Aracena [4, Corollary 3]. For the converse, suppose that $G = (V, A, \sigma)$ has p initial strongly connected components $H_1, \ldots, H_p$. For all $k \in [p]$, if H_k is trivial then i_k denotes the unique vertex it contains, and otherwise we select a positive cycle C_k in H_k and an arc (j_k, i_k) inside. Then, G can be spanned by a forest of p vertex disjoint trees $T_1, \ldots, T_p$ rooted in $i_1, \ldots, i_p$ such that if H_k is not trivial then the path from i_k to j_k contained in T_k is the one contained in C_k. For all $k \in [p]$ and all vertices j in T_k, we denote by P_{kj} the path from i_k to j contained in T_k (if $j = i_k$ this path is of length zero and positive by convention).

Now, we define $f \in F(G)$ as follows. First, for all $k \in [p]$, if H_k is trivial then f_{i_k} is the constant 0 function, and otherwise f_{i_k} is the AND function. Second, for all $k \in [p]$ and all vertices $j \neq i_k$ in T_k, f_j is the AND function if P_{kj} is positive and the OR function otherwise. Next, we define $x \in \{0, 1\}^V$ as follows: for all $j \in V$, $x_j = 0$ if and only if P_{kj} is positive (thus $x_{i_k} = 0$ for all $k \in [p]$).

We claim that $x \in \Phi(f)$. Indeed, given $k \in [p]$ and a vertex $j \neq i_k$ in T_k, it is easy to prove that $f_j(x) = x_j$ by induction on the length of P_{kj}. Next, if H_k is trivial then $f_{i_k}(x) = 0$. Otherwise, (j_k, i_k) is an arc of H_k. Let s be the sign of the path P_{kj_k}, which is in C_k by construction. Since C_k is positive, $s = \sigma_{j_k i_k}$. So if $\sigma_{j_k i_k} = 1$ then $x_{j_k} = 0$ and thus $f_{i_k}(x) = 0$, and if $\sigma_{j_k i_k} = -1$ then $x_{j_k} = 1$ and thus $f_{i_k}(x) = 0$. In all cases, $f_{i_k}(x) = 0 = x_{i_k}$. We deduce that $x \in \Phi(f)$. $\square$

Thus, to decide if $\phi(G) \geq 1$, it is sufficient to compute the non-trivial initial strongly connected components of G (this can be done in linear time [29]) and to check if they contain a positive cycle. As described below, this checking can be done in polynomial time using the following difficult theorem independently proved by Robertson, Seymour and Thomas [27] and McCuaig [20].

Theorem 1 ([20,27]). *There exists a polynomial time algorithm for deciding if a given digraph contains a cycle of even length.*

Let G be a signed digraph with n vertices, and let $\tilde{G}$ be obtained from G by replacing each positive arc by a path of length two, with two negative arcs, where the internal vertex is new. Then $\tilde{G}$ has at most $n + n^2$ vertices, and it is easy to see that G has a positive cycle if and only if $\tilde{G}$ has a cycle of even length [21]. We then deduce the following theorem.

Theorem 2. 1-MFPP *is in* P.

4 k-Maximum Fixed Point Problem for $k \geq 2$

Theorem 3. *For any $k \geq 2$, k-MFPP is NP-complete, even with $\Delta(G) \leq 2$.*

Theorem 3 is obtained from Lemmas 3, 5 and 6.

Lemma 3. *For any $k \geq 2$, k-MFPP is in NP.*

Proof (sketch, see details in Appendix ??). First, consider the case where $\Delta(G) \leq d$ for some constant d. Then a certificate of $\phi(G) \geq k$ could consist in a network $f \in F(G)$ and k distinct fixed points $x^{(1)}, \ldots, x^{(k)}$. The fact that $f \in F(G)$, and $f(x^{(i)}) = x^{(i)}$ with distinct $x^{(i)}$ for all $i \in [k]$, is checked in polynomial time.

However, when $\Delta(G)$ is not bounded, $F(G)$ can be of doubly exponential size in n. Thus, some functions f require an exponential space to be encoded. Instead, one can give as a certificate a partial function $h : X \to \{0,1\}^n$ with $X \subseteq \{0,1\}^n$ such that $f(x) = h(x)$ for any $x \in X$. In the set X, we put k fixed points and configurations which assert the effectiveness of the arcs. To check the certificate it is sufficient to ensure that there are no inconsistencies (independently for each local function). As a result, the problem is in NP. □

A shorter certificate (only the k fixed points) is possible when G is simple (see Appendix ??). This result from the following theorem. Note that the extending partial Boolean functions is a well established topic [8,9].

Theorem 4. *Let G be a simple SIG with vertex set V and consider a partial BN $h : X \to \{0,1\}^V$ with $X \subseteq \{0,1\}^V$. There is a $\mathcal{O}(|X|^2|V|^2)$-time algorithm to decide if there exists an extension of h in $F(G)$.*

We now prove that 2-MFPP is NP-hard. We will use observations from [4].

Lemma 4 ([4]). *Let $G = (V, A, \sigma)$ be a simple signed digraph, $f \in F(G)$ and x, y two distinct fixed points of f. Then there exists a positive cycle C in G such that, for any arc (i, j) in C, we have $x_i \oplus \tilde{\sigma}_{ij} = x_j \neq y_j = y_i \oplus \tilde{\sigma}_{ij}$.*

Remark 1. If the positive cycle C in Lemma 4 has only positive arcs, then either $x_i < y_i$ for all vertex i in C, or $x_i > y_i$ for all vertex i in C.

Remark 2. Given $f \in F(G)$ and x, y two distinct fixed points of f, for any feedback vertex set I of G we have $x_I \neq y_I$.

Lemma 5. *The problem 2-MFPP is NP-hard, even with $\Delta(G) \leq 2$.*

Proof. We reduce 3SAT to our problem. Let us consider a 3SAT instance ψ with n variables $\lambda_1, \ldots, \lambda_n$ and m clauses $\mu_1, \ldots, \mu_m$. We define the signed digraph $G_\psi = (V, A, \sigma)$, where $|V| = 4n + 2m + 1$, as follows (see Fig. 2).

First, $V = R \cup P \cup L \cup \bar{L} \cup S \cup T$ with $R = \{r_i \mid i \in [n]\}$, $P = \{p_i \mid i \in [0, n]\}$, $L = \{\ell_i \mid i \in [n]\}$, $\bar{L} = \{\bar{\ell}_i \mid i \in [n]\}$, $S = \{s_i \mid i \in [m]\}$, and $T = \{t_i \mid i \in [m]\}$. To simplify the notation let $s_0 = p_0$ and $s_{m+1} = p_n$. Second,

$$A := \bigcup_{i \in [n]} \{(p_{i-1}, \ell_i), (p_{i-1}, \bar{\ell}_i), (\ell_i, p_i), (\bar{\ell}_i, p_i), (r_i, \ell_i), (r_i, \bar{\ell}_i)\}$$

$$\cup \bigcup_{j \in [m]} \{(t_i, s_i), (s_i, s_{i-1})\} \cup \{(p_n, s_m)\}$$

$$\cup \{(\ell_i, t_j) \mid i \in [n], \ j \in [m] \text{ if } \lambda_i \text{ appears positively in } \mu_j\}$$

$$\cup \{(\bar{\ell}_i, t_j) \mid i \in [n], \ j \in [m] \text{ if } \lambda_i \text{ appears negatively in } \mu_j\}.$$

Arcs in $\{(s_i, t_i) \mid i \in [m]\} \cup \{(r_i, \ell_i) \mid i \in [n]\}$ are negative, all others are positive.

Let us first prove that if ψ is satisfiable then there exists a BN $f \in F(G_\psi)$ with has at least two fixed points. Consider a valid assignment $v : \{\lambda_1, \ldots \lambda_n\} \to \{\bot, \top\}$. Let $I^\bot = \{i \in [n] \mid v(\lambda_i) = \bot\}$ and $I^\top = \{i \in [n] \mid v(\lambda_i) = \top\}$. We define $f \in F(G_\psi)$ as follows.

- For all $i \in I^{\perp}$ (resp. $I^{\top}$), f_{r_i} is the constant 0 (resp. 1) function.
- For all $i \in [n]$, f_{ℓ_i} and $f_{\bar{\ell}_i}$ are both AND functions.
- For all $i \in [0, n]$, f_{p_i} is the OR function.
- For all $i \in [m]$, f_{s_i} and f_{t_i} are the AND functions.

The two following configurations x and y are distinct fixed points of f, and therefore $\phi(G_\psi) \geq 2$: for all $j \in V$,

$$x_j = \begin{cases} 1 & \text{if } j \in \{r_i \mid i \in I^{\top}\} \\ 0 & \text{otherwise} \end{cases}$$

$$y_j = \begin{cases} 1 & \text{if } j \in \{r_i \mid i \in I^{\top}\} \cup P \cup S \cup \{\ell_i \mid i \in I^{\perp}\} \cup \{\bar{\ell}_i \mid \in I^{\top}\} \\ 0 & \text{otherwise.} \end{cases}$$

Now, we prove that if $\phi(G_\psi) \geq 2$ then ψ is satisfiable. Consider a BN $f \in F(G_\psi)$ with two distinct fixed points x and y. Remark that $\{p_0\}$ is a feedback vertex set of G_ψ. In other words, all cycles of G_ψ contain p_0. We deduce from Remark 2 that $x_{p_0} \neq y_{p_0}$ and that $\phi(G_\psi) \leq 2$. Without loss of generality, suppose that $x_{p_0} < y_{p_0}$. Remark also that any cycle containing one of the vertices $t_1, \ldots, t_m$ is negative, and that no positive cycle in G_ψ contains any negative arc. Thus, according to Remark 1, there exists a cycle C such that $x_j < y_j$ for every vertex j in C. In other words, $x_P < y_P$ and $x_S < y_S$ and for every $i \in [n]$ either C contains ℓ_i and we have $x_{\ell_i} < y_{\ell_i}$, or it contains $\bar{\ell}_i$ and we have $x_{\bar{\ell}_i} < y_{\bar{\ell}_i}$. We construct the following assignment v from C.

$$v(\lambda_i) = \begin{cases} \perp & \text{if } C \text{ contains } \ell_i, \\ \top & \text{if } C \text{ contains } \bar{\ell}_i. \end{cases}$$

For the sake of contradiction, suppose that v does not satisfy the formula. As a consequence, there is a clause μ_j which is false with assignment v. In other words, any variable which appears positively in the clause is assigned to false and any variable which appears negatively is assigned to true.

Let us prove that $x_{t_j} < y_{t_j}$. Since any incoming arc of t_j is positive, and since x and y are fixed points, it is sufficient to prove that, for every in-neighbor ℓ of t_j, we have $x_\ell < y_\ell$. By definition of G_ψ, any in-neighbor of t_j corresponds to a variable λ_i of the clause. If λ_i appears positively (resp. negatively) in clause μ_j then the in-neighbor of t_j corresponding to λ_i is ℓ_i (resp. $\bar{\ell}_i$). Since $v(\lambda_i) = \perp$ (resp. $\top$) because the clause is false then C contains ℓ_i (resp. $\bar{\ell}_i$) and we have $x_{\ell_i} < y_{\ell_i}$ (resp. $x_{\bar{\ell}_i} < y_{\bar{\ell}_i}$). As a result, $x_{t_j} < y_{t_j}$.

Now, the vertex s_j has two in-neighbors. One of them is s_{j+1} and we have $\sigma_{s_{j+1}s_j} = 1$ and $x_{s_{j+1}} < y_{s_{j+1}}$. The other is t_j with $\sigma_{t_j s_j} = -1$ and $x_{t_j} < y_{t_j}$. Hence, there are two possible local functions for f_{s_i}:

- $f_{s_j}(z) = z_{s_{j+1}} \vee \neg z_{t_j}$, and then $x_{s_j} = f_{s_j}(x) = x_{s_{j+1}} \vee \neg x_{t_j} = 0 \vee \neg 0 = 1$.
- $f_{s_i}(z) = z_{s_{i+1}} \wedge \neg z_{t_j}$, and then $y_{s_j} = f_{s_j}(y) = y_{s_{j+1}} \wedge \neg y_{t_j} = 1 \wedge \neg 1 = 0$.

In both cases, we do not have $x_{s_j} < y_{s_j}$, which is a contradiction since s_j is in C. As a result, the 3SAT instance ψ is satisfiable. Additionally, remark that

$\phi(G_\psi) \geq 1$ because, with the constant 1 function for the vertices in R, and the OR local function everywhere else, the configuration $z_i = 1$ for all i is a fixed point. We can conclude that $\phi(G_\psi) = 1$ when ψ is unsatisfiable.

To get a bounded degree $\Delta(G_\psi) \leq 2$, notice that only vertices in T have in-degree three, which can be decreased by adding an intermediate vertex (see the right picture in Fig. 2) while preserving the correctness of the reduction. □

We can extend the NP-hardness reduction to any $k \geq 2$.

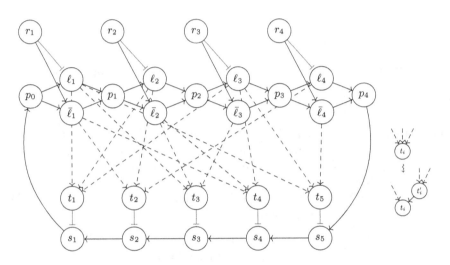

Fig. 2. Example of construction in the reduction from 3SAT to k-MFPP (Lemma 5). This signed digraph G_ψ implements the following 3SAT instance ψ: $(\lambda_1 \vee \lambda_2 \vee \lambda_3) \wedge (\neg\lambda_1 \vee \lambda_2 \vee \lambda_4) \wedge (\lambda_1 \vee \neg\lambda_2 \vee \neg\lambda_3) \wedge (\neg\lambda_1 \vee \neg\lambda_2 \vee \lambda_3) \wedge (\lambda_1 \vee \lambda_3 \vee \neg\lambda_4)$ which is satisfiable if and only if $\phi(G_\psi) \geq 2$, otherwise $\phi(G_\psi) = 1$.

Lemma 6. *For any $k \geq 2$, k-MFPP is NP-hard, even with $\Delta(G) \leq 2$.*

Proof. Let $\ell = \lfloor \log_2(k-1) \rfloor$, *i.e.* $2^\ell < k \leq 2^{\ell+1}$. Given a formula, consider the digraph G from Lemma 5, and add ℓ new isolated vertices with positive loops. Then 1 or 2 fixed points on G_ψ become respectively 2^ℓ or $2^{\ell+1}$ fixed points. □

Remark 3. For $\Delta(G) \leq 1$, $|F(G)| = 1$ since each local function is the identity or the negation, and computing $\phi(G)$ is in $\mathcal{O}(|G|)$, hence k-MFPP $\in$ P.

5 MAXIMUM FIXED POINT PROBLEM

Theorem 5. *When $\Delta(G) \leq d$, MFPP is $NP^{\#P}$-complete.*

In this first part of the section, we prove Theorem 5, from Lemmas 7 and 8.

Lemma 7. *When $\Delta(G) \leq d$, MFPP is in* $\mathsf{NP}^{\#P}$.

Proof. An algorithm in $\mathsf{NP}^{\#P}$ to solve MFPP is, on input G, k:

1. guess local functions f_i for $i \in [n]$ (polynomial from $\Delta(G) \leq d$),
2. construct $\psi = (f_1(x) = x_1) \wedge \cdots \wedge (f_n(x) = x_n)$ on variables $x_1, \ldots, x_n$,
3. compute the number of solutions of ψ with the $\#P$ oracle, that is $\phi(f)$,
4. accept if and only if $\phi(f) \geq k$.

A non-deterministic branch accepts if and only if $\phi(G) \geq k$. □

Lemma 8. *When $\Delta(G) \leq d$, MFPP is* $\mathsf{NP}^{\#P}$*-hard.*

Proof (sketch, see details in Appendix ??). We consider the following problem.

EXISTENTIAL-MAJORITY-3SAT (E-MAJ3SAT)
Input: A 3SAT formula ψ on $\{\lambda_1, \ldots, \lambda_n\}$ and $s \in [n]$
Question: Is there an assignment v of $\lambda_1, \ldots, \lambda_s$ such that the majority of assignments of $\lambda_{s+1}, \ldots, \lambda_n$ satisfy ψ?

We know that E-MAJ3SAT is $\mathsf{NP}^{\mathsf{PP}}$-complete [19] and that $\mathsf{NP}^{\#P} = \mathsf{NP}^{\mathsf{PP}}$ (direct extension of $\mathsf{P}^{\#P} = \mathsf{P}^{\mathsf{PP}}$ [22]). Consequently, it is sufficient to prove that we can reduce E-MAJ3SAT to MFPP. To represent an instance (ψ, s) of E-MAJ3SAT, we construct a digraph $G_{\psi,s}$ similar to the digraph G_ψ constructed in Lemma 5 except that we add a positive loop to the $q = n - s$ vertices $r_{s+1}, \ldots, r_n$. We claim that $\phi(G_{\psi,s}) = \alpha + 2^q$, with

$$\alpha = \max_{v:\{\lambda_1,\ldots,\lambda_s\}\to\{\bot,\top\}} |\{u : \{\lambda_{s+1}, \ldots, \lambda_n\} \to \{\bot, \top\} \mid v \cup u \text{ satisfies } \psi\}|.$$

Indeed, consider $f \in F(G_{\psi,s})$ with $\phi(f) = \phi(G_{\psi,s})$. As in Lemma 5, the functions f_i for $i \in \{\ell_1, \bar{\ell}_1, \ldots, \ell_s, \bar{\ell}_s\}$ correspond to an assignment v of $\lambda_1, \ldots, \lambda_s$. Moreover, each valuation u of $\lambda_{s+1}, \ldots, \lambda_n$ corresponds to one (resp. two) fixed points if the assignment $v \cup u$ makes ψ false (resp. true). As a consequence, the reduction is correct by setting $k = \frac{3}{2}2^q$. □

In this second part, we study MFPP with unbounded maximum degree.

Theorem 6. *When $\Delta(G)$ is unbounded, MFPP is* NEXPTIME*-complete.*

Proof (sketch, see details in Appendix ??). It is easy to see that the problem MFPP with unbounded degree is in NEXPTIME. Indeed, to know if $\phi(G) \geq k$ it is sufficient to guess a function $f \in F(G)$ (encoded in exponential space), to compute $\phi(f)$ (in exponential time) and then accept if $\phi(f) \geq k$, reject otherwise. A non-deterministic branch accepts if and only if $\phi(G) \geq k$.

For the hardness, we reduce from SUCCINT-3SAT [22], which is 3SAT where ψ has $n = 2^{\tilde{n}}$ variables, $m = 2^{\tilde{m}}$ clauses, and is given by a circuit D with:

- $\tilde{m}$ input bits for the clauses, and 2 for the three literal positions,
- $\tilde{n}$ output bits to give the corresponding variable, and 1 for its polarity.

D is acyclic, has in-degree at most 2, and has simple OR, AND, NOT, identity or constant functions. The idea is to generalize the construction from the proof of Theorem 3, with one literal for each node of the circuit D (top part), and additional clauses implementing the circuit (bottom part). With non-trivial additional elements, choosing local functions correspond to choosing an assignment. There will be a maximum of one (resp. two) fixed point for each non-satisfied (resp. satisfied) clause. As a result, ψ is satisfiable if and only if $\phi(G) \geq 2m$. □

6   Conclusion

This first work raises many open questions. First, is the problem 1-MFPP P-complete? We proved that it is equivalent to the problem of finding an even cycle in a digraph, for which the P versus NP-complete status remained open until [20, 27]. Now we know that the problem is in P, but is it a tight bound?

Several natural extensions of the present results may be addressed. What happens to the complexity when we study the minimum number of fixed points instead of the maximum? And for digraphs with only positive arcs? What about limit cycles of period greater than one instead of fixed points? Understanding the complexity of computing bounds on dynamical properties of BNs respecting a given interaction digraph is a new and promising approach, both on the theoretical and practical points of view.

Acknowledgments. The authors would like to thank for their support the *Young Researcher* project ANR-18-CE40-0002-01 "FANs", project ECOS-CONICYT C16E01, and project STIC AmSud CoDANet 19-STIC-03 (Campus France 43478PD).

References

1. The Online Encyclopedia of Integer Sequences, founded in 1964 by N. J. A. Sloane. Sequence A006126. https://oeis.org/A006126
2. Akutsu, T., Miyano, S., Kuhara, S.: Identification of genetic networks from a small number of gene expression patterns under the boolean network model. In: Biocomputing'99, pp. 17–28. World Scientific (1999)
3. Albert, R.: Boolean modeling of genetic regulatory networks. In: Ben-Naim, E., Frauenfelder, H., Toroczkai, Z. (eds.) Complex Networks. LNP, vol. 650, pp. 459–481. Springer, Heidelberg (2004). https://doi.org/10.1007/978-3-540-44485-5_21
4. Aracena, J.: Maximum number of fixed points in regulatory Boolean networks. Bull. Math. Biol. **70**(5), 1398–1409 (2008)
5. Aracena, J., Demongeot, J., Goles, E.: Fixed points and maximal independent sets in and-or networks. Discrete Appl. Math. **138**(3), 277–288 (2004)
6. Aracena, J., Richard, A., Salinas, L.: Maximum number of fixed points in AND-OR-NOT networks. J. Comput. Syst. Sci. **80**(7), 1175–1190 (2014)
7. Aracena, J., Richard, A., Salinas, L.: Number of fixed points and disjoint cycles in monotone Boolean networks. SIAM J. Discrete Math. **31**(3), 1702–1725 (2017)
8. Boros, E., Ibaraki, T., Makino, K.: Error-free and best-fit extensions of partially defined Boolean functions. Inf. Comput. **140**(2), 254–283 (1998)

9. Crama, Y., Hammer, P.L.: Boolean Functions: Theory, Algorithms, and Applications. Cambridge University Press, Cambridge (2011)
10. Feder, T.: A new fixed point approach for stable networks and stable marriages. J. Comput. Syst. Sci. **45**(2), 233–284 (1992)
11. Gadouleau, M., Richard, A., Riis, S.: Fixed points of Boolean networks, guessing graphs, and coding theory. SIAM J. Discrete Math. **29**(4), 2312–2335 (2015)
12. Gadouleau, M., Riis, S.: Graph-theoretical constructions for graph entropy and network coding based communications. IEEE Trans. Inf. Theory **57**(10), 6703–6717 (2011)
13. Goles, E., Salinas, L.: Sequential operator for filtering cycles in Boolean networks. Adv. Appl. Math. **45**(3), 346–358 (2010)
14. Kauffman, S.A.: Metabolic stability and epigenesis in randomly connected nets. J. Theor. Biol. **22**, 437–467 (1969)
15. Kauffman, S.A.: Origins of Order Self-Organization and Selection in Evolution. Oxford University Press, Oxford (1993)
16. Kaufman, M., Soulé, C., Thomas, R.: A new necessary condition on interaction graphs for multistationarity. J. Theoret. Biol. **248**(4), 675–685 (2007)
17. Kosub, S.: Dichotomy results for fixed-point existence problems for Boolean dynamical systems. Math. Comput. Sci. **1**(3), 487–505 (2008)
18. Le Novère, N.: Quantitative and logic modelling of molecular and gene networks. Nat. Rev. Genet. **16**, 146–158 (2015)
19. Littman, M.L., Goldsmith, J., Mundhenk, M.: The computational complexity of probabilistic planning. J. Artif. Intell. Res. **9**, 1–36 (1998)
20. McCuaig, W.: Pólya's permanent problem. Electron. J. Comb. **11**(1), 79 (2004)
21. Montalva, M., Aracena, J., Gajardo, A.: On the complexity of feedback set problems in signed digraphs. Electron. Not. Discrete Math. **30**, 249–254 (2008)
22. Papadimitriou, C.H.: Computational Complexity. Addison-Wesley, Reading (1994)
23. Paulevé, L., Richard, A.: Topological fixed points in Boolean networks. Comptes Rendus de l'Académie des Sci.-Ser. I-Math. **348**(15–16), 825–828 (2010)
24. Remy, E., Ruet, P., Thieffry, D.: Graphic requirements for multistability and attractive cycles in a Boolean dynamical framework. Adv. Appl. Math. **41**(3), 335–350 (2008)
25. Richard, A.: Positive and negative cycles in Boolean networks. J. Theor. Biol. **463**, 67–76 (2019)
26. Robert, F.: Discrete Iterations: A Metric Study. Springer Series in Computational Mathematics, vol. 6, p. 198. Springer, Heidelberg (1986). https://doi.org/10.1007/978-3-642-61607-5
27. Robertson, N., Seymour, P., Thomas, R.: Permanents, pfaffian orientations, and even directed circuits. Ann. Math. **150**(3), 929–975 (1999)
28. Soulé, C.: Mathematical approaches to differentiation and gene regulation. C.R. Paris Biol. **329**, 13–20 (2006)
29. Tarjan, R.: Depth-first search and linear graph algorithms. SIAM J. Comput. **1**(2), 146–160 (1972)
30. Thomas, R.: Boolean formalization of genetic control circuits. J. Theor. Biol. **42**(3), 563–585 (1973). https://doi.org/10.1016/0022-5193(73)90247-6
31. Thomas, R., d'Ari, R.: Biological Feedback. CRC Press, Boca Raton (1990)
32. Thomas, R., Kaufman, M.: Multistationarity, the basis of cell differentiation and memory. II. Logical analysis of regulatory networks in terms of feedback circuits. Chaos Interdisc. J. Nonlinear Sci. **11**(1), 180–195 (2001)

A Note on the Ordinal Analysis
of $\mathbf{RCA_0} + \mathrm{WO}(\sigma)$

Lorenzo Carlucci[1]([✉]), Leonardo Mainardi[1], and Michael Rathjen[2]

[1] Department of Computer Science, University of Rome I, Rome, Italy
carlucci@di.uniroma1.it, l.mainardi.90@gmail.com
[2] School of Mathematics, University of Leeds, Leeds LS2 9JT, UK
M.Rathjen@leeds.ac.uk

Abstract. We fill an apparent gap in the literature by giving a short and self-contained proof that the ordinal of the theory $\mathbf{RCA_0} + \mathrm{WO}(\sigma)$ is σ^ω, for any ordinal σ satisfying $\omega \cdot \sigma = \sigma$ (e.g., ω^ω, ω^{ω^ω}, ε_0). Theories of the form $\mathbf{RCA_0} + \mathrm{WO}(\sigma)$ are of interest in Proof Theory and Reverse Mathematics because of their connections to a number of well-investigated combinatorial principles related to various subsystems of arithmetic.

1 Introduction

Well-ordering statements are commonly used in Proof Theory and Reverse Mathematics as measures of strength of a theorem or a theory. For example a number of interesting theorems is known to be equivalent to $\mathrm{WO}(\omega^\omega)$ or $\mathrm{WO}(\omega^{\omega^\omega})$ over the theory $\mathbf{RCA_0}$ (see, e.g., [3,7,9,15]). It is then natural to ask what is the proof-theoretic ordinal of the theories $\mathbf{RCA_0} + \mathrm{WO}(\omega^\omega)$, $\mathbf{RCA_0} + \mathrm{WO}(\omega^{\omega^\omega})$ and, in general, $\mathbf{RCA_0} + \mathrm{WO}(\alpha)$. While it is well-known that the ordinal of $\mathbf{RCA_0}$ is ω^ω, the answer for the other theories is not so immediate and occasionally some confusion arises.[1] Even the standard argument for showing that ω^ω is an upper bound on the proof-theoretic ordinal of $\mathbf{RCA_0}$ is somewhat indirect in that it hinges on the characterization of the provably recursive functions of $\mathbf{RCA_0}$ rather than only on the computation of its proof-theoretical ordinal. A proper direct treatment approach to determining the proof-theoretic ordinal of theories of the form $\mathbf{RCA_0} + \mathrm{WO}(\alpha)$ seems to be missing from the literature. The closest match is Sommer's [17] model-theoretical treatment of first-order theories with transfinite induction restricted to various formula-classes and ordinals strictly below ε_0.

In this paper we show that, if σ is an ordinal satisfying $\omega \cdot \sigma = \sigma$, the proof-theoretic ordinal of the theory $\mathbf{RCA_0} + \mathrm{WO}(\sigma)$ is σ^ω. Examples of relevant σs

---

[1] For example, in proving that a $\varPi_1^1$-version of Ramsey's Theorem called the Adjacent Ramsey Theorem is equivalent to $\mathrm{WO}(\varepsilon_0)$ over $\mathbf{RCA_0}$, [4] Lemma 2.2 makes use of the false equivalence, over $\mathbf{RCA_0}$, between $\mathrm{WO}(\varepsilon_0)$ and the $\varPi_1^1$-soundness of $\mathbf{ACA_0}$. The presentation in the later [5] avoids this pitfall but establishes a slightly different result.

© Springer Nature Switzerland AG 2019
F. Manea et al. (Eds.): CiE 2019, LNCS 11558, pp. 144–155, 2019.
https://doi.org/10.1007/978-3-030-22996-2_13

are ω^ω, ω^{ω^ω} etc. and ε_0. This should be contrasted with the fact that the ordinal of $\mathbf{ACA}_0 + \mathrm{WO}(\varepsilon_0)$ is the much larger ε_1, as can be gleaned from the proof-theoretical analysis of transfinite induction over Peano Arithmetic (the original proof seems to be in [8]).

Essentially, we show that the first-order part of $\mathbf{RCA}_0 + \mathrm{WO}(\sigma)$ is the theory $\mathbf{I}\Sigma_1$ plus the scheme of transfinite induction up to σ restricted to Π_1 formulas, which we denote by $\mathrm{TI}(\sigma, \Pi_1)$. We give an ordinal analysis of the latter theory augmented by a generic unary predicate symbol U and then show that the theories prove the same Π_1^1-statements, where a Π_1^1-sentence $\forall X F(X)$ with $F(X)$ being arithmetic is identified with $F(\mathsf{U})$ in the first-order context of $\mathrm{TI}(\sigma, \Pi_1)$.

For the remainder of the paper, we fix an ordinal σ such that $\omega \cdot \sigma = \sigma$. The ordinal σ is assumed to be represented in a natural ordinal representation system. We denote by $\lhd$ the primitive recursive ordering on the ordinals smaller that σ to distinguish it from the usual ordering on the naturals.

2   Ordinal Analysis of $\mathbf{I}\Sigma_1 + \mathrm{TI}(\sigma, \Pi_1)$

The language of $\mathbf{T}_\sigma := \mathbf{I}\Sigma_1 + \mathrm{TI}(\sigma, \Pi_1)$ is the language of Primitive Recursive Arithmetic, $\mathbf{PRA}$, augmented by a unary predicate symbol U. Especially we assume that there is a binary surjective coding function $\langle \cdot, \cdot \rangle$ with inverses $(\cdot)_0$, $(\cdot)_1$. The order relation for the ordering on σ will be denoted by the same symbol $\lhd$ used to denote the corresponding primitive recursive relation. Bounded quantifiers $\forall x \leq t$ and $\exists x \leq t$ will be treated as quantifiers in their own right. Formulas containing only bounded quantifiers are called Δ_0-formulas. For our proof-theoretic purposes, $\mathbf{T}_\sigma$ will be formalized in a one-sided sequent calculus, using negation normal forms following [14] (this is also known as the Tait-calculus [18]). $\mathbf{T}_\sigma$ has the usual axioms pertaining to primitive recursive functions and predicates. A noteworthy feature is that transfinite induction on σ for Π_1-formulas is expressed via the rule

$$\frac{\Theta, \; \exists z \, ((z)_0 \lhd a \land \neg F((z)_1, (z)_0)), \; \forall x F(x, a)}{\Theta, \; F(t, s)} \tag{1}$$

where a is an eigenvariable, $F(x, a)$ is Δ_0, t, s are arbitrary terms, and Θ is an arbitrary finite set of formulas.

Observe that we do not need Σ_1-induction as an extra induction principle as it follows from $\mathrm{TI}(\sigma, \Pi_1)$, since $\mathbf{I}\Pi_1$ entails $\mathbf{I}\Sigma_1$.

In order to perform partial cut eliminations, we define the degree, $|A|$, of a formula A as follows:

- $|A| = |\neg A| = 0$ if A is Δ_0.

If A is not Δ_0 and of one of the forms below, then:

- $|A_0 \land A_1| = |A_0 \lor A_1| = \max(|A_0|, |A_1|) + 1$;
- $|\forall x F(x)| = |\exists x F(x)| = |F(0)| + 1$;
- $|\forall x \leq t \, F(x)| = |\exists x \leq t \, F(x)| = |F(0)| + 2$.

As the rule (1) introduces a Δ_0-formula and the main formulas of axioms are Δ_0 as well, we can easily eliminate cuts of degree greater than 0. We use the notation $\mathbf{T}_\sigma \vdash^m_k \Gamma$ to convey that Γ is deducible in $\mathbf{T}_\sigma$ by a deduction of length at most m such that all cuts occurring in this deduction are with cut formulas of a degree $< k$. Thus $\mathbf{T}_\sigma \vdash^m_1 \Gamma$ means that there is deduction in which all cut formulas (if any) are Δ_0-formulas.

Theorem 1. $\mathbf{T}_\sigma \vdash^n_{r+1} \Gamma \Rightarrow \exists m \, \mathbf{T}_\sigma \vdash^m_1 \Gamma$.

Proof. By the usual cut elimination method of Gentzen's Hauptsatz.

2.1 Embedding $\mathbf{T}_\sigma$ in an Infinitary System

Next we embed $\mathbf{T}_\sigma$ into an infinitary system, called $\mathbf{PA}_\omega$, with ω-rule (basically the same as the system Z_∞ in [14]; a definition of $\mathbf{PA}_\omega$ in a two-sided Gentzen calculus can be found in [10]). The formulas of $\mathbf{PA}_\omega$ are the closed formulas of $\mathbf{T}_\sigma$, i.e. formulas without free variables. We shall assign a rank, $|A|_{\Delta_0}$ to a formula A of $\mathbf{PA}_\omega$ as follows:

(i) $|A|_{\Delta_0} = 0$ if A is atomic or a negated atom.
(ii) $|A_0 \wedge A_1|_{\Delta_0} = |A_0 \vee A_1|_{\Delta_0} = \max(|A_0|_{\Delta_0}, |A_1|_{\Delta_0}) + 1$.
(iii) $|\exists x \le t\, F(x)|_{\Delta_0} = |\forall x \le t\, F(x)|_{\Delta_0} = |F(0)|_{\Delta_0} + 1$.
(iv) $|\exists x\, F(x)|_{\Delta_0} = |\forall x\, F(x)|_{\Delta_0} = \max(\omega, |F(0)|_{\Delta_0} + 1)$.

Note that $|A|_{\Delta_0} < \omega$ exactly when A is Δ_0, and $|\exists x\, F(x)|_{\Delta_0} = |\forall x\, F(x)|_{\Delta_0} = \omega$ when $F(0)$ is Δ_0.

Definition 1. For a natural number n we use $\bar{n}$ to denote the nth numeral, that is the term obtained from the term $\bar{0}$ for zero by adding the successor function symbol n-times in front of it. The terms of $\mathbf{PA}_\omega$ are closed and thus can be evaluated to a number. For a term t let $t^\mathbb{N}$ be the number n such that t evaluates to n (in the following we occasionally refer to $t^\mathbb{N}$ by t).

The axioms of $\mathbf{PA}_\omega$ are sequents of two kinds. Let Γ be a finite set of formulas of $\mathbf{PA}_\omega$.

(i) Let $R(t_1, \ldots, t_r)$ be an atomic formula, where R is a relation symbol for a primitive recursive relation $R^\mathbb{N}$. If $R^\mathbb{N}(t_1^\mathbb{N}, \ldots, t_r^\mathbb{N})$ is true, then

$$\Gamma, R(t_1, \ldots, t_r)$$

is an axiom. If $R^\mathbb{N}(t_1^\mathbb{N}, \ldots, t_r^\mathbb{N})$ is false, then

$$\Gamma, \neg R(t_1, \ldots, t_r)$$

is an axiom.
(ii) If $s^\mathbb{N} = t^\mathbb{N}$ holds for terms s and t, then

$$\Gamma, U(s), \neg U(t)$$

is an axiom.

The ω-rule is the following rule: If $\Gamma, F(\bar{n})$ is deducible for all n, then $\Gamma, \forall x\, F(x)$ is the conclusion.

Similarly to derivations in $\mathbf{T}_\sigma$, we will use the notation $\mathbf{PA}_\omega \vdash^{\alpha}_{\beta} \Gamma$ to convey that Γ is deducible in $\mathbf{PA}_\omega$ by a deduction of height at most α such that all cuts occurring in this deduction are with cut formulas of $|\cdot|_{\Delta_0}$-rank $< \beta$.

Lemma 1 (Reduction Lemma). *If* $|B|_{\Delta_0} = \omega$, $\mathbf{PA}_\omega \vdash^{\alpha}_{\omega} \Gamma, B$ *and* $\mathbf{PA}_\omega \vdash^{\beta}_{\omega} \Gamma, \neg B$, *then*

$$\mathbf{PA}_\omega \vdash^{\alpha \# \beta}_{\omega} \Gamma$$

where $\alpha \# \beta$ *denotes the* natural *or* Hessenberg sum *of* α *and* β.

Proof. Standard.

Theorem 2 (Embedding Theorem). *If* $\mathbf{T}_\sigma \vdash^{m}_{1} \Gamma$, *then* $\mathbf{PA}_\omega \vdash^{\sigma^m}_{\omega} \Gamma^*$, *where* Γ^* *is the result of assigning closed terms to all free variables in* Γ *(the same term to the same variable).*

Proof. We proceed by induction on m. We only need to pay attention to the case where the last inference is an instance of the rule (1). So let $\Gamma = \Theta, F(t, s)$ and assume $\mathbf{T}_\sigma \vdash^{m_0}_{1} \Lambda$ with $m_0 < m$ and $\Lambda = \Theta, \exists z\ ((z)_0 \lhd a \wedge \neg F((z)_1, (z)_0)), \forall x\, F(x, a)$.

Let $*$ be an assignment. Inductively we have for all closed terms q that

$$\mathbf{PA}_\omega \vdash^{\sigma^{m_0}}_{\omega} \Theta^*, \exists z\ ((z)_0 \lhd q \wedge \neg F^*((z)_1, (z)_0)), \forall x\, F^*(x, q). \tag{2}$$

We use transfinite induction on α for α in the field of $\lhd$ to show that:

$$\mathbf{PA}_\omega \vdash^{\sigma^{m_0} \cdot \omega \cdot (\alpha+1)}_{\omega} \Theta^*, \forall x\, F^*(x, \bar{\alpha}) \tag{3}$$

By the induction hypothesis, we have:

$$\mathbf{PA}_\omega \vdash^{\sigma^{m_0} \cdot (\omega \cdot (\eta+1))}_{\omega} \Theta^*, F^*(s', \bar{\eta})$$

for every $\eta \lhd \alpha$ and arbitrary closed term s', yielding

$$\mathbf{PA}_\omega \vdash^{\sigma^{m_0} \cdot (\omega \cdot \alpha)+1}_{\omega} \Theta^*, \bar{\eta} \lhd \bar{\alpha} \to F^*(s', \bar{\eta})$$

via an inference ($\vee$). If r is a closed term such that $r^{\mathbb{N}}$ is different from all η preceding α, then $\neg r \lhd \bar{\alpha}$ is an axiom, and thus, via an inference ($\vee$), we arrive at $\mathbf{PA}_\omega \vdash^{1}_{0} \Theta^*, r \lhd \bar{\alpha} \to F^*(s', r)$. Thus from the above we conclude that

$$\mathbf{PA}_\omega \vdash^{\sigma^{m_0} \cdot (\omega \cdot \alpha)+1}_{\omega} \Theta^*, (\bar{k})_0 \lhd \bar{\alpha} \to F^*((\bar{k})_1, (\bar{k})_0)$$

holds for all k, so that, via an application of the ω-rule, we get:

$$\mathbf{PA}_\omega \vdash^{\sigma^{m_0} \cdot (\omega \cdot \alpha)+2}_{\omega} \Theta^*, \forall z\ ((z)_0 \lhd \bar{\alpha} \to F^*((z)_1, (z)_0)). \tag{4}$$

Applying the Reduction Lemma 1 to (2) and (4) yields:

$$\mathbf{PA}_\omega \vdash_\omega^{\sigma^{m_0} \#(\sigma^{m_0}\cdot(\omega\cdot\alpha))+2} \Theta^*, \forall x\, F^*(x, \bar{\alpha}). \tag{5}$$

From (5) we finally get:

$$\mathbf{PA}_\omega \vdash_\omega^{\sigma^{m_0}\cdot(\omega\cdot(\alpha+1))} \Theta^*, \forall x\, F^*(x, \bar{\alpha}),$$

confirming (3).

If a term q has the property that $q^\mathbb{N}$ is not in the field of $\lhd$ then one can directly infer from (2) that

$$\mathbf{PA}_\omega \vdash_\omega^{\sigma^{m_0}} \Theta^*, \forall x\, F^*(x, q). \tag{6}$$

The reason for this is that if the formula $\exists z\,((z)_0 \lhd q \wedge \neg F^*((z)_1, (z)_0))$ figures as the main formula of an inference in this derivation its minor formula is of the form $(p)_0 \lhd q \wedge \neg F^*((p)_1, (p)_0)$. The latter formula conjunctively contains a false atomic formula. Such a formula can always be erased from the derivation. Formally, of course, this has to be proved by a separate induction on the ordinal of the derivation.

(3) and (6) now yield

$$\mathbf{PA}_\omega \vdash_\omega^{\sigma^m} \Theta^*, F^*(t, s)$$

for all closed terms t and s, since $\omega \cdot (\alpha + 1) \lhd \sigma$ on account of $\omega \cdot \sigma = \sigma$.

2.2   Eliminating Cuts with Δ_0-Formulas

The next step is to eliminate cuts with Δ_0-formulas that are not atomic.

Lemma 2. *Let* $0 < n < \omega$ *and suppose* $\mathbf{PA}_\omega \vdash_{n+1}^{\alpha} \Gamma$. *Then* $\mathbf{PA}_\omega \vdash_n^{\omega\cdot\alpha} \Gamma$.

Proof. We proceed by induction on α. The crucial case is when the last inference was a cut of rank n with cut formulas $A, \neg A$. Note that A is not an atomic formula. We then have $\mathbf{PA}_\omega \vdash_{n+1}^{\alpha_0} \Gamma, A$ and $\mathbf{PA}_\omega \vdash_{n+1}^{\alpha_0} \Gamma, \neg A$ for some $\alpha_0 < \alpha$. The induction hypotheses furnishes us with

$$\mathbf{PA}_\omega \vdash_n^{\omega\cdot\alpha_0} \Gamma, A \quad \text{and} \quad \mathbf{PA}_\omega \vdash_n^{\omega\cdot\alpha_0} \Gamma, \neg A. \tag{7}$$

Let A be of the form $\exists x \leq t\, F(x)$. Then $\neg A$ is the formula $\forall x \leq t\, \neg F(x)$. From (7) we obtain

$$\mathbf{PA}_\omega \vdash_n^{\omega\cdot\alpha_0} \Gamma, F(\bar{0}), \ldots, F(\bar{p}) \quad \text{and} \quad \mathbf{PA}_\omega \vdash_n^{\omega\cdot\alpha_0} \Gamma, \neg F(\bar{k}) \tag{8}$$

for all $k \leq p$, where p is the numerical value of t. As the formulas $F(\bar{k}), \neg F(\bar{k})$ have rank $< n$, we can employ $(p + 1)$-many cuts to (8) to arrive at $\mathbf{PA}_\omega \vdash_n^{\omega\cdot\alpha_0+p+1} \Gamma$. Thus we have $\mathbf{PA}_\omega \vdash_n^{\omega\cdot\alpha} \Gamma$ as $\omega \cdot \alpha_0 + p + 1 < \omega \cdot \alpha$. A similar argument works when A is of either form $A_0 \wedge A_1$ or $A_0 \vee A_1$.

Corollary 1. *If* $\mathbf{PA}_\omega \vert\frac{\alpha}{\omega}\ \Gamma$ *then* $\mathbf{PA}_\omega \vert\frac{\omega^\omega \cdot \alpha}{1}\ \Gamma$.

Proof. We use induction on α. The only interesting case arises when the last inference is a cut with a formula A of rank $k > 0$. Then we have $\mathbf{PA}_\omega \vert\frac{\alpha_0}{\omega}\ \Gamma, A$ and $\mathbf{PA}_\omega \vert\frac{\alpha_0}{\omega}\ \Gamma, \neg A$ for some $\alpha_0 < \alpha$. The induction hypothesis yields $\mathbf{PA}_\omega \vert\frac{\omega^\omega \cdot \alpha_0}{1}\ \Gamma, A$ and $\mathbf{PA}_\omega \vert\frac{\omega^\omega \cdot \alpha_0}{1}\ \Gamma, \neg A$. Hence $\mathbf{PA}_\omega \vert\frac{\omega^\omega \cdot \alpha_0 + 1}{k+1}\ \Gamma$. Applying Lemma 2 k times we arrive at $\mathbf{PA}_\omega \vert\frac{\omega^k \cdot (\omega^\omega \cdot \alpha_0 + 1)}{1}\ \Gamma$. As $\omega^k \cdot (\omega^\omega \cdot \alpha_0 + 1) = \omega^\omega \cdot \alpha_0 + \omega^k \le \omega^\omega \cdot \alpha$ we also have $\mathbf{PA}_\omega \vert\frac{\omega^\omega \cdot \alpha}{1}\ \Gamma$ as desired.

Note that $\sigma \ge \omega^\omega$ since $\omega \cdot \sigma = \sigma$.

Corollary 2. *Let* $m > 0$. *If* $\mathbf{PA}_\omega \vert\frac{\sigma^m}{\omega}\ \Gamma$ *then* $\mathbf{PA}_\omega \vert\frac{\sigma^{m+1}}{1}\ \Gamma$.

Proof. Corollary 1 yields $\mathbf{PA}_\omega \vert\frac{\omega^\omega \cdot \sigma^m}{1}\ \Gamma$. Thus the desired conclusion follows as $\omega^\omega \cdot \sigma^m \le \sigma \cdot \sigma^m = \sigma^{m+1}$.

3 Upper Bounds for the Provable Well-Orderings of $\mathbf{T}_\sigma$

The results of the previous section can be utilized to determine the ordinal rank of provable well-orderings of $\mathbf{T}_\sigma$. Let $\prec$ be a primitive recursive ordering. $\prec$ is said to be a *provable well-ordering* of $\mathbf{T}_\sigma$ if $\mathbf{T}_\sigma$ proves that $\prec$ is a total linear ordering and

$$\mathbf{T}_\sigma \vdash \mathrm{WO}(\prec)$$

where $\mathrm{WO}(\prec)$ stands for the formula

$$\forall v[\forall u \prec v \mathsf{U}(u) \rightarrow \mathsf{U}(v)] \rightarrow \forall v \mathsf{U}(v).$$

Assuming $\mathbf{T}_\sigma \vdash \mathrm{WO}(\prec)$, as a consequence of Theorems 1, 2 and Corollary 2 we then have

$$\mathbf{PA}_\omega \vert\frac{\sigma^m}{1}\ \forall v[\forall u \prec v \mathsf{U}(u) \rightarrow \mathsf{U}(v)] \rightarrow \forall v \mathsf{U}(v) \tag{9}$$

for some $m > 0$. There are several ways of obtaining an upper bound for the order-type of $\prec$ in terms of the length of a cut-free deduction of $\mathrm{WO}(\prec)$ (see e.g. [13, Theorem 23.1], [19, Theorem 3.6], [6, Theorem 2.27]) which ultimately go back to Gentzen. Schütte [13, Theorem 23.1] obtains particularly sharp bounds. He shows that the length α of a cut-free derivation of transfinite induction along an ordering $\prec$ provides an upper bound for the ordinal rank of $\prec$ if $\omega \cdot \alpha = \alpha$. For our purpose, however, we need to extract bounds from deductions that still have cuts with formulas $\mathsf{U}(s), \neg\mathsf{U}(s)$.[2] We could first eliminate these remaining cuts, however, we would get bounds of the form 2^{σ^m}, and these are too high for our purpose of showing that σ^ω is the proof-theoretic ordinal of $\mathbf{T}_\sigma$. To overcome

---

[2] They may also contain cuts with formulas $R(t_1, \ldots, t_k), \neg R(t_1, \ldots, t_k)$, where R is a symbol for a primitive recursive predicate. But these are entirely harmless.

this obstacle we shall draw on a technique that the third author has used for many years. To this end we extend $\mathbf{PA}_\omega$ by yet another infinitary rule $\text{Prog}_\prec$ due to Schütte [11, p. 384] called *Progressionsregel* ($\text{Prog}_\prec$ was also used in [12, p. 214] and in [14]):

$$\frac{\Gamma, \mathsf{U}(\bar{m}) \text{ for all } m \prec n}{\Gamma, \mathsf{U}(s)} \tag{10}$$

whenever s is a closed term with value n.

Let $\text{PROG}_\prec$ be an abbreviation for $\forall v[\forall u \prec v\, \mathsf{U}(u) \rightarrow \mathsf{U}(v)]$. The rule $\text{Prog}_\prec$ has the effect of making $\text{PROG}_\prec$ provable. We shall refer by $\mathbf{PA}^*_\infty$ to the extension of $\mathbf{PA}_\omega$ by the rule $\text{Prog}_\prec$.

Lemma 3

$$\mathbf{PA}^*_\infty \left|\frac{\alpha}{1}\right. \neg\text{PROG}_\prec, \Gamma \Rightarrow \mathbf{PA}^*_\infty \left|\frac{3\cdot\alpha}{1}\right. \Gamma. \tag{11}$$

Proof. We proceed by induction on α. If $\neg\text{PROG}_\prec$ was not the main formula of the last inference then the desired result follows immediately by applying the inductive assumption to its premises and subsequently reapplying the same inference. Thus suppose that $\neg\text{PROG}_\prec$ was the main formula of the last inference. Then

$$\mathbf{PA}^*_\infty \left|\frac{\alpha_0}{1}\right. \neg\text{PROG}_\prec, \forall u \prec s\, \mathsf{U}(u) \wedge \neg\mathsf{U}(s), \Gamma \tag{12}$$

for some $\alpha_0 \lhd \alpha$. The induction hypothesis yields

$$\mathbf{PA}^*_\infty \left|\frac{3\cdot\alpha_0}{1}\right. \forall u \prec s\, \mathsf{U}(u) \wedge \neg\mathsf{U}(s), \Gamma. \tag{13}$$

for some s. Using inversion for ($\wedge$), ($\forall$) and ($\vee$) we arrive at

$$\mathbf{PA}^*_\infty \left|\frac{3\cdot\alpha_0}{1}\right. \Gamma, \neg\bar{n} \prec s, \mathsf{U}(\bar{n}) \tag{14}$$

for all n, and

$$\mathbf{PA}^*_\infty \left|\frac{3\cdot\alpha_0}{1}\right. \neg\mathsf{U}(s), \Gamma. \tag{15}$$

Since $\mathbf{PA}^*_\infty \left|\frac{0}{0}\right. \Gamma, \bar{n} \prec s$ holds for all n with $n \prec s^\mathbb{N}$, we can apply cuts and the rule $\text{Prog}_\prec$ to (14) to arrive at

$$\mathbf{PA}^*_\infty \left|\frac{3\cdot\alpha_0+2}{1}\right. \Gamma, \mathsf{U}(s). \tag{16}$$

Applying Cut to (16) and (15) yields

$$\mathbf{PA}^*_\infty \left|\frac{3\cdot\alpha_0+3}{1}\right. \Gamma \tag{17}$$

and hence

$$\mathbf{PA}^*_\infty \left|\frac{3\cdot\alpha}{1}\right. \Gamma. \tag{18}$$

Corollary 3

$$\mathbf{PA}^*_\infty \vdash_1^{\sigma^m} U(\bar{n})$$

for all n.

Proof. Follows from (9) and Lemma 3. Note that $m > 0$.

For a closed numerical term s we denote by $|s|_\prec$ the ordinal $\{|\bar{n}|_\prec \mid \bar{n} \prec s \text{ is true}\}$.

Proposition 1. *Assume that the sequent* $\neg U(t_1), \ldots, \neg U(t_r), U(s_1), \ldots, U(s_q)$ *is not an axiom and* $s_1 \preceq \ldots \preceq s_q$ *holds. Then*

$$\mathbf{PA}^*_\infty \vdash_1^\alpha \neg U(t_1), \ldots, \neg U(t_r), U(s_1), \ldots, U(s_q)$$

implies

$$|s_1|_\prec < \omega \cdot \alpha. \tag{19}$$

Proof. Let $\neg U(t)$ be an abbreviation for $\neg U(t_1), \ldots, \neg U(t_r)$. In the above we allow $r = 0$ in which case $\neg U(t)$ is the empty sequent.

We proceed by induction on α. As the sequent is not an axiom it must have been inferred. The only two possibilities are applications of $\mathrm{Prog}_\prec$ or cuts with atomic formulas.

Case 1: The last inference was $\mathrm{Prog}_\prec$. Then there is a term s_j and $\alpha_0 \lhd \alpha$ such that $\mathbf{PA}^*_\infty \vdash_1^{\alpha_0} \neg U(t), U(s_1), \ldots, U(s_q), U(\bar{n})$ for all $\bar{n} \prec s_j$. As $s_1 \preceq s_j$ this also holds for all $\bar{n} \prec s_1$. The induction hypothesis yields that

$$|\bar{n}|_\prec < \omega \cdot \alpha_0$$

holds for those $\bar{n} \prec s_1$ for which the sequent is not an axiom. Since by Definition 1 (ii)

$$\neg U(t), U(s_1), \ldots, U(s_q), U(\bar{n}) \tag{20}$$

is an axiom only if $\bar{n}$ has the same value as some $t_1, \ldots, t_r$, then there are only finitely many n for which (20) is an axiom. Thus $|s_1|_\prec < \omega \cdot \alpha_0 + \omega$, whence $|s_1|_\prec < \omega \cdot \alpha$.

Case 2: The last inference was a cut with cut formulas $U(p), \neg U(p)$, i.e., we have

$$\mathbf{PA}^*_\infty \vdash_1^{\alpha_0} \neg U(t), U(s_1), \ldots, U(s_q), U(p) \tag{21}$$

$$\mathbf{PA}^*_\infty \vdash_1^{\alpha_0} \neg U(t), U(s_1), \ldots, U(s_q), \neg U(p) \tag{22}$$

for some $\alpha_0 < \alpha$ and closed term p. If the sequent from (22) is not an axiom, the induction hypothesis applied to that derivation yields $|s_1|_\prec < \omega \cdot \alpha_0$. If it is an axiom, there is an s_j such that p and s_j evaluate to the same numeral, and hence $s_1 \preceq p$. So in this case the induction hypothesis applied to (21) yields $|s_1|_\prec < \omega \cdot \alpha_0$.

Case 3: The last inference was a cut with cut formulas $R(u_1, \ldots, u_p), \neg R(u_1, \ldots, u_p)$ for a symbol R for a primitive recursive relation. Then we have

$$\mathbf{PA}^*_\infty \vdash^{\alpha_0}_1 \neg U(t), U(s_1), \ldots, U(s_q), R(u_1, \ldots, u_p) \tag{23}$$

$$\mathbf{PA}^*_\infty \vdash^{\alpha_0}_1 \neg U(t), U(s_1), \ldots, U(s_q), \neg R(u_1, \ldots, u_p) \tag{24}$$

for some $\alpha_0 < \alpha$. If $R(u_1, \ldots, u_p)$ is true it follows from (24) that we also have

$$\mathbf{PA}^*_\infty \vdash^{\alpha_0}_1 \neg U(t), U(s_1), \ldots, U(s_q)$$

and hence the induction hypothesis yields $|s_1|_\prec < \omega \cdot \alpha_0$.

Likewise, if $R(u_1, \ldots, u_p)$ is false it follows from (23) that we also have

$$\mathbf{PA}^*_\infty \vdash^{\alpha_0}_1 \neg U(t), U(s_1), \ldots, U(s_q)$$

and hence the induction hypothesis yields $|s_1|_\prec < \omega \cdot \alpha_0$.

Corollary 4

(i)

$$\mathbf{PA}^*_\infty \vdash^{\alpha}_1 U(s) \Rightarrow |s|_\prec < \omega \cdot \alpha.$$

(ii)

$$\mathbf{PA}_\omega \vdash^{\beta}_1 \mathrm{WO}(\prec) \Rightarrow |\prec| \leq \omega \cdot 3 \cdot \beta$$

where $|\prec|$ stands for the ordinal rank of $\prec$.

Proof. (i) is an immediate consequence of Proposition 1.
(ii) follows from (i) and Lemma 3.

In sum, it follows that the ordinal rank of $\prec$ is not bigger than σ^m, and hence σ^ω is an upper bound for the proof-theoretic ordinal of $\mathbf{T}_\sigma$.

Proposition 1 can also be shown via techniques in A. Beckmann's dissertation, notably his [1, 5.2.5 Boundedness Theorem] that also features in [2].

Turning to lower bounds, one can easily show, using external induction on n, that $\mathbf{T}_\sigma \vdash \mathrm{WO}(\sigma^n)$. This is a folklore result; details can be found in [17, Lemma 4.3]. As a consequence of the results gathered so far we have

Theorem 3. *The proof-theoretic ordinal of* $\mathbf{I\Sigma}_1 + \mathrm{TI}(\sigma, \Pi_1)$ *is* σ^ω.

It remains to transfer this result to our target theory $\mathbf{RCA}_0 + \mathrm{WO}(\sigma)$.

4 Π_1^1-Conservativity

We here prove that $\mathbf{RCA}_0 + \mathrm{WO}(\sigma)$ is Π_1^1-conservative over $\mathbf{T}_\sigma$. More precisely, a Π_1^1-sentence $\forall X F(X)$ (with $F(X)$ being arithmetic) is identified with $F(\mathsf{U})$ in the first-order context of $\mathrm{TI}(\sigma, \Pi_1)$. This is enough to apply our results from the previous section to conclude that the ordinal of $\mathbf{RCA}_0 + \mathrm{WO}(\sigma)$ is σ^ω.

To prove the conservativity result, we proceed as follows. We start by showing that any model of $\mathrm{I}\Sigma_1 + \mathrm{TI}(\sigma, \Pi_1)$ can be extended to a model of $\mathbf{RCA}_0 + \mathrm{WO}(\sigma)$. The argument is essentially contained in Simpson [16], IX.1. By writing that $\mathbf{M}_1$ is an ω-submodel of $\mathbf{M}_2$ we mean that $\mathbf{M}_1 = (M_1, \mathcal{S}_1)$ and $\mathbf{M}_2 = (M_1, \mathcal{S}_2)$ where $\mathcal{S}_1 \subseteq \mathcal{S}_2$. In other words, the two models share the same first-order part M_1.

Lemma 4. *Let* $\mathbf{M}$ *be an* L_2-*structure which satisfies the axioms of* $\mathrm{I}\Sigma_1 + \mathrm{TI}(\sigma, \Pi_1)$. *Then* $\mathbf{M}$ *is an* ω-*submodel of some model of* $\mathbf{RCA}_0 + \mathrm{WO}(\sigma)$.

Proof. We first show that $\mathbf{M}$ can be extended to a model $\mathbf{M}'$ satisfying $\mathbf{RCA}_0$ and $\mathrm{TI}(\sigma, \Delta_0^0)$ with the same first-order domain as $\mathbf{M}$. Then we show that such an extension also satisfies $\mathrm{WO}(\sigma)$.

The ω-extension $\mathbf{M}'$ is defined exactly as in Simpson [16] Lemma IX.1.8, i.e., the second-order part is given by the Δ_1^0-definable sets of the base model $\mathbf{M}$. By Lemma IX.1.8 of [16] we have that $\mathbf{M}'$ satisfies $\mathbf{RCA}_0$.

Then, in order to check that $\mathrm{TI}(\sigma, \Delta_0^0)$ is also satisfied, we use the first claim in Simpson's Lemma IX.1.8. Let φ be a Σ_0^0 formula with no free set variables and parameters in $\mathbf{M}'$. Then, there exists a Π_1^0-formula φ_Π with the same free variables and parameters only in M such that φ and φ_Π are equivalent over M'. Thus, $\mathrm{TI}(\sigma, \Pi_1^0)$ in $\mathbf{M}$ implies $\mathrm{TI}(\sigma, \Delta_0^0)$ in $\mathbf{M}'$.

Finally, we show that M' also satisfies $\mathrm{WO}(\sigma)$. If this were not the case, letting S be a set witnessing $\neg\mathrm{WO}(\sigma)$, we would have that $\bar{S}$, i.e. the complement of S, witnesses the failure of an instance of $\mathrm{TI}(\sigma, \Delta_0^0)$. More precisely: suppose that S is non-empty and has no $\lhd$-minimal element. Then $\exists x(x \in S)$. On the other hand, $\bar{S}$ is in M' (since any model of $\mathbf{RCA}_0$ is closed under Turing reducibility hence under complement) and $\forall x(\forall y(y \lhd x \to y \in \bar{S}) \to x \in \bar{S})$. Suppose in fact that for some x, $\forall y(y \lhd x \to y \in \bar{S})$ but $x \in S$. Then all $y \lhd x$ are not in S but x is in S and thus x is the minimum of S, contra our hypothesis. □

Remark 1. The proof of Lemma 4 above shows that if $\mathbf{RCA}_0 + \mathrm{WO}(\sigma)$ proves $\forall X F(X)$ with $F(X)$ arithmetic, then $\mathbf{T}_\sigma$ proves the formula $F'(\mathsf{U})$ obtained from $F(X)$ by replacing expressions of the form '$t \in X$' by '$\mathsf{U}(t)$'.

Then, we proceed by showing that Lemma 4 gives a sufficient condition for Π_1^1-conservativity.

Lemma 5. *If* T_1 *and* T_2 *are theories in the language of second order arithmetic and every model of* T_1 *is an* ω-*submodel of a model of* T_2 *then* T_2 *is* Π_1^1-*conservative over* T_1.

Proof. If ψ is Π_1^1 and T_1 does not prove ψ, let $\mathbf{M}_1$ be a model of $T_1 + \neg\psi$. Then $\mathbf{M}_1$ is an ω-submodel of a model $\mathbf{M}_2$ of T_2. Then $\mathbf{M}_2$ is a model of $T_2 + \neg\psi$ and thus T_2 does not prove ψ. □

Theorem 4. $\mathbf{RCA}_0 + \mathrm{WO}(\sigma)$ *is* Π_1^1-*conservative over* $\mathbf{I}\Sigma_1 + \mathrm{TI}(\sigma, \Pi_1)$.

Proof. Follows immediately from Lemmas 4 and 5.

Theorem 5. *The proof-theoretic ordinal of* $\mathbf{RCA}_0 + \mathrm{WO}(\sigma)$ *is* σ^ω.

Proof. The upper bound follows from Theorems 3 and 4. The lower bound follows from the observation that for each n $\mathbf{RCA}_0 + \mathrm{WO}(\sigma) \vdash \mathrm{WO}(\sigma^n)$. The proof, which we omit, is analogous to the proof that $\mathbf{RCA}_0 \vdash \mathrm{WO}(\omega^n)$, for each n. □

Remark 2. The Π_1^1-conservativity of $\mathbf{I}\Sigma_1 + \mathrm{TI}(\sigma, \Pi_1)$ over $\mathbf{RCA}_0 + \mathrm{WO}(\sigma)$ also holds and can be established by standard arguments. In particular one can prove that if $A(\mathsf{U})$ is provable in $\mathbf{T}_\sigma$ then $\forall X A^*(X)$ is provable in $\mathbf{RCA}_0 + \mathrm{WO}(\sigma)$, where $A^*(X)$ is the result of first replacing '$\mathsf{U}(t)$' by '$t \in X$' and then translating the primitive recursive function and predicate symbols not belonging to the language of $\mathbf{RCA}_0$ as in [16], Definition IX.3.4.

Acknowledgements. This publication was made possible through the support of a grant from the John Templeton Foundation ("A new dawn of intuitionism: mathematical and philosophical advances," ID 60842). The opinions expressed in this publication are those of the authors and do not necessarily reflect the views of the John Templeton Foundation.

References

1. Beckmann, A.: Separating fragments of bounded arithmetic. Ph.D. thesis, Universität Münster (1996)
2. Beckmann, A., Pohlers, W.: Applications of cut-free infinitary derivations to generalized recursion theory. Ann. Pure Appl. Logic **94**, 7–19 (1995)
3. Carlucci, L., Dehornoy, P., Weiermann, A.: Unprovability results involving braids. Proc. London Math. Soci. **102**(1), 159–192 (2011)
4. Friedman, H.: Adjacent Ramsey theory. Draft, August 2010. https://u.osu.edu/friedman.8/
5. Friedman, H., Pelupessy, F.: Independence of Ramsey theorem variants using ε_0. Proc. Am. Math. Soc. **144**, 853–860 (2016)
6. Friedman, H., Sheard, S.: Elementary descent recursion and proof theory. Ann. Pure Appl. Logic **71**, 1–45 (1995)
7. Hatzikiriakou, K., Simpson, S.G.: Reverse mathematics, Young diagrams, and the ascending chain condition. J. Symbolic Logic **82**, 576–589 (2017)
8. Kreisel, G., Lévy, A.: Reflection principles and their use for establishing the complexity of axiomatic systems. Zeitschrift für mathematische Logik und Grundlagen der Mathematik **14**, 97–142 (1968)
9. Kreuzer, A., Yokoyama, K.: On principles between Σ_1- and Σ_2-induction, and monotone enumerations. J. Math. Logic **16**, 1650004 (2016)

10. Rathjen, M.: The art of ordinal analysis. In: Sanz-Solé, M., Soria, J., Varona, J.L., Verdera, J. (eds.) Proceedings of the International Congress of Mathematicians, Madrid, 22–30 August 2006, pp. 45–69. European Mathematical Society (2006)
11. Schütte, K.: Beweistheoretische Erfassung der unendlichen Induktion in der Zahlentheorie. Mathematische Annalen **122**, 369–389 (1951)
12. Schütte, K.: Beweistheorie, 1st ed. Springer, Berlin (1960)
13. Schütte, K.: Proof Theory. Grundlehren der mathematischen Wissenschaften, vol. 225, 1st edn. Springer, Heidelberg (1977). https://doi.org/10.1007/978-3-642-66473-1
14. Schwichtenberg, H.: Proof theory: some applications of cut-elimination. In: Barwise, J. (ed.) Handbook of Mathematical Logic, pp. 867–895. North Holland, Amsterdam (1977)
15. Simpson, S.: Ordinal numbers and the Hilbert basis theorem. J. Symbolic Logic **53**, 961–974 (1988)
16. Simpson, S.: Subsystems of Second Order Arithmetic, 2nd edn. Cambridge University Press, New York (2009). Association for Symbolic Logic
17. Sommer, R.: Transfinite induction within Peano arithmetic. Ann. Pure Appl. Logic **76**, 231–289 (1995)
18. Tait, W.W.: Normal derivability in classical logic. In: Barwise, J. (ed.) The Syntax and Semantics of Infinitary Languages. LNM, vol. 72, pp. 204–236. Springer, Heidelberg (1968). https://doi.org/10.1007/BFb0079691
19. Takeuti, G.: Proof Theory, 2nd edn. North Holland, Amsterdam (1987)

Study of Stepwise Simulation Between ASM

Patrick Cégielski and Julien Cervelle[✉]

LACL, Université Paris-Est Créteil, 94010 Créteil Cedex 2, France
{patrick.cegielski,julien.cervelle}@u-pec.fr

Abstract. In this paper we study the notion of stepwise simulation between Abstract State Machines, to explore if some natural change on the original definition would keep it sound. We prove that we have to keep the classical notion and give results about the computability of the simulation itself.

1 Introduction

After one or two centuries of discussion, Richard Dedekind has given a definition of "function": A *function f* from a set A to a set B is a relation $R \subseteq A \times B$ such that if (x, y) and (x, y') belongs to the relation then $y = y'$.

From an informal point of view, a *function* is *computable* if there exists a "mechanical" process which, being given an element x of A, provides the (unique) element $y = f(x)$ of B after some finite "time of computation" (or equivalently some finite number of "steps") if $f(x)$ exists, and runs indefinitely otherwise. A formal definition was given by Alan Turing in 1936, exhibiting a non computable function.

The definition of Turing is universally accepted but other "models of computation" were and are still exhibited for various reasons.

A model of computation is a set of elements, each of them being called a *machine* or a *program*, depending on the model. Here, we do not define a program as a word over some alphabet because such a pattern is not convenient for ASM. To each machine is associated a finite set of *variables*, each variable taking values in a well defined set. Some variables are used for the *input* and possibly some variables for the *output*; the other variables are called *auxiliary variables*.

Given a finite list $v_0, v_1, \ldots, v_n$ of variables and an associated list $D_0, D_1, \ldots, D_n$ of sets, called *value sets*, an *element of trace* is an assignment for every variable which gives to v_i an assignment in D_i. A *trace* is a finite or infinite sequence of trace elements over the same variables and value sets.

For a given machine M of a given deterministic model of computation $\mathcal{M}$, the *run* of M on a given input is the unique trace whose first element is the initial assignment of the variables: The input variables are initialized with the input and the auxiliary and output variables are initialized as specified by the model. The element following an element e corresponds to the states of the variables after one step of M starting with the variable values as in the element e.

© Springer Nature Switzerland AG 2019
F. Manea et al. (Eds.): CiE 2019, LNCS 11558, pp. 156–167, 2019.
https://doi.org/10.1007/978-3-030-22996-2_14

Obviously, a run is a trace but the converse is not necessarily true: Indeed, a trace is not necessarily recursive while a run from a constructive model is always recursive.

Let $v_0, v_1, \cdots, v_n$ be a finite list of variables, $v_0, v_1, \cdots, v_p$ be a sublist of input variables, and $D_0, D_1, \cdots, D_n$ be the associated value sets. For a machine M of a model of computation $\mathcal{M}$, the *runs' log* is the set of runs of M for all the possible inputs $\mathcal{I} \in D_0 \times \cdots \times D_p$. Thus, a machine is a finite description of a runs' log, which is an infinite set.

<u>Problem.</u> *Given two models of computation, is a runs' log in one equal to a runs' log in the other?*

The answer to this problem is in general NO but there exists a model of computation, ASM, which has the property that each runs' log of any machine of any model of computation is the runs' log of a machine of the ASM model. Let us make this statement more precise. Yuri GUREVICH has given a schema of languages which is not only a *Turing-complete language* (a language allowing to program each computable function), but which also allows to describe step-by-step the behavior of all algorithms for each computable function (it is an *algorithmically complete language*); this schema of languages was first called *dynamic structures*, then *evolving algebras*, and finally ASM (for *Abstract State Machines*) [2]. In 2000, he proposed the now-called Gurevich's thesis *"the notion of algorithm is entirely captured by the model"* in [3]. A consequence is

For every runs' log, there exists an ASM with that exact runs' log.

ASM is the only known model of computation to have this property. However, authors have considered some other models of computation which are interesting candidates to have a weaker variant of the property. For some of them, it depends on the granularity of a step: Instead of requiring equality of runs, one allows the run of the model to be a specific subsequence of the simulated run. For instance, one keeps only an element every k elements, where the integer k is fixed. This leads us to the notion of *k-simulation*.

Some authors (see for instance [4]), considering and proving this weaker property for their model, insist on the strict regularity: an element every k elements (for instance 2, 4, 6, ... for $k = 2$) and not just allowing to discard at most $k - 1$ elements (for instance 1, 3, 4, 5, 7, ...). This implies adding steps which do nothing (often called "skip" of "nop") in the programs. But these authors give no explanation to justify such a strict constraint.

Then a natural question arises on the behavior of models of computation: Do we need to force the regularity? One way to get enlightenment about the question is to see if we can build two ASM A and B whose traces are all equal up to irregular dilatation but in such a way that the set of points to be removed is, somehow, a non recursive set. This means that the two computations are equivalent but that one of the ASM computes something more in its trace than the other one.

The main result of this paper proves that we have built such a pair of ASM and furthermore that these two ASM are such that one simulates the other but for a simpler notion: removing at most one point to each trace is sufficient to get the simulation. The result states that the point is not computable given the input.

The paper is organized as follows. The next section gives insight about what is a trace of the execution in the case of a general model of computation and introduces formally ASM and their traces. Section 3 analyzes some possible definitions of equivalent traces and how they are related to computability. It also includes the proofs of announced results.

2   Definitions

2.1   Traces in a General Setting of Model of Computation

Let $\mathcal{M}$ be a (discrete time) model of computation and M an instance of $\mathcal{M}$. We suppose that $\mathcal{M}$ is such that the "state" of M at some time is entirely described by values stored in a finite number of *variables*. Let $v_0, \ldots, v_n$ be these variables. For all i, the variable v_i takes its values from the set D_i (in a more general setting, it is sufficient for D_i to be just a class).

Definition 1. *A element of trace for M is an element of $D_0 \times \cdots \times D_n$.*

A trace is a sequence of elements of trace indexed by $\mathbb{N}$ or by a finite interval $I_k = \{0, \ldots, k\}$ of $\mathbb{N}$. In case of a finite interval I_k, we call k the length *of the trace.*

For a given M, some variables are distinguished and called input variables. *Without loss of generality, we suppose that $v_0, \ldots, v_m$ for some $m \leq n$ (i.e. the first $m + 1$ ones) are the input variables. An* input *for M is an element of $D_0 \times \cdots \times D_m$.*

A run *on input $\imath$ for machine M is the trace $(t_i)_{i \in I}$ where*

- *The trace element t_0 is initialized with $\imath$ i.e. the input variables are set as in $\imath$ and the remaining variables are set depending on the definition of $\mathcal{M}$ (see Remark 1).*
- *For all $i \in I$, applying one step of M to t_i leads to t_{i+1} unless M halts and in this case the interval I is $\{0, \ldots, i\}$.*

Remark 1. We suppose that the values of variables v_i for $m < i \leq n$ in t_0 are either fixed by the definition of M or can take any values and in this case, they must have no incidence on the computation of M.

2.2   Definition of ASM

We first introduce ASM, making precise our point of view on ASM, because several definitions exist.

ASM were defined formally in [2]. For this paper, we choose to use only ASM in some normal form (see [1]). We refer the reader to the aforementioned references for the general ASM definition; we just give the formal definition of ASM we use, a variant which is simpler though more verbose and equivalent in power.

Syntax

Definition 2. *An* ASM vocabulary, *or* signature, *is a first-order signature* $\mathcal{L}$ *with a finite number of static function symbols, a finite number of dynamic function symbols, a finite number of predicate symbols (among which the two boolean constant symbols* true *and* false*), and an additional symbol, of arity 0, (denoted by* undef*), logical connectives (*¬*,* ∧*, and* ∨*), and the* equality *predicate denoted by* =*.*

Terms of $\mathcal{L}$ are defined by:

- if c is a nullary function symbol (a constant, dynamic or static) of $\mathcal{L}$, then c is a term,
- if $t_1, \ldots, t_n$ are terms and f is an n-ary function symbol (dynamic or static) of $\mathcal{L}$ then $f(t_1, \ldots, t_n)$ is a term.

Definition 3. Boolean terms *of $\mathcal{L}$ are defined inductively by:*

- *if p is an n-ary predicate and $t_1, \ldots, t_n$ are terms of $\mathcal{L}$ then $p(t_1, \ldots, t_n)$ is a boolean term;*
- *if t and t' are terms of $\mathcal{L}$, then $t = t'$ is a boolean term;*
- *if F, F' are boolean terms of $\mathcal{L}$, then $\neg F$, $F \wedge F'$, $F \vee F'$ are boolean terms.*

Definition 4. *Let $\mathcal{L}$ be an ASM signature.* ASM rules *are defined inductively as follows:*

- *An* update *is an expression of the form $f(t_1, \ldots, t_n) := t_0$, where f is a n-ary dynamic functional symbol and $t_0, t_1, \ldots, t_n$ are terms of $\mathcal{L}$.*
- *If $R_1, \ldots, R_k$ are updates of signature $\mathcal{L}$, where $k \geq 1$, then the expression $R_1 || \cdots || R_k$ is called a* block *and means parallel execution of the updates.*
- *Finally, if R is a block and φ is a boolean term, the ordered pair $\langle \varphi, R \rangle$ is called a* conditional rule *which must be seen as the instruction* if φ then R. *In this paper, we call φ the* guard *and R the* block *of the conditional rule.*

Definition 5. *Let $\mathcal{L}$ be an ASM signature. A* program *on signature $\mathcal{L}$, or $\mathcal{L}$-program, is a finite set of conditional rules of that signature.*

We can now define ASM.

Definition 6. *An ASM A is a tuple $\langle \mathcal{L}, P \rangle$ where $\mathcal{L}$ is an ASM signature and P is an $\mathcal{L}$-program. We denote by $A_{\mathcal{D}}$ the set of dynamic symbols of A.*

Semantics

Definition 7. *Let $\mathcal{L}$ be an ASM signature. An ASM* abstract state, *or more precisely an $\mathcal{L}$-state, is a synonym for a first-order structure $\mathcal{A}$ of signature $\mathcal{L}$ (an $\mathcal{L}$-structure). We denote by $[t]^{\mathcal{A}}$ the value of the term t in the structure $\mathcal{A}$.*

The universe of $\mathcal{A}$, denoted by $A^{\perp}$, consists of the elements of the *data set A* and a special value $\perp$ (supposedly not in A). The interpretation of the symbol undef in $A^{\perp}$ is always $\perp$.

Definition 8. *Let $\mathcal{L}$ be an ASM signature and A a nonempty set. A set of modifications (more precisely an $(\mathcal{L}, A)$-modification set) is any finite set of triples $(f, \bar{a}, a)$, where f is an n-ary function symbol of $\mathcal{L}$, $\bar{a} = (a_1, \ldots, a_n)$ is an n-tuple of $A^{\perp}$, and a is an element of $A^{\perp}$.*

Definition 9. *Let $\mathcal{L}$ be an ASM signature, let $\mathcal{A}$ be an $\mathcal{L}$-state and let Π be an $\mathcal{L}$-program. Let $\Delta_{\Pi}(\mathcal{A})$ denote the set defined by as follows:*

1. *If u is the update rule $f(t_1, \ldots, t_n) := t_0$ then*

$$\Delta_u(\mathcal{A}) = \{(f, ([t_1]^{\mathcal{A}}, \ldots, [t_n]^{\mathcal{A}}), [t_0]^{\mathcal{A}})\}.$$

2. *If B is the block $R_1 || \cdots || R_k$ then:*

$$\Delta_B(\mathcal{A}) = \{\Delta_{R_1}(\mathcal{A}), \ldots, \Delta_{R_n}(\mathcal{A})\}.$$

3. *If T is the conditional rule $\langle \varphi, R \rangle$, we first have to evaluate the expression $t = [\varphi]^{\mathcal{A}}$. We define:*

$$\Delta_T(\mathcal{A}) = \begin{cases} \emptyset & \text{if } t \text{ is false,} \\ \Delta_R(\mathcal{A}) & \text{otherwise.} \end{cases}$$

4. *Finally, if Π is a program consisting in rules $T_1, \ldots, T_n$, then we define:*

$$\Delta_{\Pi}(\mathcal{A}) = \bigcup_{i=1}^{n} \Delta_{T_i}(\mathcal{A}).$$

We defined $\Delta_{\Pi}(\mathcal{A})$ as an $(\Pi, \mathcal{L}, A)$-set of modifications.

Definition 10. *A set of modifications is* incoherent *if it contains two elements $(f, \bar{a}, a)$ and $(f, \bar{a}, b)$ with $a \neq b$. It is* coherent *otherwise.*

Definition 11. *Let $\mathcal{L}$ be an ASM signature, Π an $\mathcal{L}$-program, and $\mathcal{A}$ an $\mathcal{L}$-state. The machine's definition must ensure $\Delta_{\Pi}(\mathcal{A})$ is coherent (otherwise, the machine's definition is invalid[1]). The transform $\tau_{\Pi}(\mathcal{A})$ of $\mathcal{A}$ by Π is the $\mathcal{L}$-structure $\mathcal{B}$ defined by:*

- *the base set of $\mathcal{B}$ is the base set $A^{\perp}$ of $\mathcal{A}$;*
- *for any n-ary element f of $\mathcal{L}$ and any element $\bar{a} = (a_1, \ldots, a_n)$ of A^n:*
  - *If there exists some (necessarily unique) a such that $(f, \bar{a}, a) \in \Delta_{\Pi}(\mathcal{A})$, then: $[f]^{\mathcal{B}}(\bar{a}) = a$.*
  - *Otherwise: $[f]^{\mathcal{B}}(\bar{a}) = [f]^{\mathcal{A}}(\bar{a})$.*

Definition 12. *Let $\mathcal{L}$ be an ASM signature, Π an $\mathcal{L}$-program, and $\mathcal{A}$ an $\mathcal{L}$-state. The* computation *is the sequence of $\mathcal{L}$-states $(\mathcal{A}_n)_{n \in \mathbb{N}}$ defined by:*

---

[1] Given an ASM, the well definition of an ASM is undecidable. However, one could add some rules to detect incoherence at runtime and behave accordingly.

- $\mathcal{A}_0 = \mathcal{A}$ *(called the* initial algebra *of the computation);*
- $\mathcal{A}_{n+1} = \tau_\Pi(\mathcal{A}_n)$ *for* $n \in \mathbb{N}$.

For ASM, a computation halts *if there exists a fixed point* $\mathcal{A}_{n+1} = \mathcal{A}_n$. *In this case, this fixed point* $\mathcal{A}_n$ *is the* result *of the computation.*

Definition 13. *The formal* semantics *is the partial class function which transforms* $\langle \Pi, \mathcal{A} \rangle$, *where* Π *is an ASM program and* $\mathcal{A}$ *a state, in the fixed point* $\mathcal{A}_n$ *obtained by iterating* τ_Π *starting from* $\tau_\Pi(\mathcal{A})$ *until a fixed point is reached if such a fixed point exists, otherwise it is undefined.*

Finally, though it is not directly mentioned in the original paper defining ASM, as we need to deal with simulation, we need a formal definition of how input is treated.

For some ASM signature $\mathcal{L}$, some of the dynamic symbols will be used for the input. We call them *input* symbols. To construct the initial algebra $\mathcal{I}(e)$ of an ASM computation on input e, we use the following:

- Some set E to use $E^\perp$ as universe for the algebra.
- For all static variables of arity n of $\mathcal{L}$, we assign a function from $(E^\perp)^n \to E$.
- The input is stored in input symbols (so input can be some values of $E^\perp$ or functional).
- The rest of the dynamic symbols are set to the constant functions equal to $\perp$ on all their inputs.
- The set E is called the *data set* of the ASM.
- We say that an ASM is m-ary when it has only one input symbol, this symbol having arity m.

2.3   Trace

We restate here the definition of trace introduced in Sect. 2.1 in the special case of ASM.

Definition 14. *For some ASM, a* trace element *is the values of all the dynamic symbols (input or not).*

The trace *of some ASM A on input e is the sequence of elements of trace* $(t_i)_{i \in I}$ *where I is either $\mathbb{N}$ or $\{0, \dots, \ell\}$ for some ℓ where t_i is the restriction of* $\mathcal{A}_i$ *to dynamic symbols if $(\mathcal{A}_n)_{n \in \mathbb{N}}$ is the computation of the ASM starting from* $\mathcal{I}(e)$. *Moreover, if the computation does not halt, then $I = \mathbb{N}$, and otherwise,* $I = \{0, \dots, \ell\}$ *when ℓ is the number of steps before halting (that is the smallest with $\tau_\Pi(\mathcal{A}_\ell) = \mathcal{A}_\ell$). We denote I by* dom(t).

If y is a trace element and S is a subset of the dynamic symbols, $y \restriction S$ is the trace element which assign values only to symbols of S as in y. If t is a trace, we denote by $t \restriction S$ the trace $((t_i \restriction S)_{i \in \text{dom}(t)})$.

2.4   Self-detection of a Halting ASM

It is possible (well-known result of the folklore) in any ASM to write a boolean expression (subsequently called HasHalted) which evaluates to true when it is in a state which is a fixed point. The idea is to test if all assignments do not change the assigned value.

3   Analysis

We first introduce a preliminary notion of subtrace.

Definition 15. *Let t_1 and t_2 be two traces. We say that t_2 is a* subtrace *of t_1 if there exists a strictly increasing function $s : \mathrm{dom}(t_1) \to \mathrm{dom}(t_2)$ such that:*

1. $\forall i \in \mathrm{dom}(t_1), t_1(i) = t_2(s(i))$,
2. $s(0) = 0$
3. $\mathrm{dom}(t_1)$ is finite if and only if $\mathrm{dom}(t_2)$ is finite
4. if $\mathrm{dom}(t_1)$ is finite, then $s(\max \mathrm{dom}(t_1)) = \max \mathrm{dom}(t_2)$.

The function s and condition 1. ensure that t_1 is extracted from the elements of t_2 keeping the order. The next conditions state that the two traces have the same start 2. both have an end or both are infinite 3. and in the finite case, have the same end 4.

We define the notion of k-weak regular subtracing which is a more constrained notion of subsequence.

Definition 16 (Weak regular subtrace). *Let t_1 and t_2 be two traces and k some positive integer. We say that t_2 is a k-weak regular subtrace of t_1 if t_2 is a subtrace of t_1 and furthermore if $\forall i \in \mathrm{dom}(t_1), k \geq s(i+1) - s(i) > 0$.*

The last condition ensures that to build the subtrace of t_2, one cannot skip more than $k-1$ elements: two consecutive terms taken from t_2 always belong to some windows of length k.

We say that s is a guide *of the k-weak regular subtracing of t_1 by t_2.*

Finally, we say that the set $w = \mathrm{dom}(t_2) \smallsetminus \{s(i) \mid i \in \mathrm{dom}(t_1)\}$ is a witness *of the k-weak regular subtracing of t_1 by t_2.*

Remark 2. Witnesses and guides are not unique: if the same trace element x occurs twice in t_2 ($t_2(a) = t_2(b) = x$) and corresponds to only one trace element in t_1 ($t_1(c) = x$), the function s can possibly choose either ($s(c) = a$ or $s(c) = b$) provided both choices comply to the weak regular subtracing condition.

Remark 3. Note that guides and witnesses are linked and uniquely defined and computable one from another. Indeed, the guide is the increasing enumeration of the complement of the witness.

For instance, consider the trace $t_1 = [x = 0], [x = 2], [x = 4], \ldots$ (x is incremented by 2 at each step) and the trace $t_2 = [x = 0], [x = 1], [x = 2], \ldots$ (x is incremented by 1 at each step). The only guide of the 2-weak regular subtracing of t_1 by t_2 is $s : t \mapsto 2t$. The only witness is $\{2k + 1 \mid k \in \mathbb{N}\}$.

We now define the stronger notion of k-regular subtrace.

Definition 17 (Regular subtrace). *Let t_1 and t_2 be two traces (finite or infinite) and k be some integer. We say that t_2 is a k-regular subtrace of t_1 if t_2 is a subtrace of t_1 and furthermore if $\forall i \in \mathrm{dom}(t_1)$, $s(i) = ki$.*

For regular subtracing, the chosen elements are picked up one every exactly k elements.

Remark 4. If t_2 is a k-regular subtrace of t_1, it is also a k-weak regular subtrace of t_1.

For an ASM A and e some input, we denote by t_e^A the run of A on initial algebra $\mathcal{I}(e)$.

Definition 18 (Weak simulation). *Some ASM B k-weakly simulates some other A if $A_\mathcal{D} \subseteq B_\mathcal{D}$ and for all input e, the run t_e^A is a k-weak regular subtrace of $t_e^B \upharpoonright A_\mathcal{D}$. We denote by $\mathcal{W}_k^e(A, B)$ the set of witnesses for this weak regular subtracing and by $\mathcal{G}_k^e(A, B)$ the set of guides.*

Definition 19. *An ASM A is arithmetic if its data set is $\mathbb{N}$ and all the static functions are Turing computable.*

Theorem 1. *Let A and B be two arithmetic ASM such that B k-weakly simulates A. Then, given any computable input e on this input, there is a computable element in $\mathcal{G}_k^e(A, B)$.*

Proof. If both traces are finite, the guides of $\mathcal{G}_k^e(A, B)$ are all finite and therefore computable. Without loss of generality, we may assume in the sequel that both traces are infinite.

Firstly, note that one can simulate the execution of A and B on input e. This can be done since the modifications of the dynamic values only concern a finite number of elements of the domains which are integers. More precisely, for all dynamic symbols f of domain D (D is $\mathbb{N}^\ell$ where ℓ is the arity of f), the simulator saves a table T_f from $\mathfrak{P}(D \times \mathbb{N})$. The simulator evaluates the program's guard with a call by value scheme: when the simulator is computing the value $f(x)$ for some x in D, the simulator first checks if there exists y such that $\langle x, y \rangle \in T_f$. In this case, the result is y. Otherwise, the result is taken from e for input symbols and $\perp$ for other dynamic symbols. When $f(x)$ is assigned a value v, the simulator adds $\langle x, v \rangle$ in T_f possibly removing an ancient value associated to x in the table. With this way of representing the state of an arithmetic ASM, it is decidable whether two trace elements corresponding to the same ASM are equal. We denote by $(\!|q|\!)$ the minimal representation of the state q using such tables (minimal meaning removing entries $\langle x, \perp \rangle$ for dynamic symbols and entries $\langle x, y \rangle$ where y is the same value as in e from input symbols).

We conclude that the functions t_1 and t_2 defined as $t_1(i) = (\!|t_e^A(i)|\!)$ and $t_2(i) = (\!|t_e^B \upharpoonright A_\mathcal{D}(i)|\!)$ are computable. Moreover, by definition of an ASM, the state of an ASM after one step only depends on the current state. Therefore, there is a computable function a such that $t_1(i + 1) = a(t_1(i))$.

We now construct an enumerable rooted DAG (directed acyclic graph) D whose nodes are labeled by integers. The DAG D is built such that, for all paths from the root p, the sequence $(x_i)_{i \in \mathbb{N}}$, where x_i is the integer labelling p_i, is a guide of $\mathcal{G}_k^e(A, B)$.

The DAG D is built inductively. We firstly label the root by 0. For each node n labeled by i, we look consider the possibly empty set $S_i = \{i' \mid i + 1 \leq i' \leq i + k \wedge t_2(i') = a(t_2(i))\}$. For all $i' \in S_i$, we add a child to n labeled by i'. If two nodes have the same label, they are merged. Since the label is only increasing, at some point, the algorithm performing the construction knows when all the ancestors of a given node have been produced.

Let us prove by induction that paths from the root are labeled by guides. The initialization comes from the fact that $t_e^A(0) = t_e^B \upharpoonright A_{\mathcal{D}}(0)$. For the induction step, consider the first n elements of a path of D: $x_0, \ldots, x_{n-1}$ and such that $\forall i < n$, $t_e^A(i) = t_e^B \upharpoonright A_{\mathcal{D}}(x_i)$. For any child x_n of x_{n+1}, one has $(\!(t_e^A(n))\!) = t_1(n) = a(t_1(n-1)) = a(t_2(x_{n-1})) = t_2(x_n) = (\!(t_e^B \upharpoonright A_{\mathcal{D}}(x_n))\!)$. Therefore, $t_e^A(n) = t_e^B \upharpoonright A_{\mathcal{D}}(x_n)$.

In order to end the proof, we need to show that D has a computable infinite path. Firstly, we consider D', subgraph of D, which is a tree, keeping, for each node, only the edge to its smaller labeled ancestor. By construction of D, each path of D', labeled by x, is such that $\forall i$, $|x_{i+1} - x_i| \leq k$.

Let us show, by contradiction, that there are at most k infinite paths. Suppose that there are $k + 1$ infinite paths labeled by $x^{(0)}, \ldots, x^{(k)}$. Let n be such that the paths have no common node labeled by an integer greater then n. Since $\forall i, j$, $|x_{i+1}^{(j)} - x_i^{(j)}| \leq k$, it means that the set $\{n+1, \ldots, n+k\}$ of cardinality k intersects the $k+1$ sequences $x^{(0)}, \ldots, x^{(k)}$ which contradicts the definition of n. Hence, D' is a computable tree with at most k infinite paths. Then the sequence of labels of each of these paths are computable. Indeed, to enumerate them increasingly: the programs knows the (finite) beginning of the path up to the first node q labeled by an integer greater than n. From node q, the tree has only one infinite branch. Then to enumerate the successor of some node r, the program starts in parallel searches in D' from all the children of r. All of the children but one, r', belong to finite subtrees. Once they are all found (i.e. all searches have halted but one), the program output r' and continues the enumeration from r'.

Proposition 1. *There exist 0-ary ASM A and B such that B 2-simulates A and such that no element of $\mathcal{W}_2^\emptyset(A, B)$ is recursive.*

Proof. Let A be the ASM, with the successor function as static symbol and dynamic 0-ary symbols m and n with program:

```
if (n = undef)
    m := 0
    n := 1
if (n ≠ undef)
    n := n+1
```

This ASM simply keeps m=0 and increments n from 1 to infinity.

Let f be the characteristic function of a non-recursive set (that is a total function from $\mathbb{N}$ to $\{0,1\}$ such that $f^{-1}(1)$ is non-recursive). Let B be the ASM, with f, the division by 2, parity test, and the integer successor as static symbols and dynamic 0-ary symbols m and n, with program:

```
if (n = undef)
   m := 0
   n := 1
if (n is even ∧ f(n/2) = 1)
   m := 1
if (n is odd ∨ m = 1)
   m := 0
   n := n+1
```

This ASM does what A does but delays the incrementation of n when $f(n/2) = 1$. The ASM B 2-weakly simulates A by simply omitting the steps where $f(n/2) = 1$. These steps occurs at time t where $\exists k \neq 0$, such that $f(k) = 1$ and $t = \sum_{i=1}^{k} 2 + \delta_{f(i)1}$ (where δ is the Kronecker delta). Hence the only witness for B 2-weak simulating A is $W = \{\sum_{i=1}^{k} 2 + \delta_{f(i)1} \mid k \in \mathbb{N} \setminus \{0\} \wedge f(k) = 1\}$. Since $f^{-1}(1)$ is a non-recursive set and can be computed from W, we conclude that W is non-recursive.

Theorem 2. *There exist some arithmetic ASM A and B such that:*

- *The ASM B 2-weakly simulates A and the set $W_2(A, B)$ contains only finite sets.*
- *The set $W_2(A, B)$ is non-recursive.*

Proof. Let $E \subset \mathbb{N}$ be a recursively enumerable, non-recursive set. Let f be a recursive function such that $E = \{x \mid f(x) \text{ halts}\}$. Let F be an arithmetic ASM computing f. Without loss of generality, assume that the input for F is in the variable x.

We design two ASM A and B as follows.

- The ASM A and B have all the symbols of F.
- We add to A and B the new variables s, c, d, and m (initialized to **undef**). The purpose of these variables is as follows:
  - s is used to have an initial step which copies x into c.
  - c is a counter which, after the initial step, decreases from x to 0 and afterwards remains equal to 0.
  - d is a counter which increases from 0 after c has reached 0. It increases while the simulation of A has not halted.
  - m is turned to 1 for one step at the beginning and at the end of the simulation of F but only in B. Otherwise it remains equal to 0. It is also the case during the whole run in A.
- To allow the expected behavior of A and B, for all rule $\langle g, R \rangle$ of F, one puts into A and B the rule $\langle g \wedge \text{s} = 1 \wedge \text{c} = 0, R \rangle$.

– In A, we add the three following rules (`HasHalted` is the halt time detection expression for the ASM F):

```
if (s = undef)
    c := x
    s := 1
    d := 0
if(s = 1 ∧ c > 0)
    c := c-1
if(c = 0 ∧ ¬HasHalted)
    d := d+1
```

– In B, we do almost the same but, using the variable m, we add differences in the execution of the ASM:

```
if(s = undef)
    c := x
    s := 1
    d := 0
if(s = 1 ∧ c > 1)
    c := c-1
if(s = 1 ∧ c = 1 ∧ m = undef)
    m := 1
if(s = 1 ∧ c = 1 ∧ m = 1)
    m := undef
    c := 0
if (c = 0 ∧ ¬HasHalted)
    d := d+1
if (HasHalted)
    m := 1
```

The runs for A and B on input x start with:

$$\imath + [\mathtt{m} = \bot, \mathtt{s} = \bot, \mathtt{x} = x, \mathtt{c} = \bot, \mathtt{d} = \bot],$$
$$(\imath + [\mathtt{m} = \bot, \mathtt{s} = 1, \mathtt{x} = x, \mathtt{c} = i, \mathtt{d} = 0])_{i \text{ from } x \text{ to } 1}$$

where $\imath$ contains the initial values of all other variables.

Then, the next element of the run for B, is

$$\imath + [\mathtt{m} = 1, \mathtt{s} = 1, \mathtt{x} = x, \mathtt{c} = 1, \mathtt{d} = 0]$$

Afterwards, the elements of the runs for A and B are again identical and contain the computation of $f(x)$ by F and d which increases by 1 at each step.

If the computation of $f(x)$ does not halt, the only witness of $\mathcal{W}_2^x(A, B)$ is the singleton $\{x\}$.

If the computation of $f(x)$ halts, the run of A ends after the computation as does the run of B which has one more element, of index r_x, equal to $\jmath + [\mathtt{m} = 1]$ where $\jmath$ is the state of both A and B at the previous step. In this case, the only witness of $\mathcal{W}_2^x(A, B)$ is $\{x, r_x\}$.

We conclude that $W = \{\mathcal{W}_k^e(A, B) \mid e \text{ input}\} = \{\{x\} \mid f(x) \text{ does not halts}\} \cup \{\{x, r_x\} \mid f(x) \text{ halts}\}$. If one can enumerate W, one can enumerate $\mathbb{N} \setminus E$ and therefore, being enumerable, E is computable which is a contradiction. We conclude that W is not enumerable and consequently non-recursive.

4   Conclusion

In this paper, we studied the simulation of computation models by ASM where one step of simulation is executed by several steps of the ASM. We showed that the correct simulation must ensure that each simulated step corresponds to the same number of steps of the ASM. Indeed, if the number can be variable, the simulated model could be computing more information, not directly in its variables but looking at which steps the content of some variables is modified.

References

1. Cégielski, P., Guessarian, I.: Normalization of some extended abstract state machines. In: Blass, A., Dershowitz, N., Reisig, W. (eds.) Fields of Logic and Computation. LNCS, vol. 6300, pp. 165–180. Springer, Heidelberg (2010). https://doi.org/10.1007/978-3-642-15025-8_9
2. Gurevich, Y.: Reconsidering Turing's Thesis: Toward More Realistic Semantics of Programs. University of Michigan, Technical report CRL-TR-38-84, EECS Department (1984)
3. Gurevich, Y.: A new thesis. Abstracts Am. Math. Soc. **6**, 317 (1985)
4. Marquer, Y., Valarcher, P.: An imperative language characterizing PTIME algorithms. In: Cégielski, P., Enayat, A., Kossak, R. (eds.) Studies in Weak Arithmetics: Volume 3. Lecture Note 217. CSLI Publications, Stanford (2016)

Cohesive Powers of Linear Orders

Rumen Dimitrov[1]([✉]), Valentina Harizanov[2], Andrey Morozov[3], Paul Shafer[4],
Alexandra Soskova[5], and Stefan Vatev[5]

[1] Department of Mathematics and Philosophy, Western Illinois University, Macomb,
IL 61455, USA
rd-dimitrov@wiu.edu
[2] Department of Mathematics, George Washington University,
Washington, DC 20052, USA
harizanv@gwu.edu
[3] Sobolev Institute of Mathematics, Novosibirsk 630090, Russia
morozov@math.nsc.ru
[4] School of Mathematics, University of Leeds, Leeds LS2 9JT, UK
P.E.Shafer@leeds.ac.uk
[5] Faculty of Mathematics and Informatics, Sofia University, 5 James Bourchier blvd.,
1164 Sofia, Bulgaria
{asoskova,stefanv}@fmi.uni-sofia.bg

Abstract. Cohesive powers of computable structures can be viewed as effective ultraproducts over effectively indecomposable sets called cohesive sets. We investigate the isomorphism types of cohesive powers $\Pi_C \mathcal{L}$ for familiar computable linear orders $\mathcal{L}$. If $\mathcal{L}$ is isomorphic to the ordered set of natural numbers $\mathbb{N}$ and has a computable successor function, then $\Pi_C \mathcal{L}$ is isomorphic to $\mathbb{N} + \mathbb{Q} \times \mathbb{Z}$. Here, $+$ stands for the sum and $\times$ for the lexicographical product of two orders. We construct computable linear orders $\mathcal{L}_1$ and $\mathcal{L}_2$ isomorphic to $\mathbb{N}$, both with noncomputable successor functions, such that $\Pi_C \mathcal{L}_1$ is isomorphic to $\mathbb{N} + \mathbb{Q} \times \mathbb{Z}$, while $\Pi_C \mathcal{L}_2$ is not. While cohesive powers preserve the satisfiability of all Π_2^0 and Σ_2^0 sentences, we provide new examples of Π_3^0 sentences Φ and computable structures $\mathcal{M}$ such that $\mathcal{M} \vDash \Phi$ while $\Pi_C \mathcal{M} \vDash \neg\Phi$.

1 Introduction and Preliminaries

Skolem was the first to construct a countable nonstandard model of true arithmetic. Various countable nonstandard models of (fragments of) arithmetic have been later studied by Feferman, Scott, Tennenbaum, Hirschfeld, Wheeler, Lerman, McLaughlin and others (see [6–9]). The following definition, and other notions from computability theory can be found in [10].

The first three and the last two authors acknowledge partial support of the NSF grant DMS-1600625. The second author acknowledges support from the Simons Foundation Collaboration Grant, and from CCFF and Dean's Research Chair GWU awards. The last two authors acknowledge support from BNSF, MON, DN 02/16. The fourth author acknowledges the support of the *Fonds voor Wetenschappelijk Onderzoek – Vlaanderen* Pegasus program.

© Springer Nature Switzerland AG 2019
F. Manea et al. (Eds.): CiE 2019, LNCS 11558, pp. 168–180, 2019.
https://doi.org/10.1007/978-3-030-22996-2_15

Definition 1. *(i) An infinite set $C \subseteq \omega$ is* cohesive *(r-cohesive) if for every c.e. (computable) set W, either $W \cap C$ or $\overline{W} \cap C$ is finite.*

(ii) A set M is maximal *(r-maximal) if M is c.e. and $\overline{M}$ is cohesive (r-cohesive).*

(iii) If M is maximal, then $\overline{M}$ is called co-maximal.

(iv) A set B is quasimaximal *if it is the intersection of finitely many maximal sets.*

In the definition above, ω denotes the set of natural numbers. We will use $=^*$ (and $\subseteq^*$) to refer to equality (inclusion) of sets up to finitely many elements. Let A be a fixed r-cohesive set. For computable functions f and g, Feferman, Scott, and Tennenbaum (see [6]) defined an equivalence relation $f \sim_A g$ if $A \subseteq^* \{n : f(n) = g(n)\}$. They then proved that the structure $\mathcal{R}/\sim_A$, with domain the set of recursive functions modulo $\sim_A$, is a model of only a fragment of arithmetic. They constructed a particular Π_3^0 sentence Φ such that for the standard model of arithmetic, $\mathcal{N}$, we have $\mathcal{N} \vDash \Phi$ but $\mathcal{R}/\sim_A \nvDash \Phi$. The sentence Φ provided in [6] essentially uses Kleene's T predicate.

Cohesive powers of computable structures are effective versions of ultrapowers. They have been introduced in [2] in relation to the study of automorphisms of the lattice $\mathcal{L}^*(V_\infty)$ of effective vector spaces. Cohesive powers of the field of rational numbers were used in [1] to characterize certain principal filters of $\mathcal{L}^*(V_\infty)$. Their isomorphism types and automorphisms were further studied in [4]. They were also used in [1] and [3] to find interesting orbits in $\mathcal{L}^*(V_\infty)$.

The goal of this paper is to show that the presentation of a computable structure matters for the isomorphism type of its cohesive power. We give computable presentations of the ordered set of natural numbers such that their cohesive powers are not elementary equivalent. Furthermore, we provide examples of computable structures $\mathcal{M}$ and Π_3^0 sentences Ψ, which do not use Kleene's T predicate, such that $\mathcal{M} \vDash \Psi$ while the cohesive power $\Pi_C \mathcal{M} \vDash \neg \Psi$. We will now present some additional definitions and known results.

Definition 2 *[2]. Let $\mathcal{A}$ be a computable structure for a computable language L and with domain A. Let $C \subseteq \omega$ be a cohesive set. The* cohesive power *of $\mathcal{A}$ over C, denoted by $\Pi_C \mathcal{A}$, is a structure $\mathcal{B}$ for L defined as follows:*

(i) Let $D = \{\varphi \mid \varphi : \omega \to A$ is a partial computable function, and $C \subseteq^* \mathrm{dom}(\varphi)\}$.

   For $\varphi_1, \varphi_2 \in D$, define $\varphi_1 =_C \varphi_2$ iff $C \subseteq^* \{x : \varphi_1(x)\downarrow = \varphi_2(x)\downarrow\}$.

   Let $B = (D/=_C)$ be the domain of $\mathcal{B} = \Pi_C \mathcal{A}$.

(ii) If $f \in L$ is an n-ary function symbol, then $f^\mathcal{B}$ is an n-ary function on B such that for every $[\varphi_1], \dots, [\varphi_n] \in B$, $f^\mathcal{B}([\varphi_1], \dots, [\varphi_n]) = [\varphi]$, where for every $x \in A$,

$$\varphi(x) \simeq f^\mathcal{A}(\varphi_1(x), \dots, \varphi_n(x)),$$

   and $\simeq$ stands for equality of partial functions.

(iii) If $P \in L$ is an m-ary predicate symbol, then $P^\mathcal{B}$ is an m-ary relation on B such that for every $[\varphi_1], \dots, [\varphi_m] \in B$,

$$P^\mathcal{B}([\varphi_1], \dots, [\varphi_m]) \Leftrightarrow C \subseteq^* \{x \in A \mid P^\mathcal{A}(\varphi_1(x), \dots, \varphi_m(x))\}.$$

(iv) If $c \in L$ is a constant symbol, then $c^{\mathcal{B}}$ is the equivalence class of the (total) computable function on A with constant value $c^{\mathcal{A}}$.

The following is the fundamental theorem of cohesive powers due to Dimitrov (see [2]). We view cohesive powers as effective ultrapowers and Theorem 3 as an effective version of Łoś's theorem.

Theorem 3. *Let C be a cohesive set and let $\mathcal{A}$ and $\mathcal{B}$ be as in the definition above.*

1. *If $\tau(y_1, \ldots, y_n)$ is a term in L and $[\varphi_1], \ldots, [\varphi_n] \in B$, then $[\tau^{\mathcal{B}}([\varphi_1], \ldots, [\varphi_n])]$ is the equivalence class of a partial computable function such that*

$$\tau^{\mathcal{B}}([\varphi_1], \ldots, [\varphi_n])(x) = \tau^{\mathcal{A}}(\varphi_1(x), \ldots, \varphi_n(x)).$$

2. *If $\Phi(y_1, \ldots, y_n)$ is a formula in L that is a Boolean combination of Σ_1^0 and Π_1^0 formulas and $[\varphi_1], \ldots, [\varphi_n] \in B$, then*

$$\mathcal{B} \vDash \Phi([\varphi_1], \ldots, [\varphi_n]) \text{ iff } C \subseteq^* \{x : \mathcal{A} \vDash \Phi(\varphi_1(x), \ldots, \varphi_n(x))\}.$$

3. *If Φ is a Π_2^0 (or Σ_2^0) sentence in L, then $\mathcal{B} \vDash \Phi$ iff $\mathcal{A} \vDash \Phi$.*
4. *If Φ is a Π_3^0 sentence in L, then $\mathcal{B} \vDash \Phi$ implies $\mathcal{A} \vDash \Phi$.*

Note that $\mathcal{A}$ is a substructure of $\mathcal{B} = \Pi_C \mathcal{A}$. For $c \in A$ let $[\varphi_c] \in B$ be the equivalence class of the total function φ_c such that $\varphi_c(x) = c$ for every $x \in \omega$. The map $d : A \to B$ such that $d(c) = [\varphi_c]$ is called the *canonical embedding* of $\mathcal{A}$ into $\mathcal{B}$.

2   Cohesive Powers of Linear Orders

We will now investigate various algebraic and computability-theoretic properties of cohesive powers of linear orders. We first provide some definitions and notational conventions we will use. Let $C \subseteq \omega$ be a cohesive set. Let $\langle \cdot, \cdot \rangle : \omega^2 \to \omega$ be a fixed computable bijection, and let (computable) functions π_1 and π_2 be such that $\pi_1(\langle m, n \rangle) = m$ and $\pi_2(\langle m, n \rangle) = n$.

Definition 4. *Let $\mathcal{L}_0 = \langle L_0, <_{\mathcal{L}_0} \rangle$ and $\mathcal{L}_1 = \langle L_1, <_{\mathcal{L}_1} \rangle$ be linear orders. Then*

(1) $\mathcal{L}_0 + \mathcal{L}_1 = \langle \{\langle 0, l \rangle : l \in L_0\} \cup \{\langle 1, l \rangle : l \in L_1\}, <_{\mathcal{L}_0 + \mathcal{L}_1} \rangle$, *where*

$$\langle i, l \rangle <_{\mathcal{L}_0 + \mathcal{L}_1} \langle j, m \rangle \text{ iff } (i < j) \vee (i = j \wedge l <_{\mathcal{L}_i} m).$$

(2) $\mathcal{L}_0 \times \mathcal{L}_1 = \langle L_0 \times L_1, <_{\mathcal{L}_0 \times \mathcal{L}_1} \rangle$, *where*

$$\langle k, m \rangle <_{\mathcal{L}_0 \times \mathcal{L}_1} \langle l, n \rangle \text{ iff } (k <_{\mathcal{L}_0} l) \vee (k =_{\mathcal{L}_0} l \wedge m <_{\mathcal{L}_1} n).$$

Remark 5.(1) By $\mathbb{N}$, $\mathbb{Z}$, and $\mathbb{Q}$ we denote the usual presentations of the ordered sets of natural numbers, integers, and rational numbers. The order types of $\mathbb{N}$, $\mathbb{Z}$, and $\mathbb{Q}$ are denoted by ω, ζ, and η.

(2) In Definition 4, we use $\mathcal{L}_0 \times \mathcal{L}_1$ to denote the lexicographical product of the linear orders $\mathcal{L}_0$ and $\mathcal{L}_1$. This product is also denoted by $\mathcal{L}_1 \cdot \mathcal{L}_0$. (For example, $\mathbb{Q} \times \mathbb{Z}$ is also denoted by $\mathbb{Z} \cdot \mathbb{Q}$, and its order type is denoted $\zeta \cdot \eta$.)

(3) We will use $\mathcal{L}^{rev}$ to denote the reverse linear order of $\mathcal{L}$. (In the literature it is also denoted by $\mathcal{L}^*$.)

(4) Let the quantifier $\forall^\infty n$ stand for "for almost every n." Note that for a cohesive C and a Σ_1^0 formula $\phi(x)$, $(\forall^\infty n \in C)[\phi(n)]$ is equivalent to "there are infinitely many $n \in C$ such that $\phi(n)$."

Before we state the next theorem, we would like to remind the reader that $\mathbb{N} + \mathbb{Q} \times \mathbb{Z}$ is the order type of a countable non-standard model of PA.

Theorem 6. *Let $\mathcal{L}_0$ and $\mathcal{L}_1$ be computable linear orders and let C be a cohesive set. Then*

(1) $\Pi_C(\mathcal{L}_0 + \mathcal{L}_1) \cong \Pi_C\mathcal{L}_0 + \Pi_C\mathcal{L}_1$

(2) $\Pi_C(\mathcal{L}_0 \times \mathcal{L}_1) \cong \Pi_C\mathcal{L}_0 \times \Pi_C\mathcal{L}_1$

(3) $\Pi_C\mathcal{L}_0^{rev} \cong (\Pi_C\mathcal{L}_0)^{rev}$

(4) Let $\mathcal{A}$ be a computable presentation of the linear order $\mathbb{N}$ with a computable successor function. Then $\Pi_C\mathcal{A} \cong \mathbb{N} + \mathbb{Q} \times \mathbb{Z}$.

(5) If $\mathcal{L}$ is a computable dense linear order without endpoints, then $\mathcal{L} \cong \Pi_C\mathcal{L}$.

Proof. (1) Let $\mathcal{A} = \Pi_C(\mathcal{L}_0 + \mathcal{L}_1)$ and $\mathcal{B} = \Pi_C\mathcal{L}_0 + \Pi_C\mathcal{L}_1$. We will define an isomorphism $\Phi : \mathcal{A} \to \mathcal{B}$. Suppose $[\varphi]_C \in \Pi_C(\mathcal{L}_0 + \mathcal{L}_1)$ for a partial computable function φ.

If $(\forall^\infty n \in C)[\varphi(n) \in \{0\} \times L_1]$, then let $\Phi([\varphi]_C) =_{def} \langle 0, [\pi_2 \circ \varphi]_C \rangle$.

If $(\forall^\infty n \in C)[\varphi(n) \in \{1\} \times L_2]$, then let $\Phi([\varphi]_C) =_{def} \langle 1, [\pi_2 \circ \varphi]_C \rangle$.

Since C is cohesive, exactly one of the two cases above applies, so it follows that Φ is well defined. It is then easy to check that Φ is an isomorphism.

(2) Let $\mathcal{A} = \Pi_C(\mathcal{L}_0 \times \mathcal{L}_1)$ and $\mathcal{B} = \Pi_C\mathcal{L}_0 \times \Pi_C\mathcal{L}_1$. We will define an isomorphism $\Phi : \mathcal{A} \to \mathcal{B}$. Suppose $[\varphi]_C \in \Pi_C(\mathcal{L}_0 \times \mathcal{L}_1)$, and let $\Phi([\varphi]_C) =_{def} \langle [\pi_1 \circ \varphi]_C, [\pi_2 \circ \varphi]_C \rangle$. We will prove that

$$[\varphi]_C <_{\mathcal{A}} [\psi]_C \Leftrightarrow \langle [\pi_1 \circ \varphi]_C, [\pi_2 \circ \varphi]_C \rangle <_{\mathcal{B}} \langle [\pi_1 \circ \psi]_C, [\pi_2 \circ \psi]_C \rangle.$$

By definition, $[\varphi]_C <_{\mathcal{A}} [\psi]_C$ iff $C \subseteq^* \{n : \varphi(n) < \psi(n)\}$. By cohesiveness of C, we will have either

$(\forall^\infty n \in C)[(\pi_1 \circ \varphi)(n) < (\pi_1 \circ \psi)(n)]$, or

$(\forall^\infty n \in C)[(\pi_1 \circ \varphi)(n) = (\pi_1 \circ \psi)(n) \wedge (\pi_2 \circ \varphi)(n) < (\pi_2 \circ \psi)(n)]$. In the first case, $[\pi_1 \circ \varphi]_C <_{\Pi_C\mathcal{L}_0} [\pi_1 \circ \psi]_C$. In the second case, $[\pi_1 \circ \varphi]_C =_{\Pi_C\mathcal{L}_0} [\pi_1 \circ \psi]_C$ and $[\pi_2 \circ \varphi]_C <_{\Pi_C\mathcal{L}_1} [\pi_2 \circ \psi]_C$. Therefore,

$$\langle [\pi_1 \circ \varphi]_C, [\pi_2 \circ \varphi]_C \rangle <_{\mathcal{B}} \langle [\pi_1 \circ \psi]_C, [\pi_2 \circ \psi]_C \rangle.$$

(3) Let $\mathcal{A} = \Pi_C\mathcal{L}_0^{rev}$ and $\mathcal{B} = (\Pi_C\mathcal{L}_0)^{rev}$. We will define an isomorphism $\Phi : \mathcal{A} \to \mathcal{B}$. If $[\varphi]_C \in \Pi_C\mathcal{L}_0^{rev}$, then let $\Phi([\varphi]_C) = [\varphi]_C$. We will prove that $[\varphi]_C <_{\mathcal{A}} [\psi]_C$ iff $[\varphi]_C <_{\mathcal{B}} [\psi]_C$. By definition, we have

$$[\varphi]_C <_{\mathcal{B}} [\psi]_C \Leftrightarrow [\psi]_C <_{\Pi_C\mathcal{L}_0} [\varphi]_C \Leftrightarrow$$

$$(\forall^\infty n \in C)\,(\psi(n) <_{\mathcal{L}_0} \varphi(n)) \Leftrightarrow$$

$$(\forall^\infty n \in C)\,(\varphi(n) <_{\mathcal{L}_0^{rev}} \psi(n)) \Leftrightarrow [\varphi]_C <_A [\psi]_C\,.$$

(4) The proof of this fact is omitted because it is a simplified version of the proof of Theorem 8.

(5) The theory of dense linear orders without endpoints is Π_2^0 axiomatizable and countably categorical. By Theorem 3 (part 4), $\Pi_C \mathcal{L}$ is also a dense linear order without endpoints. Since $\Pi_C \mathcal{L}$ is countable, we have $\mathbb{Q} \cong \mathcal{L} \cong \Pi_C \mathcal{L}$.

Part (5) in the previous theorem provides an example of an infinite structure $\mathcal{L}$ such that $\mathcal{L} \cong \Pi_C \mathcal{L}$. The linear order $\mathbb{Q}$ is an ultrahomogeneous structure; it is the Fraïssé limit of the class of finite linear orders. The relationship between Fraïssé limits and cohesive powers is considered in [5]. We now provide two more examples of structures isomorphic to their cohesive powers.

Example 7. (1) $\Pi_C (\mathbb{Q} \times \mathbb{Z}) \cong \mathbb{Q} \times \mathbb{Z}$
(2) $\Pi_C (\mathbb{N} + \mathbb{Q} \times \mathbb{Z}) \cong \mathbb{N} + \mathbb{Q} \times \mathbb{Z}$

Proof. (1) $\Pi_C \mathbb{Q} \times \Pi_C \mathbb{Z} \cong \mathbb{Q} \times \Pi_C(\mathbb{N}^{rev}+\mathbb{N}) \cong \mathbb{Q} \times (\Pi_C \mathbb{N}^{rev} + \Pi_C \mathbb{N})$
$\cong \mathbb{Q} \times [(\mathbb{N} + \mathbb{Q} \times \mathbb{Z})^{rev} + (\mathbb{N} + \mathbb{Q} \times \mathbb{Z})] \cong \mathbb{Q} \times [\mathbb{Q} \times \mathbb{Z} + \mathbb{N}^{rev} + \mathbb{N} + \mathbb{Q} \times \mathbb{Z}]$
$\cong \mathbb{Q} \times [\mathbb{Q} \times \mathbb{Z} + \mathbb{Z} + \mathbb{Q} \times \mathbb{Z}] \cong \mathbb{Q} \times [\mathbb{Q} \times \mathbb{Z}] \cong \mathbb{Q} \times \mathbb{Z}$
(2) $\Pi_C (\mathbb{N} + \mathbb{Q} \times \mathbb{Z}) \cong \Pi_C \mathbb{N} + \Pi_C (\mathbb{Q} \times \mathbb{Z}) \cong \mathbb{N} + \mathbb{Q} \times \mathbb{Z} + \mathbb{Q} \times \mathbb{Z} \cong \mathbb{N} + \mathbb{Q} \times \mathbb{Z}$

Part (4) of Theorem 6 demonstrates that having a computable successor function is a sufficient condition for the cohesive power of a computable linear order of type ω to be isomorphic to $\mathbb{N} + \mathbb{Q} \times \mathbb{Z}$. The next theorem shows that this condition is not necessary.

Theorem 8. *There is a computable linear order $\mathcal{L}$ of order type ω with a non-computable successor function such that for every cohesive set C we have $\Pi_C \mathcal{L} \cong \mathbb{N} + \mathbb{Q} \times \mathbb{Z}$.*

Proof. Fix a non-computable c.e. set A, and let f be a total computable injection on the set of natural numbers with range A. Let $\mathcal{L} = (\omega, <_{\mathcal{L}})$ be the linear order obtained by ordering the even numbers according to their natural order, and by setting $2a <_{\mathcal{L}} 2k + 1 <_{\mathcal{L}} 2a + 2$ if and only if $f(k) = a$. Specifically, we set

$2c <_{\mathcal{L}} 2d$	$\Leftrightarrow$	$2c < 2d$
$2c <_{\mathcal{L}} 2k + 1$	$\Leftrightarrow$	$c \leq f(k)$
$2k + 1 <_{\mathcal{L}} 2c$	$\Leftrightarrow$	$f(k) < c$
$2k + 1 <_{\mathcal{L}} 2\ell + 1$	$\Leftrightarrow$	$f(k) < f(\ell).$

Then $\mathcal{L}$ is a computable linear order of type ω. Let $S^{\mathcal{L}}$ denote the successor function of $\mathcal{L}$. Then $A \leq_T S^{\mathcal{L}}$ (indeed, $A \equiv_T S^{\mathcal{L}}$) because $a \in A$ if and only if $S^{\mathcal{L}}(2a) \neq 2a + 2$. Thus $S^{\mathcal{L}}$ is not computable.

Let C be cohesive, and let $\mathcal{P} = \Pi_C \mathcal{L}$. We show that $\mathcal{P} \cong \mathbb{N} + \mathbb{Q} \times \mathbb{Z}$. To do this, we begin by establishing the following properties of $\mathcal{P}$.

(a) $\mathcal{P}$ has an initial segment of type ω.
(b) Every element of $\mathcal{P}$ has a $<_\mathcal{P}$-immediate successor.
(c) Every element of $\mathcal{P}$ that is not the least element has an $<_\mathcal{P}$-immediate predecessor.

For (a), note that the range of the canonical embedding of $\mathcal{L}$ into $\mathcal{P}$ is an initial segment of $\mathcal{P}$ of type ω.

For (b), consider a $[\psi] \in \mathcal{P}$. We define a partial computable φ such that, for almost every $n \in C$, $\varphi(n)$ is the $<_\mathcal{L}$-immediate successor of $\psi(n)$. It then follows that $[\varphi]$ is the $<_\mathcal{P}$-immediate successor of $[\psi]$. To define φ, observe that, by the cohesiveness of C, exactly one of the following three cases occurs.

(1) $(\forall^\infty n \in C)(\psi(n) \text{ is odd})$
(2) $(\forall^\infty n \in C)(\exists a \in A)(\psi(n) = 2a)$
(3) $(\forall^\infty n \in C)(\exists a \notin A)(\psi(n) = 2a)$

Note that we cannot effectively decide which case occurs, but in each case we can define a particular φ_i such that $[\varphi_i]$ is the $<_\mathcal{P}$-immediate successor of $[\psi]$.

If case (1) occurs, define

$$\varphi_1(n) = \begin{cases} 2a + 2 & \text{if } \psi(n)\downarrow, \psi(n) = 2k + 1, \text{ and } f(k) = a; \\ \uparrow & \text{otherwise.} \end{cases}$$

If case (2) occurs, define

$$\varphi_2(n) = \begin{cases} 2k + 1 & \text{if } \psi(n)\downarrow, \psi(n) = 2a, a \in A, \text{ and } f(k) = a; \\ \uparrow & \text{otherwise.} \end{cases}$$

If case (3) occurs, define

$$\varphi_3(n) = \begin{cases} 2a + 2 & \text{if } \psi(n)\downarrow \text{ and } \psi(n) = 2a; \\ \uparrow & \text{otherwise.} \end{cases}$$

In each case (i) ($i = 1, 2, 3$), we have that for almost every $n \in C$, $\varphi_i(n)$ is the $<_\mathcal{L}$-immediate successor of $\psi(n)$.

The proof of (c) is analogous to the proof of (b).

For $[\psi], [\varphi] \in \mathcal{P}$, write $[\psi] \ll_\mathcal{P} [\varphi]$ if $[\psi] <_\mathcal{P} [\varphi]$ and the interval $([\psi], [\varphi])_\mathcal{P}$ in $\mathcal{P}$ is infinite. Using the cohesiveness of C, we check that $[\psi] \ll_\mathcal{P} [\varphi]$ if and only if $[\psi] <_\mathcal{P} [\varphi]$ and $\limsup_{n \in C} |(\psi(n), \varphi(n))_\mathcal{L}| = \infty$, where $|(a, b)_\mathcal{L}|$ denotes the cardinality of the interval $(a, b)_\mathcal{L}$ in $\mathcal{L}$. Note that for even numbers $2a$ and $2b$, $2a <_\mathcal{L} 2b$ if and only if $2a < 2b$. Therefore, if $2a < 2b$, then $|(2a, 2b)_\mathcal{L}| \geq b - a - 1$.

To finish the proof, we show the following.

(d) If $[\psi], [\varphi] \in \mathcal{P}$ satisfy $[\psi] \ll_\mathcal{P} [\varphi]$, then there is a $[\theta] \in \mathcal{P}$ such that $[\psi] \ll_\mathcal{P} [\theta] \ll_\mathcal{P} [\varphi]$.
(e) If $[\psi] \in \mathcal{P}$, then there is a $[\varphi] \in \mathcal{P}$ with $[\psi] \ll_\mathcal{P} [\varphi]$.

For (d), suppose that $[\psi], [\varphi] \in \mathcal{P}$ satisfy $[\psi] \ll_\mathcal{P} [\varphi]$. Again, by considering the cases (1)–(3) above, either $\psi(n)$ is odd for almost every $n \in C$, or $\psi(n)$ is even for almost every $n \in C$. In the case where $\psi(n)$ is odd for almost every $n \in C$, $\widehat{\psi}(n)$ is even for almost every $n \in C$, where $[\widehat{\psi}]$ is the $<_\mathcal{P}$-immediate successor of $[\psi]$. Thus, we may assume that $\psi(n)$ and $\varphi(n)$ are even for almost every $n \in C$ by replacing $[\psi]$ and $[\varphi]$ by their $<_\mathcal{P}$-immediate successors if necessary. The condition $\limsup_{n \in C} |(\psi(n), \varphi(n))_\mathcal{L}| = \infty$ is now equivalent to $\limsup_{n \in C}(\varphi(n) - \psi(n)) = \infty$.

Define a partial computable θ by

$$\theta(n) = \begin{cases} \left\lfloor \frac{\psi(n)+\varphi(n)}{2} \right\rfloor & \text{if } \left\lfloor \frac{\psi(n)+\varphi(n)}{2} \right\rfloor \text{ is even;} \\[2mm] \left\lfloor \frac{\psi(n)+\varphi(n)}{2} \right\rfloor + 1 & \text{if } \left\lfloor \frac{\psi(n)+\varphi(n)}{2} \right\rfloor \text{ is odd.} \end{cases}$$

By the definition of θ, we have that $\limsup_{n \in C}(\theta(n) - \psi(n)) = \infty$ and that $\limsup_{n \in C}(\varphi(n) - \theta(n)) = \infty$. Since $\psi(n)$, $\varphi(n)$, and $\theta(n)$ are even for almost all $n \in C$, we have that:

$$\limsup_{n \in C} |(\psi(n), \theta(n))_\mathcal{L}| = \infty \quad \text{and} \quad \limsup_{n \in C} |(\theta(n), \varphi(n))_\mathcal{L}| = \infty.$$

Thus, $[\psi] \ll_\mathcal{P} [\theta] \ll_\mathcal{P} [\varphi]$, as desired.

For (e), consider $[\psi] \in \mathcal{P}$. As argued above, we may assume that $\psi(n)$ is even for almost every $n \in C$ by replacing $[\psi]$ by its $<_\mathcal{P}$-immediate successor, if necessary. If $\limsup_{n \in C} \psi(n)$ is finite, then by the cohesiveness of C, the function ψ must be eventually constant on C. In this case, $[\psi] \ll_\mathcal{P} [2\mathrm{id}]$. If $\limsup_{n \in C} \psi(n) = \infty$, then $[\psi] \ll_\mathcal{P} [2\psi]$.

This completes the proof since the properties (a)–(e) ensure that $\mathcal{P} \cong \mathbb{N} + \mathbb{Q} \times \mathbb{Z}$.

3   Non-isomorphic Cohesive Powers of Isomorphic Structures

Theorem 9. *For every co-maximal set $C \subseteq \omega$ there exist two isomorphic computable structures $\mathcal{A}$ and $\mathcal{B}$ such the cohesive powers $\prod_C \mathcal{A}$ and $\prod_C \mathcal{B}$ are not isomorphic.*

Proof. Note that it suffices to prove the theorem for an arbitrary co-maximal set consisting of even numbers only. Indeed, if C is an arbitrary co-maximal set, then $C_1 = \{2s \mid s \in C\}$ is also a co-maximal set, and for any computable structure $\mathcal{M}$, we have $\prod_C \mathcal{M} \cong \prod_{C_1} \mathcal{M}$ (via $[\phi(x)] \mapsto [\phi(\lfloor \frac{x}{2} \rfloor)]$). Then, if $\mathcal{M}_0$ and $\mathcal{M}_1$ are isomorphic computable structures such that $\prod_{C_1} \mathcal{M}_0 \not\cong \prod_{C_1} \mathcal{M}_1$, then $\prod_C \mathcal{M}_0 \not\cong \prod_C \mathcal{M}_1$.

Let $S = \{2s \mid s \in \omega\}$. Let $A \subseteq S$ be such that $A_1 = S - A$ is infinite and c.e. For every such A we will define a computable structure $\mathcal{M}_A$ with a single ternary relation.

Let $F = \{4s + 1 \mid s \in \omega\}$ and $B = \{4s + 3 \mid s \in \omega\}$. Fix a computable bijection f from the set $\{\langle i, j \rangle \in S \mid i < j\}$ onto F. Let also b be a computable bijection from the set $\{\langle j, i \rangle \in S \mid i < j \wedge (i \in A_1 \vee j \in A_1)\}$ onto B. For the function f, we write f_{ij} instead of $f(i, j)$, and similarly for the function b. Define a ternary relation P as follows:

$$P = \{(x, f_{xy}, y) \mid x, y \in S \wedge x < y\} \cup$$
$$\{(y, b_{yx}, x) \mid x, y \in S \wedge x < y \wedge (x \in A_1 \vee y \in A_1)\}.$$

Finally, let $\mathcal{M}_A = \langle \omega; P \rangle$. Informally, we can view the triples x, w, y with the property $P(x, w, y)$ as labelled arrows (e.g., $x \xrightarrow{w} y$). We start with a structure consisting of the set $S \cup F$ with arrows $i \xrightarrow{f_{ij}} j$ that connect i with j for all $i, j \in S$ such that $i < j$. These arrows can be viewed as a way of redefining the natural ordering $<$ on S. Elements of S can be thought of as "stem elements" and elements of F can be thought of as "forward witnesses." Next, we start enumerating the c.e. set $A_1 = S - A$. At every stage a new element k is enumerated into A_1, we add new arrows together with appropriate elements from B, the "backward witnesses," which intend to exclude k from the initial ordering on S. More precisely, we add arrows $k \xrightarrow{b_{ki}} i$ for all i with $i < k$, and arrows $j \xrightarrow{b_{jk}} k$, for each j with $j > k$. Eventually, exactly the elements of A_1 will be excluded from the ordering, and the final ordering will be an ordering on the set A.

In the resulting structure, every element $x \in A_1$ is connected with every element $y \in S$ such that $x \neq y$ with exactly two arrows: $x \xrightarrow{w} y$ and $y \xrightarrow{w_1} x$. If $x, y \in A$ are such that $x \neq y$ then they are connected with arrows of the type $x \xrightarrow{w} y$ exactly when $x < y$. In other words, the formula

$$\Phi(x, y) =_{def} \exists w P(x, w, y) \wedge \neg \exists w_1 P(y, w_1, x)$$

will be satisfied by exactly those $x, y \in A$ such that $x < y$. The formula Φ will not be satisfied by any pair (x, y) for which at least one of x or y has been excluded.

The following properties of the structure $\mathcal{M}_A$ follow immediately from the definition above.

(1) For every w there is at most one pair x, y such that $P(x, w, y)$.
(2) If $x \in S - A$, then for any $y \in S$, $y \neq x$, there is a unique w_1 such that $P(x, w_1, y)$ and a unique w_2 such that $P(y, w_2, x)$.
(3) If $x, y \in A$, then $x < y \Leftrightarrow \exists w P(x, w, y)$.
(4) $\mathcal{M}_A$ is computable.

To prove (4) note that the relation P is computable because

$$P(x, z, y) \Leftrightarrow x, y \in S \wedge \left[(x < y \wedge z = f_{xy}) \vee (x > y \wedge z \in B \wedge b^{-1}(z) = \langle x, y \rangle) \right].$$

(5) Let $D, E \subseteq S$ be infinite and such that $S - D$ and $S - E$ are infinite and c.c. Then $\mathcal{M}_D \cong \mathcal{M}_E$.

Since D and E are infinite, the orders $(D, <)$ and $(E, <)$, where $<$ is the natural order, are isomorphic to $\mathbb{N}$. The isomorphism between these orders, extended by any bijection between $S - D$ and $S - E$, has a unique natural extension to a map from the domain of $\mathcal{M}_D$ to the domain of $\mathcal{M}_E$. That is, the arrows in $\mathcal{M}_D$ (the elements of F and B) can be uniquely mapped to corresponding arrows in $\mathcal{M}_E$.

To continue with the proof, we let

$$\Theta(x) =_{def} (\exists t) [\Phi(x, t) \vee \Phi(t, x)].$$

The formula $\Theta(x)$ defines the set A in $\mathcal{M}_A$.

For any structure $\mathcal{M} = (M, P)$ in the language with one ternary predicate symbol we will use the following notation:

$L_\mathcal{M} =_{def} \{x \in M | \mathcal{M} \vDash \Theta(x)\}$, and

$<_{L_\mathcal{M}} =_{def} \{(x, y) \in M \times M | \mathcal{M} \vDash \Phi(x, y)\}$.

Fix $A \subseteq S$ such that $S - A$ is infinite and c.e.

It follows from the discussion above that the formula $\Phi(x, y)$ defines in $\mathcal{M}_A$ the restriction of the natural order $<$ to A. Clearly, $\left(L_{\mathcal{M}_A}, <_{L_{\mathcal{M}_A}}\right)$ has order type ω.

Let $\mathcal{M}_A^\sharp = \prod_C \mathcal{M}_A$. For partial computable functions g and h such that $[g], [h] \in \text{dom}(\mathcal{M}_A^\sharp)$ we have:

(i) $\mathcal{M}_A^\sharp \vDash \Phi([g], [h]) \Leftrightarrow C \subseteq^* \{i| (g(i) \in A) \wedge (h(i) \in A) \wedge (g(i) < h(i))\}$

(ii) $L_{\mathcal{M}_A^\sharp} = \{[g] \in \mathcal{M}_A^\sharp| g(C) \subseteq^* A\}$ and $\left(L_{\mathcal{M}_A^\sharp}, <_{L_{\mathcal{M}_A^\sharp}}\right)$ is a linear order.

Note that (i) follows from Theorem 3, part (2), since $\Phi(x, y)$ is a Boolean combination of Σ_1^0 and Π_1^0 formulas.

For the proof of (ii) note that for any $[g] \in \mathcal{M}_A^\sharp$ we have either $C \subseteq^* \{i|g(i) \in A\}$ or $C \subseteq^* \{i|g(i) \in \omega - A\}$ because C is cohesive and $\omega - A$ is c.e. Since

$$[g] \in L_{\mathcal{M}_A^\sharp} \Leftrightarrow (\exists x) [\Phi([g], x) \vee \Phi(x, [g])],$$

the equivalence in part (i) implies that $L_{\mathcal{M}_A^\sharp} = \{[g] \in \mathcal{M}_A^\sharp| g(C) \subseteq^* A\}$. It is easy to show that the relation $<_{L_{\mathcal{M}_A^\sharp}}$ is a linear order on $L_{\mathcal{M}_A^\sharp}$.

For any $a \in A$ let $h_a(i) = a$ for all $i \in \omega$. We will call the element $[h_a]$ in $\mathcal{M}_A^\sharp$ a constant in $\mathcal{M}_A^\sharp$.

(6) The set of constants $\{[h_a] | a \in A\}$ in the structure $\mathcal{M}_A^\sharp$ forms an initial segment of $\left(L_{\mathcal{M}_A^\sharp}, <_{L_{\mathcal{M}_A^\sharp}}\right)$ of order type ω.

Clearly, if $a_0, a_1 \in A$, then $\Phi([h_{a_0}], [h_{a_1}])$ if and only if $a_0 < a_1$. Therefore, $\{[h_a] | a \in A\}$ is an ordered set of type ω. It remains to check that $\{[h_a] | a \in A\}$

is an initial segment. Suppose $[h] \in \mathcal{M}_A^\sharp$ and $a \in A$ are such that $\mathcal{M}_A^\sharp \models \Phi([h], [h_a])$. Then

$$C \subseteq^* \{i | \mathcal{M}_A \models \Phi(h(i), a)\} = \{i | h(i) \in A \wedge h(i) < a\} = \bigcup_{k \in A \wedge k < a} \{i | h(i) = k\}.$$

The last expression is a union of a finite family of mutually disjoint c.e. sets. Since C is cohesive, there exists a $k \in A$ such that $C \subseteq^* \{i | h(i) = k\}$, which means that $[h] = [h_k]$ is a constant.

We now define the following Σ_3^0 sentence

$$\Psi =_{def} (\exists x) [\Theta(x) \wedge (\forall y) [\Theta(y) \Rightarrow \Phi(y, x)]].$$

The intended interpretation of Ψ is that when $\Phi(x, t)$ defines a linear order $(L_\mathcal{M}, <_{L_\mathcal{M}})$, then the order has a greatest element. Note that $\mathcal{M}_A \models \neg \Psi$. This is because $\left(L_{\mathcal{M}_A}, <_{L_{\mathcal{M}_A}}\right)$ has order type ω and hence has no greatest element.

Before we continue with the proof we recall Proposition 2.1 from [8].

Proposition 10 *(Lerman [8]). Let R be a co-r-maximal set, and let f be a computable function such that $f(R) \cap R$ is infinite and $f(R) \subset R$ almost everywhere. Then the restriction $f \upharpoonright R$ differs from the identity function at only finitely many points.*

We now fix a co-maximal (hence co-r-maximal) set $C \subseteq S$ and an infinite co-infinite computable set $D \subseteq S$. By property (5) above, we have $\mathcal{M}_C \cong \mathcal{M}_D$. Let $\mathcal{M}_C^\sharp = \prod_C \mathcal{M}_C$ and $\mathcal{M}_D^\sharp = \prod_C \mathcal{M}_D$.

It is not hard to show that, since C is co-maximal, for every partial computable function φ for which $C \subseteq^* dom(\varphi)$, there is a computable function f_φ such that $[\varphi] = [f_\varphi]$ (see [4]).

To finish the proof we will establish the following facts:

(7) $\mathcal{M}_C^\sharp \models \Psi$

(8) $\mathcal{M}_D^\sharp \models \neg \Psi$

To prove (7) recall that $L_{\mathcal{M}_C^\sharp} = \{[f] \in \mathcal{M}_C^\sharp | f(C) \subseteq^* C\}$. By Proposition 10, if $[f] \in \mathcal{M}_C^\sharp$ is such that $f(C) \subseteq^* C$ and $f(C)$ is infinite, then $[f] = [id]$. If $f(C)$ is finite, then f is eventually a constant on C, because C is cohesive. Therefore, $L_{\mathcal{M}_C^\sharp} = \{[f_c] \mid c \in C\} \cup \{[id]\}$. It is easy to see that if $c \in C$, then $\Phi([f_c], [id])$. Thus, $\left(L_{\mathcal{M}_C^\sharp}, <_{L_{\mathcal{M}_C^\sharp}}\right)$ has order type $\omega + 1$ with the greatest element $[id]$. Therefore, $\mathcal{M}_C^\sharp \models \Psi$.

To prove (8), let $D = \{d_0 < d_1 < \cdots\}$. The function g defined as $g(d_i) = d_{i+1}$ is computable. Suppose that $\mathcal{M}_D^\sharp \models \Psi$ and let $[f]$ be the greatest element in $\left(L_{\mathcal{M}_D^\sharp}, <_{L_{\mathcal{M}_D^\sharp}}\right)$. Since $[f] <_{L_{\mathcal{M}_D^\sharp}} [g \circ f]$, it follows that $\mathcal{M}_D^* \models \neg \Psi$.

In conclusion, we defined computable isomorphic structures $\mathcal{M}_C$ and $\mathcal{M}_D$ such that $\prod_C \mathcal{M}_C$ and $\prod_C \mathcal{M}_D$ are not even elementary equivalent. The structure $\mathcal{M}_C$ also provides a sharp bound for the fundamental theorem of cohesive powers. Namely, for the Σ_3^0 sentence Ψ, $\mathcal{M}_C \models \neg \Psi$ but $\prod_C \mathcal{M}_C \models \Psi$.

4   Orders of Type ω with Cohesive Powers Not Isomorphic to $\mathbb{N} + \mathbb{Q} \times \mathbb{Z}$

We prove that if C is co-maximal, then there is a computable linear order $\mathcal{L}$ of type ω (necessarily with a non-computable successor function) such that $\Pi_C \mathcal{L} \not\cong \mathbb{N} + \mathbb{Q} \times \mathbb{Z}$.

Lemma 11. *Let $C \subseteq \omega$ be co-c.e., infinite, and co-infinite. Then there is a computable linear order $\mathcal{L} = (\omega, <_\mathcal{L})$ of type ω such that for every partial computable function φ_e,*

$$\forall^\infty n \in C(\varphi_e(n)\!\downarrow \Rightarrow \varphi_e(n) \text{ is not the } \mathcal{L}\text{-immediate successor of } n). \qquad (*)$$

Proof. Fix an infinite computable set $R \subseteq \overline{C}$. We define $<_\mathcal{L}$ in stages. By the end of stage s, $<_\mathcal{L}$ will have been defined on $X_s \times X_s$ for some finite $X_s \supseteq \{0, 1, \ldots, s\}$. At stage 0, set $X_0 = \{0\}$ and define $0 \not<_\mathcal{L} 0$. At stage $s > 0$, start with $X_s = X_{s-1}$ and update X_s and $<_\mathcal{L}$ according to the following procedure.

1. If $<_\mathcal{L}$ has not yet been defined on s (i.e., if $s \notin X_s$), then update X_s to $X_s \cup \{s\}$ and extend $<_\mathcal{L}$ to make s the $<_\mathcal{L}$-greatest element of X_s.
2. Consider each $\langle e, n \rangle < s$ in order. If
   (a) $\varphi_{e,s}(n)\!\downarrow \in X_s$,
   (b) $\varphi_e(n)$ is currently the $<_\mathcal{L}$-immediate successor of n in X_s,
   (c) $n \notin R$, and
   (d) n is not $<_\mathcal{L}$-below any of $0, 1, \ldots, e$,
   then let m be the least element of $R - X_s$. Update X_s to $X_s \cup \{m\}$, and extend $<_\mathcal{L}$ so that $n <_\mathcal{L} m <_\mathcal{L} \varphi_e(n)$.

This completes the construction.

We claim that for every k, there are only finitely many elements $<_\mathcal{L}$-below k. It follows that $\mathcal{L}$ is of type ω. Say that φ_e *acts for n and adds m* when $<_\mathcal{L}$ is defined on an $m \in R$ to make $n <_\mathcal{L} m <_\mathcal{L} \varphi_e(n)$ as in (2). Let s_0 be a stage with $k \in X_{s_0}$. Suppose at stage $s > s_0$ we add an m to X_s and define $m <_\mathcal{L} k$. This can only be due to a φ_e acting for an $n \notin R$ and adding m at stage s. At stage s, we must have $n <_\mathcal{L} k$ because $n <_\mathcal{L} m <_\mathcal{L} k$. Therefore, we must also have $e < k$, for otherwise k would be among $0, 1, \ldots e$, and condition (2d) would prevent the action of φ_e. Furthermore, m is chosen so that $m \in R$ and thus only elements of R are added $<_\mathcal{L}$-below k after stage s_0. Hence an m can only be added $<_\mathcal{L}$-below k after stage s_0 when a φ_e with $e < k$ acts for an $n <_\mathcal{L} k$ with $n \notin R$. Each φ_e acts at most once for every n, and no new $n \notin R$ appears $<_\mathcal{L}$-below k after stage s_0. Thus, after stage s_0, only finitely many m are ever added $<_\mathcal{L}$-below k.

We claim that for every e, $(*)$ holds. Given e, let ℓ be the $<_\mathcal{L}$-greatest element of $\{0, 1, \ldots, e\}$. Suppose that $n >_\mathcal{L} \ell$ and $n \in C$. If $\varphi_e(n)\!\downarrow$, let s be large enough so that $\langle e, n \rangle < s$, $\varphi_{e,s}(n)\!\downarrow$, $n \in X_s$, and $\varphi_e(n) \in X_s$. Then either $\varphi_e(n)$ is already not the $\mathcal{L}$-immediate successor of n at stage $s+1$, or at stage $s+1$ the conditions of (2) are satisfied for $\langle e, n \rangle$, and an m is added such that $n <_\mathcal{L} m <_\mathcal{L} \varphi_e(n)$.

Theorem 12. *Let C be a co-maximal set. Then there is a computable linear order $\mathcal{L}$ of type ω such that* [id] *does not have a successor in $\Pi_C\mathcal{L}$. Therefore, $\Pi_C\mathcal{L} \not\cong \mathbb{N} + \mathbb{Q} \times \mathbb{Z}$.*

Proof. Let $\mathcal{L}$ be a computable linear order for C as in Lemma 11. Suppose that φ is a partial computable function such that [id] $<_{\Pi_C\mathcal{L}}$ [φ]. We show that [φ] is not the $<_{\Pi_C\mathcal{L}}$-immediate successor of [id]. The inequality [id] $<_{\Pi_C\mathcal{L}}$ [φ] means that $(\forall^\infty n \in C)\,(n <_\mathcal{L} \varphi(n))$. However, by Lemma 11,

$$(\forall^\infty n \in C)\,(\varphi(n) \text{ is not the } \mathcal{L}\text{-immediate successor of } n).$$

Define a partial computable ψ so that, for every n,

$$\psi(n) = \begin{cases} \text{the least } m \text{ such that } n <_\mathcal{L} m <_\mathcal{L} \varphi(n) & \text{if there is such an } m; \\ \uparrow & \text{otherwise.} \end{cases}$$

Then $(\forall^\infty n \in C)\,(n <_\mathcal{L} \psi(n) <_\mathcal{L} \varphi(n))$. Thus, [id] $<_{\Pi_C\mathcal{L}}$ [ψ] $<_{\Pi_C\mathcal{L}}$ [φ]. So, [φ] is not the $<_{\Pi_C\mathcal{L}}$-immediate successor of [id].

It follows that $\Pi_C\mathcal{L} \not\cong \mathbb{N} + \mathbb{Q} \times \mathbb{Z}$ because every element of $\mathbb{N} + \mathbb{Q} \times \mathbb{Z}$ has an immediate successor, but [id] $\in \Pi_C\mathcal{L}$ does not have an immediate successor.

Note that the sentence Ψ that states that every element has an immediate successor is Π_3^0. Then for the computable linear order $\mathcal{L}$ of type ω constructed above, $\mathcal{L} \models \Psi$ but $\Pi_C\mathcal{L} \models \neg\Psi$. Note that when working on this paper we proved Theorem 9 first, and even though it is subsumed by Theorem 12 as stated, we wanted to include our first proof in case others also find the method useful.

References

1. Dimitrov, R.D.: A class of Σ_3^0 modular lattices embeddable as principal filters in $\mathcal{L}^*(V_\infty)$. Arch. Math. Logic **47**, 111–132 (2008)
2. Dimitrov, R.D.: Cohesive powers of computable structures, vol. 99, pp. 193–201. Annuaire De L'Universite De Sofia "St. Kliment Ohridski", Fac. Math. and Inf. (2009)
3. Dimitrov, R.D., Harizanov, V.: Orbits of maximal vector spaces. Algebra Logic 54, 680–732 (2015) (Russian) 440–477 (2016) (English translation)
4. Dimitrov, R., Harizanov, V., Miller, R., Mourad, K.J.: Isomorphisms of non-standard fields and ash's conjecture. In: Beckmann, A., Csuhaj-Varjú, E., Meer, K. (eds.) CiE 2014. LNCS, vol. 8493, pp. 143–152. Springer, Cham (2014). https://doi.org/10.1007/978-3-319-08019-2_15
5. Dimitrov, R., Harizanov, V., Morozov, A., Shafer, P., Soskova, A., and Vatev, S.: Cohesive powers, linear orders and Fraïssé limits, Unpublished manuscript
6. Feferman, S., Scott, D.S., Tennenbaum, S.: Models of arithmetic through function rings. Not. Amer. Math. Soc. **6**, 173 (1959). Abstract #556-31
7. Hirschfeld, J., Wheeler, W.H.: Forcing, Arithmetic, Division Rings. LNM, vol. 454. Springer, Heidelberg (1975). https://doi.org/10.1007/BFb0064082
8. Lerman, M.: Recursive functions modulo co-r-maximal sets. Trans. Am. Math. Soc. **148**, 429–444 (1970)

9. McLaughlin, T.: Sub-arithmetical ultrapowers: a survey. Ann. Pure Appl. Logic **49**, 143–191 (1990)
10. Soare, R.I.: Recursively Enumerable Sets and Degrees: A Study of Computable Functions and Computably Generated Sets. Springer, Heidelberg (1987)

An algorithmic approach
to characterizations of admissibles

Bruno Durand[(✉)] and Grégory Lafitte[(✉)]

LIRMM, CNRS, Université de Montpellier, 161 rue Ada, 34090 Montpellier, France
{bruno.durand,gregory.lafitte}@lirmm.fr

Abstract. Sacks proved that every admissible countable ordinal is the first admissible ordinal relatively to a real. We give an algorithmic proof of this result for constructibly countable admissibles. Our study is completed by an algorithmic approach to a generalization of Sacks' theorem due to Jensen, that finds a real relatively to which a countable sequence of admissibles, having a compatible structure, constitutes the sequence of the first admissibles. Our approach deeply involves infinite time Turing machines. We also present different considerations on the constructible ranks of the reals involved in coding ordinals.

Introduction

One motivation for ordinal computability (see, *e.g.*, [5]) is to find new proofs for theorems in constructible and descriptive set theory. Such proofs may yield extra information and give a further perspective. Examples so far include the proof by Koepke and Seyfferth of the existence of incomparable α-degrees using α-Turing machines in [26], Koepke's proof of the continuum hypothesis in L using ordinal Turing machines in [25] and the proof by Schlicht and Seyfferth of Shoenfield's absoluteness theorem in [34], also via ordinal Turing machines.

In this paper, we prove a constructive version of a theorem of Sacks, along with a strengthening thereof due to Jensen, the latter using *infinite time Turing machines* (ITTMs). These were invented by Hamkins and Kidder and first introduced in [17]. A topic that has received particular attention is the issue of *clockability*. This subject is of special concern for us as it shows some deep relations with admissible ordinals.

Admissible ordinals correspond to levels of the constructible hierarchy that are closed enough to carry computabilities, in other words closed under Σ_1 definability; the reference text is Barwise's book [1]. This notion has several equivalent definitions stated in rather different terms and involves many deep properties witnessed by ITTMs.

The research for this paper has been done thanks to the support of the *Agence nationale de la recherche* through the RaCAF ANR-15-CE40-0016-01 grant.

F. Manea et al. (Eds.): CiE 2019, LNCS 11558, pp. 181–192, 2019.
https://doi.org/10.1007/978-3-030-22996-2_16

Sacks' theorem is very important to understand countable admissible ordinals. It states that for any countable admissible ordinal α, there exists a real r such that α is the first non-recursive ordinal relatively to r (written $\omega_1^{CK,r}$, this ordinal is admissible and the first admissible $> \omega$ relatively to r). For the first admissibles, this statement is rather clear: the first admissible is ω_1^{CK}, the next one is ω_1^{CK,r_1} where r_1 is a code of lowest constructible rank for ω_1^{CK}, and so on. The situation is more complex with the first *recursively inaccessible* ordinal (an admissible which is a limit of admissibles). Indeed a computability construction is required to transform a sequence of admissibles co-final in this admissible into an adequate real. The situation is much more complex for larger ordinals because of coding problems. For successor admissibles, those that are not recursively inaccessible, our idea is to use a code for the admissible just below, but questions arise to know whether such a code exists and if it exists, in which constructibility level. For recursively inaccessibles, the situation is even more subtle since the ω-sequence of admissibles cofinal in this ordinal must also be definable, and various cases arise depending on the different gaps in which the ordinals of the sequence might appear.

When one has proven Sacks' theorem, it is rather natural to try to obtain that any two admissibles become the first and the second admissibles relatively to a real. Once again the case where the ordinals are small is not difficult (although it requires to work both with Turing and hyperarithmetic reducibilities). For transfinite sequences of admissibles, the situation is more complex since our goal is that this sequence coincides with an initial segment of the admissibles using only an oracle r—which one can see as some kind of translation function. We provide an elementary proof of Jensen's generalization to Sacks' theorem for sequences of constructibly countable admissibles, the ordinal type of which is bounded by the *supremum* of the clockable times of ITTMs.

We begin our paper by describing various gaps and ranks concerning definability/coding issues, and then describe our algorithmic and computability approach to proving versions of these two theorems.

1 Gaps and ranks

In the different results of this paper, in particular in Theorem 3, we are interested in finding the *simplest* reals, simplest in the sense that they appear at the lowest possible rank in the constructible hierarchy. This approach is naturally linked to various gaps and ranks described in this section.

1.1 The constructible hierarchy

We recall Gödel's constructible hierarchy $(L_\alpha : \alpha \in \mathrm{On})$ and an accompanying *ad hoc* hierarchy $\mathcal{J}_\alpha$ in which we can find Jensen's master codes for levels where new reals appear. The reader is referred to [22,10,21] for more on this hierarchy.

Definition 1. $L_0 = \mathcal{J}_0 = \varnothing$, $L_{\alpha+1} = \mathrm{Def}(L_\alpha)$, $L_\lambda = \bigcup_{\alpha < \lambda} L_\alpha$, $L = \bigcup_{\alpha \in \mathrm{On}} L_\alpha$, $\mathcal{J}_{\omega \cdot \xi} = L_\xi$, $\mathcal{J}_{\omega \cdot \xi + n} = \Delta_n(L_\xi)$, where λ is a limit ordinal and $\mathrm{Def}(X)$, resp.

$\Delta_n(X)$, is the set of all subsets of X definable with parameters in $\langle X, \in \rangle$ by a first order formula, resp. by both Σ_n and Π_n first order formulae.

We are interested in finding the lowest level of Gödel's constructible hierarchy where certain sets appear, in particular certain reals (subsets of ω). The least level α of L where a certain set A appears ($A \in L_{\alpha+1}$) is called the L-rank of A.

1.2 Gaps in the constructible hierarchy

H. W. Putnam and G. S. Boolos identified levels of L where no new reals appear.

Theorem 1. *There are arbitrarily long gaps in L where no new reals appear.*[1]

This leads to the notion of gaps in the constructible hierarchy: an ordinal α is in a *Putnam gap* (or L-gap) if no new reals appear in L_α. The proof of Theorem 1 uses the fact that if M is a countable elementary subset of $L_{\omega_2^L}$, then the image of ω_1^L under the Mostowski collapse of M is a very long Putnam gap. The idea behind such a Putnam gap is that an ordinal $\alpha < \omega_1^L$ starts a long gap if it is very similar to ω_1^L. When new reals appear at level $\alpha + 1$, a real coding all of L_α is one of them.

Lemma 1 ([2,3]). *If new reals appear at $\alpha + 1$, then among them is an arithmetical copy[2] E_α of L_α.*

If α is a Putnam-gap ordinal and if $r \in L_\alpha$ is a real coding a well-order on ω, then the order type of that well-order is less than α. Thus if α starts a Putnam-gap, then α is a limit ordinal and Marek and Srebrny [30] actually showed that an ordinal α starts a Putnam-gap if and only if $L_\alpha \models \text{ZFC}^- + V = \text{HC}$.

Gaps naturally appear both in the L and the $\mathcal{J}$ hierarchies. When new reals appear at a level, we call this level *an index*. Let $\ell_{\mathcal{J}}(\alpha)$ be the maximum β such that $[\alpha, \alpha + \beta)$ is a $\mathcal{J}$-gap. Thus α is a $\mathcal{J}$-gap ordinal if and only if $\ell_{\mathcal{J}}(\alpha) \neq 0$, and if α starts an $\mathcal{J}$-gap, $\ell_{\mathcal{J}}(\alpha)$ is the length of that $\mathcal{J}$-gap.

Special reals in $\mathcal{J}$ were identified by Jensen: a real r is a *master code* for $\mathcal{J}_\xi$ (or, for ξ) if $\{x \subseteq \omega : x \leqslant_T r\} = \mathcal{J}_{\xi+1} \cap \mathcal{P}(\omega)$.

Theorem 2 (Jensen). *ξ is a $\mathcal{J}$-index if and only if there is a master code for ξ. Furthermore, if r is a master code for ξ, then r' is the master code for $\xi + 1$.*

The gaps in the $\mathcal{J}$ hierarchy can be described in the following way (for more on master codes and these gaps, see [20,18,19,22,23]): let $\varsigma(\alpha)$ be the least strict upper bound on $\{\text{Ind}_{\mathcal{J}}(\xi) : \xi < \alpha\}$, where $\text{Ind}_{\mathcal{J}} : \omega_1^L \to \omega_1^L$ enumerates the $\mathcal{J}$-indices in increasing order; obviously $\alpha \leqslant \text{Ind}_{\mathcal{J}}(\alpha)$. $\alpha \leqslant \varsigma(\alpha)$ and $\alpha < \varsigma(\alpha)$

---

[1] Let $\beta > \alpha$ be countable ordinals such that there is an elementary embedding $j : L_\beta \to L_{\omega_2}$ with *critical point* $cr(j) \geqslant \alpha$. For every $\gamma < cr(j)$, $L_{\omega_2} \models$ "No new reals appear between ranks ω_1 and $\omega_1 + \gamma$." No new reals thus appear between $cr(j)$ and $cr(j) + \gamma$, by elementarity and absoluteness. Cf. [31,2,3,28].

[2] E_α is an *arithmetical copy* of L_α if there is one-one function f from L_α to ω (and onto the field of E_α) such that $\forall x, y \in L_\alpha, x \in y \iff \langle f(x), f(y) \rangle \in E_\alpha$.

iff $\mathrm{Ind}_{\mathcal{J}}(\alpha) > \alpha$ and α does not start a $\mathcal{J}$-gap. $\varsigma(\alpha) \neq \mathrm{Ind}_{\mathcal{J}}(\alpha)$ iff α starts a $\mathcal{J}$-gap. $\mathrm{Ind}_{\mathcal{J}}(\alpha) = \varsigma(\alpha) + \ell_{\mathcal{J}}(\varsigma(\alpha))$. $\varsigma(\lambda)$ is a $\mathcal{J}$-gap ordinal iff $\varsigma(\lambda)$ is admissible iff λ is admissible and locally countable ($\models \forall a\,(a$ is countable$)$). If α starts an L-gap, then α also starts a $\mathcal{J}$-gap. If α starts a $\mathcal{J}$-gap, then α is the *supremum* of L-indices and α starts an L-gap iff $\ell_{\mathcal{J}}(\alpha) \geqslant \omega$. Moreover if α starts a $\mathcal{J}$-gap and $\ell_{\mathcal{J}}(\alpha) \geqslant n$, then α is Σ_n-admissible.

1.3   Definability and coding

There are some other gaps that we call *definability gaps*, which are obviously linked to Putnam gaps: there are countable ordinals α, υ such that α is not definable in L_υ.[3] It is possible to characterize the least such definability gap: the ordinal υ_0, which is the least υ such that there is an ordinal α not definable in L_υ, can be characterized as the least η such that there exists an ordinal $\delta < \eta$ such that $L_\delta \prec L_\eta$ ($\prec$ means *being an elementary submodel*).[4]

Now we would like to know for a particular real where it first appears, especially for a real coding a countable ordinal.

Definition 2. *Let α be a countable ordinal. α is definable at (level)γ if α is definable without parameters in L_γ. α is codable at (level)γ if a real appears in $L_{\gamma+1}$ coding α. The code-rank of α ($< \omega_1^L$), code-rk(α), is the least γ such that α is codable at γ. α is countable at (level)γ if $L_\gamma \models$ "α is countable".*

If $\alpha < \omega_1^L$, then it is codable at some level. And if α is countable at β, then α is codable and definable at β.

Lemma 2. *For every countable ordinal α, there exists a countable β such that α is definable at β.*[5]

By Löwenheim-Skolem and a combination of footnotes 5 and 3, there exists an ordinal $\alpha < \omega_1^L$ which is definable at a β, then not definable at a $\beta' > \beta$, etc. There is actually an upper bound for the ordinals that remain definable from

---

[3] There exists α such that $L_\alpha \prec L_{\omega_1}$, α is thus not definable in L_{ω_1}. There is a countable $\upsilon > \alpha$ such that $L_\upsilon \prec L_{\omega_1}$, and α is already not definable in L_υ.

[4] υ_0 is clearly $\leqslant$ the least such η, η_0, since whenever one has $L_\alpha \prec L_\beta$, α is not definable in L_β. Now, suppose that $\upsilon_0 < \eta_0$, in other words, for all $\delta < \upsilon_0$, $L_\delta \not\prec L_{\upsilon_0}$. Now, by Löwenheim-Skolem there is a countable elementary submodel of L_{υ_0}. Take the $\subseteq$-least such model M. By the Condensation Lemma, there is an $\alpha < \upsilon_0$ and an isomorphism j such that the Mostowski collapse of M is isomorphic to L_α via j. j cannot be trivial as this would mean that $L_\alpha \prec L_\delta$, although $\delta < \upsilon_0$ and υ_0 is the least such ordinal. We can thus consider κ, the critical point[1] of j. Since $L_\alpha \cong M \prec L_{\upsilon_0}$, $L_\kappa \prec L_{j(\kappa)}$. But then κ cannot be definable in $L_{j(\kappa)}$, and thus $\upsilon_0 \leqslant j(\kappa)$. But $j(\kappa) < \upsilon_0$, contradiction.

[5] Consider $\kappa = \aleph_\alpha$. κ is definable as the greatest cardinal in L_{κ^+}. (Here κ^+ denotes the least ordinal of cardinality greater than κ.) And thus α is also definable in L_{κ^+}. Löwenheim-Skolem's theorem, in conjunction with Mostowski's lemma and the Condensation Lemma, provides the countable β such that α is definable in L_β.

some point on (they have been called *memorable*[6] ordinals): ordinals α for which there exists β such that for any countable $\gamma \geq \beta$, α is still definable at γ.[7]

1.4   Clockability

Computations by ITTMs are intimately related to admissibles since they are in some sense universal tools for Σ_1 functions. In particular, an ordinal α is said to be *clockable* if there exists an ITTM that halts exactly in time α on input 0. ITTMs can also write a real coding an ordinal. In this case the ordinal is said to be *writable*. See [17] for more on these notions. A very powerful theorem by Welch [36] asserts that the *supremum* of clockable ordinals is the same as the *supremum* of the writable ordinals. In our paper, we denote this ordinal by λ_∞.[8]

Writable ordinals have no gaps: α is writable if and only if $\alpha < \lambda_\infty$. The situation is different with clockability: there are gaps inside clockable ordinals. The study of these gaps is very interesting and has been carried out through many papers (see in particular the seminal [17,37]), among them we refer to [6] since it contains all the considerations on gaps needed in the present paper. We can summarize the situation in terms of clockable gaps as follows. The situation resembles Putnam gaps with a major difference: for a clockable gap, the starting point is also a limit ordinal, but its size is always a limit ordinal. Furthermore, they are deeply related to admissibles. The properties we most often use in the present paper are that all starting points of gaps are admissibles, that no admissible is clockable, but some admissibles can be strictly inside a gap (see [6]). Moreover, if α starts a gap, the ordinal type of clockable ordinals below α is exactly α.

The structure of admissible ordinals can be relativized to a real without any major change. With ITTMs we can use the standard oracle definition of Turing machines (with a special oracle tape), or even see the oracle real as an input.

Some other considerations are important for us. If an ordinal can be written by an ITTM in time γ then it is codable at level γ. The writing time of a recursive ordinal is exactly ω and if the ordinal α is not recursive, its writing time is exactly the *supremum* of the ends of those gaps that start no later than α. For admissibles, the situation is simple: their writing time is exactly the end of the gap they belong to. A detailed proof of these results can be found in [27,11]. A consequence of these results is that for every clockable $\alpha < \lambda_\infty$ which does not end a clockable gap, we have that the code-rank of α is $< \alpha$ and bounded by its writing time.

---

[6] *Cf.* https://mathoverflow.net/questions/259100/memorable-ordinals.

[7] Any countable τ such that $L_\tau \prec L_{\omega_1}$ is such an upper bound: if α is definable at β, take δ above τ and β such that $L_\delta \prec L_{\omega_1}$. We then have $L_\tau \prec L_\delta \prec L_{\omega_1}$. α is thus definable at δ, since δ is above β, and also at τ. τ is therefore above α and any other definable ordinal. In fact, the least non-memorable ordinal τ_0 is the least ordinal τ with uncountably many elementary extensions $L_\tau \prec L_\gamma$. (*Cf.* footnote 6).

[8] We use Barwise's convention for admissibles: $\tau_0 = \omega$, $\tau_1 = \omega_1^{CK}$, ..., τ_α is the α-th admissible. Note that there exist admissibles α such that $\alpha = \tau_\alpha$. Such is the case for λ_∞, but it is not the first one.

If we set a (sufficiently closed) bound for computation times of the ITTMs, we obtain nice computation models. If the bound is chosen as ω, we obtain classical Turing machines; if the bound is chosen to be any admissible ordinal, then the model is well defined, keeping the same universal machine for all bounds. Therefore, if α is admissible, we can define $\leqslant_\alpha$ as the ITTM reducibilities with computations bounded by admissible time α. The reducibilities $\leqslant_\alpha$ and $\leqslant_\beta$ are the same over reals if and only if α and β belong to the same clockable gap. If α and β do not belong to the same clockable gap and $\alpha < \beta$ then $\leqslant_\alpha$ is a refinement of $\leqslant_\beta$.

2   Sacks' theorem revisited

Theorem 3 (Sacks). *For every admissible countable ordinal α, there exists a real r such that $\alpha = \omega_1^{\mathrm{CK},r}$.*

Gerald E. Sacks first proved Theorem 3 in [32] by a forcing argument. Friedman and Jensen [13] gave an alternative proof, which does not make use of forcing but involves infinitary logic. See also Chong and Yu [8, Theorem 5.4.12] for a proof using Steel forcing.

In the proof of our version of Sacks' theorem, Theorem 4, we need to be able to construct reals with *ad hoc* properties. We choose to isolate this construction in the following lemma.

Lemma 3. *For any countable set of reals $\{r_i : i \in \omega\}$ such that for every i, $r'_i \leqslant_T r_{i+1}$, there exists a real r such that for all i, $r_i \leqslant_T r$, but[9] $\bigoplus_i r_i \not\leqslant_T r$. Moreover $r \leqslant_T (\bigoplus_i r_i)'$.*

Proof. We first prove the lemma when for every i, $r''_i \leqslant_T r_{i+1}$, then we adapt the proof to the hypothesis $r'_i \leqslant_T r_{i+1}$. Please note that it is not straightforward (we cannot just take every other real in the sequence) since it might be possible that $\bigoplus_i r_{2i} <_T \bigoplus_i r_i$. This cannot be the case for the sequences constructed in our use of this lemma but we prefer to formulate the lemma and prove it in its most general form.

The real r needs to code the r_i's in such a way that for every i, r_i can be computed from r but not uniformly, as $\bigoplus_i r_i$ would then be $\leqslant_T$-below r. We consider that r is a mapping from ω^2 to $\{0,1\}$ that contains r_i in the column $\mathfrak{b}(i)$. At the same time, we make sure that $\bigoplus_i r_i \not\leqslant_T r$ by adding just the needed supplementary information to hide the r_i's.

Construction: r is constructed as $\bigcup_i o_i$ where the o_i's are compatible (infinite) oracles that represent the left part of r up to the column $\mathfrak{b}(i)$, the rest is empty (at 0). We now assume that o_i has been built and that we know $\mathfrak{b}(i)$, and we give the construction for o_{i+1} and $\mathfrak{b}(i+1)$.

We now consider all the computations $\varphi_i^\tau(\langle i+1, j \rangle)$ for every j and every finite extension τ of o_i. We look for the first (τ, j) such that it either *(i)* converges

---

[9] The infinite join, $\bigoplus_i r_i$, of the r_i's is defined as $\{\langle i, j \rangle : j \in r_i\}$.

and is $\neq r_{i+1}(j)$, or *(ii)* for every extension of τ, it diverges. $\mathfrak{b}(i+1)$ is then taken to be greater $(+1)$ than [case *(i)*] the maximum between $\mathfrak{b}(i)$ and the greatest column reached during the computation, [case *(ii)*] the greatest column reached in the enumeration of the extensions of o_i; which is the column from which every extension will make the computation diverge on j.

o_{i+1} is the found τ, in which we add r_{i+1} at column $\mathfrak{b}(i+1)$. Thus o_{i+1} contains o_i, r_{i+1} in column $\mathfrak{b}(i+1)$, and some extra *finite* information. $\square_{Construction}$

We now show, by induction, that the construction works: we suppose that o_i has been constructed and that $o_i \leqslant_T r'_i$. We show that o_{i+1} is built such that $o_{i+1} \leqslant_T r'_{i+1}$. In the construction of o_{i+1} from o_i, we end up either in case *(i)* or *(ii)*: by *reductio ad absurdum*, suppose that both cases are false. We then

–a– either have convergence for every j such that they all converge to $r_{i+1}(j)$. But then we would have $r_{i+1} \leqslant_T o_i$, which is impossible since $o_i \leqslant_T r'_i$,
–b– or there are some j's where it diverges and all extensions end up making it converge to $r_{i+1}(j)$, but then r_{i+1} would be recursively enumerable in o_i, which is also impossible since $r_{i+1} \geqslant_T r''_i$.

Looking closely at the construction, one observes that $o_{i+1} \leqslant_T r'_{i+1}$.

Now, for every i, $r_i \leqslant_T r$ (from a finite information, $\mathfrak{b}(i)$). And if we had $\bigoplus_i r_i \leqslant_T r$, there would be an e such that for all i, j, $\varphi^r_e(\langle i, j \rangle) = r_i(j)$. Take $i = e + 1$, we observe a contradiction:

Case *(i)*: $\varphi^r_e(\langle e+1, j \rangle) \neq r_{e+1}(j)$ for the found j,
Case *(ii)*: $\varphi^r_e(\langle e+1, j \rangle)$ diverges for every extension. But r is an extension of o_{i+1}, which implies that φ^r_e diverges on the found j.

As we have that for every i, $o_i \leqslant_T r'_i$, we get that $r = \bigcup_i o_i \leqslant_T (\bigoplus_i r_i)'$.

Now let us adapt the proof to the hypothesis $r'_i \leqslant_T r_{i+1}$: instead of coding r_i in the column $\mathfrak{b}(i)$ we code r_{2i} in $\mathfrak{b}(i)$ and r_{2i+1} in $\mathfrak{b}(i) + 1$. The oracle o_i is thus defined until the $\mathfrak{b}(i) + 1$ column and the finite extension which is defined uses r_{2i+3} instead of r_{i+1} since we have that $r_{2i+3} \geqslant_T o'_i$. Thus we have non-recursivity relative to oracle r only on the r_{2i+1} terms of $\bigoplus_i r_i$ but it is sufficient for our lemma. $\qquad\square$

The construction can be slightly enhanced to make r and $\bigoplus_i r_i$ incomparable, by introducing witnesses for $r \not\leqslant_T \bigoplus_i r_i$. We just need to be careful to introduce only a finite number of witnesses on every column. The following theorem provides an explicit construction of a weaker version of Sacks' result.

Theorem 4. *For every admissible constructibly countable ordinal α, there exists a real r such that $\alpha = \omega_1^{CK,r}$.*

Proof. First, we solve the easy case, where α is a *successor admissible*, i.e., admissible but not a limit of admissibles: $\alpha = \beta^+$ for some admissible β.

We first assume that α and β are not codable at the same level. There exist (many) reals that code β. For any such real r, we have that $\alpha \leqslant \omega_1^{CK,r}$ and $\omega_1^{CK,r}$ is admissible. Among those reals, we choose r_β of least constructible rank γ.

If $\alpha < \omega_1^{CK,r_\beta}$, then α is recursive in r_β and codable at level γ. Therefore, β and α are codable at the same level, which contradicts the hypothesis.

We now assume that α and β have the same code-rank (in L). Let γ then be the least level of $\mathcal{J}$ where a code for α appears, and thus also where one appears for β. New reals appear at this level and among them, there is a master code for γ. From this master code, we can extract a code r for β of minimal Turing degree. α is not recursive in r, since α is admissible. And thus, $\alpha = \omega_1^{CK,r}$.

Now assume that α is *recursively inaccessible*. As $\alpha < \omega_1^L$, we have $\alpha = \lim_{n<\omega} \alpha_n$, where the α_n's are admissible, the α_n's are codable at the code-rank of α, and α is admissible relative[10] to $\{\alpha_n : n < \omega\}$. For each n, r_n is chosen as a "simplest" code for α_n. To precise the meaning of "simplest", there are two cases: either they are all of the same code-rank than α, or they can be chosen to have strictly increasing code-ranks.

In the latter case, by the admissibility hypothesis on the structure, we choose the r_n's of least code-rank so that the $\bigoplus_n r_n \equiv_T r_\alpha$. In other terms, we do not add extra information in the precise chosen sequence.

In the former case, α and all α_n's are of code-rank γ. As a code for α, we choose a real r_α extracted from a master code of $\mathcal{J}$ for the least level where a code for α appears. From r_α we can define integer indices $i_0, i_1, \ldots$ such that in the order that codes α in r_α, α_n is at index i_n. Thus, the order for α truncated at level i_n is a code for α_n and is represented by a real r_n. We can remark that 1's in r_n are also 1's in r_{n+1} while some 0's in r_n become 1's in r_{n+1}. By the admissibility hypothesis on the structure, the infinite join of the r_n's is Turing-equivalent to r_α.

In both cases, we also have that $r'_n \leqslant_T r_{n+1}$, since the α_n's are admissible. The real r we look for is obtained directly by Lemma 3. As r_α is not recursive in r, and because of the hypothesis on the sequence of the α_n's, r_n is recursive in r and α is the least ordinal which is not recursive in r. α is thus $\omega_1^{CK,r}$. □

This proof provides properties that go beyond the statement of Theorem 4. For instance, if $\alpha = \omega_1^{CK,r}$ for $r \in L_\alpha$ then $\alpha = \beta^+$, i.e., is a successor admissible, and there is γ, such that $\beta < \gamma < \alpha$ and new reals appear in L_γ, i.e., the interval (β, α) is not inside a coding gap. An analogous result can be found in [7].

Note that since Lemma 3 gives $r \leqslant_T (\bigoplus_i r_i)'$, our proof provides a certain minimality property of the constructed real. In general, it can be expressed in terms of hyperarithmetic[11] minimality, exactly as the refinement that Sacks obtained in [33, Theorem 4.26] of his original theorem. But in the case where the admissible is a successor admissible, we get an improved optimality result:

Theorem 5. *For every admissible $\alpha < \omega_1^L$, there exists a real r such that $\omega_1^{CK,r} = \alpha$, and for every real $s <_h r$, $\omega_1^{CK,s} < \alpha$. Moreover if α is a successor admissible, for every real $s <_T r$, $\omega_1^{CK,s} < \alpha$.*

---

[10] An ordinal α is *admissible relative to a set of ordinals A* if $\langle L_\alpha [A] ; A \cap \alpha \rangle$ is an admissible structure.

[11] Recall that $Y \leqslant_h X$ if Y is hyperarithmetic in X, that is if Y is ITTM-computable in some bounded recursive ordinal length of time ($< \omega_1^{CK,X}$).

This naturally leads to the study of the least constructible rank of the real that defines a countable admissible ordinal *via* Sacks' result. The *Sacks-rank* of a countable ordinal α, Sacks-rk(α), is the least γ such that there is a real r in $L_{\gamma+1}$ such that $\omega_1^{CK,r} = \alpha$. Thanks to our construction, some structural properties can be proved concerning Sacks ranks: for every countable ordinal $\alpha < \omega_1^L$, if α is a successor admissible $(=\beta^+)$, we have that Sacks-rk(α) = code-rk(β), and otherwise (α is recursively inaccessible) that Sacks-rk(α) = code-rk(α).

3   Jensen's theorem revisited

Ronald B. Jensen and Harvey Friedman gave in [13] a model-theoretic proof of Sacks' theorem (*cf.* Theorem 3). Jensen had formulated the *model existence* theorem and applied it to provide an alternative proof. This method could not be applied to prove Jensen's theorem, which is a generalization of Sacks' theorem to a sequence of admissibles. He had to use proper class forcing over admissible sets. A proof of Jensen's theorem can be found in his unpublished manuscript [23, Chapter 6, Theorem 4] and also in Simpson and Weitkamp's [35].

It has to be noted that one needs to be careful when stating the hypothesis of this theorem: choose for example a recursively inaccessible for α_ω and $\{\alpha_i : i \in \omega\}$ a sequence of admissibles cofinal in α_ω. Relative to any oracle, the *supremum* of the first ω admissibles cannot be admissible: to see that, design an ITTM looking for clockable gaps and make it halt at the *sup* of the ω first gaps (*cf.* [6]). The hypothesis proposed by Jensen, which solves this difficulty, asserts that every admissible in the considered sequence is still admissible relative[10] to admissibles of the sequence below it.[12] This is the hypothesis that we use in our version of the theorem (Theorem 6).

We first present a computability lemma that provides a real that verifies some computability specifications in terms of ITTM reducibilities ($\leqslant_{\tau_\alpha}$ means ITTM-computable in time $< \tau_\alpha$). The proof of this lemma is a direct generalization of the proof of Lemma 3.

Lemma 4. *For any ordinal $\gamma < \lambda_\infty$, for any sequence of reals $(r_\alpha : \alpha < \omega \cdot \gamma)$, such that for every α, $r'_\alpha <_T r_{\alpha+1}$, there exists $r : \gamma \to \mathcal{P}(\omega)$ such that for all $\beta < \gamma$, for all $i < \omega$, $r_{\omega\cdot\beta+i} \leqslant_{\tau_\beta} r(\beta)$, but $\bigoplus_i r_{\omega\cdot\beta+i}$ and $r(\beta)$ are $\leqslant_{\tau_\beta}$-incomparable, and $\bigoplus_i r_{\omega\cdot\beta+i}$ and $r(\beta + 1)$ are $\leqslant_{\tau_\beta}$-incomparable; and for any limit ordinal $\lambda < \omega \cdot \gamma$ and any $\alpha < \lambda$, $r_\alpha \leqslant_{\tau_\lambda} r(\lambda)$, $\bigoplus_{\alpha<\lambda} r_\alpha$ and $r(\lambda)$ are $\leqslant_{\tau_\lambda}$-incomparable. Moreover for all $\beta < \gamma$, $r(\beta) \leqslant_T (\bigoplus_i r_{\omega\cdot\beta+i})'$ and for any limit ordinal $\lambda < \omega \cdot \gamma$, $r(\lambda) \leqslant_T (\bigoplus_{\alpha<\lambda} r_\alpha)'$.*

$\bigoplus_{\alpha<\lambda} r_\alpha$ is defined as follows: first note that if λ is a recursive ordinal, we can use a standard bi-recursive encoding of ω^2 in ω. Otherwise, we form the

---

[12] This hypothesis carries the ideas of progressivity of the sequence and indiscernibility by first order properties: in the list of admissibles $\langle \tau_\beta : \beta < \lambda_\infty \rangle$, the sequence of indices that correspond to the considered sequence does not contain too much information in itself.

equivalence classes of those ITTMs (represented by their indices in a standard enumeration) that halt in a given ordinal time δ. In each class we have ω integers. We encode $r_{\omega+\delta}$ at the abscisse given by the first ITTM that halts in time δ. Note that this encoding covers all clockable ordinals. When we enter gaps, we proceed in the same way but on the second ITTM that halts in time η, for the η-th non-clockable ordinal. Then we continue the construction for ω imbrications of gaps of gaps of gaps, *etc*. But we could have more than ω such imbrications. Thus, we use once again ITTM indices inside the sequence of ITTMs that halt at a given time to find the proper abscisse. Please note that the lengths and the ranks of the clockable gaps are co-final in λ_∞. Thus, for any $\lambda < \lambda_\infty$, we get a coding that is recursive for ITTMs bounded by time *supremum* of length of the possible gap and starting point. This construction of a transfinite join is compatible with Jensen's hypothesis on relative admissibility of the admissible ordinals of the considered sequence.

Our version of Jensen's theorem provides an explicit construction: we keep Jensen's hypothesis, but add λ_∞ as an upper bound on the *length* of the sequence and also require that the admissibles of the sequence are below ω_1^L.

Theorem 6. *Let $\gamma < \lambda_\infty$. If $\langle \alpha_\beta : \beta < \gamma \rangle$ is a sequence of constructibly countable admissibles such that for every $\delta < \gamma$, α_δ is admissible relative[10] to $\{\alpha_\beta : \beta < \delta\}$, then there is a real r such that α_β is the β-th r-admissible ordinal.*

Proof sketch. To start with, we would like to construct r such that $\alpha_0 = \omega_1^{\mathrm{CK},r}$. And of course, we would like to add much more things in r in order to get that α_1 becomes the next admissible, and so on. The situation is analogous to that of Sacks' theorem: if α_0 is a successor admissible, then we consider r_0 a code of its predecessor that we write in r (if we are in a definability gap, we proceed as we have done in the proof of Theorem 4). If α_0 is recursively inaccessible, then we encode in a special way an ω-sequence of codes $\langle r_{\beta_i} : i < \omega \rangle$ for admissibles $\langle \beta_i : i < \omega \rangle$ cofinal in α_0. We use Lemma 3 to get the *ad hoc* real r_0.

Now we would like to add some new information to encode also the real r_1 that makes α_1 the second admissible. But while doing this, we should not make α_0 computable. Thus we can make a special version of Lemma 3 where the reduction used is not the Turing reducibility, but the ITTM reducibility with time bounded by the first admissible (which is exactly the hyperarithmetic reducibility, $\leqslant_h$) and while doing this, we make sure that we still have $\alpha_0 \nleq_T r_0 \oplus r_1$.

This construction gives the induction step when β is a successor ordinal. When β is limit, let us first assume that β is not strictly inside an ITTM-gap; thanks to the relative admissibility hypothesis of the sequence, α_β is recursively inaccessible if and only if τ_β is. We propose a single construction for the two cases concerning α_β, being recursively inaccessible or a successor admissible. We proceed as above, with the help of a more complex lemma. We consider ITTM-reducibilities bounded by the β-th admissible, and apply this to the ω-sequence extracted from admissibles below α_β if inaccessible, or to the real coding its predecessor if α_β is a successor admissible. The hypotheses of our improved lemma (Lemma 4) that provides r_β are verified thanks to the relative admissibility hypothesis and the constructibly countability of the considered ordinals.

The problem now is when the β-th admissible is inside an ITTM clockable gap. Indeed, if τ_η starts a clockable gap which contains τ_β, then $\leqslant_{\tau_\eta}$ is exactly the same reducibility than $\leqslant_{\tau_\beta}$. Note that in this case, α_η starts a clockable gap which contains α_β. We use a slightly modified version of Lemma 4 that uses α_η as an oracle from rank η on. If α_β is in a gap of a gap of a gap ... (of rank δ), then we modify analogously Lemma 4 adding as an oracle the starting points of the gaps of rank $< \delta$.   □

We propose an application of our theorem to a bounded version of Solovay's problem, namely finding a real relatively to which the admissibles are the recursively inaccessibles. The solution to Solovay's problem by Sy D. Friedman (*cf.* [15]) proves that the sequence of recursively inaccessible ordinals below ω_1^L verify the relative admissibility hypothesis of Jensen, thus our construction works as is.

Theorem 7 (Solovay's problem below λ_∞). *If $\langle \iota_\beta : \beta < \lambda_\infty \rangle$ is the sequence of recursively inaccessible ordinals below λ_∞, then there is a real r such that ι_β is the β-th r-admissible ordinal.*

References

1. Barwise, J.: Admissible Sets and Structures: An Approach to Definability Theory, Perspectives in Mathematical Logic, vol. 7. Springer, Heidelberg (1975)
2. Boolos, G.S.: The hierarchy of constructible sets of integers. Ph.D. thesis, Massachusetts Institute of Technology, Cambridge, Mass. (1966)
3. Boolos, G.S., Putnam, H.: Degrees of unsolvability of constructible sets of integers. J. Symb. Log. **33**, 497–513 (1968)
4. Boyd, R., Hensel, G., Putnam, H.: A recursion-theoretic characterization of the ramified analytical hierarchy. Trans. Am. Math. Soc. **141**, 37–62 (1969)
5. Carl, M.: Ordinal Computability. De Gruyter Series in Logic and Its Applications, vol. 9. De Gruyter, July 2019
6. Carl, M., Durand, B., Lafitte, G., Ouazzani, S.: Admissibles in gaps. In: Kari et al. [24], pp. 175–186
7. Chong, C.T.: A recursion-theoretic characterization of constructible reals. Bull. Lond. Math. Soc. **9**, 241–244 (1977)
8. Chong, C.T., Yu, L.: Recursion theory. De Gruyter Series in Logic and Its Applications, vol. 8. De Gruyter (2015)
9. David, R.: A functorial Π_2^1 singleton. Adv. Math. **74**, 258–268 (1989)
10. Devlin, K.: Constructibility. Springer, Heidelberg (1984)
11. Durand, B., Lafitte, G.: A constructive Swiss knife for infinite time Turing machines (2016)
12. Friedman, H.: Minimality in the Δ_2^1-degrees. Fundamenta Mathematicae **81**(3), 183–192 (1974)
13. Friedman, H., Jensen, R.: Note on admissible ordinals. In: Barwise, J. (ed.) The Syntax and Semantics of Infinitary Languages. LNM, vol. 72, pp. 77–79. Springer, Heidelberg (1968). https://doi.org/10.1007/BFb0079683

The authors would like to thank the anonymous referees for their constructive comments which helped a lot to improve the manuscript.

14. Friedman, S.D.: An introduction to the admissibility spectrum. In: Marcus, R.B., Dorn, G.J., Weingartner, P. (eds.) Logic, Methodology and Philosophy of Science VII, Proceedings of the Seventh International Congress of Logic, Methodology and Philosophy of Science (Salzburg, 1983). Studies in Logic and the Foundations of Mathematics, vol. 114, pp. 129–139. North-Holland (1986)
15. Friedman, S.D.: Strong coding. Ann. Pure Appl. Log. **35**, 1–98 (1987)
16. Grilliot, T.: Omitting types: applications to recursion theory. J. Symb. Log. **37**, 81–89 (1972)
17. Hamkins, J.D., Lewis, A.: Infinite time Turing machines. J. Symb. Log. **65**(2), 567–604 (2000)
18. Hodes, H.T.: Jumping through the transfinite: the master code hierarchy of Turing degrees. J. Symb. Log. **45**(2), 204–220 (1980)
19. Hodes, H.T.: Upper bounds on locally countable admissible initial segments of a Turing degree hierarchy. J. Symb. Log. **46**, 753–760 (1981)
20. Hodes, H.T.: Jumping through the transfinite. Ph.D. thesis, Harvard University (1977)
21. Jech, T.: Set Theory: The Third Millennium Edition, Revised and Expanded. SMM. Springer, Heidelberg (2003). https://doi.org/10.1007/3-540-44761-X
22. Jensen, R.B.: The fine structure of the constructible hierarchy. Ann. Math. Log. **4**, 229–308 (1972). Erratum, ibid. 4, 443 (1972)
23. Jensen, R.B.: Admissible sets, December 2010. https://www.mathematik.hu-berlin.de/~raesch/org/jensen.html. Handwritten notes (1969)
24. Kari, J., Manea, F., Petre, I. (eds.): CiE 2017. LNCS, vol. 10307. Springer, Cham (2017). https://doi.org/10.1007/978-3-319-58741-7
25. Koepke, P.: Turing computations on ordinals. Bull. Symb. Log. **11**, 377–397 (2005)
26. Koepke, P., Seyfferth, B.: Ordinal machines and admissible recursion theory. Ann. Pure Appl. Log. **160**, 310–318 (2009)
27. Le Scornet, P.: Les machines de Turing en temps transfini. Rapport de stage, Ecole Normale Supérieure de Rennes, June 2017
28. Leeds, S., Putnam, H.: An intrinsic characterization of the hierarchy of constructible sets of integers. In: Gandy, R.O., Yates, C.M.E. (eds.) Logic Colloquium '69 (Proceedings of the Summer School and Colloquium in Mathematical Logic, Manchester, August 1969), pp. 311–350. North-Holland (1971)
29. Lukas, J.D., Putnam, H.: Systems of notations and the ramified analytical hierarchy. J. Symb. Log. **39**, 243–253 (1974)
30. Marek, W., Srebrny, M.: Gaps in the constructible universe. Ann. Math. Log. **6**, 359–394 (1974)
31. Putnam, H.: A note on constructible sets of integers. Notre Dame J. Formal Log. **4**(4), 270–273 (1963)
32. Sacks, G.E.: Forcing with perfect closed sets. In: Proceedings of the Symposia Pure Math, vol. XIII, pp. 331–355. American Mathematical Society (1971)
33. Sacks, G.E.: Countable admissible ordinals and hyperdegrees. Adv. Math. **19**, 213–262 (1976)
34. Seyfferth, B.: Three models of ordinal computability. Ph.D. thesis, Rheinischen Friedrich-Wilhelms-Universitat Bonn (2012)
35. Simpson, S.G., Weitkamp, G.: High and low Kleene degrees of coanalytic sets. J. Symb. Log. **48**(2), 356–368 (1983)
36. Welch, P.D.: Eventually infinite time Turing degrees: infinite time decidable reals. J. Symb. Log. **65**(3), 1193–1203 (2000)
37. Welch, P.D.: Characteristics of discrete transfinite time Turing machine models: halting times, stabilization times, and normal form theorems. Theoret. Comput. Sci. **410**, 426–442 (2009)

Destroying Bicolored P_3s by Deleting Few Edges

Niels Grüttemeier, Christian Komusiewicz[iD], Jannik Schestag,
and Frank Sommer[(✉)][iD]

Fachbereich Mathematik und Informatik, Philipps-Universität Marburg,
Marburg, Germany
{niegru,komusiewicz,jschestag,fsommer}@informatik.uni-marburg.de

Abstract. We introduce and study the BICOLORED P_3 DELETION problem defined as follows. The input is a graph $G = (V, E)$ where the edge set E is partitioned into a set E_b of blue edges and a set E_r of red edges. The question is whether we can delete at most k edges such that G does not contain a bicolored P_3 as an induced subgraph. Here, a bicolored P_3 is a path on three vertices with one blue and one red edge. We show that BICOLORED P_3 DELETION is NP-hard and cannot be solved in $2^{o(|V|+|E|)}$ time on bounded-degree graphs if the ETH is true. Then, we show that BICOLORED P_3 DELETION is polynomial-time solvable when G does not contain a bicolored K_3, that is, a triangle with edges of both colors. Moreover, we provide a polynomial-time algorithm for the case where G contains no induced blue P_3, red P_3, blue K_3, and red K_3. Finally, we show that BICOLORED P_3 DELETION can be solved in $\mathcal{O}(1.85^k \cdot |V|^5)$ time and that it admits a kernel with $\mathcal{O}(\Delta k^2)$ vertices, where Δ is the maximum degree of G.

1 Introduction

Graph modification problems are a popular topic in computer science. In these problems, one is given a graph and one wants to apply a minimum number of modifications, for example edge deletions, to obtain a graph that fulfills some graph property Π.

One important reason for the popularity of graph modification problems is their usefulness in graph-based data analysis. A classic problem in this context is CLUSTER EDITING where we may insert and delete edges and Π is the set of cluster graphs. These are exactly the graphs that are disjoint unions of cliques and it is well-known that a graph is a cluster graph if and only if it does not contain a P_3, a path on three vertices, as induced subgraph. CLUSTER EDITING has many applications [4], for example in clustering gene interaction networks [3] or protein sequences [23]. The variant where we may only delete edges is known as CLUSTER DELETION [20]. Further, graph-based data analysis problems that

Some of the results of this work are contained in the third author's Bachelor thesis [19].
F. Sommer was supported by the DFG, project MAGZ (KO 3669/4-1).

© Springer Nature Switzerland AG 2019
F. Manea et al. (Eds.): CiE 2019, LNCS 11558, pp. 193–204, 2019.
https://doi.org/10.1007/978-3-030-22996-2_17

lead to graph modification problems for some graph property Π defined by small forbidden induced subgraphs arise in the analysis of biological [7,12] or social networks [5,18].

Besides the application, there is a more theoretical reason why graph modification problems are very important in computer science: Often these problems are NP-hard [17,24] and thus represent interesting case studies for algorithmic approaches to NP-hard problems. For example, by systematically categorizing graph properties based on their forbidden subgraphs one may outline the border between tractable and hard graph modification problems [2,16,24].

In recent years, multilayer graphs have become an increasingly important tool for integrating and analyzing network data from different sources [15]. Formally, multilayer graphs can be viewed as edge-colored (multi-)graphs, where each edge color represents one layer of the input graph. With the advent of multilayer graphs in network analysis it can be expected that graph modification problems for edge-colored graphs will arise in many applications as it was the case in uncolored graphs.

One example for such a problem is MODULE MAP [21]. Here, the input is a graph with red and blue edges and the aim is to obtain by a minimum number k of edge deletions and edge insertions a graph that contains no induced P_3 with two blue edges, no induced P_3 with a red and a blue edge, and not a triangle, called K_3, with two blue edges and one red edge. MODULE MAP arises in computational biology [1,21]; the red layer represents genetic interactions and the blue layer represents physical protein interactions [1]. MODULE MAP is NP-hard, even if G contains only blue edges, and can be solved in $\mathcal{O}(2^k \cdot |V|^3)$ time [21].

Motivated by the practical application of MODULE MAP, an edge deletion problem with bicolored forbidden induced subgraphs, we aim to study such problems from a more systematic and algorithmic point of view. Given the importance of P_3-free graphs in the uncolored case, we focus on the problem where we want to destroy all induced *bicolored* P_3s, that is, all induced P_3s with one blue and one red edge, by edge deletions.

BICOLORED P_3 DELETION (BPD)
Input: A two-colored graph $G = (V, E_r, E_b)$ and an integer $k \in \mathbb{N}$.
Question: Can we delete at most k edges from G such that the remaining graph contains no bicolored P_3 as induced subgraph?

We use $E := E_r \uplus E_b$ to denote the set of all edges of G, n to denote the number of vertices in G, and m to denote the number of edges in G.

Our Results. We show that BPD is NP-hard and that, assuming the Exponential-Time Hypothesis (ETH) [14], it cannot be solved in a running time that is subexponential in the instance size. We then study two different aspects of the computational complexity of the problem.

First, we consider special cases that can be solved in polynomial time, motivated by similar studies for problems on uncolored graphs [6]. We are in particular interested in whether or not we can exploit structural properties of input graphs that can be expressed in terms of *bicolored* forbidden subgraphs. We show that BPD can be solved in polynomial time on graphs that do not contain a certain type of bicolored K_3s as induced subgraphs, where bicolored K_3s are triangles with edges of both colors. Moreover, we show that BPD can be solved in polynomial time on graphs that contain no K_3s with one edge color and no P_3s with one edge color as induced subgraphs.

Second, we consider the parameterized complexity of BPD with respect to the natural parameter k. We show that BPD can be solved in $\mathcal{O}(1.85^k \cdot nm^2)$ time and that it admits a problem kernel with $\mathcal{O}(\Delta k^2)$ vertices, where Δ is the maximum degree in G. As a side result, we show that BPD admits a trivial problem kernel with respect to $\ell := m - k$. Due to lack of space several proofs are deferred to a long version of the article.

2   Preliminaries

We consider undirected simple graphs $G = (V, E)$ where the edge set E is partitioned into a set E_b of *blue* edges and a set E_r of *red* edges, denoted by $G = (V, E_r, E_b)$. For a vertex v, $N_G(v) := \{u \mid \{u, v\} \in E\}$ denotes the *open neighborhood* of v and $N_G[v] := N_G(v) \cup \{v\}$ denotes the *closed neighborhood* of v. For a vertex set W, $N_G(W) := \bigcup_{w \in W} N(w) \setminus W$ denotes the *open neighborhood* of W and $N_G[W] := N_G(W) \cup W$ denotes the *closed neighborhood* of W. The *degree* $\deg(v) := |N_G(v)|$ of a vertex v is the size of its open neighborhood. Moreover, we define $\deg_b(v) := |\{u \mid \{u, v\} \in E_b\}|$ as the *blue degree* of v and $\deg_r(v) := |\{u \mid \{u, v\} \in E_r\}|$ as the *red degree* of v, respectively. We let $N_G^2(v) := N_G(N_G(v)) \setminus \{v\}$ and $N_G^3(v) := N_G(N_G^2(v)) \setminus N_G(v)$ denote the *second* and *third neighborhood* of v.

For any two vertex sets $V_1, V_2 \subseteq V$, we denote by $E_G(V_1, V_2) := \{\{v_1, v_2\} \in E \mid v_1 \in V_1, v_2 \in V_2\}$ the set of edges between V_1 and V_2 in G and write $E(V') := E(V', V')$. For any $V' \subseteq V$, $G[V'] := (V', E(V') \cap E_r, E(V') \cap E_b)$ denotes the *subgraph induced by* V'. We say that some graph $H = (V^H, E_r^H, E_b^H)$ is an *induced subgraph* of G if there is a set $V' \subseteq V$, such that $G[V'] = H$, otherwise G is called H-*free*. Two vertices u and v are *connected* if there is a path from u to v in G. A *connected component* is a maximal vertex set S such that each two vertices are connected in $G[S]$. A *clique* in a graph G is a set $K \subseteq V$ of vertices such that in $G[K]$ each pair of vertices is adjacent. The graph $(\{u, v, w\}, \{\{u, v\}\}, \{\{v, w\}\})$ is called bicolored P_3. We say that a vertex $v \in V$ is *part of* a bicolored P_3 in G if there is a set $V' \subseteq V$ with $v \in V'$ such that $G[V']$ is a bicolored P_3. Furthermore, we say that two edges $\{u, v\}$ and $\{v, w\}$ *form* a bicolored P_3 if $G[\{u, v, w\}]$ is a bicolored P_3. For any set E' of edges we denote by $G - E' := (V, E_r \setminus E', E_b \setminus E')$ the graph we obtain by deleting all edges in E'. As a shorthand, we write $G - e := G - \{e\}$ for an edge e. An edge deletion set S is a *solution* for an instance (G, k) of BPD if $G - S$ is bicolored P_3-free and $|S| \leq k$. In each context we may omit the subscript G if the graph is clear from the context.

A *reduction rule* for some problem L is a computable function that maps an instance w of L to an instance w' of L such that w is a yes-instance if and only if w' is a yes-instance and $|w'| \leq |w|$. For the relevant notions of parameterized complexity we refer to the standard monographs [8,9].

3 Bicolored P_3 Deletion is NP-Hard

In this section, we prove the NP-hardness of BPD. This hardness result motivates our study of polynomial-time solvable cases and the parameterized complexity in Sects. 4 and 5, respectively.

Theorem 1. BPD *is NP-hard even if the maximum degree of G is eight.*

Proof. To show the NP-hardness we give a polynomial-time reduction from the NP-hard (3,4)-SAT problem which is given a 3-CNF formula ϕ where each variable occurs in at most four clauses, and asks if there is a satisfying assignment for ϕ [22].

Let ϕ be a 3-CNF formula with variables $X = \{x_1, \ldots, x_n\}$ and clauses $\mathcal{C} = \{C_1, \ldots, C_m\}$ with four occurrences per variable. For a given variable x_i that occurs in a clause C_j we define the *occurrence number* $\Psi(C_j, x_i)$ as the number of clauses in $\{C_1, C_2, \ldots, C_j\}$ where x_i occurs. Intuitively, $\Psi(C_j, x_i) = r$ means that the rth occurrence of variable x_i is the occurrence in clause C_j. Since each variable occurs in at most four clauses, we have $\Psi(C_j, x_i) \in \{1, 2, 3, 4\}$.

Construction: We describe how to construct an equivalent instance $(G = (V, E_r, E_b), k)$ of BPD from ϕ.

For each variable $x_i \in X$ we define a *variable gadget* as follows. The variable gadget of x_i consists of a *central vertex* v_i and two sets $T_i := \{t_i^1, t_i^2, t_i^3, t_i^4\}$ and $F_i := \{f_i^1, f_i^2, f_i^3, f_i^4\}$ of vertices. We add a blue edge from v_i to every vertex in T_i and a red edge from v_i to every vertex in F_i.

For each clause $C_j \in \mathcal{C}$ we define a *clause gadget* as follows. The clause gadget of C_j consists of three vertex sets $A_j := \{a_j^1, a_j^2, a_j^3\}$, $B_j := \{b_j^1, b_j^2, b_j^3\}$, and $W_j := \{w_j^1, w_j^2, w_j^3, w_j^4\}$. We add blue edges such that the vertices in $B_j \cup W_j$ form a clique with only blue edges in G. Moreover, we add blue edges $\{a_j^p, b_j^p\}$ and red edges $\{a_j^p, u\}$ for every $p \in \{1, 2, 3\}$ and $u \in W_j \cup B_j \setminus \{b_j^p\}$.

We connect the variable gadgets with the clause gadgets by identifying vertices in $T_i \cup F_i$ with vertices in A_j as follows. Let C_j be a clause containing variables x_{i_1}, x_{i_2}, and x_{i_3}. For each $p \in \{1, 2, 3\}$ we set

$$a_j^p = \begin{cases} t_{i_p}^{\Psi(C_j, x_{i_p})} & \text{if } x_{i_p} \text{ occurs as a positive literal in } C_j, \text{ and} \\ f_{i_p}^{\Psi(C_j, x_{i_p})} & \text{if } x_{i_p} \text{ occurs as a negative literal in } C_j. \end{cases}$$

Now, for every variable $x_i \in X$ each vertex in $T_i \cup F_i$ is identified with at most one vertex a_j^p. Figure 1 shows an example of a clause gadget and its connection with the variable gadgets. To complete the construction of the BPD instance (G, k) we set $k := 4n + 14m$.

Correctness: The formal proof of the correctness is deferred to a long version. Instead, we briefly describe its idea. For each variable x_i we have to delete all

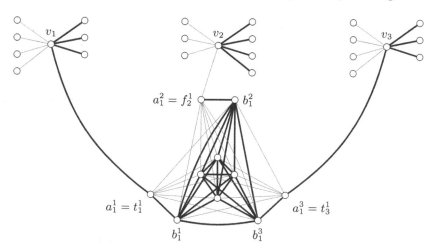

Fig. 1. The lower part of the figure shows the clause gadget of a clause $C_1 = (x_1 \lor \overline{x_2} \lor x_3)$. The upper part of the figure shows variable gadgets representing variables x_1, x_2, and x_3. The vertices $a_1^1, a_1^2,$ and a_1^3 from the clause gadget are identified with vertices from the variable gadgets. The bold lines represent blue edges and the thin lines represent red edges. (Color figure online)

blue edges in $E(\{v_i\}, T_i)$ or all red edges in $E(\{v_i\}, F_i)$ in the corresponding variable gadget. Deleting the edges in $E(\{v_i\}, T_i)$ models a true-assignment of the variable x_i while deleting the edges in $E(\{v_i\}, F_i)$ models a false-assignment of x_i. Since we identify vertices in $T_i \cup F_i$ with vertices in A_j the information of the truth assignment is transmitted to the clause gadgets. This ensures that we can make a clause-gadget bicolored P_3-free with 14 edge deletions if and only if there is at least one vertex in A_j which is incident with a deleted edge of its variable gadget.   □

Note that in the proof of Theorem 1 we constructed a graph with $8n + 42m \in \mathcal{O}(n)$ edges, $k = 4n + 14m \in \mathcal{O}(n)$ and therefore $\ell = 4n + 28m \in \mathcal{O}(n)$. Considering the ETH [14] and the fact that there is a reduction from 3-SAT to (3,4)-SAT with a linear blow up in the number of variables [22] this implies the following.

Corollary 1. *If the ETH is true, then* BPD *cannot be solved in* $2^{o(|V|+|E|+k+\ell)}$ *time even if the maximum degree is eight.*

4   Polynomial-Time Solvable Cases

Since BPD is NP-hard in general, there is little hope to find a polynomial-time algorithm that solves BPD on arbitrary instances. In this section, we provide polynomial-time algorithms for two special cases of BPD.

Our first result is a polynomial-time algorithm for BPD, when $G = (V, E_r, E_b)$ does not contain a certain type of K_3s.

Definition 1. *Three vertices* u, v, w *form a* bicolored K_3 *if* $G[\{u, v, w\}]$ *contains exactly three edges such that exactly two of them have the same color. A bicolored* K_3 *is* endangered *in* G *if at least one of the two edges with the same color is part of a bicolored* P_3 *in* G.

Theorem 2. BPD *can be solved in polynomial time if* G *contains no endangered bicolored* K_3*s.*

Proof. We prove the theorem by reducing BPD to VERTEX COVER on bipartite graphs which can be solved in polynomial time since it is equivalent to computing a maximum matching.

VERTEX COVER
Input: A graph $G = (V, E)$ and an integer $k \in \mathbb{N}$.
Question: Is there a *vertex cover* of size at most k in G, that is, a set $S \subseteq V$ with $|S| \leq k$ such that every edge $e \in E$ has at least one endpoint in S.

Let $(G = (V, E_b, E_r), k)$ be an instance of BPD where G contains no endangered bicolored K_3. We define an instance (G', k') of VERTEX COVER as follows. Let $G' = (V', E')$ be the graph with vertex set $V' := E_r \cup E_b$ and edge set $E' := \{\{e_1, e_2\} \subseteq E_b \cup E_r \mid e_1 \text{ and } e_2 \text{ form a bicolored } P_3 \text{ in } G\}$. That is, G' contains a vertex for each edge of G and edges are adjacent if they form a P_3 in G. Moreover, let $k' = k$. The graph G' is obviously bipartite with partite sets E_b and E_r.

We now show that (G, k) is a yes-instance for BICOLORED P_3-DELETION if and only if (G', k') is a yes-instance for VERTEX COVER.

($\Rightarrow$) Let S be a solution for (G, k). Note that the edges of G are vertices of G' by construction and therefore $S \subseteq V'$. We show that S is a vertex cover in G'.

Assume towards a contradiction that there is an edge $\{x, y\} \in E'$ with $x, y \notin S$. By the definition of E', the edges x and y form a bicolored P_3 in G. This contradicts the fact that $G - S$ is bicolored P_3-free. Hence, S is a vertex cover of size at most k in G'.

($\Leftarrow$) Let $C \subseteq V'$ with $|C| \leq k$ be a minimal vertex cover of G'. Note that the vertices of G' are edges of G by construction and therefore $C \subseteq E$. We show that $G - C$ is bicolored P_3-free.

Assume towards a contradiction that there are $x = \{u, v\} \in E_b \setminus C$, and $y = \{v, w\} \in E_r \setminus C$ forming an induced bicolored P_3 in $G - C$. Then, x and y do not form an induced bicolored P_3 in G since otherwise there is an edge $\{x, y\} \in E'$, which has no endpoint in the vertex cover C. It follows that $\{u, w\} \in C$. Obviously, the vertices u, v, w form a bicolored K_3. Since x and y form an induced bicolored P_3 in $G - C$, one of these edges has the same color as $\{u, w\}$. Since $\{u, w\} \in C$ and C is minimal, it follows that $\{u, w\} \in V'$ is an endpoint of an edge in G' and thus $\{u, w\}$ is part of an induced bicolored P_3 in G. Therefore, $G[\{u, v, w\}]$ forms an endangered bicolored K_3 in G which contradicts the fact that G contains no endangered bicolored K_3s. This proves the correctness of the reduction.

For a given instance (G, k) of BPD, the VERTEX COVER instance (G', k') can be computed in $\mathcal{O}(nm)$ time by computing all induced bicolored P_3s of G. Since VERTEX COVER can be solved in $\mathcal{O}(|E| \cdot \sqrt{|V|})$ time on bipartite graphs [13] and since $|V'| = m$ and $|E'| \leq nm$, we have that BPD can be solved in $\mathcal{O}(nm^{\frac{3}{2}})$ time on graphs without endangered K_3s.   □

Theorem 2 implies that BPD can be solved in polynomial time on bipartite graphs and thus also on triangle-free graphs.

We now provide a second polynomial-time solvable special case that is characterized by four colored forbidden induced subgraphs: the two *monochromatic* K_3s, these are the K_3s where all three edges have the same color, and the two *monochromatic* P_3s, these are the P_3s where both edges have the same color. Observe that a graph that does not contain these forbidden induced subgraphs may still contain K_3s or P_3s as induced subgraphs. The algorithm exploits the following observation.

Lemma 1. *Let G be a graph that contains no monochromatic K_3 and no monochromatic P_3 as induced subgraphs. Then the maximum blue degree and the maximum red degree in G are two.*

Proof. We show the proof only for the blue degree, the bound for the red degree can be shown symmetrically.

Assume towards a contradiction that G contains a vertex t with at least three blue neighbors u, v, and w. Since G contains no induced blue P_3, $G[\{u, v, w\}]$ has three edges. Moreover, since G contains no monochromatic K_3 not all of the three edges in $G[\{u, v, w\}]$ are red. Assume without loss of generality that $\{u, v\}$ is blue. Then, $G[\{u, v, t\}]$ is a blue K_3, a contradiction.   □

The algorithm now applies the following two reduction rules exhaustively.

Reduction Rule 1. *(a) Remove all bicolored P_3-free components from G.*
(b) If G contains a connected component C of size at most five, then compute the minimum number of edge deletions k_C to make $G[C]$ bicolored P_3-free, remove C from G, and decrease k by k_C.

The second reduction rule involves certain bridges that may be deleted greedily. An edge e is a *bridge* if the graph $G - e$ has more connected components than G.

Reduction Rule 2. *If G contains an induced bicolored P_3 formed by $\{u, v\}$ and $\{v, w\}$ such that $\{u, v\}$ is a bridge of G and the connected component C containing v and w in $G - \{u, v\}$ is bicolored P_3-free, then remove C from G and decrease k by one.*

On graphs without monochromatic K_3s and P_3s we can apply both reduction rules exhaustively in $\mathcal{O}(n)$ time. Moreover, the remaining instance, which has maximum degree two, can be solved in $\mathcal{O}(n)$ time as well.

Theorem 3. *BPD can be solved in $\mathcal{O}(n)$ time if G contains no monochromatic K_3 and no monochromatic P_3 as induced subgraph.*

5   Parameterized Complexity

In this section, we study the parameterized complexity of BPD parameterized by k, $\ell := m - k$, and (k, Δ), where Δ denotes the maximum degree of G. First, we provide the following running time bound for BPD.

Theorem 4. BPD *can be solved in* $\mathcal{O}(1.85^k \cdot nm^2)$ *time.*

Next, we consider problem kernelization for BPD parameterized by (k, Δ) and $\ell := m - k$. We show that BPD admits problem kernels with at most $\mathcal{O}(k^2 \Delta)$ vertices or 2ℓ edges, respectively.

In the following, we provide two reduction rules leading to an $\mathcal{O}(k^2 \Delta)$ vertex kernel for BPD. The first reduction rule deletes all edges which form more than k bicolored P_3s.

Reduction Rule 3. *If G contains an edge $\{u, v\}$ such that there exist vertices $w_1, \ldots, w_t$ with $t > k$ such that $G[\{u, v, w_i\}]$ is an induced bicolored P_3 for each i, then remove $\{u, v\}$ from G and decrease k by one.*

Lemma 2. *Reduction Rule 3 is correct and it can be applied exhaustively in* $\mathcal{O}(nm)$ *time.*

Proof. First, we prove the correctness of Reduction Rule 3. Let S be a solution for (G, k). Without loss of generality consider an edge $\{u, v\} \in E_r$ such that there exist vertices $w_1, \ldots, w_t$ such that for each i the graph $G[\{u, v, w_i\}]$ is an induced bicolored P_3. At least one edge of each $G[\{u, v, w_i\}]$ is an element of S. Assume towards a contradiction $\{u, v\} \notin S$. In each $G[\{u, v, w_i\}]$, the blue edge has to be removed. In other words, $\{\{v, w_i\} | 1 \leq i \leq t\} \subseteq S$. Since $t > k$, $|S| > k$, a contradiction to $|S| \leq k$. Hence, $\{u, v\} \in S$.

Second, we bound the running time of applying Reduction Rule 3 exhaustively. In a first step, for each edge $e \in E$ compute the number of bicolored P_3s containing e. This can be done in $\mathcal{O}(nm)$ time. In a second step, check if an edge $e = \{u, v\}$ is part of more than k bicolored P_3s, then remove e. After, the removal of e, only new bicolored P_3s can arise which contain vertices u and v. Hence, for each remaining vertex $w \in V$ check if $G[\{u, v, w\}]$ is a new induced bicolored P_3 in $G - \{u, v\}$. Afterwards, update the number of bicolored P_3s for edges $\{u, w\}$ and $\{v, w\}$. This can be done in $\mathcal{O}(n)$ time. Since at most k edges can be removed, this step can be done in $\mathcal{O}(kn)$ time. Since $k < m$, the overall running time of Reduction Rule 3 is $\mathcal{O}(nm)$. $\square$

Let $\mathcal{P}$ denote the set of all vertices of G which are part of bicolored P_3s. Then, the set $N[\mathcal{P}]$ contains all vertices which are either part of an induced bicolored P_3 or which are adjacent to a vertex in an induced bicolored P_3. In the following, we present a reduction rule to remove all vertices in $V \setminus N[\mathcal{P}]$. Note that a vertex v is contained in $V \setminus N[\mathcal{P}]$ if and only if each vertex $u \in N[v]$ is not part of any induced bicolored P_3.

Reduction Rule 4. *If G contains a vertex $v \in V$ such that each vertex $u \in N[v]$ is not part of any induced bicolored P_3, then remove vertex v from G.*

Lemma 3. *Reduction Rule 4 is correct and it can be applied exhaustively in $\mathcal{O}(nm)$ time.*

Proof. Let $H := G \setminus \{v\}$. We prove that there exist a solution S for (G, k) if and only if S is also a solution for (H, k)

($\Rightarrow$) Let S be a solution for (G, k). Since H is an induced subgraph of G, S is also a solution for (H, k).

($\Leftarrow$) Let S be a solution for (H, k). Further, assume without loss of generality that for each $S' \subsetneq S$ the graph $H - S'$ is *not* bicolored P_3-free. Assume towards a contradiction that $G - S$ is not bicolored P_3-free. Then, each induced bicolored P_3 in $G - S$ has to contain vertex v. Since vertex v is not part of any induced bicolored P_3 in G, at least one edge deletion is incident with v. We will use the following claim to obtain a contradiction.

Claim 1. *There exists an ordering $(e_1, \ldots, e_{|S|})$ of the elements of S such that for each i the edge e_i is part of an induced bicolored P_3 in $G - \{e_1, \ldots, e_{i-1}\}$.*

Proof. Assume towards a contradiction that such an ordering does not exist. Since there is at least one edge deletion incident with vertex v, S is not empty. Thus, there exists a maximal index $1 \leq i < |S|$ such that there is a finite sequence $(e_1, \ldots, e_i)$ of elements of S where for each j the edge e_j is part of an induced bicolored P_3 in $G - \{e_1, \ldots, e_{j-1}\}$. According to our choice of i, there exists no edge of S which is part of an induced bicolored P_3 in $G - \{e_1, \ldots, e_i\}$. If $G - \{e_1, \ldots, e_i\}$ contains an induced bicolored P_3 formed by $\{u, v\}$ and $\{v, w\}$, then $\{u, v\} \in S$ or $\{v, w\} \in S$. This contradicts the fact that no edge of S is part of an induced bicolored P_3 in $G - \{e_1, \ldots, e_i\}$. Otherwise, $G - \{e_1, \ldots, e_i\}$ is bicolored P_3-free. Then, $\{e_1, \ldots, e_i\}$ is a solution for (G, k). This contradicts the fact that no $S' \subsetneq S$ is a solution for (G, k).   ◊

We assumed that there exists an edge deletion incident with vertex v. Since v is not part of any induced bicolored P_3 in G and according to Claim 1 there exists an ordering $F = (e_1, \ldots, e_{|S|})$ of edge deletions such that for each i the edge e_i is part of an induced bicolored P_3 in $G - \{e_1, \ldots, e_{i-1}\}$. Let $W \subseteq N(v)$ be the set of neighbors of v which are incident with an edge deletion in S. For each $x \in W$ define $i(x) := \min_{x \in e_j} j$. Intuitively, $i(x)$ denotes the minimal index of an edge deletion of F incident with x.

Let $w \in W$ be the vertex with lowest value $i(w)$ and let z be the other endpoint of edge $e_{i(w)}$. Intuitively, edge $e_{i(w)}$ is the first edge deletion incident with a neighbor w of v. Observe that $z \in N_G^2(v)$ since no vertex of $N_G(v)$ is part of any induced bicolored P_3 in G.

Without loss of generality assume that the edge $\{v, w\}$ is red. The edge $\{w, z\}$ is red since otherwise $G[\{v, w, z\}]$ is an induced bicolored P_3. According to Claim 1 the edge $\{w, z\}$ is part of an induced bicolored P_3 in $G - \{e_1, \ldots, e_{i(w)-1}\}$. Hence, there exists a vertex $y \in V$ with $y \neq v$ such that $G[\{w, y, z\}]$ is an induced bicolored P_3 in $G - \{e_1, \ldots, e_{i-1}\}$. First, assume $y \in N_G(v)$. According to our assumption that no vertex in $N_G[v]$ is part of an induced bicolored P_3 in G, vertex y is not part of an induced bicolored P_3 in G. Hence, in the ordering F there exists a minimal index j such that the edge e_j is the first edge deletion incident with vertex y. Since $j < i(w)$ this contradicts the maximality of $i(w)$. Second, assume $y \in N_G^2(v)$. Since all edges between w and $N_G^2(v) \cap N_G(w)$ are red, the edge $\{w, y\}$ is red. Hence, in the graph $G - \{e_1, \ldots, e_{i-1}\}$ the edge $\{w, y\}$ is contained in S. This contradicts the choice of $i(w)$. Hence, vertex v is not incident with an edge deletion. Third, assume towards a contradiction $y \in N_G^3(v)$. Since $\{w, z\} \in E_r$ and $\{w, y\} \notin E$, we conclude $\{y, z\} \in E_r$. Since vertex w is in no induced bicolored P_3 in G, $\{w, y\} \in E_G$. Hence, $y \in N_G^2(v)$, a contradiction.

Next, consider the running time of applying Reduction Rule 4 exhaustively. In a first step, determine all bicolored P_3s in G. Afterwards, determine for each vertex $v \in V$ if it is part of some induced bicolored P_3. This needs $\mathcal{O}(nm)$ time. Now, check for each vertex $v \in V$ if each vertex $u \in N[v]$ is not part of any induced bicolored P_3. This can be done in $\mathcal{O}(m)$ time. The claimed running time follows.   □

Theorem 5. BPD *admits a* $\mathcal{O}(k^2\Delta)$-*vertex kernel that can be computed in* $\mathcal{O}(nm)$ *time.*

Proof. First, apply Reduction Rule 3 exhaustively. Second, apply Reduction Rule 4 exhaustively. This needs $\mathcal{O}(nm)$ time. Let $\mathcal{P}$ be the set of vertices which are contained in an induced bicolored P_3 in G. We prove that (G, k) is a yes-instance if G contains at most $\mathcal{O}(k^2\Delta)$ vertices. If (G, k) is a yes-instances for BPD the graph $G[\mathcal{P}]$ can contain at most k vertex-disjoint bicolored P_3s. Since Reduction Rule 3 was applied exhaustively, each edge is contained in at most k bicolored P_3s. Hence, $|\mathcal{P}| \leq 2k^2$. Since Reduction Rule 4 was applied exhaustively, $V \setminus N[\mathcal{P}] = \emptyset$. In other words, set $\mathcal{P}$ has no second neighborhood in G. Since each vertex has degree at most Δ we have $|N(\mathcal{P})| \leq 2k^2\Delta$. Hence, the overall number of vertices in G is at most $\mathcal{O}(k^2\Delta)$.   □

Note that a kernelization by Δ alone is unlikely since BPD is NP-hard even if $\Delta = 8$ by Theorem 1. Since BPD is fixed-parameter tractable with respect to parameter k, we can trivially conclude that it admits an exponential-size problem kernel. It is open if there is a polynomial kernel depending only on k while CLUSTER DELETION has a relatively simple $4k$-vertex kernel [11]. Summarizing, BPD seems to be somewhat harder than CLUSTER DELETION for parameter k.

In contrast, BPD seems to be easier than CLUSTER DELETION if parameterized by the dual parameter $\ell := m - k$: there is little hope that CLUSTER DELETION admits a problem kernel of size $\ell^{\mathcal{O}(1)}$ [10] while BPD has a trivial linear size problem kernel as we show below.

Theorem 6. BPD *admits a problem kernel with* $\mathcal{O}(2\ell)$ *edges and vertices which can be computed in* $\mathcal{O}(n + m)$ *time.*

Proof. We show that instances with more than 2ℓ edges are trivial yes-instances. Let (G, k) with $|E| \geq 2\ell$ be an instance of BPD. Then, since E_r and E_b forms a partition of E, we conclude $|E_r| \geq \ell$ or $|E_b| \geq \ell$. Without loss of generality let $|E_r| \geq \ell$. Since $|E_b| = m - |E_r| \leq m - \ell = k$, E_b is a solution for (G, k).   □

6   Outlook

We have left open many questions for future work. First, it would be interesting to further investigate the structure and usefulness of bicolored P_3-free graphs. Since each color class may induce an arbitrary graph it seems difficult to obtain a concise and nontrivial structural characterization of these graphs. Nevertheless, one could aim to identify graph problems that are NP-hard on general two-edge colored graphs but polynomial-time solvable on bicolored P_3-free graphs.

Second, there are many open questions concerning BICOLORED P_3 DELE-TION. Does BICOLORED P_3 DELETION admit a polynomial kernel for k? Can BICOLORED P_3 DELETION be solved in $2^{O(n)}$ time? Can BICOLORED P_3 DELE-TION be solved in polynomial time on graphs that contain no induced monochromatic P_3s? Can BICOLORED P_3 DELETION be solved in polynomial time on graphs that contain no cycle consisting only of blue edges? Even simpler is the following question: Can BICOLORED P_3 DELETION be solved in polynomial time if the subgraph induced by the red edges and the subgraph induced by the blue edges are each a disjoint union of paths?

Moreover, it would be interesting to perform a similar study on BICOL-ORED P_3 EDITING where we may also insert blue and red edges. Finally, we were not able to resolve the following question: Can we find bicolored P_3s in linear time?

References

1. Amar, D., Shamir, R.: Constructing module maps for integrated analysis of heterogeneous biological networks. Nucleic Acids Res. **42**(7), 4208–4219 (2014)
2. Aravind, N.R., Sandeep, R.B., Sivadasan, N.: Dichotomy results on the hardness of H-free edge modification problems. SIAM J. Discrete Math. **31**(1), 542–561 (2017)
3. Ben-Dor, A., Shamir, R., Yakhini, Z.: Clustering gene expression patterns. J. Comput. Biol. **6**(3–4), 281–297 (1999)
4. Böcker, S., Baumbach, J.: Cluster editing. In: Bonizzoni, P., Brattka, V., Löwe, B. (eds.) CiE 2013. LNCS, vol. 7921, pp. 33–44. Springer, Heidelberg (2013). https://doi.org/10.1007/978-3-642-39053-1_5
5. Brandes, U., Hamann, M., Strasser, B., Wagner, D.: Fast quasi-threshold editing. In: Bansal, N., Finocchi, I. (eds.) ESA 2015. LNCS, vol. 9294, pp. 251–262. Springer, Heidelberg (2015). https://doi.org/10.1007/978-3-662-48350-3_22
6. Brandstädt, A., Le, V.B., Spinrad, J.P.: Graph Classes: A Survey. SIAM, Philadelphia (1999)
7. Bruckner, S., Hüffner, F., Komusiewicz, C.: A graph modification approach for finding core-periphery structures in protein interaction networks. Algorithms Mol. Biol. **10**, 16 (2015)
8. Cygan, M., et al.: Parameterized Algorithms. Springer, Cham (2015). https://doi.org/10.1007/978-3-319-21275-3
9. Downey, R.G., Fellows, M.R.: Fundamentals of Parameterized Complexity. TCS. Springer, London (2013). https://doi.org/10.1007/978-1-4471-5559-1
10. Grüttemeier, N., Komusiewicz, C.: On the relation of strong triadic closure and cluster deletion. In: Brandstädt, A., Köhler, E., Meer, K. (eds.) WG 2018. LNCS, vol. 11159, pp. 239–251. Springer, Cham (2018). https://doi.org/10.1007/978-3-030-00256-5_20

11. Guo, J.: A more effective linear kernelization for cluster editing. Theoret. Comput. Sci. **410**(8–10), 718–726 (2009)
12. Hellmuth, M., Wieseke, N., Lechner, M., Lenhof, H.-P., Middendorf, M., Stadler, P.F.: Phylogenomics with paralogs. Proc. Nat. Acad. Sci. **112**(7), 2058–2063 (2015)
13. Hopcroft, J.E., Karp, R.M.: An $n^{5/2}$ algorithm for maximum matchings in bipartite graphs. SIAM J. Comput. **2**(4), 225–231 (1973)
14. Impagliazzo, R., Paturi, R., Zane, F.: Which problems have strongly exponential complexity? J. Comput. Syst. Sci. **63**(4), 512–530 (2001)
15. Kivelä, M., Arenas, A., Barthelemy, M., Gleeson, J.P., Moreno, Y., Porter, M.A.: Multilayer networks. J. Complex Netw. **2**(3), 203–271 (2014)
16. Komusiewicz, C.: Tight running time lower bounds for vertex deletion problems. ACM Trans. Comput. Theory **10**(2), 6:1–6:18 (2018)
17. Lewis, J.M., Yannakakis, M.: The node-deletion problem for hereditary properties is NP-complete. J. Comput. Syst. Sci. **20**(2), 219–230 (1980)
18. Nastos, J., Gao, Y.: Familial groups in social networks. Soc. Netw. **35**(3), 439–450 (2013)
19. Schestag, J.: Liechtenstein-P_3s in two-colored graphs. Bachelorarbeit, Philipps-Universität Marburg (2019)
20. Shamir, R., Sharan, R., Tsur, D.: Cluster graph modification problems. Discrete Appl. Math. **144**(1–2), 173–182 (2004)
21. Sommer, F., Komusiewicz, C.: Parameterized algorithms for module map problems. In: Lee, J., Rinaldi, G., Mahjoub, A.R. (eds.) ISCO 2018. LNCS, vol. 10856, pp. 376–388. Springer, Cham (2018). https://doi.org/10.1007/978-3-319-96151-4_32
22. Tovey, C.A.: A simplified NP-complete satisfiability problem. Discrete Appl. Math. **8**(1), 85–89 (1984)
23. Wittkop, T., Baumbach, J., Lobo, F.P., Rahmann, S.: Large scale clustering of protein sequences with force-a layout based heuristic for weighted cluster editing. BMC Bioinform. **8**(1), 396 (2007)
24. Yannakakis, M.: Edge-deletion problems. SIAM J. Comput. **10**(2), 297–309 (1981)

Degree Spectra for Transcendence in Fields

Iskander Kalimullin[1], Russell Miller[2,4]($\boxtimes$), and Hans Schoutens[3,4]

[1] Kazan Federal University, Kremlyovskaya Street 18, 420008 Kazan, Russia
ikalimul@gmail.com
[2] Queens College, 65-30 Kissena Blvd., Queens, NY 11367, USA
Russell.Miller@qc.cuny.edu
[3] New York City College of Technology, 300 Jay Street, Brooklyn, NY 11201, USA
hschoutens@citytech.cuny.edu
[4] C.U.N.Y. Graduate Center, 365 Fifth Avenue, New York, NY 10016, USA

Abstract. We show that for both the unary relation of transcendence and the finitary relation of algebraic independence on a field, the degree spectra of these relations may consist of any single computably enumerable Turing degree, or of those c.e. degrees above an arbitrary fixed Δ_2^0 degree. In other cases, these spectra may be characterized by the ability to enumerate an arbitrary Σ_2^0 set. This is the first proof that a computable field can fail to have a computable copy with a computable transcendence basis.

Keywords: Computability · Computable structure theory · Degree spectrum · Field · Transcendence basis

1 Introduction

It has been known since the work of Metakides and Nerode in [7] that a computable field need not have a computable transcendence basis. This result, readily established, is fundamental to the study of effectiveness for fields. Under the usual definition of computable structure, a computable field is simply a field whose domain is a computable subset of ω (usually just ω itself, the set of all nonnegative integers) and whose atomic diagram, in the language with addition and multiplication, is computable. The theorem of Metakides and Nerode shows that working with an arbitrary computable field will be difficult, as one cannot

The research of the first author is supported by RSF Grant no. 18-11-00028; he is also funded by the Russian Ministry of Education and Science (project 1.451.2016/1.4) as a federal professor in mathematics. The second author was partially supported by Grant # 581896 from the Simons Foundation, and the second and third authors were both supported by grants from the City University of New York PSC-CUNY Research Award Program. The authors wish to acknowledge useful conversations with Dr. Kenneth Kramer.

© Springer Nature Switzerland AG 2019
F. Manea et al. (Eds.): CiE 2019, LNCS 11558, pp. 205–216, 2019.
https://doi.org/10.1007/978-3-030-22996-2_18

in general distinguish the algebraic elements of the field (relative to the prime subfield, either $\mathbb{Q}$ or $\mathbb{Z}/(p)$) from those transcendental over the prime subfield.

We recently realized that the following very natural question had not been addressed: must every computable field be isomorphic to a computable field with a computable transcendence basis? It is well known that there need not exist a computable isomorphism between two isomorphic computable fields, and so it is plausible that the answer might be affirmative: two computable fields, one with a computable transcendence basis and the other without any such basis, can certainly be isomorphic. (We normally refer to isomorphic computable fields as *computable copies* of each other.) In case of an affirmative answer, one would be justified in always assuming a computable transcendence basis, as this would only require choosing a "nice" computable copy of the field in question.

Initially we were optimistic that the answer would indeed be affirmative, and even that a single Turing procedure might produce such a copy uniformly, using the original field's atomic diagram as an oracle. (This could also then be extended to noncomputable fields.) However, conversations with our colleague Ken Kramer disabused us of that notion, and in fact we will demonstrate here that the answer is negative. For uniform procedures, the negative answer is proven in Sect. 2, which introduces and illustrates the use of algebraic curves of positive genus for this purpose. The remainder of the article shows that there is not even any nonuniform procedure: certain computable fields have no computable copy with any computable transcendence basis. Indeed, the spectrum of the transcendence relation on a field has many possible configurations, plenty of which do not include the degree $\mathbf{0}$. For each computably enumerable Turing degree c, it is possible for transcendence to be intrinsically of degree c, or for it intrinsically to compute c, or for it intrinsically to enumerate a given Σ_2^0 set. The proofs here make substantial use of results on algebraic curves developed during earlier work by two of us in [11].

Transcendence bases are not in general definable, and a single field of infinite transcendence degree will have continuum-many different transcendence bases. This makes it difficult to define "the" Turing degree for transcendence bases. To address this, we use two $L_{\omega_1\omega}$-definable relations on fields. The transcendence relation T, which is unary, holds of those elements not algebraic over the prime subfield Q of a field F:

$$x \in T \iff (\forall f \in Q[X]^*) f(x) \neq 0.$$

(Here $Q[X]^*$ is the set of nonzero polynomials over Q.) The algebraic independence relation I is a generalization of this to tuples of all arities n:

$$(x_1, \ldots, x_n) \in I \iff (\forall f \in Q[X_1, \ldots, X_n]^*) f(x_1, \ldots, x_n) \neq 0.$$

A computable field of infinite transcendence degree will possess transcendence bases having each Turing degree $\geq_T \deg(I)$, but not of any other Turing degree: given any basis as an oracle, one can decide the independence relation on the field, and conversely, from an I-oracle, one can compute a transcendence basis for F. Thus the Turing degree of I can stand in for the set of Turing degrees of transcendence bases, as this set is the upper cone above $\deg(I)$. In turn, the unary relation T is always computable from I, although sometimes strictly below I under Turing reducibility. In the fields we consider here, we will always have $I \equiv_T T$. We remark the following useful property.

Lemma 1. *In a computable field F, for each Turing degree $\boldsymbol{d}$, every $\boldsymbol{d}$-computably enumerable transcendence basis B over the prime subfield Q is $\boldsymbol{d}$-computable.*

Proof. Given any $x \in F$, use a $\boldsymbol{d}$-oracle to list the elements $b_0, b_1, \ldots$ of B, and search for an n and a polynomial $f \in Q[X, Y_0, \ldots, Y_n]^*$ with $f(x, b_0, b_1, \ldots, b_n) = 0$ in F. This search must terminate, and $x \in B$ just if $x \in \{b_0, \ldots, b_n\}$.   □

Our notation is generally standard. The articles [8,9] form good introductions to computable field theory, and myriad other articles have contributed to the area: [3,4,7,12,16] all have historical importance, while [5,10,11] describe related recent work in the discipline.

2   Curves of Positive Genus

Proposition 1. *Let $f(Y, Z) = 0$ define a curve over a field k of characteristic 0. If the genus of this curve is positive, then $f = 0$ has no solutions in the purely transcendental field extension $K = k(t_1, t_2, \ldots)$ except those solutions within k itself. (We say that $f = 0$ has no* nonconstant *solutions in K.)*

Proof. A solution in K would lie within some subfield $k(t_1, \ldots, t_n)$, so we prove by induction on n that no such subfield contains a nonconstant solution. For $n = 1$, we can view the extension $k(t_1)$ as an algebraic curve of genus 0 over k. If an extension $k(y, z)$ (where $f(y, z) = 0$) lies within $k(t_1)$, with $\{y, z\} \not\subseteq k$, then the Riemann-Hurwitz formula dictates that $f = 0$ must also have genus 0, contradicting the hypothesis of the proposition.

For the inductive step, suppose $y, z \in k(t_1, \ldots, t_{n+1})$ satisfy $f(y, z) = 0$. We express $y = \frac{g_1}{h_1}$ and $z = \frac{g_2}{h_2}$ as rational functions of $t_1, \ldots, t_n$ over the field $k(t_{n+1})$ Of course, each of g_1, g_2, h_1, h_2 has finitely many nonzero coefficients in that field, and the pairs (g_1, h_1) and (g_2, h_2) may be taken to have no common factor. Having characteristic 0, k is infinite, so it must contain an element a such that, when t_{n+1} is replaced by a, all of these coefficients remain nonzero and no common factors are introduced. Substituting a for t_{n+1} in y and z yields a solution to $f = 0$ in $k(t_1, \ldots, t_n)$. By inductive hypothesis this solution lies in k, meaning that the original y and z did not involve any of $t_1, \ldots, t_n$ (since no common factors arose to be cancelled when we made the substitution). But then y and z were a solution to $f = 0$ in $k(t_{n+1})$, hence must be a constant solution, according to the base case of the induction.   □

Corollary 1. *There is no uniform procedure for transforming a countable field into an isomorphic countable field that decides its own transcendence relation T.*

That is, there does not exist any Turing functional Ψ such that, for every atomic diagram F of a countable field with domain ω (in the signature with just

$+$ and $\cdot$), Ψ^F computes the atomic diagram, in the larger signature with $+$, $\cdot$ and T, of a structure with reduct F and for which T is the (unary) transcendence relation. The same holds with the ($<\omega$-ary) algebraic independence relation I in place of T.

Proof. Suppose Ψ were such a functional. Fix an irreducible curve with affine equation $f(Y, Z) = 0$ of positive genus over $\mathbb{Q}$, and let F be a presentation of the field $\mathbb{Q}(y_1, y_2, y_3, \ldots)(z_1)$, with $\{y_1, y_2, \ldots\}$ algebraically independent over $\mathbb{Q}$ and $f(y_1, z_1) = 0$. Then Ψ^F must compute the atomic diagram of an isomorphic field L with a transcendence relation T, say with isomorphism $h : F \rightarrow L$. Therefore T will hold of the 1-tuple $h(y_1)$ in the field L. Let σ be an initial segment of the atomic diagram of F such that Ψ^σ ensures that T holds of $h(y_1)$ and that $f(h(y_1), h(z_1)) = 0$ in L.

Now let E be a presentation of the field whose atomic diagram begins with σ. However, the atomic diagram of E (beyond σ) specifies that y_1 is in fact rational itself, in some way consistent with σ, and thus that z_1 is algebraic over $\mathbb{Q}$ in E. (It may not be possible to make z_1 rational too, as $f = 0$ will have only finitely many solutions in $\mathbb{Q}$, by Faltings' proof of the Mordell Conjecture. However, there is no difficulty in making z_1 algebraic over $\mathbb{Q}$.) The rest of E is then generated by this portion and by elements $y_2, y_3, \ldots$ algebraically independent over $\mathbb{Q}$, just as in F.

Thus Ψ^E will build a field in which the domain element $h(y_1)$ is transcendental over $\mathbb{Q}$ (being thus specified by Ψ^E) and $f(h(y_1), h(z_1)) = 0$. However, E is a purely transcendental extension of the field $k = \mathbb{Q}(z_1)$, which is algebraic over $\mathbb{Q}$. By Proposition 1, E does not contain any solution to $f(Y, Z) = 0$ outside of k, so every solution in E consists of elements algebraic over $\mathbb{Q}$. This ensures that E and the field with atomic diagram Ψ^E are not isomorphic as fields, proving the Corollary. (The result for the relation I follows directly.) $\square$

3   Background on Algebraic Curves

Corollary 1 proved that there is no uniform method of taking a computable field and producing a computable copy with a computable transcendence basis. Now we wish to show that a single computable field can entirely fail to have a computable copy with a computable transcendence basis. Indeed, we will establish far more specific results, with detailed descriptions of the possible degrees of transcendence bases in computable copies of the field. To do this, however, we need to work with infinitely many curves of positive genus at once, as a single curve will only allow our field to avoid being isomorphic to a single computable field with computable transcendence basis.

Fortunately, an appropriate collection of curves has already been built. We recall the following result from [11], as stated there.

Theorem 1 (Miller and Schoutens, Thm. 3.1 of [11]). *There exists a non-covering collection $\mathcal{C}$ of curves with the effective Faltings property, over a computable ground field k.*

That is, $\mathcal{C} = \{f_0, f_1, \dots\}$ is an infinite set of smooth projective curves C_i with corresponding affine equations $f_i \in k[X, Y]$ such that:

- for each i, the function field $k(f_i)$ does not embed into $k(\mathcal{C} - \{f_i\})$; and
- the function $i \mapsto f_i(k)$ giving (a strong index for the finite set of) all solutions of $f_i(X, Y) = 0$ in k^2 is computable. (This is the *effective Faltings property*.)

In particular, the second item requires that each equation $f_i = 0$ should have only finitely many solutions by elements of k.

The specific example $\mathcal{C}$ given in [11] is in fact a collection of Fermat curves $f_i = X^{q_i} + Y^{q_i} - 1$, for a fixed increasing computable sequence $q_0 < q_1 < \cdots$ of odd prime numbers. By Fermat's Theorem, each has exactly two solutions in $k = \mathbb{Q}$, and the non-covering property for this $\mathcal{C}$ is established in [11]. It is believed that many other computable sets of curves have the same property, but rather than pursuing that question here, we will use this same set $\mathcal{C}$. It should be borne in mind that not all odd primes belong to the sequence $\langle q_i \rangle_{i \in \omega}$. Indeed, this sequence is quite sparse within the primes: each element is the least prime $q_{i+1} > (4(q_i - 1)(q_i - 2))^2$, with q_0 chosen to be 5 (or any other odd prime except 3, which is ruled out because the genus of $(X^d + Y^d - 1)$ is $\frac{(d-2)(d-1)}{2}$ and we need genera > 1).

For the rest of this article we fix these curves $C_0, C_1, \dots$ with affine equations $f_0, f_1, \dots$ exactly as given here. The usefulness of Theorem 1 lies in the fact that it enables us to adjoin to a ground field k (such as $\mathbb{Q}$ or $\overline{\mathbb{Q}}$) a transcendental element x and then an element y satisfying $f_i(x, y) = 0$ (so y is also transcendental) without creating any transcendental solutions to any other f_j in the new field $K = k(x)[y]/(f_i)$. Indeed, our k might already have been built this way, with pairs $(x_0, y_0), \dots, (x_{i-1}, y_{i-1})$ of transcendental solutions to $f_0, \dots, f_{i-1}$, say, and the new (x, y) will not generate any solutions to any of $f_0, \dots, f_{i-1}$ that were not already in k. This allows us to work independently with the distinct polynomials f_i and their solutions, and avoids the need for priority arguments and the like.

It should be noted that $\mathbb{Q}(x_i)[y_i]/(f_i)$ actually contains eight solutions to $f_i(X, Y) = 0$. Two are the trivial solutions $(0, 1)$ and $(1, 0)$, which we can always recognize and ignore. Then the solution (x_i, y_i) generates $(\frac{-x_i}{y_i}, \frac{1}{y_i})$, $(\frac{-y_i}{x_i}, \frac{1}{x_i})$, and the transpositions of these three. Moreover, we will sometimes work over the algebraic closure $\overline{\mathbb{Q}}$ instead of $\mathbb{Q}$, and in this case we get $6q_i^2$ nontrivial solutions from (x_i, y_i), since for a primitive q_i-th root θ of unity, $(x_i \theta^j, y_i \theta^k)$ will be another solution. (There will also be plenty of non-transcendental solutions within $\overline{\mathbb{Q}}$, of course.) However, it was shown by Leopoldt [6] and Tzermias [15] that no further transcendental solutions will exist; this result is also used in [11], appearing there as Theorem 4.4.

4   Examples of Degree Spectra

Our initial goal was to produce a computable field such that no computable copy of the field has a computable transcendence basis. In fact, we will give a much more specific answer to the question, using the well-established notion of the degree spectrum of a relation.

Definition 1. *For a computable structure $\mathcal{A}$ and a relation R on $\mathcal{A}$, the* Turing degree spectrum *of R on $\mathcal{A}$ is the set of all Turing degrees of images of R in computable structures isomorphic to $\mathcal{A}$:*

$$DgSp_{\mathcal{A}}(R) = \{deg(g(R)) : g : \mathcal{A} \to \mathcal{B} \text{ is an isomorphism onto a computable } \mathcal{B}\}.$$

In many contexts this definition is restricted to n-ary relations R, but it applies equally well to finitary relations, i.e., those defined on all finite tuples from $(\mathcal{A})^{<\omega}$, of arbitrary length.

Notice first that both the (unary) transcendence relation T on a computable field and the (finitary) relation I of algebraic independence are definable in the field by computable infinitary Π_1^0 formulas. Therefore, in every computable field F, both T and I will be Π_1^0 sets, hence of c.e. Turing degree. This places an upper bound on the complexity of the degrees in $DgSp_F(T)$ and $DgSp_F(I)$, since all such degrees must be $\leq_T \mathbf{0}'$. Even below that bound, it also rules out a number of further candidates, namely those degrees that do not contain any c.e. set. (We call these *properly Δ_2^0 degrees*, meaning that they are Δ_2^0 but not Σ_1^0.)

Our ultimate goal is to know the degrees of the transcendence bases in the various computable copies of F. Recall, however, that these are precisely the degrees $\geq_T \deg(I)$. Thus, once one knows the Turing degree of I in a particular computable copy of the field, one knows all the degrees of transcendence bases in that copy, and so we view $DgSp_F(I)$ as a reasonable answer to the question.

Our first example shows that, for a fixed computably enumerable Turing degree $\mathbf{c}$, the relations T and I can both be *intrinsically of degree $\mathbf{c}$*. This term was used in [2], in which Downey and Moses showed that the relation of adjacency in a computable linear order can be intrinsically of degree $\mathbf{0}'$. Subsequently, Downey, Lempp, and Wu showed in [1] that the only degrees $\mathbf{c}$ for which the adjacency relation can be intrinsically of degree $\mathbf{c}$ are $\mathbf{c} = \mathbf{0}'$ and (if the adjacency relation is finite) $\mathbf{c} = \mathbf{0}$. Therefore Theorem 2 distinguishes the situation for transcendence and for independence in fields from that of adjacency in linear orders.

Theorem 2. *For each computably enumerable Turing degree $\mathbf{c}$, there exists a computable field F for which the spectrum of the transcendence relation T and of the independence relation I are both the singleton $\{\mathbf{c}\}$.*

Proof. Fixing a computable enumeration $\langle C_s \rangle_{s \in \omega}$ of a c.e. set $C \in \mathbf{c}$, we can describe the isomorphism type of our field quickly. For every $i \notin C$, it will contain a transcendental element x_i (over the ground field $\mathbb{Q}$) and an additional element y_i with $f_i(x_i, y_i) = 0$. Moreover, these elements x_i will form a transcendence basis, as i ranges over $\overline{C}$. For each $i \in C$, the field will contain elements called x_i and y_i, again satisfying $f_i(x_i, y_i) = 0$, but this x_i will lie within $\mathbb{Q}$, making y_i algebraic over $\mathbb{Q}$. These x_i and y_i (for all $i \in \omega$) will generate the field. (The choice of which rational number equals x_i, for $i \in C$, will depend on the least s with $i \in C_s$.)

Next we give a computable presentation F of this field. At stage 0, F_0 consists of a finite substructure of the field $\mathbb{Q}$ (with the operations viewed as relations, so that it makes sense for $\mathbb{Q}$ to have a finite substructure). At stage $s + 1$, we add elements x_s and y_s to F_s, along with as many new elements as are needed in order for the relational atomic diagram of F_{s+1} to specify that $f_s(x_s, y_s) = 0$ (but without making x_s itself algebraic over F_s). Then, for the least $i \leq s$ (if any) such that $i \in C_s$ and we have not yet acted on behalf of i, we add enough new elements to F_{s+1} and define the operations on them to make x_i lie within $\mathbb{Q}$ (in a way consistent with F_s, of course: nothing in the atomic diagram should ever be redefined). This must be possible, since x_i has been treated as a transcendental up until this stage. Finally, we take another step to close F under the field operations, adding another element and extending the relations in F_{s+1} in a way consistent with the principle that the set

$$\{x_i : i \leq s \ \& \ \text{we have not yet acted on behalf of } i\}$$

should form a transcendence basis for F_{s+1}. That is, we make sure not to create any algebraic relations involving these x_i's, and all new elements added to the field are generated by $\{x_i, y_i : i \leq s\}$. This is the entire construction, and it is clear that it does ultimately build a computable field $F = \cup_s F_s$. Furthermore, $\{x_i : i \notin C\}$ will indeed be a transcendence basis for F, and every x_i with $i \in C$ will lie within $\mathbb{Q}$ in the field F.

We now argue that for every computable field $E \cong F$, the transcendence relation T and the independence relation I on E have $T \equiv_T I \equiv_T C$. First, given a C-oracle, we enumerate a transcendence basis for F by collecting, for each $i \notin C$, the first coordinate of the first pair (x, y) that we find in E^2 for which $f_i(x, y) = 0 \neq xy$. (This pair is not unique, as mentioned in Sect. 3, but the six possible x-values are pairwise interalgebraic.) By the construction, this enumerates a transcendence basis B for F, hence computes one, by Lemma 1, and from B we can compute I and T.

To show that $C \leq_T T$, we claim that $i \in \overline{C}$ just if there exists a pair (x, y) of elements of E with $x \in T$ and $f_i(x, y) = 0$ in E. (Thus C is Π_1^T, as well as Σ_1.) Indeed, for $i \notin C$, the isomorphic image in E of the elements (x_i, y_i) from F will be such a pair. For the converse, suppose $i \in C$. Then x_i and y_i were made algebraic at some stage in the construction of F, and by Theorem 1 with $k = \mathbb{Q}(y_i : i \in C) \subseteq F$ (which is the subfield containing all elements algebraic over $\mathbb{Q}$), the function field of the collection $\{f_j : j \notin C\}$ over k does not contain any nontrivial solution to $f_i = 0$. This function field is isomorphic to F itself, so we have $C \leq_T T$.   □

Theorem 2 answers the initial question posed above. Theorems 3 and 4 will provide further examples.

Corollary 2. *There exists a computable field F such that no computable field E isomorphic to F has a computable transcendence basis, nor even a computable transcendence relation.*   □

Our next example shows that the relations T and I can also *intrinsically compute* a c.e. degree $\mathbf{c}$, in the sense that the spectra of T and I can equal the upper cone above c (subject to the restriction that these spectra only contain c.e. degrees). Once again, this parallels a result of Downey and Moses in [2] for linear orders. In Corollary 3 below, we will generalize this result to all Δ_2^0 degrees c, which is not known (to us) to be possible for adjacency on linear orders.

Theorem 3. *For each computably enumerable Turing degree c, there exists a computable field F for which the spectrum of the transcendence relation T and of the independence relation I are both*

$$DgSp_F(T) = DgSp_F(I) = \{d \geq c : d \text{ is a c.e. degree}\}.$$

Proof. Fix some c.e. set $C \in c$, with a computable enumeration $\langle C_t \rangle_{t \in \omega}$ by finite nested sets C_t. The field F for this degree is the field

$$F = \overline{\mathbb{Q}}(x_k)[y_k]/(f_k(x_k, y_k)),$$

with k ranging over the set $\overline{C} \oplus \omega$. That is, F has the algebraic closure $\overline{\mathbb{Q}}$ as its ground field, and contains an algebraically independent set $\{x_{2i} : i \in \overline{C}\} \cup \{x_{2j+1} : j \in \omega\}$ of elements, along with corresponding elements y_{2i} and y_{2j+1} that "tag" the individual x-elements by forming solutions to $f_{2i} = 0$ or $f_{2j+1} = 0$. (The reason for the odd-indexed elements x_{2j+1} will become clear below: they will give us the upward closure we desire.)

To see that F has a computable presentation, start building a computable copy of $\overline{\mathbb{Q}}$, with only finitely many elements added at each stage. At stage $s + 1$, we add new elements x_s and y_s to the field, with $f_s(x_s, y_s) = 0$, and treat x_s as a transcendental over all previously existing field elements. For odd values $s = 2j + 1$, we simply continue at each subsequent stage to build the field, with x_s remaining transcendental. For even $s = 2i$, at each subsequent stage $t > s + 1$, we check whether $i \in C_t$. As long as $i \notin C_t$, we simply add to the field the next element generated by x_s, continuing to treat x_s as transcendental over the preceding elements. However, for the first t (if any) with $i \in C_t$, we switch strategies and make x_s a rational number, finding some way to do this that is consistent with the finite portion of the atomic diagram of F that has already been defined. Of course, this also makes y_s algebraic over $\mathbb{Q}$, though not rational. This enlarges our presentation of the ground field $\overline{\mathbb{Q}}$, of course, but since only finitely much of $\overline{\mathbb{Q}}$ had been built so far, it is easy to incorporate x_s and y_s into it and to continue building $\overline{\mathbb{Q}}$, including them, at each subsequent stage.

Now for any computable field $E \cong F$, with transcendence relation T, we can compute C from T. Indeed, by Theorem 1, a number i lies in $\overline{C}$ if and only if E contains transcendental elements x and y such that $f_{2i}(x, y) = 0$, so $\overline{C}$ is Σ_1^T, while C is Σ_1. Thus $DgSp_F(T)$ contains only degrees above c, and these must all be c.e. degrees, as the relation T is definable in F by a computable infinitary Π_1^0 formula. The same analysis applies to the independence relation I.

To prove the reverse inclusion, let d be any c.e. degree that computes c, and fix some c.e. set $D \in d$ with computable enumeration $\langle D_s \rangle_{s \in \omega}$. We build a specific computable copy E of F in which $T \equiv_T I \equiv_T D$, by a process quite similar to the above construction of F itself. E includes a copy of $\overline{\mathbb{Q}}$, built slowly, with only finitely many elements added at each stage. Once again, the even-indexed x_{2i} and y_{2i} are added at stage $2i$ and treated as transcendental until i enters C, at which point x_{2i} becomes rational. The odd-indexed elements x_{2j+1} and y_{2j+1} are added at stage $2j+1$ and treated as transcendental until (if ever) we reach a stage s with $j \in D_s$. If such a stage occurs, then this x_{2j+1} is made rational at that stage (in the same way as with x_{2i} if i enters C), and we adjoin to the field new elements x'_{2j+1} and y'_{2j+1}, again with x'_{2j+1} transcendental over all existing elements of F and with $f_{2j+1}(x'_{2j+1}, y'_{2j+1}) = 0$. These new elements will forever remain transcendental over the ground field $\overline{\mathbb{Q}}$, and the original x_{2j+1} and y_{2j+1} have now been "swallowed up" by $\overline{\mathbb{Q}}$. Thus the E built here is indeed isomorphic to F, and is a computable field. However, from the transcendence relation T on F, we can compute D, since $j \in \overline{D}$ if and only if the original x_{2j+1} lies in T. Conversely, from a D-oracle we can decide whether x_{2j+1} will ever be swallowed up by $\overline{\mathbb{Q}}$ or not, and also (since $C \leq_T D$) whether x_{2i} will remain transcendental in E or not. Thus $T \equiv_T D$, and so $\boldsymbol{d} \in \mathrm{DgSp}_F(T)$.

The same argument also shows that $\boldsymbol{d} \in \mathrm{DgSp}_F(I)$, since the elements x_s that stay transcendental forever form a transcendence basis for E, from which we can compute the independence relation. It should be remarked here, as in Sect. 3, that the first transcendental solution to $f_k = 0$ that one finds in E will only be one of the $6q_i^2$ such solutions, but in enumerating a transcendence basis, it is safe to choose the first coordinate of the first transcendental solution we find, and then to ignore all other solutions to the same f_k, as their coordinates are all either in $\overline{\mathbb{Q}}$ or interalgebraic with the coordinate we chose. In fact, since the automorphism group of E acts transitively on these solutions, there is nothing to distinguish one such choice from another.   □

Our next result suggests that many spectra of transcendence relations can be viewed as upper cones of enumeration degrees. To be clear, the spectrum is still a set of Turing degrees, by definition, but the defining property of the spectrum may be the ability to enumerate a particular set. (It remains true that only c.e. degrees may lie in $\mathrm{DgSp}_F(T)$, although other Turing degrees may enumerate the same set. So the spectrum will never truly be an upper cone of e-degrees.)

Theorem 4. *Let S be any Σ_2^0 subset of ω. Then there exists a computable field F such that*

$$DgSp_F(T) = DgSp_F(I) = \{c.e.\ degrees\ \boldsymbol{d} : S \in \Sigma_1^{\boldsymbol{d}}\}.$$

That is, $DgSp_F(T)$ contains exactly those c.e. degrees that have the ability to enumerate S.

Proof. Since S is Σ_2^0, there exists a computable total "chip function" $h : \omega \to \omega$ such that $S = \{n : h^{-1}(n) \text{ is finite}\}$. The field F we use for this set is the field

$$F = \overline{\mathbb{Q}}(x_k)[y_k]/(f_i(x_k, y_k)),$$

with k ranging over the set $S \oplus \omega$, much as in Theorem 3 but using the set S itself instead of its complement.

To give a computable presentation of F, we start building a copy of the field $\overline{\mathbb{Q}}(x_k)[y_k]/(f_k(x_k, y_k))$ with k ranging over all of ω, so that every x_k is initially treated as a transcendental. For odd $k = 2j + 1$, x_k stays transcendental throughout this construction. For even $k = 2i$, we write $x_{2i,0} = x_{2i}$ for the initial element described above. At each stage $s + 1$, we check whether $h(s) = i$. If not, then we keep $x_{2i,s+1} = x_{2i,s}$ and continue to treat it as a transcendental. If $h(s) = i$, however, then we suspect that i might not lie in S (since $h^{-1}(i)$ might turn out to be infinite). In this case we make the current $x_{2i,s}$ into a rational number, consistently with the finite portion of the atomic diagram of F built so far, and thus make $y_{2i,s}$ algebraic. We then adjoin new elements $x_{2i,s+1}$ and $y_{2i,s+1}$ to F, treating $x_{2i,s+1}$ as transcendental and setting $f_{2i}(x_{2i,s+1}, y_{2i,s+1}) = 0$. We continue building $\overline{\mathbb{Q}}$ as the ground field, now incorporating the old $x_{2i,s}$ and $y_{2i,s}$ into it, and continuing closing F itself under the field operations, but always adding only finitely many new elements at each stage. This completes the construction, and it is clear that $f_{2i}(X, Y) = 0$ will have a solution by transcendental elements in F just if $h^{-1}(i)$ is finite, which is to say, just if $i \in S$. The rest of the construction then makes it clear that the field we have built is a computable copy of the field F described above.

Given any computable field $E \cong F$, let T be the transcendence relation on E. Then, given a T-oracle, we may search in E for a solution to $f_{2i}(X, Y) = 0$ using transcendental elements x and y. If we find one, then by the definition of F we know that $i \in S$. Conversely, if $i \in S$, then such a solution exists, and we will eventually find it. Thus S is c.e. relative to the degree $\boldsymbol{d}$ of T, as required.

Conversely, fix any c.e. degree $\boldsymbol{d}$ such that S is $\Sigma_1^{\boldsymbol{d}}$, and fix a c.e. set $D \in \boldsymbol{d}$ and a computable enumeration of it. Also fix an index e such that $S = W_e^D = \mathrm{dom}(\Phi_e^D)$; we will use this below to give a computable chip function for S, similar to that used in the original computation of F but specific to this D. To build a computable copy E of F whose transcendence relation T satisfies $T \equiv_T D$, we use the strategy from Theorem 3. The elements $x_{2i,0}$ and $y_{2i,0}$ are defined and initially treated as transcendentals. However, at each stage $s + 1$, the current $x_{2i,s}$ and $y_{2i,s}$ are made into algebraic elements and replaced by new elements $x_{2i,s+1}$ and $y_{2i,s+1}$ unless $\Phi_{e,s}^{D_s}(i) \downarrow$ with some use u such that $D_{s+1} \upharpoonright u = D_s \upharpoonright u$. This is our new chip function for S: if $i \in S = \mathrm{dom}(\Phi_e^D)$, then there will be some s_0 such that we keep x_{2i,s_0} transcendental at all stages $\geq s_0$; whereas if $i \notin S$, then for every stage $s + 1$ at which $\Phi_{e,s}^{D_s}(i) \downarrow$ with a use u, there must be some $t > s$ with $D_t \upharpoonright u \neq D_s \upharpoonright u$, so that $x_{2i,s}$ will be made algebraic at stage $t + 1$ and replaced by a new $x_{2i,t+1}$.

We also revamp the construction for the odd-indexed elements x_{2j+1} and y_{2j+1}, using exactly the same process as in the proof of Theorem 3. If we ever reach a stage at which j enters D, then we turn x_{2j+1} into a rational number, consistently with the construction so far, and adjoin a new transcendental x'_{2j+1} and corresponding y'_{2j+1} with $f_{2j+1}(x'_{2j+1}, y'_{2j+1}) = 0$ in E. This completes the construction of E, which is clearly a computable field and isomorphic to F.

Now from an oracle for the transcendence relation T on E, we can determine whether x_{2j+1} is algebraic in E or not, thus deciding whether or not $j \in D$.

Thus $D \leq_T T$. For the reverse reduction, we claim that with a D-oracle we can enumerate a transcendence basis B for E, thus deciding the independence relation I on E, which in turn computes T. This will prove $I \equiv_T T \equiv_T D$ as required. The D-oracle allows us to decide, for each j, whether $j \in D$, from which we determine either that x_{2j+1} lies in B (if $j \notin D$) or that x'_{2j+1} does (if $j \in D$, in which case we identify x'_{2j+1} by waiting for a stage at which j has entered D). Next, for each i, we watch for a stage s at which $\Phi^{D_s}_{e,s}(i) \downarrow$ with a use u such that $D_s \upharpoonright u = D \upharpoonright u$. The D-oracle allows us to check this, and if we ever find such an s, then we enumerate $x_{2i,s+1}$ into our basis, since the computable enumeration of D will never again change below u. (This is where our argument would fail if d were a properly Δ^0_2 degree, rather than a c.e. degree. With only a computable approximation to D, we could not be sure whether $D_s \upharpoonright u$ would ever again change, even knowing that $D_s \upharpoonright u = D \upharpoonright u$.) Thus we have enumerated exactly the set of elements x_k given when we first defined the isomorphism type of F above, and this set is a transcendence basis for E.   □

Corollary 3. *Let c be any Δ^0_2 Turing degree. Then there exists a computable field F such that*

$$DgSp_F(T) = DgSp_F(I) = \{c.e.\ degrees\ d : c \leq_T d\}.$$

In particular, both $DgSp_F(T)$ and $DgSp_F(I)$ can fail to contain a least degree.

The condition of not containing a least degree also holds for many of the spectra given in Theorem 4. In particular, if no set A' (with A c.e.) is 1-equivalent to S, then the spectrum has no least degree, as $\deg(D) \in DgSp_F(T)$ if and only if $S \leq_1 D'$. The proof uses the Sacks Jump Theorem (see [13], or [14, Thm. VII.3.1]), to avoid the upper cone above a hypothetical least degree.

Proof. Fix a set $C \in c$, and apply Theorem 4 to the Σ^0_2 set $S = C \oplus \overline{C}$. The ability to enumerate S is exactly the ability to compute C, so the corollary follows. (To avoid having a least degree in the spectra, just choose a degree $c \leq 0'$ that is not c.e. and has no least c.e. degree above it.)   □

Corollary 3 extends Theorem 3 to the Δ^0_2 degrees. The result can be viewed as an upper-cone result, but in a somewhat odd way. If the c in the corollary is c.e., then the corollary merely repeats Theorem 3. If c is Δ^0_2 but not c.e., then the degree spectrum is the restriction of the upper cone above c to the c.e. degrees, and therefore does not contain the base degree c itself, nor any other non-c.e. degree $\geq_T c$.

References

1. Downey, R.G., Lempp, S., Wu, G.: On the complexity of the successivity relation in computable linear orderings. J. Math. Log. **10**(01n02), 83–99 (2010)
2. Downey, R.G., Moses, M.F.: Recursive linear orders with incomplete successivities. Trans. Am. Math. Soc. **326**(2), 653–668 (1991)

3. Ershov, Yu.L.: Theorie der Numerierungen. Zeits. Math. Logik Grund. Math. **23**, 289–371 (1977)
4. Frohlich, A., Shepherdson, J.C.: Effective procedures in field theory. Phil. Trans. R. Soc. Lond. Ser. A **248**(950), 407–432 (1956)
5. Frolov, A., Kalimullin, I., Miller, R.: Spectra of algebraic fields and subfields. In: Ambos-Spies, K., Löwe, B., Merkle, W. (eds.) CiE 2009. LNCS, vol. 5635, pp. 232–241. Springer, Heidelberg (2009). https://doi.org/10.1007/978-3-642-03073-4_24
6. Leopoldt, H.-W.: Über die Automorphismengrupper des Fermatkorpers. J. Number Theory **56**(2), 256–282 (1996)
7. Metakides, G., Nerode, A.: Effective content of field theory. Ann. Math. Log. **17**, 279–320 (1979)
8. Miller, R.: Computable fields and Galois theory. Not. Am. Math. Soc. **55**(7), 798–807 (2008)
9. Miller, R.: An introduction to computable model theory on groups and fields. Groups Complex. Cryptol. **3**(1), 25–46 (2011)
10. Miller, R., Poonen, B., Schoutens, H., Shlapentokh, A.: A computable functor from graphs to fields. J. Symbol. Log. **83**(1), 326–348 (2018)
11. Miller, R., Schoutens, H.: Computably categorical fields via Fermat's Last Theorem. Computability **2**, 51–65 (2013)
12. Rabin, M.: Computable algebra, general theory, and theory of computable fields. Trans. Am. Math. Soc. **95**, 341–360 (1960)
13. Sacks, G.E.: Recursive enumerability and the jump operator. Trans. Am. Math. Soc. **108**, 223–239 (1963)
14. Soare, R.I.: Recursively Enumerable Sets and Degrees. Springer, New York (1987)
15. Tzermias, P.: The group of automorphisms of the Fermat curve. J. Number Theory **53**(1), 173–178 (1995)
16. van der Waerden, B.L.: Algebra, volume I. Springer, New York (1970 hardcover, 2003 softcover). Trans. Blum, F., Schulenberger, J.R.

More Intensional Versions of Rice's Theorem

Jean-Yves Moyen[1] and Jakob Grue Simonsen[2(✉)]

[1] LIPN, Université Paris 13, Villetaneuse, France
`Jean-Yves.Moyen@lipn.univ-paris13.fr`
[2] Department of Computer Science, University of Copenhagen (DIKU),
Copenhagen, Denmark
`simonsen@diku.dk`

Abstract. Classic results in computability theory concern extensional results: the behaviour of partial recursive functions rather than the programs computing them. We prove a generalisation of Rice's Theorem concerning equivalence classes of programs and show how it can be used to study intensional properties such as time and space complexity. While many results that follow from our general theorems can - and have - been proved by more involved, specialised methods, our results are sufficiently simple that little work is needed to apply them.

1 Introduction

Rice's Theorem [17] states that any non-trivial extensional set of programs is undecidable. This generic formulation allows to prove undecidability of a variety of sets *e.g.* "programs that, on input 0, return 42", "programs that compute a bijection" or "programs that compute a non-total function".

Rice's Theorem showcases a fundamental dichotomy between programs and the partial functions they compute: it gives an undecidability criterion for sets of programs, but these sets are defined by the functions computed by the programs. Underlying this dichotomy and the Theorem is the notion of *extensional equivalence*: "two programs are equivalent iff they compute the same function". Rice's Theorem is, quintessentially, that this equivalence relation is undecidable.

After 60 years, scant research has investigated *intensional* analogues of Rice's Theorem: undecidability results concerning *how* programs compute rather than *what* they compute. One exception is Asperti's work on *complexity cliques* [1] that, roughly, considers two programs equivalent if they compute the same function with comparable (up to big-Θ) complexity, and proceeds to prove Rice-like theorems for this equivalence relation. In a different vein, several classic results such as the Rice-Shapiro [14,18] and Kreisel-Lacombe-Shoenfield/Tseitin-Moschovakis Theorems [10,11,20,21] consider what is essentially continuity properties on appropriate metric spaces to characterize the semi-decidable properties of partial recursive functions, and decidable properties of total recursive

---

A very preliminary version of this paper first appeared at the DICE '16 workshop [12].

F. Manea et al. (Eds.): CiE 2019, LNCS 11558, pp. 217–229, 2019.
https://doi.org/10.1007/978-3-030-22996-2_19

functions, respectively. In recent work, Hoyrup et al. [4–6] have investigated topological characterizations of (semi-)decidable sets under different representations, remarkably including results for the class of primitive recursive functions [4]. Our current work considers *programs* as the representations of choice, rather than the (partial) functions computed by programs (i.e., we use what is sometimes called the "Markov" approach in the literature); while some of our techniques are clearly topological in nature, and thus related to Hoyrup's recent results, we are more interested in (a) the interplay between equivalence relations on the set of programs–not just extensional equivalence–, and (b) (semi-)decidable *over*-approximations of such equivalence classes.

Contribution: We generalise Rice's Theorem in two ways: The first generalisation (Theorem 2) only asks that the studied sets accept all programs computing one given function, but has no condition on programs computing other functions (which may or not be in the set); The second generalisation (Theorem 3) imposes very general conditions on equivalence relations (via so-called *switching families*), and can be used to prove several well-known results traditionally proved by specialized methods. While Theorem 2 is a corollary of Theorem 3, Theorem 2 is perhaps more accessible when wanting to prove simple (un)decidability results, whence we give separate proofs.

Our results shed light on the problems with over-approximations in the field of implicit computational complexity: For example, one may want a procedure that accepts all polynomial-time programs and "a few" exponential-time programs. But our results can be applied to show that over-approximations are patently not viable: there will, necessarily, always be *many* bad eggs and there will always be *extremely* bad eggs.

2   Preliminaries and Notation

We assume an unspecified, Turing-complete, programming language and denote by $\mathcal{P}$ the set of all programs in the language[1]). We assume the language has a single datatype $\mathcal{D}$, typically $\mathcal{D} = \mathcal{N}$ or $\mathcal{D} = \{0,1\}^*$, and assume appropriate computable encodings of elements of $\mathcal{N}$, $\mathcal{P}$, finite sets, and finite strings of symbols as elements of $\mathcal{D}$. We denote by $\phi_{\mathrm{p}} : \mathcal{D} \rightharpoonup \mathcal{D}$ the partial function computed by $\mathrm{p} \in \mathcal{P}$, and for any partial function $f : \mathcal{D} \rightharpoonup \mathcal{D}$, we denote by $\mathcal{P}_f$ the set of all programs computing function f. We write $\phi_{\mathrm{p}}(x) = \bot$ to denote that p does not terminate on input x. By Turing completeness of $\mathcal{P}$, we assume computable pairing and unpairing functions such that partial functions ϕ_{p} may informally be considered to have type $\mathcal{D}^m \rightharpoonup \mathcal{D}^k$ for any $m, k \in \mathcal{N}$. By Turing completeness, we further assume wlog. the existence of a fixed universal program that allows simulation of all other programs in $\mathcal{P}$, and that the s-m-n Theorem holds. As usual [7] in examples, we shall freely use informal vernacular such as "loop" and appropriate pseudocode with the understanding that this is to be interpreted in

---

[1] For the technical development, we need only assume $\mathcal{P}$ to be r.e. but morally we are interested in concrete programming languages where $\mathcal{P}$ will always be a decidable set.

$\mathcal{P}$. When S is a countable set (suitably encoded by elements of $\mathcal{D}$, we denote by REC_S and RE_S the sets of decidable, resp. r.e. subsets of S, usually suppressing S in the notation if S is clear from the context).

We refer to [3] for basic definitions of Blum complexity. With the notation of the current paper (assuming a standard encoding of elements of $\mathcal{N}$ as elements of $\mathcal{D}$), a map Φ from $\mathcal{P}$ to $\mathcal{P}$ is (the complexity function of) a Blum complexity measure if (i), for every $\mathbf{p} \in \mathcal{P}$ and every $n \in \mathcal{N}$, $\phi_{\Phi(\mathbf{p})}(n) \neq \bot$ iff $\phi_{\mathbf{p}}(n) \neq \bot$, and (ii) the ternary predicate $\Phi(\mathbf{p})(n) = m$ on $\mathcal{P} \times \mathcal{N} \times \mathcal{N}$ is dedidable. As usual, for two partial functions $f, g : \mathcal{N} \rightharpoonup \mathcal{N}$, we write $f \in O(g)$ if there are $m, c \in \mathcal{N}$ such that for all $n \geq m$, $g(n) \neq \bot$ implies $f(n) \leq cg(n)$; and we write $f \in \Theta(g)$ if $f \in O(g)$ and $g \in O(f)$.

We shall use a straightforward abstraction of Smullyan's notion of recursively separable sets [19]: Let $\mathcal{S}$ be a family of subsets on a set $\mathcal{W}$, and let $A, B \subseteq \mathcal{W}$. Set A is said to be $\mathcal{S}$-*separated* from B if there is a set $C \in \mathcal{S}$ such that $A \subseteq C$ and $B \cap C = \emptyset$. If no such C exists, A is said to be $\mathcal{S}$-*inseparable* from B. A and B are said to be $\mathcal{S}$-*separable* if A is $\mathcal{S}$-separated from B, or B is $\mathcal{S}$-separated from A. If A and B are not $\mathcal{S}$-separable, they are said to be $\mathcal{S}$-*inseparable*. The standard example of a family $\mathcal{S}$ and $\mathcal{S}$-inseparable sets is $\mathcal{S} = \text{REC}$–the family of all decidable sets; A and B are $\mathcal{S}$-(in)separable iff they are recursively (in)separable. The classic example of two specific recursively inseparable sets are $A = \{\mathbf{p} : \phi_{\mathbf{p}}(0) = 0\}$ and $B = \{\mathbf{p} : \phi_{\mathbf{p}}(0) \notin \{0, \bot\}\}$ (see, e.g., [16, Sec. 3.3]) for proof of the following.

3   Rice's Theorem and the Rice Equivalence Relation

In the below definition, the notions of intensionally complete, extensionally sound, complete and universal are new, but completely natural and straightforward.

Definition 1. *Let $\mathfrak{P} \subseteq \mathcal{P}$ and F be a non-empty set of partial recursive functions $f : \mathcal{D} \rightharpoonup \mathcal{D}$. We say that $\mathfrak{P}$ is (i)* non-trivial *if $\mathfrak{P} \notin \{\emptyset, \mathcal{P}\}$; (ii)* extensional *if $(\mathbf{p} \in \mathfrak{P}) \wedge (\mathbf{q} \notin \mathfrak{P})$ implies $\phi_{\mathbf{p}} \neq \phi_{\mathbf{q}}$; (iii)* intensionally complete *for F if $\mathcal{P}_F \subseteq \mathfrak{P}$ (i.e., $\mathfrak{P}$ contains all programs computing a function in F and is thus an over-approximation of $\mathcal{P}_F$); (iv)* extensionally complete *for F if $F \subseteq \phi_{\mathfrak{P}}$ (i.e. $\mathfrak{P}$ contains at least one program computing each function in F: (v)* extensionally sound *for F if $\phi_{\mathfrak{P}} \subseteq F$ (i.e. $\mathfrak{P}$ contains only programs computing functions in F and is thus an under-approximation of $\mathcal{P}_F$); (vi)* extensionally universal *if it is extensionally complete for the set of all partial recursive functions.*

Note that a set $\mathfrak{P}$ intensionally complete for F does not need to be extensional, and thus functions not in F may be computed by both programs in, and not in, $\mathfrak{P}$. An example: "the set of programs that contain a loop" is intensionally complete for the set $F = \{x \mapsto \bot\}$ where $x \mapsto \bot$ is the nowhere defined partial function (because programs without loops always terminate). Note that this set of programs is non-trivial and decidable.

Typical examples of extensionally sound and complete $\mathfrak{P}$ found in the literature are for subrecursive classes F, e.g., Bellantoni-Cook style characterizations of the set of polynomial-time computable functions [2]; extensionally sound and complete $\mathfrak{P}$ are informally known as *characterizations* of F in the field of implicit computational complexity. Rice's Theorem is:

Theorem 1 (Rice [17]). *Any (non-trivial) extensional set of programs is undecidable.*

Theorem 1 is limited to extensional properties: it is does not yield any information about *intensional* sets of programs, that depend on the actual program behaviour. That is, it provides undecidability results on "what is computed?", but is unable to provide undecidability results on "how is it computed?".

The extensionality central to Rice's Theorem defines an equivalence relation between programs:

Definition 2 (The Rice Equivalence Relation). *The Extensional Equivalence Relation, or* Rice Equivalence Relation *is* $\mathfrak{R} \subseteq \mathcal{P} \times \mathcal{P}$*, defined by:* $p\mathfrak{R}q \Leftrightarrow \phi_p = \phi_q$.

Thus "two programs are $\mathfrak{R}$-equivalent iff they compute the same function", an extensional set of programs is exactly the union of equivalence classes of the Extensional Equivalence Relation $\mathfrak{R}$.

The set of equivalence relations (or, equivalently, partitions) on a set has a useful order structure, the refinement ordering. The refinement order $\preceq$ is defined by $\equiv_0 \preceq \equiv_1$ iff $x \equiv_0 y$ implies $x \equiv_1 y$ (i.e., each class of $\equiv_0$ is a subset of a class of $\equiv_1$). Rice's Theorem may hence be restated as: any (non-trivial) equivalence relation $\equiv \subsetneq \mathcal{D} \times \mathcal{D}$ with $\mathfrak{R} \preceq \equiv$ is undecidable.

We shall momentarily generalise Rice's Theorem by utilizing the view of Rice's Theorem as a result on equivalence classes. First, we shall require the target set to simply *contain* some classes of $\mathfrak{R}$, and not necessarily to be exactly the union of these (partial extensionality); second, we shall consider what happens if we completely change the equivalence relation and use some other equivalence than $\mathfrak{R}$. We first re-cast Asperti's result from [1] purely in terms of equivalence relations.

Definition 3 (The Asperti Equivalence Relation). *The Asperti Equivalence Relation* $\mathfrak{A} \subseteq \mathcal{P} \times \mathcal{P}$ *is defined by* $p\mathfrak{A}q \Leftrightarrow \phi_p = \phi_q \wedge \Phi(p) \in \Theta(\Phi(q))$.

That is, "two programs are equivalent iff they compute the same function with the same complexity". $\mathfrak{A}$ is exactly Asperti's "similarity" [1, Def. 3, Thm. 4], and unions of equivalence classes of $\mathfrak{A}$ are exactly Asperti's "complexity cliques" [1, Def. 5, Thm. 8]. We can then reformulate the main result of [1] (the "Asperti-Rice Theorem") as: *any equivalence relation* $\equiv \subsetneq \mathcal{D} \times \mathcal{D}$ *with* $\mathfrak{A} \preceq \equiv$ *is undecidable.*

The set of equivalence relations on $\mathcal{P}$ equipped with $\preceq$ forms a complete lattice [15] whose minimal element, $\perp$, is (syntactic) equality, and whose maximal element, $\top$, is the trivial equivalence with a single class (where every program is equivalent to every other program). In the vernacular of lattice theory, Rice's Theorem is then: an equivalence relation $\equiv_0$ in the principal filter of $\mathfrak{R}$ is decidable iff $\equiv_0 = \top$, and the Asperti-Rice Theorem is: *an equivalence relation in the principal filter of* $\mathfrak{A}$ *is decidable iff* $\equiv_0 = \top$.

The fact that these results are so cleanly expressed with order-theoretic vocabulary is a clue that the study of the lattice of equivalence relations on programs can shed some light on the extensional/intensional dichotomy [13]. However, to obtain a more general result, one must prove undecidability of equivalence relations *not* in the principal filter at either $\mathfrak{R}$ or $\mathfrak{A}$. We now proceed to do so.

4 Intensionality I: A Simple Generalisation of Rice's Theorem

We now consider sets $\mathfrak{P}$ that are not unions of classes of $\mathfrak{R}$ (i.e., extensional sets), but instead *contain* some equivalence classes (i.e., $\mathfrak{P}$ will be an intensionally complete set). Our first main result is:

Theorem 2. *Any non-empty, decidable set of programs that is intensionally complete for a non-empty set F of partial recursive functions, is extensionally universal.*

Proof. Let f be a computable function and let $\mathfrak{P}$ be a decidable set of programs that is intensionally complete for $\{f\}$. Let p be a program computing f: $\phi_p = f$, thus $p \in \mathfrak{P}$. Assume, for contradiction, that $\mathfrak{P}$ is not extensionally universal. Then there is a program q such that no program computing the same function is in $\mathfrak{P}$: $\phi_s = \phi_q \Rightarrow s \notin \mathfrak{P}$. Now, for every program r, construct program r′ which, on input x, works as follows: r′ first simulates program r on input 0; if it terminates, then (i) if the result is 0, r′ simulates p on input x; otherwise (ii) if the result is not 0, r′ simulates q on input x (end of description of r′).

Note that by Turing-completeness of $\mathcal{P}$, given r, r′ is effectively computable (in informal syntax, r′ is essentially the program if r(0)=0 then p(x) else q(x)).

- If $\phi_r(0) = 0$, then $\phi_{r'} = \phi_p$, hence because $\mathfrak{P}$ is intensionally complete for $f = \phi_p$, r′ $\in \mathfrak{P}$.
- If $\phi_r(0) \neq 0$ but it does terminates, then $\phi_{r'} = \phi_q$, and hence r′ $\notin \mathfrak{P}$ by hypothesis.

Now, consider the set $C = \{\, r \ : \ r' \in \mathfrak{P} \,\}$. As $\mathfrak{P}$ is decidable (by hypothesis) and r′ is effectively computable from r, then C is decidable. But by the above observations $\{\, r \ : \ \phi_r(0) = 0 \,\} \subseteq C$, and $\{\, r \ : \ \phi_r(0) \notin \{0, \bot\} \,\} \cap C = \emptyset$. However, these sets cannot be recursively separated. Hence, C cannot be decidable, and we obtain a contradiction. □

Note that, contrary to Rice's Theorem, Theorem 2 only requires the set of programs to be non-empty rather than non-trivial—indeed, the trivial set of all programs is certainly decidable, intensionally complete and extensionally universal. Also observe that Theorem 2 holds more information than simply "the set of all programs computing some computable partial function is undecidable": every decidable *over-approximation* of such a set must, for every partial recursive function f, contain at least one (but not necessarily all) program(s) computing f.

Theorem 2 creates a "deadlock" for attempts to devise programming languages that attempt to over-approximate subrecursive classes, e.g., the polynomial-time computable functions. This is illustrated in Fig. 1, where any over-approximation will contain programs computing the Ackermann function, or winning the Hydra game [9].

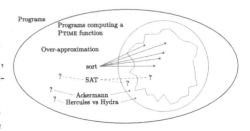

Fig. 1. Extensionally universal over-approximation.

Theorem 2 has several corollaries:

Corollary 1. *The following hold:*

Rice's Theorem: *Any (non-trivial) extensional set of programs is undecidable.*

The Rice-Asperti Theorem: *Any equivalence relation $\equiv \subsetneq \mathcal{P} \times \mathcal{P}$ with $\mathfrak{A} \preceq\equiv$ is undecidable.*

All ICC characterisations are intensionally incomplete: *Any proper subset of $\mathcal{P}$ that is intensionally complete and extensionally sound for some non-empty set F of partial functions is undecidable.*

Proof. We prove each part in turn:

- Let $\mathfrak{P}$ be an extensional decidable set of programs. If $\mathfrak{P} = \emptyset$, then it is trivial. Otherwise, by definition, any extensional set is intensionally complete. Hence, by Theorem 2, $\mathfrak{P}$ is extensionally universal. Let $p \in \mathfrak{P}$ and q be any program. Because $\mathfrak{P}$ is extensionally universal, there exists q' with $\phi_q = \phi_{q'}$ and $q' \in \mathfrak{P}$. Because $\mathfrak{P}$ is extensional, any program that computes the same function as q' is also in it. Especially $q \in \mathfrak{P}$.
  Hence, $\mathfrak{P} = \mathcal{P}$, and $\mathfrak{P}$ is thus trivial.
- Observe that "extensional" in the previous argument is equivalent to stating $\mathfrak{P}$ is union of equivalence classes of $\mathfrak{R}$. The argument can now be repeated, replacing $\mathfrak{R}$ by $\mathfrak{A}$, *mutatis mutandis* (observing that r' has the same complexity as either p or q up to the constant factor $\Phi(r)(0)$).
- Let $\mathfrak{P} \subseteq \mathcal{P}$ be intensionally complete and extensionally sound for F. If $\mathfrak{P}$ is decidable, so is the complement $\mathcal{P} \setminus \mathfrak{P}$. As $\mathfrak{P}$ is a proper subset of $\mathcal{P}$, then $\mathcal{P} \setminus \mathfrak{P} \neq \emptyset$, and as $\mathfrak{P}$ is extensionally sound for F, $\mathcal{P} \setminus \mathfrak{P}$ is extensionally complete for the set of partial recursive functions not in F, and this set is non-empty (otherwise, $\mathfrak{P}$ would be extensionally complete for the set of all partial recursive functions, and as it is also intensionally complete, we would have $\mathfrak{P} = \mathcal{P}$). Then, Theorem 2 applied to $\mathcal{P} \setminus \mathfrak{P}$ yields that $\mathcal{P} \setminus \mathfrak{P}$ is extensionally universal, and hence contains at least one p such that $\phi_p \in F$, contradicting intensional completeness of $\mathfrak{P}$. □

5   Intensionality II: A General Theorem

Theorem 2 concerns extensionality: it speaks of sets of programs determined by the function they compute, no matter how they compute it. This limitation is set in stone by the use of $\mathfrak{R}$, whence one should properly parameterise the result over the equivalence relation used. However, changing the equivalence relation must be done cautiously as some equivalence relations on programs are decidable (*e.g.*, "having the same number of lines of code").

In order to replace $\mathfrak{R}$ by other suitable equivalence relations, we adapt our vocabulary as follows:

Definition 4. *Let $\mathfrak{F}$ be an equivalence relation on a set S and let $\{\mathfrak{F}_1, \mathfrak{F}_2, \ldots\}$ be the set of its equivalence classes. A subset $\mathfrak{P}$ of S is* (i) partially compatible *with an equivalence class $\mathfrak{F}_k$ if $\mathfrak{F}_k \subseteq \mathfrak{P}$;* (ii) complete *for a set $\mathcal{F}$ of equivalence classes if, for each class $\mathfrak{F}_k \in \mathcal{F}$, we have $\mathfrak{P} \cap \mathfrak{F}_k \neq \emptyset$;* (iii) sound *for a set $\mathcal{F}$ of equivalence classes if $\mathfrak{P} \subseteq \bigcup_{\mathfrak{F}_k \in \mathcal{F}} \mathfrak{F}_k$;* (iv) universal *(for $\mathfrak{F}$) if it is complete for $\mathfrak{F}$.*

The primary means of using our second main result (Theorem 3 below) is to construct a family of (typically, but not necessarily computable) maps called *switching functions* that interact appropriately with two sets S and T (where typically $T = \mathcal{P}$) and an equivalence relation on T.

Definition 5. *Let S and T be sets and $\approx$ an equivalence relation on T. Let $\mathcal{S} \subseteq \mathscr{P}(S)$ and $\mathcal{T} \subseteq \mathscr{P}(T)$ be families of subsets of S and T. A switching function is a total function $\pi : S \to T$. It is $\mathcal{S}$-$\mathcal{T}$-continuous if the reverse image of any set in $\mathcal{T}$ is in $\mathcal{S}$: $\forall T' \in \mathcal{T}, \pi^{-1}(T') \in \mathcal{S}$. If F, G are sets of equivalence classes of $\approx$, An F-G-switching family is a family of switching functions $I = (\pi_{a,b})_{(a,b) \in (\cup F) \times (\cup G)}$.*
Let $(a, b) \in (\cup F) \times (\cup G)$. Define the sets $A_{a,b} = \{x \in S : \pi_{a,b}(x) \approx a\}$ and $B_{a,b} = \{x \in S : \pi_{a,b}(x) \approx b\}$. An F-G-switching family is $\mathcal{S}$-F-G-intricated with $\approx$ if, for each pair of elements $(a, b) \in (\cup F) \times (\cup G)$ such that there is a $T' \in \mathcal{T}$ with $A_{a,b} \subseteq \pi_{a,b}^{-1}(T')$, $A_{a,b}$ is $\mathcal{S}$-inseparable from $B_{a,b}$.

Observe that if $A_{a,b} \cap B_{a,b} \neq \emptyset$, then $A_{a,b}$ and $B_{a,b}$ are $\mathcal{S}$-inseparable. Thus, it suffices to check a-b-$\mathcal{S}$-intrication for a, b where $A_{a,b}$ and $B_{a,b}$ are disjoint. Note in particular that if $a \approx b$ then $A_{a,b}$ and $B_{a,b}$ are the same and hence cannot be $\mathcal{S}$-separated. Observe that there may be elements in S that are neither in $A_{a,b}$ nor in $B_{a,b}$, i.e., $\pi_{a,b}(x)$ is not equivalent to a and not equivalent to b.

Example 1 (Standard Switching family). Let $S = T = \mathcal{P}$. The *standard switching family* is: $\pi_{p,q}(r)(x) = r'(x) \triangleq$ if r(0)=0 then p(x) else q(x)

– Each function in the family is REC-REC-continuous. Indeed, let D be a decidable set, to decide if $x \in \pi_{p,q}^{-1}(D)$, it suffices to compute $\pi_{p,q}(x)$ and decide $\pi_{p,q}(x) \in D$, which is clearly possible as each function in the family is computable.

- For any non-empty sets F, G of equivalence classes of $\mathfrak{R}$ $(\pi_{a,b})_{(a,b)\in(\cup F)\times(\cup G)}$ is REC-F-G-intricated with $\mathfrak{R}$. Indeed, for any $\mathsf{p}, \mathsf{q} \in \mathcal{P}$ with $\phi_\mathsf{p} \neq \phi_\mathsf{q}$ we have $A_{\mathsf{p},\mathsf{q}} = \{\, \mathbf{r} \; : \; \phi_{\pi_{\mathsf{p},\mathsf{q}}(\mathbf{r})} = \phi_\mathsf{p} \,\}$. From the proof of Theorem 2 we know that $A_{\mathsf{p},\mathsf{q}} = \{\, \mathbf{r} \; : \; \mathbf{r}(0) = 0 \,\}$ and similarly $B_{\mathsf{p},\mathsf{q}} = \{\, \mathbf{r} \; : \; \mathbf{r}(0) \notin \{0, \bot\} \,\}$, and these sets are recursively inseparable. Hence, if $\mathsf{p} \in \cup F$ and $\mathsf{q} \in \cup G$, the result follows (if $\cup F = \cup G$ there is nothing to prove).
- REC-intrication also implies, for example, PTIME-intrication as every PTIME set is decidable. Similarly, for any non-empty F, G, the standard switching family is REC-intricated with $\mathfrak{A}$.

Example 2 (Hyperconnectedness in topological spaces). Let $(S, \mathcal{S})$ and $(T, \mathcal{T})$ be topological spaces, and let $\approx$ be any equivalence relation on S such that the quotient topology $\mathcal{S}/\approx$ is hyperconnected (i.e., no two non-empty open sets are disjoint). Let $g : S/\approx \rightarrow T$ be continuous. Then, for any non-empty sets F, G of equivalence classes of $\approx$, the switching family $(\pi_{a,b})_{(a,b)\in(\cup F)\times(\cup G)}$ defined by $\pi_{a,b} = g$ for any $(a, b) \in S \times S$ is $\mathcal{S}/\approx$-intricated with $\approx$ (because g is continuous, and no two non-empty open sets of $\mathcal{S}/\approx$ are disjoint).

The next example shows a switching family where $S \neq T$. We first introduce the following equivalence relations.

Definition 6. *Let f, g be partial recursive functions. We write $f \simeq g$, iff they differ only on a finite set of inputs: $f \simeq g \Leftrightarrow \{\, x \; : \; f(x) \neq g(x) \,\}$ is finite. The equivalence relation $\eqcirc \subseteq \mathcal{P} \times \mathcal{P}$ is defined by $\mathsf{p} \eqcirc \mathsf{q}$ iff $\phi_\mathsf{p} \simeq \phi_\mathsf{q}$ Let $\mathsf{s} \in \mathcal{P}$. The equivalence relation $\eqcirc_\mathsf{s} \subseteq \mathcal{P} \times \mathcal{P}$ is defined by $\mathsf{p} \eqcirc_\mathsf{s} \mathsf{q}$ iff either $\phi_\mathsf{p} = \phi_\mathsf{q} = \phi_\mathsf{s}$, or $(\, \phi_\mathsf{p}, \phi_\mathsf{q} \neq \phi_\mathsf{s}$ and $\mathsf{p} \eqcirc \mathsf{q})$.*

Example 3 (A Rice-Shapiro family). Let $S = \mathcal{P} \times \mathcal{N}$ and $T = \mathcal{P}$. Define, for each $(\mathbf{r}, n) \in \mathcal{P} \times \mathcal{N}$ and $\mathsf{p}, \mathsf{q} \in \mathcal{P}$, a switching function as follows: The program $\mathsf{s}_{\mathbf{r},n} = \pi_{\mathsf{p},\mathsf{q}}(\mathbf{r}, n)$, on input k runs $\mathbf{r}$ on input n for k steps. If $\mathbf{r}$ halts on n in at most k steps, the program returns $\mathsf{q}(k)$, and otherwise returns $\mathsf{p}(k)$.

- Each function in the switching family is RE-RE-continuous: Let T' be any r.e. subset of $\mathcal{P}$; as the switching function is clearly total and computable, we can use the enumeration of T' to enumerate the set of pairs $(\mathbf{r}, n)$ in $S = \mathcal{P} \times \mathcal{N}$ mapped to elements of T (i.e., if $\mathsf{s}_{\mathbf{r},n} \in T$, then $\mathbf{r} \in \pi_{\mathsf{p},\mathsf{q}}^{-1}(T')$). Hence, $\pi_{\mathsf{p},\mathsf{q}}^{-1}(T)$ is r.e.
- Let $\mathsf{s} \in \mathcal{P}$ and consider the equivalence relation $\eqcirc_\mathsf{s}$. Let F be the singleton set containing the equivalence class $\{\mathsf{p} : \phi_\mathsf{p} = \phi_\mathsf{s}\}$, and let G be the set of equivalence classes H of $\eqcirc_\mathsf{s}$ such that all $\mathsf{q} \in H$ satisfy $\neg(\mathsf{q} \eqcirc \mathsf{s})$ (*i.e.*, every $\mathsf{q} \in H$ differs from s on infinitely many inputs). Then, the switching family $(\pi_{a,b})_{(a,b)\in(\cup F)\times(\cup G)}$ is RE-F-G-intricated with $\eqcirc_\mathsf{s}$.

We now prove a general theorem stating that the existence of an intricated switching family of continuous functions is enough to ensure that any partially compatible set in the family is also universal. In Theorem 2, $\mathcal{S}$ was the family of all decidable sets and took $\mathfrak{F} = \mathfrak{R}$. The existence of an intricated switching family was not required in the statement of the Theorem but used in the proof (with the standard switching family). We now generalise the result.

Theorem 3. *Let S, T be sets, $\mathcal{S} \subseteq \mathscr{P}(S)$, $\mathcal{T} \subseteq \mathscr{P}(T)$ be two families of sets. Let $\approx$ be an equivalence relation on T and F, G be sets of equivalence classes of T. Let $I = (\pi_{a,b})_{(a,b) \in (\cup F) \times (\cup G)}$ be an F-G-switching family of $\mathcal{S}$-$\mathcal{T}$-continuous functions that is $\mathcal{S}$-F-G-intricated with $\approx$. Then, any $T' \in \mathcal{T}$ that is partially compatible with F is complete for G.*

Proof. Suppose, for contradiction, that T' were not complete for G, that is, there is a b such that the equivalence class $[b] \in G$ containing b satisfies $[b] \cap T' = \emptyset$. Let $a \in F$; as T' is partially compatible with F, we have $[a] = F \subseteq T'$. Let $x \in S$ be arbitrary and set $y \triangleq \pi_{a,b}(x)$. Now, if $x \in A_{a,b}$ then by definition $y = \pi_{a,b}(x) \approx a$. Hence, $y \in [a]$ and thus $y \in T'$. On the other hand, if $x \in B_{a,b}$ then by definition $y = \pi_{a,b}(x) \approx b$. Hence, $y \in [b]$ and thus $y \notin T'$.

Let $T' \subseteq T$, and define $C = \pi_{a,b}^{-1}(T') = \{ x : y = \pi_{a,b}(x) \in T' \}$. Because $T' \in \mathcal{T}$ and $\pi_{a,b}$ is $\mathcal{S}$-$\mathcal{T}$-continuous, we have $C \in \mathcal{S}$. Hence, $x \in A_{a,b}$ implies $y \in T'$, and thus $A_{a,b} \subseteq C$. On the other hand, $x \in B_{a,b}$ implies $y \notin T'$ and thus $B_{a,b} \subseteq S \setminus C$. But then $C \cap B_{a,b} = \emptyset$. In other words, $A_{a,b}$ is $\mathcal{S}$-separated from $B_{a,b}$, contradicting the assumption that the switching family is $\mathcal{S}$-intricated with $\approx$. Hence, there cannot exist a class $[b]$ of $\approx$ such that $[b] \cap T' = \emptyset$, and thus T' is complete for G. □

5.1   Examples

We now give a sequence of corollaries showing the usefulness of Theorem 3. The first states that our first generalisation of Rice's Theorem (Theorem 2 is a consequence of Theorem 3).

Corollary 2 (Theorem 2). *Any non-empty, decidable set of programs that is intensionally complete for some F containing at least one computable partial function is extensionally universal.*

Proof. By Example 1, for $S = T = \mathcal{P}$, the standard switching family is REC − REC-continuous and REC-intricated with $\mathfrak{R}$.

As $\mathcal{S} = \mathcal{T} = \text{REC}$ is closed under complements, Theorem 3 yields that any non-empty, decidable set that contains all programs in an equivalence class of $\mathfrak{R}$ is also universal for $\mathfrak{R}$, and the result follows. □

As Theorem 2 follows from Theorem 3, so do the results of Corollary 1, in particular Rice's Theorem and the Rice-Asperti Theorem.

An Intensional Rice-Shapiro Theorem. To see why it is useful that Theorem 3 is formulated using two distinct families of sets $\mathcal{S}$ and $\mathcal{T}$, we treat the classic Rice-Shapiro Theorem [14,18]. As with Rice's Theorem, the Rice-Shapiro Theorem was originally formulated as a result on partial recursive functions (and is hence purely extensional, in our terminology). Recall that if f_1 and f_2 are partial recursive functions defined on domains $D_1, D_2 \subseteq \mathcal{D}$, respectively. For two programs p, q, define $\mathsf{p} \precsim \mathsf{q}$ iff $\forall x \in \mathcal{D}.(\mathsf{p}(x) = \bot$ or $\mathsf{p}(x) = \mathsf{q}(x))$.

With the terminology of the present paper, the classic Rice-Shapiro Theorem is then:

Theorem 4 (Rice-Shapiro [14, 18]**).** *An extensional set of programs $\mathfrak{P}$ is r.e. iff : $\mathsf{p} \in \mathfrak{P} \Leftrightarrow \exists \mathsf{q} \in \mathfrak{P}, \mathsf{q}$ is defined only on a finite subset of $\mathcal{D}$, and $\mathsf{q} \precsim \mathsf{p}$.*

When moving to intensional completeness as elsewhere in this paper, we cannot hope to have a necessary *and* sufficient criterion for an extensional set being r.e. (because we can in general only ensure that sets are extensionally universal, i.e. that a set of programs contains at least one representative of each equivalence class of $\mathfrak{R}$, but not all elements of the class). However, using Theorem 3, we can obtain a result that strengthens the *necessary* ("only if") part of the Rice-Shapiro Theorem:

Corollary 3 (Rice-Shapiro, intensional version). *Let $\mathfrak{P}$ be an r.e. set of programs that is intensionally complete for some non-empty set H of partially recursive functions. Then, for any partial recursive function f (not necessarily in F), there exists a program $\mathsf{p} \in \mathfrak{P}$ such that $\phi_\mathsf{p} \eqcirc f$.*

Proof. Let $\mathsf{s} \in \mathcal{P}$ be such that $\phi_\mathsf{s} \in H$. Consider the equivalence relation $\eqcirc_\mathsf{s}$ (cf. Definition 6). It suffices to prove that $\mathfrak{P}$ is universal for $\eqcirc_\mathsf{s}$. If $\mathsf{p} \in \mathcal{P}$ satisfies $\mathsf{p} \eqcirc \mathsf{s}$, we are done, and it thus suffices to consider classes of $\eqcirc_\mathsf{s}$ where all elements of the class satisfy $\neg(\mathsf{s} \eqcirc \mathsf{p})$.

Let F be the singleton set containing the equivalence class $\{\mathsf{p} : \phi_\mathsf{p} = \phi_\mathsf{s}\}$, and let G be the set of equivalence classes such that every element p satisfies $\neg(\mathsf{p} \eqcirc \mathsf{s})$. By Example 3, the F-G-switching family $(\pi_{a,b})_{(a,b) \in (\cup F) \times (\cup G)}$ is RE-F-G-intricated with $\eqcirc_\mathsf{s}$, and the family is RE-RE-continuous. Theorem 3 then yields that any r.e. set of programs partially compatible with F is universal for G, and the result follows. □

An example application of Corollary 3: Let $\mathfrak{P}$ be an r.e. set of programs intensionally complete for a total function (*e.g.*, the constant function $x \mapsto 0$); then by choosing f as the everywhere undefined function, we conclude that $\mathfrak{P}$ must contain a program that loops on all but finitely many inputs; by choosing a different constant function (*e.g.*, $x \mapsto 1$), we can conclude that $\mathfrak{P}$ contains a program whose result is 1 for all but finitely many inputs, and so on.

Spambots. Our final example is another illustration of why Theorem 3 is formulated using two distinct families of sets $\mathcal{S}, \mathcal{T}$. The example shows that a typical prosaic formulation of program properties is treatable without much special encoding.

Let S be the set of (one tape, 2-symbol) Turing Machines, and let $T = \mathcal{P}$ be the set of all programs in a given language $\mathcal{L}$ that has I/O in the form of sending e-mails (we assume wlog. that sending an email is a special, detectable command, e.g., akin to a particular designated state in a Turing machine). Let $\mathcal{S}$ be the family of decidable sets of (Gödel encodings of) Turing Machines, and let $\mathcal{T}$ be the family of decidable sets of programs in $\mathcal{L}$. Let $\mathfrak{F}$ be the equivalence relation on T defined by $\mathsf{p}\mathfrak{F}\mathsf{q}$ iff, for any input x, $\mathsf{p}(x)$ and $\mathsf{q}(x)$ send the same number of email. Note that this is not a complexity measure in the sense of Blum—the predicate "$\mathsf{p}(x)$ sends k emails" is not decidable due to Turing equivalence of $\mathcal{L}$ (e.g., p may diverge after sending $k - 1$ emails).

By Turing completeness, let `eval` $\in \mathcal{P}$ be a universal program takes as inputs (the representation of) a Turing Machine, M, and an integer n and returns the result of M on input n: $\phi_{\texttt{eval}}(M, n) = \phi_M(n)$. Moreover, we consider that `eval` never ever sends a single email. Consider the set of switching functions defined by

$$\pi_{\mathsf{p,q}}(M)(x) = \mathbf{r}(x) = \texttt{if eval(M, 0)=0 then p(x) else q(x)} \qquad (1)$$

Because `eval` never sends an email, $\pi_{\mathsf{p,q}}(M)\mathfrak{F}\mathsf{p}$ iff $\phi_M(0) = 0$ and $\pi_{\mathsf{p,q}}(M)\mathfrak{F}\mathsf{p}$ iff $\phi_M(0) \notin \{0, \perp\}$. These sets are REC-inseparable, and if F and G are the sets of all classes of $\mathfrak{R}$ and $\mathfrak{F}$, respectively, the F-G-switching family $(\pi_{\mathsf{p,q}})_{(\mathsf{p,q})\in\mathcal{P}\times\mathcal{P}}$ is thus REC-F-G-intricated with $\mathfrak{F}$.

To show intrication it is crucial that the evaluator does not send any emails. Thus, we need the argument of the evaluator (here, the Turing Machine) to live in a world where email cannot be sent, and trying to do the same thing with $S = T = \mathcal{P}$ would be more difficult. By Theorem 3, any decidable set of programs that does not send any emails must thus contain programs that send an *arbitrarily large* number of emails. That is, if the set contains all programs that does not send any emails, it must also contain "Spambots" that will constantly send emails. Similarly, and in the same prosaic vein, one can prove that a decidable set containing all programs that do not access the Internet will also contain botnets that constantly will contact external servers and receive orders from them.

6   Future Work

The generalisations of Rice's Theorem proved in the present paper provide (un)decidability results for sets containing (all) programs whose extension is in some subclass of the partial recursive functions (e.g., the set of polynomial-time computable functions). However, undecidability hinges on the language in question being Turing-complete; for some subrecursive classes, the set of "programs" becomes much easier to wield—e.g., the set of primitive recursive functions has an r.e. presentation (morally, a set of programs computing it) without semantic repetition, i.e., every program computes a different function[2], and thus clearly has decidable non-trivial extensional properties. The techniques of the present paper patently fail for such classes—we hope that more refined undecidability results, perhaps in the vein of [4], may be procured for these in the future.

A   Material Omitted from the Main Text

Proof (Full proof of intrication from Example 3). We prove that the switching family $(\pi_{a,b})_{(a,b)\in(\cup F)\times(\cup G)}$ is RE-F-G-intricated with $\backsim_{\mathsf{s}}$:

If $\mathsf{p} \in \cup F$, $\mathsf{q} \in \cup G$ and $T' \in \mathcal{T}$ such that $A_{\mathsf{p,q}} \subseteq \pi_{\mathsf{p,q}}^{-1}(T')$, there are two cases:

---

[2] See, e.g., [8] for an easily readable account.

- If $\phi_\mathsf{p} = \phi_\mathsf{q}$, then $A_{\mathsf{s},\mathsf{p}} = B_{\mathsf{s},\mathsf{q}}$, and there is clearly no set separating $A_{\mathsf{s},\mathsf{p}}$ from $B_{\mathsf{s},\mathsf{q}}$.
- If $\phi_\mathsf{p} \neq \phi_\mathsf{q}$, there are again two cases:
  - If r does not halt on n, then $\phi_{s_{r,n}} = \phi_\mathsf{p} = \phi_\mathsf{s} \in F$, and thus $(r,n) \in A_{\mathsf{p},\mathsf{q}}$. Conversely, if $(r,n) \in A_{\mathsf{p},\mathsf{q}}$, then $\phi_{s_{r,n}} = \phi_\mathsf{s}$, and as $\neg(\mathsf{q} \leftrightsquigarrow \mathsf{s})$, this implies that r does not halt on n.
  - If r halts on n, then $\phi_{s_{r,n}} \leftrightsquigarrow \phi_\mathsf{q}$. As $\mathsf{q} \in \cup G$, we have $\neg(\mathsf{q} \leftrightsquigarrow \mathsf{s})$, and thus $\neg(\phi_{s_{r,n}} \leftrightsquigarrow \mathsf{s})$, whence in particular $\phi_{s_{r,n}} \neq \phi_\mathsf{s}$, and hence $\phi_{s_{r,n}} \leftrightsquigarrow_\mathsf{s} \mathsf{q}$. Thus, $(r,n) \in B_{\mathsf{p},\mathsf{q}}$. Conversely, if $(r,n) \in B_{\mathsf{p},\mathsf{q}}$, $s_{r,n} \leftrightsquigarrow_\mathsf{s} \mathsf{q}$ which–as $\neg(\mathsf{q} \leftrightsquigarrow \mathsf{p})$– implies $\phi_{s_{r,n}} \neq \phi_\mathsf{s}$ and hence that r halts on n.

Thus $(r,n) \in A_{\mathsf{p},\mathsf{q}}$ iff r does not halt on n, and $(r,n) \in B_{\mathsf{p},\mathsf{q}}$ iff r halts on n. Hence, $A_{\mathsf{p},\mathsf{q}} \cup B_{\mathsf{p},\mathsf{q}} = \mathcal{P} \times \mathcal{N}$, and $A_{\mathsf{p},\mathsf{s}}$ are $B_{\mathsf{p},\mathsf{s}}$ is co-r.e.-complete and r.e.-complete, respectively. If there were an an r.e. set C such that $A_{\mathsf{p},\mathsf{q}} \subseteq C$ and $B_{a,b} \cap C = \emptyset$, then $A_{\mathsf{p},\mathsf{q}} \cup B_{\mathsf{p},\mathsf{q}} = \mathcal{P} \times \mathcal{N}$ implies $A_{a,b} = C$, whence $A_{a,b}$ is r.e., contradicting that it is co-r.e.-complete. $\qquad\square$

References

1. Asperti, A.: The intensional content of Rice's theorem. In: POPL 2008 (2008)
2. Bellantoni, S., Cook, S.: A new recursion-theoretic characterization of the polytime functions. Comput. Complex. **2**, 97–110 (1992)
3. Blum, M.: A machine-independent theory of the complexity of recursive functions. J. ACM **14**(2), 322–336 (1967)
4. Hoyrup, M.: The decidable properties of subrecursive functions. In: ICALP 2016, pp. 108:1–108:13 (2016)
5. Hoyrup, M.: Topological analysis of representations. In: Manea, F., Miller, R.G., Nowotka, D. (eds.) CiE 2018. LNCS, vol. 10936, pp. 214–223. Springer, Cham (2018). https://doi.org/10.1007/978-3-319-94418-0_22
6. Hoyrup, M., Rojas, C.: On the information carried by programs about the objects they compute. Theory Comput. Syst. **61**(4), 1214–1236 (2017)
7. Jones, N.D.: Computability and Complexity, from a Programming Perspective. MIT Press, Cambridge (1997)
8. Kahrs, S.: The primitive recursive functions are recursively enumerable. Technical report, University of Kent at Canterbury
9. Kirby, L., Paris, J.: Accessible independence results for peano arithmetic. Bull. Lond. Math. Soc. **14**(4), 285–293 (1982)
10. Kreisel, G., Lacombe, D., Shoenfield, J.: Fonctionelles recursivement définissables et fonctionnelles récursives. Comptes Rendus Hebdomadaires des Séances de l'Académie des Sciences **245**, 399–402 (1957)
11. Moschovakis, Y.: Recursive metric spaces. Fundamenta Mathematicae **55**(3), 215–238 (1964)
12. Moyen, J.Y., Simonsen, J.G.: More intensional versions of Rice's Theorem. In: Mazza, D. (ed.) Developments in Implicit Computational Complexity, DICE 2016, Eindhoven, Netherlands, p. 2 (2016)
13. Moyen, J., Simonsen, J.G.: Computability in the lattice of equivalence relations. In: Proceedings 8th Workshop on Developments in Implicit Computational Complexity and 5th Workshop on Foundational and Practical Aspects of Resource Analysis (DICE-FOPARA@ETAPS 2017), pp. 38–46 (2017)

14. Myhill, J.R., Shepherdson, J.C.: Effective operations on partial recursive functions. Zeitschrift für mathematische Logik und Grundlagen der Mathematik **1**, 310–317 (1955)
15. Ore, Ø.: Theory of equivalence relations. Duke Math. J. **9**(3), 573–627 (1942)
16. Papadimitriou, C.H.: Computational Complexity. Addison-Wesley, Boston (1994)
17. Rice, H.G.: Classes of recursively enumerable sets and their decision problems. Trans. Am. Math. Soc. **74**, 358–366 (1953)
18. Shapiro, N.: Degrees of computability. Trans. AMS **82**, 281–299 (1956)
19. Smullyan, R.M.: Undecidability and recursive inseparability. Zeitschrift für mathematische Logik und Grundlagen der Mathematik **4**(7–11), 143–147 (1958)
20. Tseitin, G.: Algorithmic operators in construtive complete separable metric spaces. Dokl. Akad. Nauk **128**, 49–52 (1959). (in Russian)
21. Tseitin, G.: Algorithmic operators in constructive metric spaces. Tr. Mat. Inst. Steklov **67**, 295–361 (1962). (in Russian. English translation. In: AMS Trans. 64 (1967))

On Approximate Uncomputability
of the Kolmogorov Complexity Function

Ruslan Ishkuvatov[1,2] and Daniil Musatov[1,3,4(✉)]

[1] Moscow Institute of Physics and Technology, Dolgoprudny, Russia
musatych@gmail.com
[2] LIRMM CNRS & University of Montpellier, Montpellier, France
[3] Russian Presidential Academy of National Economy and Public Administration,
Moscow, Russia
[4] Caucasus Mathematical Center at Adyghe State University, Maykop, Russia

Abstract. Kolmogorov complexity $C(x)$ of a string x is the length of
its shortest possible description. It is well known that $C(x)$ is not computable. Moreover, any computable lower estimate of $C(x)$ is bounded
by a constant. We study the following question: suppose that we want to
compute C with some precision and some amount of errors. For which
parameters is it possible? Our main result is the following: the error
must be at least an inverse exponential function of the precision. It
gives two striking implications. Firstly, no computable function approximate Kolmogorov complexity much better than the length function does.
Secondly, time-bounded Kolmogorov complexity is sufficiently far from
unbounded Kolmogorov complexity for any particular computable time
bound.

1 Introduction

Kolmogorov complexity $C(x)$ of a string x is the minimal possible length of a
program that generates x for some universal programming language. This notion
was introduced in the 1960s and since then the area was comprehensively studied.
There are many applications in computability theory, computational complexity,
machine learning, statistics etc. Extensive expositions of the subject may be
found in books by Li and Vitányi [6] and by Shen, Uspensky and Vereshchagin [8].

A simple observation similar to the Berry paradox shows that Kolmogorov
complexity is an uncomputable function. Moreover, there are no computable
and unbounded lower estimates. In this paper we study how far is it from being
computable. Firstly, we consider two well-known notions of being close to computable. The first one is generic computability: there exists a computable function that is defined almost everywhere and equals to our function on its domain.
The second one is coarse computability: there exists a totally computable function that coincides with ours almost everywhere. Then we relax the second notion

---

The reported study was funded by RFBR according to the research project 18-31-00428.

F. Manea et al. (Eds.): CiE 2019, LNCS 11558, pp. 230–239, 2019.
https://doi.org/10.1007/978-3-030-22996-2_20

by allowing a computable function to be close to the complexity function, not necessarily be equal.

The length function is a good approximation for a majority of strings. Specifically, there are two well-known results:

Lemma 1. *There is some constant c such that for any string x of length n it holds that $C(x) \leq n + c$.*

Lemma 2. *For any constant c there are less than 2^{n-c} strings of length n and complexity less than $n - c$.*

Thus, for all but a fraction 2^{-c} of strings it holds that $|\text{len}(x) - C(x)| \leq c$. Our main question is whether this estimate may be substantially improved by another computable function instead of $\text{len}(x)$. The main (and surprising) result is that the answer is negative: the fraction 2^{-c} may be replaced by a smaller constant, but it is still a constant for any constant c. Moreover, this exponential dependency is preserved for less accurate precisions, where the difference between a computable function and the complexity function grows superconstantly.

1.1 Related Work

Algorithmic properties of the complexity function were studied in many sources. Some basic properties are listed in [8, Sect. 1.2], [6, Sect. 2.7] and [7, Sect. 2.1]. The incomputability of $C(x)$ traces back to the pioneering works by Kolmogorov [4] and Solomonoff [9]. Kummer [5] studies algorithmic properties of the set of random strings (that is, the strings for which the complexity is not less then the length) and, in particular, proves that this set is not *frequency enumerable*. This means that there is no computable function f that gets k strings and guesses for at least one of them whether it is random or not. A similar in spirit result due to Beigel et al. [2] tells about complexity function itself: if an algorithm produces a list of numbers that is guaranteed to contain $C(x)$, then the length of this list must be linear. Bauwens et al. [1] show that, given a string x, it is possible to algorithmically produce a small list of programs, one of which generates x and is optimal up to a constant additive term. Unfortunately, this list is not short enough to make any implication about the complexity itself. Fenner and Fortnow [3] show that it is possible to produce a single optimal program for all strings of length n if the generator receives a piece of advice of length approximately n.

1.2 Roadmap

The rest of the text is organized as follows. In Sect. 2 we strictly define the notions we use and state some basic properties. In Sect. 3 we show that there is no coarsely computable function that approximates $C(x)$ with a constant precision. In Sect. 4 we prove our main result about the exact relationship between the precision of approximation and the number of errors. In Sect. 5 we give a brief conclusion and present some open questions.

2   Preliminaries

In this section we give precise definitions of the notions we use.

2.1   Kolmogorov Complexity

A simple definition "Kolmogorov complexity of a string is the length of a shortest program that generates this string" lacks the specification of a programming language. To address this issue, we consider the notion of a description method, or a decompressor. We define a more general notion of conditional complexity.

Definition 1. *Let D be a computable function with two arguments. The complexity $C_D(x|y)$ of a string $x \in \{0,1\}^*$ conditional on a string $y \in \{0,1\}^*$ with respect to a decompressor D is the minimal length of a string p such that $D(p,y) = x$. If there is no such p, then $C_D(x|y) = \infty$.*

For any particular string x, one may consider a decompressor that hardwires x and outputs it, say, on the empty program p. Thus, changing the decompressor may drastically change the complexity. The following celebrated theorem shows that this change is limited.

Theorem 1 (Kolmogorov-Solomonoff [4,9]**).** *There exists a decompressor U such that for any decompressor D there exists a constant c such that for all x and y it holds that $C_U(x|y) \leq C_D(x|y) + c$.*

Such machine is called a universal decompressor. The proof idea is simple: U treats a part of its first input as a description of D and launches it on the rest of the input. Usually a particular U is fixed and the index is omitted. Most equations are valid up to some additive constant. In the paper we consider two specific complexities:

Definition 2. *Let x be some string. Its unconditional complexity $C(x)$ is just $C(x|\epsilon)$, where ϵ is the empty string. The length conditional complexity of x is $C(x|n)$, where $n = |x|$.*

The idea behind length conditional complexity is the following. Complexity may be considered as a measure of information contained in a particular string. If a string is a prefix of a computable sequence, then its complexity equals the complexity of its length plus some constant. If a string is random, then its complexity is close to its length. Length condition makes the complexity of a computable string essentially zero, but keeps the complexity of a random string close to its length. Thus length condition helps to separate the information about the length of the string from the information in the string itself.

We consider also a time-bounded version of Kolmogorov complexity.

Definition 3. *Let D be a Turing machine. The complexity $C_D^t(x|y)$ of a string $x \in \{0,1\}^*$ conditional on y in time t with respect to decompressor D is the minimal length of a string p such that $D(p,y) = x$ and $D(p,y)$ halts in at most t steps. If there are no such p, then $C_D^t(x|y) = \infty$.*

For the time-bounded version the universal decompression theorem is the following:

Theorem 2. *There exists a decompressor U such that for any decompressor D there exist constants c and d such that for all x, y and t it holds that $C_U^{dt \log t}(x|y) \le C_D^t(x|y) + c$.*

It is clear that a universal decompressor in the time-bounded framework is also a universal decompressor in the unbounded framework. In the sequel we fix some such decompressor, and thus obtain $C^t(x|y) \ge C(x|y)$ for all x, y and t without adding a constant.

2.2 Generic and Coarse Computability

Apart from computable functions, one may consider functions that are computable "almost everywhere". This notion may be formalized in two non-equivalent ways. Both of them employ the notion of asymptotic density.

Definition 4. *Let $S \subset \{0,1\}^*$. The density of S in length n is defined as the fraction of length-n strings that lie in S, i.e., $\rho_n(S) = \frac{|S \cap \{0,1\}^n|}{2^n}$. If the sequence $\rho_n(S)$ has a limit $\rho(S)$, then it is called the asymptotic density of S. If $\rho(S) = 1$, we call S a generic set. If $\rho(S) = 0$, we call S a negligible set.*

Definition 5. *A total function $h\colon \{0,1\}^* \to \{0,1\}^*$ is called generic computable if there exists a partially computable function $f\colon \{0,1\}^* \to \{0,1\}^*$ such that if $f(x)$ is defined, then $f(x) = h(x)$, and the domain of f is a generic set.*

Definition 6. *A total function $h\colon \{0,1\}^* \to \{0,1\}^*$ is called coarsely computable if there exists a total computable function $f\colon \{0,1\}^* \to \{0,1\}^*$ such that the set $\{x \mid f(x) = h(x)\}$ is a generic set.*

3 Approximating Kolmogorov Complexity with Constant Accuracy

The following theorem shows that there does not exist generic computable lower estimate of $C(x)$ in any sense.

Theorem 3. *There is no partially computable function $f(x)$ such that $f(x) \le C(x)$ in the domain of f and $f(x)$ takes on arbitrarily large values.*

Proof. The standard argument is valid here. Suppose that such function f exists. From its definition, the set $\{x \mid f(x) > n\}$ is non-empty and computably enumerable. The complexity of the first element x_0 in an enumeration is at most $\log n + O(1)$. On the other hand, $C(x_0) \ge f(x_0) > n$, hence a contradiction for a sufficiently large n.

On the other hand, $C(x)$ does have coarsely computable lower estimates. For instance, $f(x) = \min\left\{\frac{|x|}{2}, C(x)\right\}$ is not greater than $C(x)$ and coincides with computable function $\frac{|x|}{2}$ with asymptotic density 1. So, the question arises how accurately can $C(x)$ be approximated by a coarsely computable function. It turns out that any computable superconstant accuracy can be achieved: just take $f(x) = \min\{|x| - \alpha(|x|), C(x)\}$ for any computable function $\alpha(\cdot)$ that tends to infinity. Such a function is not greater than $C(x)$ and coincides with $|x| - \alpha(|x|)$ with density $O(2^{-\alpha(|x|)})$ that tends to zero. We want to show that a constant accuracy cannot be achieved. We start from showing a similar fact about the length conditional complexity $C(x|n)$.

Theorem 4. *There is no coarsely computable function $f(x)$ such that for some constant c it holds that $C(x|n) - c \le f(x) \le C(x|n)$ for all x.*

Proof. Suppose that such $f(x)$ exists for some c. Consider a computable function h that coincides with f with asymptotic density 1. Take an arbitrary number $\delta = 2^{-k}$, where $k > 2$. From the definition, there must exist N such that for all $n > N$ the fraction of strings x of length n with different values $f(x)$ and $h(x)$ is less than δ. Now fix an arbitrary $n > N$. Consider a program P that gets n and enumerates all strings of length n in a specific order. The strings are ordered by $h(x)$ in a non-increasing manner, and then lexicographically. Denote by x_i the ith string in this enumeration. It is clear that $C(x_i|n) \le C(i) + O(1)$, where the constant in $O(1)$ depends on the program that computes h. From the properties of h and the pigeonhole principle, there must exist some $j \le 2^{n-k}$ such that $f(x_j) = h(x_j)$. We fix such x_j and estimate its complexity. On the one hand, it has low complexity: $C(x_j|n) \le C(j) + O(1) \le n - k + O(1)$. On the other hand, it must have high complexity. Specifically, for at least half of the strings it holds that $C(x_i|n) \ge n - 1$. Since $f(x) \ge C(x|n) - c$, for at least half of the strings it holds that $f(x_i) \ge n - c - 1$. Because $\delta < \frac{1}{4}$, for at least a quarter of the strings it holds simultaneously that $f(x_i) = h(x_i)$ and $f(x_i) \ge n - c - 1$. Since the output of P is ordered by the value of h in a non-increasing way, for all strings in the first quarter of the list it holds that $h(x_i) \ge n - c - 1$. Thus, for the previously fixed j it holds that $C(x_j|n) \ge f(x_j) \ge n - c - 1$. Now we obtain a contradiction: on the one hand, $C(x_j|n) \le n - k + O(1)$. On the other hand, $C(x_j|n) \ge n - c - 1$. Since c is fixed and k is arbitrary, we obtain a contradiction for large enough k.

This proof cannot be literally reproduced for the unconditional complexity, because a logarithmic term should be added to the upper bound, but not to the lower one, and with this addition the contradiction vanishes. Instead, we employ the following lemma that compares the complexities $C(x)$ and $C(x|n)$.

Lemma 3. *Suppose that $C(x|n) < n - k$. Then $C(x) \le n - k + O(\log k)$.*

Proof. Firstly, we modify the description method such that it obtains the following property: if $C(x|n) < n - k$, then there exists a description of length exactly $n - k$. This is done by the following: discard the leading zeros and the first one from a description and then launch a usual decompressor. In this case any number of zeros may be attached from the beginning without changing the output, and thus the desired property is satisfied.

Secondly, the description of x now consists of a description of k and a description of x conditional on n of length exactly $n - k$. By restoring k and adding it to the length of the description we may compute n and then obtain x. The total length of the description is $n - k + O(\log k)$, as claimed.

Now we are ready to expand the result to the case of unconditional complexity. It is not a direct corollary of Theorem 4, but can be obtained by a similar argument.

Theorem 5. *There is no coarsely computable function $f(x)$ such that for some constant c it holds that $C(x) - c \leq f(x) \leq C(x)$ for all x.*

Proof. The proof proceeds like the proof of Theorem 4. We suppose that such $f(x)$ exists, denote by $h(x)$ the respective computable function and define $\delta = 2^{-k}$ as before. Consider a large enough n and the enumeration of all strings of length n ordered by $h(x)$ in the non-increasing manner, and then lexicographically. As before, we denote by x_j the earliest x such that $f(x) = h(x)$. On the one hand, we have $f(x_j) \geq n - c - 1$ and thus $C(x_j) \geq n - c - 1$. On the other hand, we have $C(x_j|n) \leq n - k + O(1)$, as before. By Lemma 3 we get $C(x_j) \leq n - k + O(\log k)$. Since c is fixed and k is arbitrary, the two bounds contradict for large enough k.

Note that now we can modify the theorem to be symmetric:

Corollary 1. *There is no coarsely computable function $f(x)$ such that for some constant d it holds that $|C(x) - f(x)| \leq d$ for all x.*

Proof. If such function $f(x)$ exists, then $f(x) - d$ contradicts Theorem 5 for $c = 2d$.

One interesting corollary deals with computable upper estimates of $C(x)$.

Corollary 2. *Suppose that $t(n)$ is some total computable function. Then for any constant d and all sufficiently large n the density of $x \in \{0,1\}^n$ such that $C^{t(n)}(x) - C(x) \geq d$ is bounded away from zero.*

This is more than a direct application of Corollary 1, because we claim not only that the density does not tend to zero but also that it is bounded away from zero. This is why we repeat a part of the proof.

Proof. Suppose that, on the contrary, there exists a constant d and an increasing sequence n_m, such that the fraction of $x \in \{0,1\}^{n_m}$ such that $C^{t(n_m)}(x) - C(x) \leq d$ tends to zero. We repeat the argument from the proof of Theorem 5 for such values of n and $h(x) = C^{t(|x|)}(x)$. All $x \in \{0,1\}^n$ are sorted by h in a non-increasing order, then lexicographically. Among the first 2^{n-k} strings there must be x_j, such that $h(x_j) \leq C(x_j) + d$. On the one hand, $h(x_j)$ must be at least $n - 1$, thus $C(x_j) \geq n - d - 1$. On the other hand, $C(x_j|n) \leq n - k$ and a contradiction follows.

This corollary is very meaningful: giving more working time allows to substantially economize on the program length. It does not matter, how much time do you already have—polynomial, exponential, power of exponents etc.—allowing unlimited time may shorten a constant fraction of programs by more than a constant number of bits.

4   Measuring the Exact Accuracy

In this section we do a more precise analysis. We consider two parameters: a threshold on the difference between a complexity function and a totally computable function, and the fraction of strings that exceed this difference. Formally, we use the following definition:

Definition 7. *Let $F\colon \{0,1\}^* \to \mathbb{N}$ be some function, $d\colon \mathbb{N} \to \mathbb{N}$ be a total computable function, and $\alpha\colon \mathbb{N} \to (0,1)$ be another function. We say that F is* approximately computable *with precision $d(n)$ and error $\alpha(n)$ if there exists a total computable function $h\colon \{0,1\}^* \to \mathbb{N}$ such that for all large enough n the fraction of $x \in \{0,1\}^n$ satisfying $|F(x) - h(x)| > d(n)$ does not exceed $\alpha(n)$. If, moreover, it holds that $h(x) \geq F(x)$ for all x (resp., $h(x) \leq F(x)$), we call F* approximately computable from above *(resp.,* approximately computable from below*).*

Note that by modifying h in a finite number of arguments we may replace "for all large enough n" by "for all n". In these terms Theorems 4 and 5 may be restated as follows: for any constant $d \in \mathbb{N}$ and any function $\alpha(n) = o(1)$ the functions $C(x|n)$ and $C(x)$ are not approximately computable with precision d and error α. Now we consider the case of a non-constant precision.

4.1   Approximating Length Conditional Complexity

In this section we prove a generalized version of Theorem 4. We start by considering approximate computability from above. This leads to a corollary about approximation of the plain complexity by the time-bounded complexity.

Theorem 6. *Let $d(n) < n - \Omega(1)$ be a total computable function. Then $C(x|n)$ is not approximately computable from above with precision $d(n)$ and error $\alpha(n) = o(2^{-d(n)})$.*

Note that if $d(n) = n - O(1)$, then the theorem is wrong, at least for some choice of the decompressor. Indeed, we can think that $c_1 < C(x|n) < n + c_2$ for arbitrarily large $c_1 - c_2$. In this case $n + c_2$ is a computable upper bound. The length function is also an approximation with an arbitrary precision $d(n)$ and error $O(2^{-d(n)})$, so this bound is tight.

Proof. Let $h(x)$ be a computable approximation of $C(x|n)$ from above with precision $d(n)$ and error $\alpha(n) = o(2^{-d(n)})$. Take an arbitrary $\delta = 2^{-k}$ and n so large that $\alpha(n) < \delta \cdot 2^{-d(n)} = 2^{-d(n)-k}$. Consider the enumeration of

$\{0,1\}^n$ by $h(x)$ in a non-decreasing order, then lexicographically. Among the first $2^{n-d(n)-k}$ elements there must exist x_j such that $h(x_j) \leq C(x_j|n) + d(n)$. On the one hand, since $h(x) \geq C(x|n)$ and x_j is in the first half of the list, it must hold that $h(x_j) \geq n-1$, and thus $C(x_j|n) \geq n-d(n)-1$. On the other hand, $C(x_j|n) \leq n-d(n)-k+O(1)$, because x_j can be specified by its ordinal number in the enumeration. If k is large enough, we obtain a contradiction.

Now consider the general case of approximate computability.

Theorem 7. *Let $d(n) < \frac{n}{2} - \Omega(1)$ be a total computable function. Then $C(x|n)$ is not approximately computable with precision $d(n)$ and error $\alpha(n) = o(2^{-2d(n)})$.*

Again the claim is wrong for $d(n) = \frac{n}{2} - O(1)$. The function $\frac{n}{2} + c$ is a good approximation. Error $O(2^{-2d(n)})$ is achieved by a simple approximation $n - d(n)$, so this bound is also tight.

Proof. Let $h(x)$ be a computable approximation of $C(x|n)$ with precision $d(n)$ and error $\alpha(n) = o(2^{-2d(n)})$. Take an arbitrary $\delta = 2^{-k}$, $k > 2$, and take n so large that $\alpha(n) < \delta \cdot 2^{-2d(n)} = 2^{-2d(n)-k}$. Consider the enumeration of $\{0,1\}^n$ by $h(x)$ in a non-increasing order, then lexicographically. Among the first $2^{n-2d(n)-k}$ elements there must exist x_j such that $|h(x_j) - C(x_j|n)| \leq d(n)$. On the one hand, for at least half of x it holds that $C(x|n) \geq n-1$, thus for at least a quarter of x it holds that $h(x) \geq n-d(n)-1$. Since $k > 2$, the chosen x_j must lie in this quarter. We obtain $C(x_j|n) \geq h(x_j) - d(n) \geq n - 2d(n) - 1$. On the other hand, $C(x_j|n) \leq n - 2d(n) - k + O(1)$. If k is large enough, we obtain a contradiction.

Thus, two-sided approximation can be obtained with an error smaller than one-sided approximation (note that approximation from below cannot be made at all). But the order of this approximation is the same: for instance, logarithmic precision may be achieved for all but inverse polynomial fraction of strings.

4.2   Approximating Plain Complexity

In order to obtain a result about approximating plain complexity $C(x)$, we need to prove an analogue of Lemma 3. Direct application produces logarithmic discrepancies and leads to a weaker theorem. We slightly change the statement and employ the fact that $d(n)$ is computable.

Lemma 4. *Suppose that $C(x|n) < n-k$. Then $C(x) \leq n-k+C(n|n-k)+O(1)$.*

Proof. The proof proceeds along the lines of the proof of Lemma 3. At the last step we replace the description of k by a description of n conditional on $n - k$, that is sufficient to restore n.

When is the complexity $C(n|n - d(n))$ constant? For instance, if $d(n)$ is growing slow. Specifically, we use the following mild condition.

Definition 8. *Let $d: \mathbb{N} \to \mathbb{N}$ be a total computable function. Say that d grows uniformly slower than linearly if there exists a constant c such that for all n and m such that $m > n + c$ it holds that $d(m) - d(n) < m - n$.*

It is clear that "usual" functions, like logarithms, polylogarithmic functions, power functions n^δ for $\delta < 1$, linear functions αn for $\alpha < 1$, etc. all possess this property.

Lemma 5. *If d is totally computable and grows uniformly slower than linearly, then $C(n|n - d(n)) = O(1)$.*

Proof. Let K be some number and n_0 be the smallest n such that $n - d(n) = K$. If $m > n_0 + c$, then $m - d(m) > n_0 - d(n_0)$ from the properties of d. Thus, any n with $n - d(n) = K$ must lie in $[n_0, n_0 + c]$. Since c is a constant, one needs a constant number of bits to specify a particular n.

Now we are ready to prove the theorems.

Theorem 8. *Let $d(n) < n - \Omega(1)$ be a total computable function that grows uniformly slower than linearly. Then $C(x)$ is not approximately computable from above with precision $d(n)$ and error $\alpha(n) = o(2^{-d(n)})$.*

Theorem 9. *Let $d(n) < \frac{n}{2} - \Omega(1)$ be a total computable function such that $2d(n)$ grows uniformly slower than linearly. Then $C(x)$ is not approximately computable with precision $d(n)$ and error $\alpha(n) = o(2^{-2d(n)})$.*

Proof. The proofs proceed along the same lines up to the condition $C(x_j|n) \le n - d(n) - k + O(1)$ (resp., $C(x_j|n) \le n - 2d(n) - k + O(1)$). By applying Lemma 4, we get $C(x_j) \le n - d(n) - k + C(n|n - d(n) - k) + O(1) \le n - d(n) - k + C(n|n - d(n)) + C(k) + O(1) \le n - d(n) - k + O(\log k)$, where the last inequality employs Lemma 5. The contradiction for large enough k still holds.

As before, we obtain a proposition about time-bounded complexity:

Corollary 3. *Suppose that $t(n)$ is some computable function. Then for any computable function $d(n) = n - \Omega(1)$ that grows slower then linearly the density of x such that $C^{t(|x|)}(x) - C(x) \ge d(|x|)$ is $\Omega(2^{-d(|x|)})$.*

The proof combines the previously used techniques and thus is omitted. The informal meaning expands that of Corollary 2: allowing unlimited time may decrease the program length by at least logarithm for an inverse polynomial fraction of the strings, by at least $n^{-\delta}$ for the fraction at least $\Omega(\frac{1}{2^{n-\delta}})$ etc.

5   Conclusion

In this paper we introduced the notion of approximate computability and analyzed it for the case of Kolmogorov complexity function. Despite its naturalness, it seems to have never been appeared in the literature. It would be interesting to study the structural properties of approximately computable functions.

For instance, what can be done with an oracle that approximately computes $C(x)$ with sufficiently small precision and error? Can one then compute $C(x)$ exactly? What other functions that are not approximately computable may be constructed? Are there natural examples that do not deal with Kolmogorov complexity? How can these examples be classified? Is there any nice hierarchy?

Acknowledgments. We want to thank Alexei Milovanov and Alexander Shen for their support and advice during the work on this paper. We want to thank three anonymous referees for their useful comments about the text and the attendants of a seminar in LIRMM for their attention and questions.

References

1. Bauwens, B., Makhlin, A., Vereshchagin, N., Zimand, M.: Short lists with short programs in short time. In: 2013 IEEE Conference on Computational Complexity (CCC), pp. 98–108. IEEE (2013)
2. Beigel, R., et al.: Enumerations of the Kolmogorov function. J. Symb. Log. **71**(2), 501–528 (2006)
3. Fenner, S., Fortnow, L.: Compression complexity. arXiv preprint arXiv:1702.04779 (2017)
4. Kolmogorov, A.N.: Three approaches to the quantitative definition of information. Probl. Inf. Transm. **1**(1), 1–7 (1965)
5. Kummer, M.: On the complexity of random strings. In: Puech, C., Reischuk, R. (eds.) STACS 1996. LNCS, vol. 1046, pp. 25–36. Springer, Heidelberg (1996). https://doi.org/10.1007/3-540-60922-9_3
6. Li, M., Vitányi, P.M.B.: An Introduction to Kolmogorov Complexity and Its Applications. Texts in Computer Science, 3rd edn. Springer, New York (2008). https://doi.org/10.1007/978-0-387-49820-1
7. Nies, A.: Computability and Randomness, vol. 51. OUP Oxford (2009)
8. Shen, A., Uspensky, V.A., Vereshchagin, N.: Kolmogorov Complexity and Algorithmic Randomness, vol. 220. American Mathematical Society (2017)
9. Solomonoff, R.J.: A formal theory of inductive inference. Part I. Inf. control **7**(1), 1–22 (1964)

Borel and Baire Sets in Bishop Spaces

Iosif Petrakis$^{(\boxtimes)}$

University of Munich, Munich, Germany
petrakis@math.lmu.de

Abstract. We study the Borel sets `Borel`(F) and the Baire sets `Baire`(F) generated by a Bishop topology F on a set X. These are inductively defined sets of F-complemented subsets of X. Because of the constructive definition of `Borel`(F), and in contrast to classical topology, we show that `Baire`(F) = `Borel`(F). We define the uniform version of an F-complemented subset of X and we show the Urysohn lemma for them. We work within Bishop's system BISH* of informal constructive mathematics that includes inductive definitions with rules of countably many premises.

1 Introduction

The set of Borel sets generated by a given family of complemented subsets of a set X, with respect to a set Φ of real-valued functions on X, was introduced in [2], p. 68. This set is inductively defined and plays a crucial role in providing important examples of measure spaces in Bishop's measure theory developed in [2]. As this measure theory was replaced in [4] by the Bishop-Cheng measure theory, an enriched version of [3] that made no use of Borel sets, the Borel sets were somehow "forgotten" in the constructive literature.

In the introduction of [3], Bishop and Cheng explained why they consider their new measure theory "much more natural and powerful theory". They do admit though that some results are harder to prove (see [3], p. v). As it is also noted though, in [20], p. 25, the Bishop-Cheng measure theory is highly impredicative, while, although we cannot explain this here, Bishop's measure theory in [2] is highly predicative. This fact makes the original Bishop-Cheng measure theory hard to implement in some functional-programming language, a serious disadvantage from the computational point of view. This is maybe why, later attempts to develop constructive measure theory were done within an abstract algebraic framework (see [7,8,21].)

Despite the history of measure theory within Bishop-style constructive mathematics, the set of Borel sets is interesting on its own, and, as we try to show here, there are interesting interconnections between the theory of Bishop spaces and the notion of Borel sets. The notion of Bishop space, Bishop used the term function space for it, was also introduced by Bishop in [2], p. 71, as a constructive and function-theoretic alternative to the notion of a topological space. The notion of a least Bishop topology generated by a given set of function from X

© Springer Nature Switzerland AG 2019
F. Manea et al. (Eds.): CiE 2019, LNCS 11558, pp. 240–252, 2019.
https://doi.org/10.1007/978-3-030-22996-2_21

to $\mathbb{R}$, together with the set of Borel sets generated by a family of complemented subsets of X, are the main two inductively defined concepts found in [2]. The theory of Bishop spaces was not elaborated by Bishop, and it remained in oblivion, until Bridges and Ishihara revived the subject in [5] and [12], respectively. In [14–18] we tried to develop their theory.

This paper is the first step towards a systematic study of Borel sets and Baire sets, that we introduce here, in Bishop spaces. A Bishop topology is a set of real-valued functions on X, all elements of which are "a priori" continuous. The study of Borel and Baire sets within Bishop spaces is a constructive counterpart to the study of Borel and Baire algebras within topological spaces.

As it is indicated here, but needs to be elaborated further somewhere else, using complemented subsets to represent pairs of basic open sets and basic closed sets has as a result that some parts of the classical duality between open and closed sets in a topological space are recovered constructively. This reinforces our conviction that the notion of a complemented subset is one of the most important positive notions introduced by Bishop to overcome the difficulties that negatively defined concepts generate in constructive mathematics.

We work within Bishop's informal system of constructive mathematics BISH*, that is BISH together with inductive definitions with rules of countably many premises. Roughly speaking, [2] is within BISH*, while [3,4] are within BISH. A formal system for BISH* is Myhill's system CST* in [13], or CZF with dependence choice and some weak form of Aczel's regular extension axiom (see [1]).

All proofs that are not included here, are omitted as straightforward.

2 F-Complemented Subsets

A Bishop space is a constructive, function-theoretic alternative to the set-theoretic notion of topological space and a Bishop morphism is the corresponding notion of "continuous" function between Bishop spaces. In contrast to topological spaces, continuity of functions is a primitive notion and a concept of open set comes a posteriori. A Bishop topology on a set can be seen as an abstract and constructive approach to the ring of continuous functions $C(X)$ of a topological space X.

Definition 1. *A Bishop space is a couple $\mathcal{F} := (X, F)$, where X is an inhabited set (i.e., a set with a given element in it) and F is a subset of $\mathbb{F}(X)$, the set of all real-valued functions on X, such that the following conditions hold:*

(BS$_1$) *The set of constant functions $\mathrm{Const}(X)$ on X is included in F.*
(BS$_2$) *If $f, g \in F$, then $f + g \in F$.*
(BS$_3$) *If $f \in F$ and $\phi \in \mathrm{Bic}(\mathbb{R})$, then $\phi \circ f \in F$, where $\mathrm{Bic}(\mathbb{R})$ is the set of all Bishop-continuous functions from $\mathbb{R}$ to $\mathbb{R}$ i.e., of all functions that are uniformly continuous on every closed interval $[-n, n]$, where $n \geq 1$.*
(BS$_4$) *If $f \in \mathbb{F}(X)$ and $(g_n)_{n=1}^{\infty}$ such that $U(f, g_n, \frac{1}{n}) :\Leftrightarrow \forall_{x \in X}\left(|f(x) - g_n(x)| \leq \frac{1}{n}\right)$, for every $n \geq 1$, then $f \in F$.*

We call F a *Bishop topology* on X. If $\mathcal{G} := (Y, G)$ *is a Bishop space, a Bishop morphism from $\mathcal{F}$ to $\mathcal{G}$ is a function* $h : X \to Y$ *such that* $\forall_{g \in G}(g \circ h \in F)$. *We denote by* $\mathrm{Mor}(\mathcal{F}, \mathcal{G})$ *the set of Bishop morphisms from $\mathcal{F}$ to $\mathcal{G}$.*

It is easy to show (see [2], p. 71) that a Bishop topology F is an algebra and a lattice, where $f \vee g$ and $f \wedge g$ are defined pointwise, and if $a, b \in \mathbb{R}$, then $a \vee b := \max\{a, b\}$ and $a \wedge b := \min\{a, b\}$. Moreover, $\mathrm{Bic}(\mathbb{R})$ is a Bishop topology on $\mathbb{R}$, $\mathrm{Const}(X)$ and $\mathbb{F}(X)$ are Bishop topologies on X. If F is a Bishop topology on X, then $\mathrm{Const}(X) \subseteq F \subseteq \mathbb{F}(X)$, and $F^* := F \cap \mathbb{F}^*(X)$ is a Bishop topology on X, where $\mathbb{F}^*(X)$ denotes the bounded elements of $\mathbb{F}(X)$.

Definition 2. *Turning the definitional clauses* $(\mathrm{BS}_1) - (\mathrm{BS}_4)$ *into inductive rules one can define the least Bishop topology* $\bigvee F_0$ *on X that includes a given subset F_0 of $\mathbb{F}(X)$. In this case F_0 is called a subbase of F. A base of F is a subset B of F such that for every $f \in F$ there is a sequence $(g_n)_{n=1}^{\infty} \subseteq B$ such that* $\forall_{n \geq 1}\left(U\left(f, g_n, \frac{1}{n}\right)\right)$.
From now on, F denotes a Bishop topology on an inhabited set X and G a Bishop topology on an inhabited set Y. For simplicity, we denote the constant function on X with value $a \in \mathbb{R}$ also by a.

A complemented subset of X is a couple (A^1, A^0) of subsets of X such that every element of A^1 is "apart" from every element of A^0, where the apartness relation $x \neq y$ on a set X is a positive and stronger version of the negatively defined inequality $\neg(x =_X y)$. Here $x \neq y$ is defined through a given set of functions from X to $\mathbb{R}$. The induced apartness between A^1 and A^0 is a positive and stronger version of the negatively defined disjointness $A^1 \cap A^0 = \emptyset$.

Definition 3. *An inequality on X is a relation $x \neq y$ such that the following conditions are satisfied:*

(Ap_1) $\forall_{x,y \in X}\left(x =_X y \ \& \ x \neq y \Rightarrow \bot\right)$.
(Ap_2) $\forall_{x,y \in X}\left(x \neq y \Rightarrow y \neq x\right)$.
(Ap_3) $\forall_{x,y \in X}\left(x \neq y \Rightarrow \forall_{z \in X}(z \neq x \ \vee \ z \neq y)\right)$.

If $a, b \in \mathbb{R}$, we define $a \neq_{\mathbb{R}} b :\Leftrightarrow |a - b| > 0$. Usually, we write $a \neq b$ instead of $a \neq_{\mathbb{R}} b$. The inequality $x \neq_F y$ on X generated by F is defined by

$$x \neq_F y :\Leftrightarrow \exists_{f \in F}\left(f(x) \neq_{\mathbb{R}} f(y)\right).$$

A complemented subset of X with respect to $\neq_F$, or an F-complemented subset of X, is a pair $\boldsymbol{A} := (A^1, A^0)$ such that $\forall_{x \in A^1} \forall_{y \in A^0}(x \neq_F y)$. In this case we write $A^1][_F A^0$, and we denote their totality by $\mathcal{P}^{][_F}(X)$. The characteristic function of $\boldsymbol{A}$ is the map $\chi_{\boldsymbol{A}} : A^0 \cup A^1 \to \mathbf{2}$ defined by

$$\chi_{\boldsymbol{A}}(x) := \begin{cases} 1, & x \in A^1 \\ 0, & x \in A^0. \end{cases}$$

If $\boldsymbol{A}, \boldsymbol{B}$ are in $\mathcal{P}^{][_F}(X)$, then $\boldsymbol{A} = \boldsymbol{B} :\Leftrightarrow A^1 = B^1 \ \& \ A^0 = B^0$, and $\boldsymbol{A} \subseteq \boldsymbol{B} :\Leftrightarrow A^1 \subseteq B^1 \ \& \ B^0 \subseteq A^0$.

Clearly, $a \neq b \Leftrightarrow a > b \vee a < b$ and $A^1][_F A^0 \Rightarrow A^1 \cap A^0 = \emptyset$. If $A, B \subseteq \mathbb{R}$, the implication $A \cap B = \emptyset \Rightarrow A][_{\mathrm{Bic}(\mathbb{R})} B$ implies Markov's principle, hence it cannot be accepted in BISH*. To see this, take $A := \{x \in \mathbb{R} \mid \neg(x = 0)\}$ and $B := \{x \in \mathbb{R} \mid x = 0\}$. If $A][_{\mathrm{Bic}(\mathbb{R})} B$, then for every $x \in A$, there is some $\phi_x \in \mathrm{Bic}(\mathbb{R})$ such that $\phi_x(x) \neq \phi_x(0)$. Every element of $\mathrm{Bic}(\mathbb{R})$ though, is strongly extensional i.e., $\phi_x(x) \neq \phi_x(0) \Rightarrow x \neq 0$ (see Proposition 5.1.2 in [14], p. 102). Actually, we have that $\forall_{x,y \in \mathbb{R}}(x \neq_{\mathrm{Bic}(\mathbb{R})} y \Leftrightarrow x \neq_{\mathbb{R}} y)$. In this way we get $\forall_{x \in \mathbb{R}}(\neg(x = 0) \Rightarrow x \neq 0)$, which is equivalent to Markov's principle (see [6], p. 20).

Corollary 1. *If* $A, B \in \mathcal{P}][_F(X)$, *then*

$$A \cup B := (A^1 \cup B^1, A^0 \cap B^0) \ \& \ A \cap B := (A^1 \cap B^1, A^0 \cup B^0),$$

$$A - B := (A^1 \cap B^0, A^0 \cup B^1) \ \& \ -A := (A^0, A^1),$$

are F-complemented subsets of X.

Clearly, $-(-A)) = A$, and $A - B = A \cap (-B)$. In [3], p. 16, and in [4], p. 73, the "union" and the "intersection" of A and B are defined in a more complex way, so that their corresponding characteristic functions are given through the characteristic functions of A and B. Since here we do not use the characteristic functions of the complemented subsets, we keep the above simpler definitions given in [2], p. 66. If $A_{2n} := A$ and $A_{2n+1} := B$, for every $n \geq 1$, the definitions of $A \cup B$ and $A \cap B$ are special cases of the following definitions.

Corollary 2. *If* $(A_n)_{n=1}^{\infty} \subseteq \mathcal{P}][_F(X)$, *then*

$$\bigcup_{n=1}^{\infty} A_n := \left(\bigcup_{n=1}^{\infty} A_n^1, \bigcap_{n=1}^{\infty} A_n^0 \right) \ \& \ \bigcap_{n=1}^{\infty} A_n := \left(\bigcap_{n=1}^{\infty} A_n^1, \bigcup_{n=1}^{\infty} A_n^0 \right),$$

are F-complemented subsets of X. *Moreover,*

$$-\bigcap_{n=1}^{\infty} A_n = \bigcup_{n=1}^{\infty} (-A_n) \ \& \ -\bigcup_{n=1}^{\infty} A_n = \bigcap_{n=1}^{\infty} (-A_n).$$

Proposition 1. *If* $h \in \mathrm{Mor}(\mathcal{F}, \mathcal{G})$, $A, B \in \mathcal{P}][_G(Y)$, $(A_n)_{n=1}^{\infty} \subseteq \mathcal{P}][_F(Y)$, *let*

$$h^{-1}(A) := \left(h^{-1}(A^1), h^{-1}(A^0) \right).$$

(i) $h^{-1}(A) \in \mathcal{P}][_F(X)$.
(ii) $h^{-1}(A \cup B) = h^{-1}(A) \cup h^{-1}(B)$, *and* $h^{-1}(A \cap B) = h^{-1}(A) \cap h^{-1}(B)$.
(iii) $h^{-1}(-A) = -h^{-1}(A)$ *and* $h^{-1}(A - B) = h^{-1}(A) - h^{-1}(B)$.
(iv) $h^{-1}\left(\bigcup_{n=1}^{\infty} A_n \right) = \bigcup_{n=1}^{\infty} h^{-1}(A_n)$ *and* $h^{-1}\left(\bigcap_{n=1}^{\infty} A_n \right) = \bigcap_{n=1}^{\infty} h^{-1}(A_n)$.

Proof. (i) Let $x \in h^{-1}(A^1)$ and $y \in h^{-1}(A^0)$ i.e., $h(x) \in A^1$ and $h(y) \in A^0$. Let $g \in G$ such that $g(h(x)) \neq g(h(y))$. Hence, $g \circ h \in F$ and $(g \circ h)(x) \neq (g \circ h)(y)$. The rest of the proof is omitted as straightforward.

3   Borel Sets

The Borel sets in a topological space (X, T) is the least set of subsets of X that includes the open (or, equivalently the closed) sets in X and it is closed under countable unions, countable intersections and relative complements. The Borel sets in a Bishop space (X, F) is the least set of complemented subsets of X that includes the *basic F-complemented* (open-closed) subsets of X that are generated by F, and it is closed under countable unions and countable intersections (the closure under relative complements is redundant in the case of Bishop spaces). The next definition is Bishop's definition, given in [2], p. 68, restricted though, to Bishop topologies.

Definition 4. *An I-family of F-complemented subsets of X is an assignment routine λ that assigns to every $i \in I$ an F-complemented subset $\lambda(i)$ of X such that $\forall_{i,j \in I}\big(i =_I j \Rightarrow \lambda(i) =_{\mathcal{P}\mathrm{ll}_F(X)} \lambda(j)\big)$. An I-family of F-complemented subsets of X is called an I-set of complemented subsets of X, if $\forall_{i,j \in I}\big(\lambda(i) =_{\mathcal{P}\mathrm{ll}_F(X)} \lambda(j) \Rightarrow i =_I j\big)$. The set $\mathtt{Borel}(\lambda)$ of Borel sets generated by λ is defined inductively by the following rules:*

(Borel_1)
$$\frac{i \in I}{\lambda(i) \in \mathtt{Borel}(\lambda)}$$

(Borel_2)
$$\frac{B(1) \in \mathtt{Borel}(\lambda), B(2) \in \mathtt{Borel}(\lambda), \dots}{\bigcup_{n=1}^{\infty} B(n) \in \mathtt{Borel}(\lambda) \ \& \ \bigcap_{n=1}^{\infty} B(n) \in \mathtt{Borel}(\lambda)}.$$

In the induction principle $\mathrm{Ind}_{\mathtt{Borel}(\lambda)}$ associated to the definition of $\mathtt{Borel}(\lambda)$ we take P to be any formula in which the set $\mathtt{Borel}(F)$ does not occur.

$$\forall_{i \in I}\big(P(\lambda(i))\big) \ \& \ \forall_{\alpha:\mathbb{N} \to \mathcal{P}\mathrm{ll}_F(X)}\bigg[\forall_{n \geq 1}\big(\alpha(n) \in \mathtt{Borel}(\lambda) \ \& \ P(\alpha(n))\big) \Rightarrow$$

$$P\bigg(\bigcup_{n=1}^{\infty} \alpha(n)\bigg) \ \& \ P\bigg(\bigcap_{n=1}^{\infty} \alpha(n)\bigg)\bigg] \Rightarrow \forall_{B \in \mathtt{Borel}(\lambda)}\big(P(B)\big).$$

Let o_F, or simply o, be the F-family of the basic F-complemented subsets of X:

$$o_F(f) := \big([f > 0], [f \leq 0]\big),$$

$$[f > 0] := \{x \in X \mid f(x) > 0\}, \ [f \leq 0] := \{x \in X \mid f(x) \leq 0\}.$$

We write $\mathtt{Borel}(F) := \mathtt{Borel}(o_F)$ and we call its elements the Borel sets in F.

Proposition 2. *(i) If we keep the pointwise equality of functions as the equality of F, then the F-family o is not a set of F-complemented subsets of X.*
(ii) $o(1) = (X, \emptyset)$ and $o(-1) = (\emptyset, X)$.
(iii) If $f, g \in F$, then $o(f) \cup o(g) = o(f \vee g)$.
(iv) If $B \in \mathtt{Borel}(F)$, then $-B \in \mathtt{Borel}(F)$.

(v) There is a Bishop space (X, F) and some $f \in F$ such that $-o(f)$ is not equal to $o(g)$ for some $g \in F$.

(vi) $o(f) = o([f \vee 0] \wedge 1)$.

Proof. (i) and (ii) If $f \in F$, then $o(f) = o(2f)$, but $\neg(f = 2f)$. (ii) is trivial. (iii) This equality is implied from the following property for reals

$$a \vee b > 0 \Leftrightarrow a > 0 \vee b > 0 \, \& \, a \vee b \leq 0 \Leftrightarrow a \leq 0 \wedge b \leq 0.$$

(iv) If $a \in \mathbb{R}$, then $a \leq 0 \Leftrightarrow \forall_{n \geq 1}\left(a < \frac{1}{n}\right)$ and $a > 0 \Leftrightarrow \exists_{n \geq 1}\left(a \geq \frac{1}{n}\right)$, hence

$$-o(f) := ([f \leq 0], [f > 0])$$

$$= \left(\bigcap_{n=1}^{\infty} [(\frac{1}{n} - f) > 0], \bigcup_{n=1}^{\infty} [(\frac{1}{n} - f) \leq 0] \right)$$

$$:= \bigcap_{n=1}^{\infty} o(\frac{1}{n} - f) \in \mathtt{Borel}(F).$$

If $P(\boldsymbol{B}) := -\boldsymbol{B} \in \mathtt{Borel}(F)$, the above equality proves the first step of the corresponding induction on $\mathtt{Borel}(F)$. The rest of the inductive proof is easy. (v) Let the Bishop space $(\mathbb{R}, \mathrm{Bic}(\mathbb{R}))$. If we take $o(\mathrm{id}_{\mathbb{R}}) := ([x > 0], [x \leq 0])$, and if we suppose that $-o(\mathrm{id}_{\mathbb{R}}) := ([x \leq 0], [x > 0]) = ([\phi > 0], [\phi \leq 0]) =: o(\phi)$, for some $\phi \in \mathrm{Bic}(\mathbb{R})$, then $\phi(0) > 0$ and ϕ is not continuous at 0, which contradicts the fact that ϕ is uniformly continuous on every bounded subset of $\mathbb{R}$. (vi) The proof is based on basic properties of $\mathbb{R}$, like $a \wedge 1 = 0 \Rightarrow a = 0$.

Since $\mathtt{Borel}(F)$ is closed under intersections and complements, if $A, B \in \mathtt{Borel}(F)$, then $A - B \in \mathtt{Borel}(F)$. As Bishop remarks in [2], p. 69, the proof of Proposition 2 (iv) rests on the property of F that $\left(\frac{1}{n} - f\right) \in F$, for every $f \in F$ and $n \geq 1$. If we define similarly the Borel sets generated by any family of real-valued functions Θ on X, then we can find Θ such that $\mathtt{Borel}(\Theta)$ is closed under complements without satisfying the condition $f \in \Theta \Rightarrow \left(\frac{1}{n} - f\right) \in \Theta$. Such a family is the set $\mathbb{F}(X, 2)$ of all functions from X to $2 := \{0, 1\}$. In this case we have that

$$o_{\mathbb{F}(X,2)}(f) := ([f = 1], [f = 0]) \, \& - o_{\mathbb{F}(X,2)}(f) = o_{\mathbb{F}(X,2)}(1 - f).$$

Hence, the property mentioned by Bishop is sufficient, but not necessary. Constructively, we cannot show, in general, that $o(f) \cap o(g) = o(f \wedge g)$. If $f := \mathrm{id}_{\mathbb{R}} \in \mathrm{Bic}(\mathbb{R})$ and $g := -\mathrm{id}_{\mathbb{R}} \in \mathrm{Bic}(\mathbb{R})$, then $o(\mathrm{id}_{\mathbb{R}}) \cap o(-\mathrm{id}_{\mathbb{R}}) = ([x > 0] \cap [x < 0], [x \leq 0] \cup [-x \leq 0]) = (\emptyset, [x \leq 0] \cup [x \geq 0])$ Since $x \wedge (-x) = -|x|$, we get $o(\mathrm{id}_{\mathbb{R}} \wedge (-\mathrm{id}_{\mathbb{R}})) = o(-|x|) = (\emptyset, [|x| \geq 0])$. The supposed equality implies that $|x| \geq 0 \Leftrightarrow x \leq 0 \vee x \geq 0$. Since $|x| \geq 0$ is always the case, we get $\forall_{x \in \mathbb{R}}(x \leq 0 \vee x \geq 0)$, which implies LLPO (see [6], p. 20). If one add the condition $|f| + |g| > 0$, then $o(f) \cap o(g) = o(f \wedge g)$ follows constructively. The condition BS_4 in the definition of a Bishop space is crucial to the next proof.

Proposition 3. *If* $(f_n)_{n=1}^{\infty} \subseteq F$, *then* $f := \sum_{n=1}^{\infty} (f_n \vee 0) \wedge 2^{-n} \in F$ *and*

$$o(f) = \bigcup_{n=1}^{\infty} o(f_n) = \left(\bigcup_{n=1}^{\infty} [f_n > 0], \bigcup_{n=1}^{\infty} [f_n \leq 0] \right).$$

Proof. The function f is well-defined by the comparison test (see [4], p. 32). If $g_n := (f_n \vee 0) \wedge 2^{-n}$, for every $n \geq 1$, then

$$\left| \sum_{n=1}^{\infty} g_n - \sum_{n=1}^{N} g_n \right| = \left| \sum_{n=N+1}^{\infty} g_n \right| \leq \sum_{n=N+1}^{\infty} |g_n| \leq \sum_{n=N+1}^{\infty} \frac{1}{2^n} \xrightarrow{N} 0,$$

the sequence of the partial sums $\sum_{n=1}^{N} g_n \in F$ converges uniformly to f, hence by BS_4 we get $f \in F$. Next we show that $[f > 0] \subseteq \bigcup_{n=1}^{\infty} [f_n > 0]$. If $x \in X$ such that $f(x) > 0$, there is $N \geq 1$ such that $\sum_{n=1}^{N} g_n(x) > 0$. By Proposition (2.16) in [4], p. 26, there is $n \geq 1$ and $n \leq N$ with $g_n(x) > 0$, hence $(f_n(x) \vee 0) \geq g_n(x) > 0$, which implies $f_n(x) > 0$. For the converse inclusion, if $f_n(x) > 0$, for some $n \geq 1$, then $g_n(x) > 0$, hence $f(x) > 0$. To show $[f \leq 0] \subseteq \bigcup_{n=1}^{\infty} [f_n \leq 0]$, let $x \in X$ such that $f(x) \leq 0$, and suppose that $f_n(x) > 0$, for some $n \geq 1$. By the previous argument we get $f(x) > 0$, which contradicts our hypothesis $f(x) \leq 0$. For the converse inclusion, let $f_n(x) \leq 0$, for every $n \geq 1$, hence $f_n(x) \vee 0 = 0$ and $g_n(x) = 0$, for every $n \geq 1$. Consequently, $f(x) = 0$. $\square$

Proposition 4. *If* $h \in \mathrm{Mor}(\mathcal{F}, \mathcal{G})$ *and* $\boldsymbol{B} \in \mathrm{Borel}(\mathcal{G})$, *then* $h^{-1}(\boldsymbol{B}) \in \mathrm{Borel}(F)$.

Proof. By the definition of $h^{-1}(\boldsymbol{B})$ in Proposition 1, if $g \in G$, then

$$\begin{aligned}
h^{-1}(o_G(g)) &:= h^{-1}([g > 0], [g \leq 0]) \\
&:= \left(h^{-1}[g > 0], h^{-1}[g \leq 0] \right) \\
&= \left([(g \circ h) > 0], [(g \circ h) \leq 0] \right) \\
&:= o_F(g \circ h) \in \mathrm{Borel}(F).
\end{aligned}$$

If $P(\boldsymbol{B}) := h^{-1}(\boldsymbol{B}) \in \mathrm{Borel}(F)$, the above equality is the first step of the corresponding inductive proof on $\mathrm{Borel}(\mathcal{G})$. The rest of the inductive proof follows immediately from Proposition 1 (iv). $\square$

Definition 5. *If* $B \subseteq F$, *let* o_B *be the B-family of F-complemented subsets of* X *defined by* $o_B(f) := o_F(f)$, *for every* $f \in B$. *We denote by* $\mathrm{Borel}(B)$ *the set of Borel sets generated by* o_B.

If F_0 is a subbase of F, then, $\mathrm{Borel}(F_0) \subseteq \mathrm{Borel}(F)$. More can be said on the relation between $\mathrm{Borel}(B)$ and $\mathrm{Borel}(F)$, when B is a base of F.

Proposition 5. *Let* B *be a base of* F.

(i) *If for every* $f \in F$, $o_F(f) \in \mathrm{Borel}(B)$, *then* $\mathrm{Borel}(F) = \mathrm{Borel}(B)$.

(ii) If for every $g \in B$ and $f \in F$, $f \wedge g \in B$, then $\mathtt{Borel}(F) = \mathtt{Borel}(B)$.
(iii) If for every $g \in B$ and every $n \geq 1$, $g - \frac{1}{n} \in B$, then $\mathtt{Borel}(F) = \mathtt{Borel}(B)$.

Proof. (i) It follows by a straightforward induction on $\mathtt{Borel}(F)$.
(ii) and (iii) Let $(g_n)_{n=1}^{\infty} \subseteq B$ such that $\forall_{n \geq 1}\left(U(f, g_n, \frac{1}{n})\right)$. We have that

$$o_F(f) \subseteq \bigcup_{n=1}^{\infty} o_B(g_n) := \left(\bigcup_{n=1}^{\infty} [g_n > 0], \bigcap_{n=1}^{\infty} [g_n \leq 0] \right)$$

i.e., by Definition 3, $[f > 0] \subseteq \bigcup_{n=1}^{\infty}[g_n > 0]$ and $\bigcap_{n=1}^{\infty}[g_n \leq 0] \subseteq [f \leq 0]$; if $x \in X$ with $f(x) > 0$ there is $n \geq 1$ with $g_n(x) > 0$, and if $\forall_{n \geq 1}(g_n(x) \leq 0)$, then for the same reason $f(x)$ cannot be > 0, hence $f(x) \leq 0$.
Because of (i), for (ii) it suffices to show that $o_F(f) \in \mathtt{Borel}(B)$. We show that

$$o_F(f) = \bigcup_{n=1}^{\infty} o_B(f \wedge g_n) := \left(\bigcup_{n=1}^{\infty} [(f \wedge g_n) > 0], \bigcap_{n=1}^{\infty} [(f \wedge g_n) \leq 0] \right) \in \mathtt{Borel}(B).$$

If $f(x) > 0$, then we can find $n \geq 1$ such that $g_n(x) > 0$, hence $f(x) \wedge g_n(x) > 0$. Hence we showed that $[f > 0] \subseteq \bigcup_{n=1}^{\infty}[(f \wedge g_n) > 0]$. For the converse inclusion, let $x \in X$ and $n \geq 1$ such that $(f \wedge g_n)(x) > 0$. Then $f(x) > 0$ and $x \in [f > 0]$. If $f(x) \leq 0$, then $\forall_{n \geq 1}(f(x) \wedge g_n(x) \leq 0)$. Suppose next that $\forall_{n \geq 1}(f(x) \wedge g_n(x) \leq 0)$. If $f(x) > 0$, there is $n \geq 1$ with $g_n(x) > 0$, hence $f(x) \wedge g_n(x) > 0$, which contradict the hypothesis $f(x) \wedge g_n(x) \leq 0$. Hence $f(x) \leq 0$.
Because of (i), for (iii) it suffices to show that $o_F(f) \in \mathtt{Borel}(B)$. We show that

$$o_F(f) = \bigcup_{n=1}^{\infty} o_B\left(g_n - \frac{1}{n}\right) := \left(\bigcup_{n=1}^{\infty} \left[\left(g_n - \frac{1}{n}\right) > 0\right], \bigcap_{n=1}^{\infty} \left[\left(g_n - \frac{1}{n}\right) \leq 0\right] \right) \in \mathtt{Borel}(B).$$

First we show that $[f > 0] \subseteq \bigcup_{n=1}^{\infty} \left[\left(g_n - \frac{1}{n}\right) > 0\right]$. If $f(x) > 0$, there is $n \geq 1$ with $f(x) > \frac{1}{n}$, hence, since $-\frac{1}{2n} \leq g_{2n}(x) - f(x) \leq \frac{1}{2n}$, we get

$$g_{2n}(x) - \frac{1}{2n} \geq \left(f(x) - \frac{1}{2n}\right) - \frac{1}{2n} = f(x) - \frac{1}{n} > 0$$

i.e., $x \in \left[\left(g_{2n} - \frac{1}{2n}\right) > 0\right]$. For the converse inclusion, let $x \in X$ and $n \geq 1$ such that $g_n(x) - \frac{1}{n} > 0$. Since $0 < g_n(x) - \frac{1}{n} \leq f(x)$, we get $x \in [f > 0]$. Next we show that $[f \leq 0] \subseteq \bigcap_{n=1}^{\infty}\left[\left(g_n - \frac{1}{n}\right) \leq 0\right]$. Let $x \in X$ with $f(x) \leq 0$, and suppose that $n \geq 1$ with $g_n(x) - \frac{1}{n} > 0$. Then $0 \geq f(x) > 0$. By this contradiction we get $g_n(x) - \frac{1}{n} \leq 0$. For the converse inclusion let $x \in X$ such that $g_n(x) - \frac{1}{n} \leq 0$, for every $n \geq 1$, and suppose that $f(x) > 0$. Since we have already shown that $[f > 0] \subseteq \bigcup_{n=1}^{\infty}\left[\left(g_n - \frac{1}{n}\right) > 0\right]$, there is some $n \geq 1$ with $g_n(x) - \frac{1}{n} > 0$, which contradicts our hypothesis, hence $f(x) \leq 0$.

4   Baire Sets

One of the definitions[1] of the set of Baire sets in a topological space $(X, \mathcal{T})$, which was given by Hewitt in [11], is that it is the least σ-algebra of subsets of X that includes the zero sets of X i.e., the sets of the form $f^{-1}(\{0\})$, where $f \in C(X)$. Clearly, a Baire set in $(X, \mathcal{T})$ is a Borel set in $(X, \mathcal{T})$, and for many topological spaces, like the metrizable ones, the two classes coincide. In this section we adopt Hewitt's notion in Bishop spaces and the framework of F-complemented subsets.

Definition 6. Let ζ_F, or simply ζ, be the F-family of the F-zero complemented subsets of X:

$$\zeta_F(f) := ([f = 0], [f \neq 0]),$$

$$[f = 0] := \{x \in X \mid f(x) = 0\}, [f \neq 0] := \{x \in X \mid f(x) \neq 0\}.$$

We write $\mathtt{Baire}(F) := \mathtt{Borel}(\zeta_F)$ and we call its elements the Baire sets in F.

Since $a \neq 0 :\Leftrightarrow |a| > 0 \Leftrightarrow a < 0 \vee a > 0$, for every $a \in \mathbb{R}$, we get

$$\zeta_F(f) = ([f = 0], [|f| > 0]) = ([f = 0], [f > 0] \cup [f < 0]).$$

Proposition 6. (i) If we keep the pointwise equality of functions as the equality of F, then the F-family ζ is not a set of F-complemented subsets of X.
(ii) $\zeta(0) = (X, \emptyset)$ and $\zeta(1) = (\emptyset, X)$.
(iii) If $f, g \in F$, then $\zeta(f) \cap \zeta(g) = \zeta(|f| \vee |g|)$.
(iv) If $B \in \mathtt{Baire}(F)$, then $-B \in \mathtt{Baire}(F)$.
(v) There is a Bishop space (X, F) and some $f \in F$ such that $-\zeta(f)$ is not equal to $\zeta(g)$ for some $g \in F$.
(vi) $\zeta(f) = \zeta(|f| \wedge 1)$.

Proof. (i) and (ii) If $f \in F$, then $\zeta(f) = \zeta(2f)$, but $\neg(f = 2f)$. (ii) is trivial.
(iii) This equality is implied from the following property for reals

$$|a| \vee |b| = 0 \Leftrightarrow |a| = 0 \wedge |b| = 0 \,\&\, |a| \vee |b| \neq 0 \Leftrightarrow |a| > 0 \vee |b| > 0.$$

(iv) If $f \in F$, then $-\zeta(f) := ([f \neq 0], [f = 0])$. For every $n \geq 1$, let

$$g_n := \left(|f| \wedge \frac{1}{n}\right) - \frac{1}{n} \in F.$$

We show that

$$\bigcup_{n=1}^{\infty} \zeta(g_n) := \left(\bigcup_{n=1}^{\infty} [g_n = 0], \bigcap_{n=1}^{\infty} [g_n \neq 0]\right) = -\zeta(f) \in \mathtt{Baire}(F).$$

First we show that $[f \neq 0] = \bigcup_{n=1}^{\infty} [g_n = 0]$. If $|f(x)| > 0$, there is $n \geq 1$ such that $|f(x)| > \frac{1}{n}$, hence $|f(x)| \wedge \frac{1}{n} = \frac{1}{n}$, and $g_n(x) = 0$. For the converse inclusion, let

---

[1] A different definition is given in [10]. See [19] for the relations between these two definitions.

$x \in X$ and $n \geq 1$ such that $g_n(x) = 0 \Leftrightarrow |f(x)| \wedge \frac{1}{n} = \frac{1}{n}$, hence $|f(x)| \geq \frac{1}{n} > 0$. Next we show that $[f = 0] = \bigcap_{n=1}^{\infty} [g_n \neq 0]$. If $x \in X$ such that $f(x) = 0$, and $n \geq 1$, then $g_n(x) = -\frac{1}{n} < 0$. For the converse inclusion, let $x \in X$ such that for all $n \geq 1$ we have that $g_n(x) \neq 0$. If $|f(x)| > 0$, there is $n \geq 1$ such that $|f(x)| > \frac{1}{n}$, hence $g_n(x) = 0$, which contradicts our hypothesis. Hence, $|f(x)| \leq 0$, which implies that $|f(x)| = 0 \Leftrightarrow f(x) = 0$. If $P(B) := -B \in \mathtt{Baire}(F)$, the above equality proves the first step of the corresponding induction on $\mathtt{Baire}(F)$. The rest of the inductive proof is immediate[2].

(v) Let the Bishop space $(\mathbb{R}, \mathrm{Bic}(\mathbb{R}))$. If we take $\zeta(\mathrm{id}_\mathbb{R}) := ([x = 0], [x \neq 0])$, and if we suppose that $-\zeta(\mathrm{id}_\mathbb{R}) := ([x \neq 0], [x = 0]) = ([\phi = 0], [\phi \neq 0]) =: \zeta(\phi)$, for some $\phi \in \mathrm{Bic}(\mathbb{R})$, then $\phi(0) > 0 \vee \phi(0) < 0$ and $\phi(x) = 0$, if $x < 0$ or $x > 0$. Hence ϕ is not continuous at 0, which contradicts the fact that ϕ is uniformly continuous on every bounded subset of $\mathbb{R}$.

(vi) This proof is straightforward.

As in the case of basic Borel sets in F, we cannot show constructively that $\zeta(f) \cup \zeta(g) = \zeta(|f| \wedge |g|)$. If we add the condition $|f| + |g| > 0$ though, this equality is constructively provable.

Proposition 7. *If* $(f_n)_{n=1}^{\infty} \subseteq F$, *then* $f := \sum_{n=1}^{\infty} |f_n| \wedge 2^{-n} \in F$ *and*

$$\zeta(f) = \bigcap_{n=1}^{\infty} \zeta(f_n) = \left(\bigcap_{n=1}^{\infty} [f_n = 0], \bigcup_{n=1}^{\infty} [f_n \neq 0] \right).$$

Proof. Working as in the proof of Proposition 3, f is well-defined and if $g_n := |f_n| \wedge 2^{-n}$, for every $n \geq 1$, then the sequence of the partial sums $\sum_{n=1}^{N} g_n \in F$ converges uniformly to f, and by BS_4 we get $f \in F$. Since $f(x) = 0 \Leftrightarrow \forall_{\geq 1}(g_n(x) = 0) \Leftrightarrow \forall_{\geq 1}(f_n(x) = 0)$, we get $[f = 0] = \bigcap_{n=1}^{\infty} [f_n = 0]$. Next we show that $[f \neq 0] \subseteq \bigcup_{n=1}^{\infty} [f_n \neq 0]$. If $|f(x)| > 0$, then there is $N \geq 1$ such that $\sum_{n=1}^{N} g_n(x) > 0$. By Proposition (2.16) in [4], p. 26, there is some $n \geq 1$ and $n \geq N$ such that $g_n(x) > 0$, hence $|f_n(x)| \geq g_n(x) > 0$. The converse inclusion follows trivially.

Theorem 1. (i) *If* $B \in \mathtt{Baire}(F)$, *then* $B \in \mathtt{Borel}(F)$.
(ii) *If* $o(f) \in \mathtt{Baire}(F)$, *for every* $f \in F$, *then* $\mathtt{Baire}(F) = \mathtt{Borel}(F)$.
(iii) *If* $f \in F$, *then* $o(f) = -\zeta((-f) \wedge 0)$.
(iv) $\mathtt{Baire}(F^*) = \mathtt{Baire}(F) = \mathtt{Borel}(F) = \mathtt{Borel}(F^*)$.

Proof. (i) By Proposition 2(iv) $-o(f) = ([f \leq 0], [f > 0]) \in \mathtt{Borel}(F)$, for every $f \in F$, hence $-o(-f) = ([f \geq 0], [f < 0]) \in \mathtt{Borel}(F)$ too. Consequently

$$-o(f) \cap -o(-f) = ([f \leq 0] \cap [f \geq 0], [f > 0] \cup [f < 0]) = \zeta(f) \in \mathtt{Borel}(F).$$

---

[2] Hence, if we define the set of Baire sets over an arbitrary family Θ of functions from X to $\mathbb{R}$, a sufficient condition so that $\mathtt{Baire}(\Theta)$ is closed under complements is that Θ is closed under $|.|$, under wedge with $\frac{1}{n}$ and under subtraction with $\frac{1}{n}$, for every $n \geq 1$. If $\Theta := \mathbb{F}(X, 2)$, then $-o_{\mathbb{F}(X,2)}(f) = o_{\mathbb{F}(X,2)}(1 - f) = \zeta_{\mathbb{F}(X,2)}(f)$, hence by Proposition 4(ii) we get $\mathtt{Borel}(\mathbb{F}(X, 2)) = \mathtt{Baire}(\mathbb{F}(X, 2))$.

If $P(\boldsymbol{B}) := \boldsymbol{B} \in \mathrm{Borel}(F)$, the above equality is the first step of the corresponding inductive proof on $\mathrm{Baire}(F)$. The rest of the inductive proof is trivial.
(ii) The hypothesis is the first step of the obvious inductive proof on $\mathrm{Borel}(F)$, which shows that $\mathrm{Borel}(F) \subseteq \mathrm{Baire}(F)$. By (i) we get $\mathrm{Baire}(F) \subseteq \mathrm{Borel}(F)$.
(iii) We show that

$$\big([f > 0], [f \leq 0]\big) = \big([(-f) \wedge 0 \neq 0], [(-f) \wedge 0 = 0]\big).$$

First we show that $[f > 0] \subseteq [(-f) \wedge 0 \neq 0]$; if $f(x) > 0$, then $-f(x) \wedge 0 = -f(x) < 0$. For the converse inclusion, let $-f(x) \wedge 0 \neq 0 \Leftrightarrow -f(x) \wedge 0 > 0$ or $-f(x) \wedge 0 < 0$. Since $0 \geq -f(x) \wedge 0$, the first option is impossible. If $-f(x) \wedge 0 < 0$, then $-f(x) < 0$ or $0 < 0$, hence $f(x) > 0$. Next we show that $[f \leq 0] = [(-f) \wedge 0 = 0]$; since $f(x) \leq 0 \Leftrightarrow -f(x) \geq 0 \Leftrightarrow -f(x) \wedge 0 = 0$ (see [6], p. 52), the equality follows.
(iv) Clearly, $\mathrm{Baire}(F^*) \subseteq \mathrm{Baire}(F)$. By Proposition 6(vi) $\zeta(f) = \zeta(|f| \wedge 1)$, where $|f| \wedge 1 \in F^*$. Continuing with the obvious induction we get $\mathrm{Baire}(F) \subseteq \mathrm{Baire}(F^*)$. By case (iii) and Proposition 6(iv) we get $o(f) \in \mathrm{Baire}(F)$, hence by case (ii) we conclude that $\mathrm{Baire}(F) = \mathrm{Borel}(F)$. Clearly, $\mathrm{Borel}(F^*) \subseteq \mathrm{Borel}(F)$. By Proposition 2(vi) $o(f) = o((f \vee 0) \wedge 1)$, where $(f \vee 0) \wedge 1 \in F^*$. Continuing with the obvious induction we get $\mathrm{Borel}(F) \subseteq \mathrm{Borel}(F^*)$.

Either by definition, as in the proof of Proposition 4, or by Theorem 1(iii) and Proposition 4, if $h \in \mathrm{Mor}(\mathcal{F}, \mathcal{G})$ and $\boldsymbol{B} \in \mathrm{Baire}(G)$, then $h^{-1}(\boldsymbol{B}) \in \mathrm{Baire}(F)$.

5   Uniformly F-Complemented Subsets

Next follows the uniform version of the notion of an F-complemented subset.

Definition 7. *If $\boldsymbol{A} := (A^1, A^0) \in \mathcal{P}^{][_F}(X)$, we say that $\boldsymbol{A}$ is uniformly F-complemented, and we write $A^1 \neq_F A^0$, if*

$$\exists_{f \in F} \forall_{x \in A^1} \forall_{y \in A^0} \big(f(x) = 1 \ \& \ f(y) = 0\big).$$

Taking $(f \vee 0) \wedge 1$ we get $A^1 \neq_F A^0 \Leftrightarrow \exists_{f \in F} \big[0 \leq f \leq 1 \ \& \ \forall_{x \in A^1} \forall_{y \in A^0} \big(f(x) = 1 \ \& \ f(y) = 0\big)\big]$. In [3], p. 55, the following relation is defined:

$$\boldsymbol{A} \leq \boldsymbol{B} :\Leftrightarrow A^1 \subseteq B^1 \ \& \ A^0 \subseteq B^0.$$

If $A^1 \neq_F A^0$, then $\boldsymbol{A} \leq o(f)$. According to the classical Urysohn lemma for $C(X)$-zero sets, the disjoint zero sets of a topological space X are separated by some $f \in C(X)$ (see [9], p. 17). We show a constructive version of this, where disjointness is replaced by a stronger, but positively defined form of it.

Theorem 2 (Urysohn lemma). *If $\boldsymbol{A} := (A^1, A^0) \in \mathcal{P}^{][_F}(X)$, then*

$$A^1 \neq_F A^0 \Leftrightarrow \exists_{f,g \in F} \exists_{c>0} \big(\boldsymbol{A} \leq \zeta(f) \ \& \ -\boldsymbol{A} \leq \zeta(g) \ \& \ |f| + |g| \geq c\big).$$

Proof. ($\Rightarrow$) Let $h \in F$ such that $0 \leq h \leq 1$, $A^1 \subseteq [h = 1]$ and $A^0 \subseteq [h = 0]$. We take $f := 1 - h \in F$, $g := h$ and $c := 1$. First we show that $\boldsymbol{A} \leq \zeta(f)$. If $x \in A^1$, then $h(x) = 1$, and $f(x) = 0$. If $y \in A^0$, then $h(y) = 0$, hence $f(y) = 1$ and $y \in [f \neq 0]$. Next we show that $-\boldsymbol{A} \leq \zeta(g)$. If $y \in A^0$, then $h(y) = 0 = g(y)$. If $x \in A^1$, then $h(x) = 1 = g(y)$ i.e., $x \in [g \neq 0]$. If $x \in X$, then $1 = |1 - h(x) + h(x)| \leq |1 - h(x)| + |h(x)|$.
($\Leftarrow$) Let $h := 1 - \left(\frac{1}{c}|f| \wedge 1\right) \in F$. If $x \in A^1$, then $f(x) = 0$, and hence $h(x) = 1$. If $y \in A^0$, then $g(y) = 0$, hence $|f(y)| \geq c$, and consequently $h(y) = 0$.

The condition BS$_3$ is crucial to the next proof.

Corollary 3. *If* $\boldsymbol{A} := (A^1, A^0) \in \mathcal{P}^{][_F}(X)$ *and* $f \in F$, *then*

$$f(A^1) \neq_{\mathrm{Bic}(\mathbb{R})} f(A^0) \Rightarrow A^1 \neq_F A^0.$$

Proof. If $f(\boldsymbol{A}) := \left(f(A^1), f(A^0)\right)$ is uniformly Bic($\mathbb{R}$)-complemented, then by Urysohn lemma there are $\phi, \theta \in \mathrm{Bic}(\mathbb{R})$ and $c > 0$ with $f(\boldsymbol{A}) \leq \zeta(\phi)$, $-f(\boldsymbol{A}) \leq \zeta(\theta)$ and $|\phi| + |\theta| \geq c$. Consequently, $\boldsymbol{A} \leq \zeta(\phi \circ f)$, $-\boldsymbol{A} \leq \zeta(\theta \circ f)$ and $|\phi \circ f| + |\theta \circ f| \geq c$. Since by BS$_3$ we have that $\phi \circ f$ and $\theta \circ f \in F$, by the other implication of the Urysohn lemma we get $A^1 \neq_F A^0$.

Acknowledgment. This paper was written during my research visit to CMU that was funded by the EU-research project "Computing with Infinite Data". I would like to thank Wilfried Sieg for hosting me at CMU.

References

1. Aczel, P., Rathjen, M.: Constructive Set Theory, Book Draft (2010)
2. Bishop, E.: Foundations of Constructive Analysis. McGraw-Hill, New York (1967)
3. Bishop, E., Cheng, H.: Constructive Measure Theory, vol. 116. Memoirs of the American Mathematical Society (1972)
4. Bishop, E., Bridges, D.S.: Constructive Analysis. Grundlehren der mathematischen Wissenschaften, vol. 279. Springer, Heidelberg (1985). https://doi.org/10.1007/978-3-642-61667-9
5. Bridges, D.S.: Reflections on function spaces. Ann. Pure Appl. Log. **163**, 101–110 (2012)
6. Bridges, D.S., Vîţă, L.S.: Techniques of Constructive Analysis. Universitext. Springer, New York (2006). https://doi.org/10.1007/978-0-387-38147-3
7. Coquand, T., Palmgren, E.: Metric Boolean algebras and constructive measure theory. Arch. Math. Logic **41**, 687–704 (2002)
8. Coquand, T., Spitters, B.: Integrals and valuations. J. Log. Anal. **1**(3), 1–22 (2009)
9. Gillman, L., Jerison, M.: Rings of Continuous Functions. Van Nostrand (1960)
10. Halmos, P.R.: Measure Theory. Springer, New York (1974)
11. Hewitt, E.: Linear functionals on spaces of continuous functions. Fund. Math. **37**, 161–189 (1950)
12. Ishihara, H.: Relating Bishop's function spaces to neighborhood spaces. Ann. Pure Appl. Log. **164**, 482–490 (2013)
13. Myhill, J.: Constructive set theory. J. Symb. Log. **40**, 347–382 (1975)

14. Petrakis, I.: Constructive topology of Bishop spaces. Ph.D. thesis, LMU (2015)
15. Petrakis, I.: Completely regular Bishop spaces. In: Beckmann, Arnold, Mitrana, Victor, Soskova, Mariya (eds.) CiE 2015. LNCS, vol. 9136, pp. 302–312. Springer, Cham (2015). https://doi.org/10.1007/978-3-319-20028-6_31
16. Petrakis, I.: The Urysohn extension theorem for Bishop spaces. In: Artemov, Sergei, Nerode, Anil (eds.) LFCS 2016. LNCS, vol. 9537, pp. 299–316. Springer, Cham (2016). https://doi.org/10.1007/978-3-319-27683-0_21
17. Petrakis, I.: A constructive function-theoretic approach to topological compactness. In: Proceedings of the 31st Annual ACM-IEEEE Symposium on Logic in Computer Science (LICS 2016), 5–8 July 2016, pp. 605–614 (2016)
18. Petrakis, I.: Constructive uniformities of pseudometrics and Bishop topologies. J. Log. Anal. 42 p. (2019)
19. Ross, K.A., Stromberg, K.: Baire sets and Baire measures. Arkiv för Matematik 6(8), 151–160 (1965)
20. Spitters, B.: Constructive and intuitionistic integration theory and functional analysis. Ph.D. thesis, University of Nijmegen (2002)
21. Spitters, B.: Constructive algebraic integration theory. Ann. Pure Appl. Logic 137(1–3), 380–390 (2006)

Nets and Reverse Mathematics
Some Initial Results

Sam Sanders[⊠]

School of Mathematics, Leeds University, Leeds, UK
sasander@me.com

Abstract. Nets are generalisations of sequences involving possibly *uncountable* index sets; this notion was introduced about a century ago by Moore and Smith, together with the generalisation to nets of various basic theorems of analysis due to Bolzano-Weierstrass, Dini, Arzelà, and others. This paper deals with the Reverse Mathematics study of theorems about nets indexed by subsets of Baire space, i.e. part of third-order arithmetic. Perhaps surprisingly, over Kohlenbach's base theory of *higher-order* Reverse Mathematics, the Bolzano-Weierstrass theorem for nets and the unit interval implies the Heine-Borel theorem *for uncountable covers*. Hence, the former theorem is *extremely hard* to prove (in terms of the usual hierarchy of comprehension axioms), but also *unifies* the concepts of sequential and open-cover compactness. Similarly, Dini's theorem for nets is equivalent to the uncountable Heine-Borel theorem.

1 Introduction

Nets, also known as *Moore-Smith sequences*, are generalisations of sequences to possibly *uncountable* index sets; following the pioneering work of Moore ([24–26]) nets were introduced about a century ago by Moore and Smith ([27]). They also established the generalisation to nets of various basic theorems due to Bolzano-Weierstrass, Dini, and Arzelà ([27, Sects. 8 and 9]).

We deal with the *Reverse Mathematics* (RM hereafter) study of theorems about nets, based on the motivation provided by Remark 2.6. Since the study of nets treats uncountable objects as first-class citizens, we will work in Kohlenbach's *higher-order* RM (see Sect. 2.1), while our nets are exclusively indexed by subsets of Baire space, i.e. part of third-order arithmetic, as discussed in Sect. 3.1. We study the Bolzano-Weierstrass theorem in Sect. 3.2 and Dini's theorem in Sect. 3.3, and our main result is that each implies the Heine-Borel theorem for *uncountable* covers, while we obtain an equivalence in the case of Dini's theorem. By [35, Sect. 3], the minimal comprehension axiom needed to proved the uncountable Heine-Borel theorem is $(\exists^3)$ (see Sect. 2.2 for the latter), which implies full

This research was supported by the John Templeton Foundation grant *a new dawn of intuitionism* with ID 60842. Opinions expressed in this paper do not necessarily reflect those of the John Templeton Foundation. I thank Thomas Streicher and Anil Nerode for their valuable advice. Finally, I thank the three referees for their valuable suggestions that have significantly improved the paper.

F. Manea et al. (Eds.): CiE 2019, LNCS 11558, pp. 253–264, 2019.
https://doi.org/10.1007/978-3-030-22996-2_22

second-order arithmetic. Moreover, the aforementioned theorems thus falls (far) outside the Gödel hierarchy, as discussed in Sect. 4. It goes without saying that this paper constitutes a spin-off from the joint project with Dag Normann that has so far led to [33–36].

Finally, since sequences are a special case of nets, the Bolzano-Weierstrass theorem *for nets* thus unifies sequential and open-cover compactness in one elegant package simply by replacing the concept of sequence by that of net. The exact results are detailed in Sect. 3.2.

2   Preliminaries

We introduce *Reverse Mathematics* in Sect. 2.1, as well as its generalisation to *higher-order arithmetic*. We introduce some essential axioms in Sect. 2.2.

2.1   Reverse Mathematics

Reverse Mathematics is a program in the foundations of mathematics initiated around 1975 by Friedman ([11,12]) and developed extensively by Simpson ([41]). The aim of RM is to identify the minimal axioms needed to prove theorems of ordinary, i.e. non-set theoretical, mathematics. We refer to [43] for a basic introduction to RM and to [40,41] for an overview of RM. We expect basic familiarity with RM, but do sketch some aspects of Kohlenbach's *higher-order* RM ([20]) essential to this paper. Since the latter is officially a type theory, we discuss how we can represent subsets of Baire space in Definition 2.4. Furthermore, in contrast to 'classical' RM based on *second-order arithmetic*, higher-order RM makes use of the richer language of *higher-order arithmetic*, as follows.

As suggested by its name, higher-order arithmetic extends second-order arithmetic. Indeed, while the latter is restricted to natural numbers and sets of natural numbers, higher-order arithmetic also has sets of sets of natural numbers, sets of sets of sets of natural numbers, et cetera. To formalise this idea, we introduce the collection of *all finite types* $\mathbf{T}$, defined by the two clauses:

$$(i)\ 0 \in \mathbf{T} \quad \text{and} \quad (ii)\ \text{If } \sigma, \tau \in \mathbf{T} \text{ then } (\sigma \to \tau) \in \mathbf{T},$$

where 0 is the type of naturals, and $\sigma \to \tau$ is the type of mappings from objects of type σ to objects of type τ. In this way, $1 \equiv 0 \to 0$ is the type of functions from numbers to numbers, and where $n + 1 \equiv n \to 0$. Viewing sets as given by characteristic functions, we note that Z_2 only includes objects of type 0 and 1.

The language L_ω includes variables $x^\rho, y^\rho, z^\rho, \ldots$ of any finite type $\rho \in \mathbf{T}$. Types may be omitted when they can be inferred from context. The constants of L_ω includes the type 0 objects $0, 1$ and $<_0, +_0, \times_0, =_0$ which are intended to have their usual meaning as operations on $\mathbb{N}$. Equality at higher types is defined in terms of '$=_0$' as follows: for any objects x^τ, y^τ, we have

$$[x =_\tau y] \equiv (\forall z_1^{\tau_1} \ldots z_k^{\tau_k})[xz_1 \ldots z_k =_0 yz_1 \ldots z_k], \tag{2.1}$$

if the type τ is composed as $\tau \equiv (\tau_1 \to \ldots \to \tau_k \to 0)$. Furthermore, L_ω includes the *recursor constant* $\mathbf{R}_\sigma$ for any $\sigma \in \mathbf{T}$, which allows for iteration on type σ-objects as in the special case (2.2). Formulas and terms are defined as usual.

Definition 2.1. The base theory RCA_0^ω consists of the following axioms.

1. Basic axioms expressing that $0, 1, <_0, +_0, \times_0$ form an ordered semi-ring with equality $=_0$.
2. Basic axioms defining the well-known Π and Σ combinators (aka K and S in [2]), which allow for the definition of λ-*abstraction*.
3. The defining axiom of the recursor constant $\mathbf{R}_0$: For m^0 and f^1:

$$\mathbf{R}_0(f, m, 0) := m \text{ and } \mathbf{R}_0(f, m, n+1) := f(\mathbf{R}_0(f, m, n)). \qquad (2.2)$$

4. The *axiom of extensionality*: for all $\rho, \tau \in \mathbf{T}$, we have:

$$(\forall x^\rho, y^\rho, \varphi^{\rho \to \tau})\big[x =_\rho y \to \varphi(x) =_\tau \varphi(y)\big]. \qquad (\mathsf{E}_{\rho,\tau})$$

5. The induction axiom for quantifier-free[1] formulas of L_ω.
6. $\mathsf{QF\text{-}AC}^{1,0}$: The quantifier-free Axiom of Choice as in Definition 2.2.

Definition 2.2. The axiom $\mathsf{QF\text{-}AC}$ consists of the following for all $\sigma, \tau \in \mathbf{T}$:

$$(\forall x^\sigma)(\exists y^\tau)A(x, y) \to (\exists Y^{\sigma \to \tau})(\forall x^\sigma)A(x, Y(x)), \qquad (\mathsf{QF\text{-}AC}^{\sigma,\tau})$$

for any quantifier-free formula A in the language of L_ω.

By [20, Sect. 2], RCA_0^ω and RCA_0 prove the same sentences 'up to language' as the latter is set-based and the former function-based. Recursion as in (2.2) is called *primitive recursion*; the class of functionals obtained from $\mathbf{R}_\rho$ for all $\rho \in \mathbf{T}$ is called *Gödel's system* T of all (higher-order) primitive recursive functionals.

We use the usual notations for natural, rational, and real numbers, and the associated functions, as introduced in [20, pp. 288–289].

Definition 2.3 (Real numbers and related notions in RCA_0^ω)

1. Natural numbers correspond to type zero objects, and we use 'n^0' and '$n \in \mathbb{N}$' interchangeably. Rational numbers are defined as signed quotients of natural numbers, and '$q \in \mathbb{Q}$' and '$<_\mathbb{Q}$' have their usual meaning.
2. Real numbers are coded by fast-converging Cauchy sequences $q_{(\cdot)} : \mathbb{N} \to \mathbb{Q}$, i.e. such that $(\forall n^0, i^0)(|q_n - q_{n+i}| <_\mathbb{Q} \frac{1}{2^n})$. We use Kohlenbach's 'hat function' from [20, p. 289] to guarantee that every f^1 defines a real number.
3. We write '$x \in \mathbb{R}$' to express that $x^1 := (q_{(\cdot)}^1)$ represents a real as in the previous item and write $[x](k) := q_k$ for the k-th approximation of x.
4. Two reals x, y represented by $q_{(\cdot)}$ and $r_{(\cdot)}$ are *equal*, denoted $x =_\mathbb{R} y$, if $(\forall n^0)(|q_n - r_n| \leq \frac{1}{2^{n-1}})$. Inequality '$<_\mathbb{R}$' is defined similarly. We sometimes omit the subscript '$\mathbb{R}$' if it is clear from context.

---

[1] To be absolutely clear, variables (of any finite type) are allowed in quantifier-free formulas of the language L_ω: only quantifiers are banned.

5. Functions $F : \mathbb{R} \to \mathbb{R}$ are represented by $\Phi^{1 \to 1}$ mapping equal reals to equal reals, i.e. $(\forall x, y \in \mathbb{R})(x =_{\mathbb{R}} y \to \Phi(x) =_{\mathbb{R}} \Phi(y))$.
6. The relation '$x \leq_\tau y$' is defined as in (2.1) but with '$\leq_0$' instead of '$=_0$'. Binary sequences are denoted '$f^1, g^1 \leq_1 1$', but also '$f, g \in C$' or '$f, g \in 2^{\mathbb{N}}$'.
7. Sets of type ρ objects $X^{\rho \to 0}, Y^{\rho \to 0}, \ldots$ are given by their characteristic functions $f_X^{\rho \to 0}$, i.e. $(\forall x^\rho)[x \in X \leftrightarrow f_X(x) =_0 1]$, where $f_X^{\rho \to 0} \leq_{\rho \to 0} 1$.

The following special case of item 7 is singled out, as it will be used frequently. Indeed, our notion of net introduced in Sect. 3.1 (only) involves index sets restricted to subsets of Baire space.

Definition 2.4 [RCA$_0^\omega$]. A 'subset D of $\mathbb{N}^{\mathbb{N}}$' is given by its characteristic function $F_D^2 \leq_2 1$, i.e. we write '$f \in D$' for $F_D(f) = 1$ for any $f \in \mathbb{N}^{\mathbb{N}}$. A 'binary relation $\preceq$ on a subset D of $\mathbb{N}^{\mathbb{N}}$' is given by the associated characteristic function $G_\preceq^{(1 \times 1) \to 0}$, i.e. we write '$f \preceq g$' for $G_\preceq(f, g) = 1$ and any $f, g \in D$. Assuming extensionality on the reals as in item 5, we obtain characteristic functions that represent subsets of $\mathbb{R}$ and relations thereon. Using pairing functions, it is clear we can also represent sets of finite sequences (of reals), and relations thereon.

Finally, we mention the highly useful **ECF**-interpretation.

Remark 2.5 (The ECF -interpretation). The technical definition of ECF may be found in [47, p. 138, Sect. 2.6]. Intuitively speaking, the ECF-interpretation $[A]_{\mathsf{ECF}}$ of a formula $A \in L_\omega$ is just A with all variables of type two and higher replaced by countable representations of continuous functionals. Such representations are also (equivalently) called 'associates' or 'codes' (see [19, Sect. 4]). The ECF-interpretation connects RCA$_0^\omega$ and RCA$_0$ (see [20, Propostion 3.1]) in that if RCA$_0^\omega$ proves A, then RCA$_0$ proves $[A]_{\mathsf{ECF}}$, again 'up to language', as RCA$_0$ is formulated using sets, and $[A]_{\mathsf{ECF}}$ is formulated using types, namely only using type zero and one objects.

2.2   Some Axioms of Higher-Order RM

We introduce some functionals which constitute the counterparts of second-order arithmetic Z_2, and some of the Big Five systems, in higher-order RM. We assume basic familiarity with the aforementioned second-order systems, while the exact definitions can be found in [41]. We use the formulation of the higher-order systems from [20,35]. First of all, ACA$_0$ is readily derived from:

$$(\exists \mu^2)(\forall f^1)\big[(\exists n)(f(n) = 0) \to [f(\mu(f)) = 0 \wedge (\forall i < \mu(f))f(i) \neq 0] \quad (\mu^2)$$
$$\wedge \, [(\forall n)(f(n) \neq 0) \to \mu(f) = 0]\big],$$

and ACA$_0^\omega \equiv$ RCA$_0^\omega + (\mu^2)$ proves the same sentences as ACA$_0$ by [16, Theorem 2.5]. The (unique) functional μ^2 in (μ^2) is called *Feferman's μ* ([2]), and is *discontinuous* at $f =_1 11\ldots$; in fact, (μ^2) is equivalent to the existence of $F : \mathbb{R} \to \mathbb{R}$ such that $F(x) = 1$ if $x >_{\mathbb{R}} 0$, and 0 otherwise ([20, Sect. 3]), and to

$$(\exists \varphi^2 \leq_2 1)(\forall f^1)\big[(\exists n)(f(n) = 0) \leftrightarrow \varphi(f) = 0\big]. \quad (\exists^2)$$

Secondly, $\Pi_1^1\text{-CA}_0$ is readily derived from the following sentence:

$$(\exists S^2 \leq_2 1)(\forall f^1)\big[(\exists g^1)(\forall n^0)(f(\bar{g}n) = 0) \leftrightarrow S(f) = 0\big], \qquad (S^2)$$

and $\Pi_1^1\text{-CA}_0^\omega \equiv \text{RCA}_0^\omega + (S^2)$ proves the same Π_3^1-sentences as $\Pi_1^1\text{-CA}_0$ by [38, Theorem 2.2]. The (unique) functional S^2 in (S^2) is also called *the Suslin functional* ([20]). By definition, the Suslin functional S^2 can decide whether a Σ_1^1-formula (as in the left-hand side of (S^2)) is true or false. We similarly define the functional S_k^2 which decides the truth or falsity of Σ_k^1-formulas; we also define the system $\Pi_k^1\text{-CA}_0^\omega$ as $\text{RCA}_0^\omega + (S_k^2)$, where (S_k^2) expresses that S_k^2 exists. Note that we allow formulas with *function* parameters, but **not** *functionals* here. In fact, Gandy's *Superjump* ([14]) constitutes a way of extending $\Pi_1^1\text{-CA}_0^\omega$ to parameters of type two.

Thirdly, second-order arithmetic Z_2 is readily derived from $\cup_k \Pi_k^1\text{-CA}_0^\omega$, or from:

$$(\exists E^3 \leq_3 1)(\forall Y^2)\big[(\exists f^1)Y(f) = 0 \leftrightarrow E(Y) = 0\big], \qquad (\exists^3)$$

and we therefore define $Z_2^\Omega \equiv \text{RCA}_0^\omega + (\exists^3)$ and $Z_2^\omega \equiv \cup_k \Pi_k^1\text{-CA}_0^\omega$, which are conservative over Z_2 by [16, Corollary 2.6]. Despite this close connection, Z_2^ω and Z_2^Ω can behave quite differently, as discussed in e.g. [35, Sect. 2.2]. The functional from $(\exists^3)$ is also called '$\exists^3$', and we use the same convention for other functionals.

Finally, the Heine-Borel theorem (aka *Cousin's lemma* [10, p. 22]) states the existence of a finite sub-cover for an open cover of certain spaces. Now, a functional $\Psi : \mathbb{R} \to \mathbb{R}^+$ gives rise to the *canonical* cover $\cup_{x \in I} I_x^\Psi$ for $I \equiv [0, 1]$, where I_x^Ψ is the open interval $(x - \Psi(x), x + \Psi(x))$. Hence, the uncountable cover $\cup_{x \in I} I_x^\Psi$ has a finite sub-cover by the Heine-Borel theorem; in symbols:

$$(\forall \Psi : \mathbb{R} \to \mathbb{R}^+)(\exists \langle y_1, \ldots, y_k \rangle)(\forall x \in I)(\exists i \leq k)(x \in I_{y_i}^\Psi). \qquad (\text{HBU})$$

By the results in [35, 36], Z_2^Ω proves HBU but $\Pi_k^1\text{-CA}_0^\omega + \text{QF-AC}^{0,1}$ cannot, and many basic properties of the *gauge integral* ([28, 44]) are equivalent to HBU.

We finish this section with a motivation for the study of nets in RM.

Remark 2.6. First of all, nets were introduced about a century ago ([26, 27]) in mathematics, and therefore should count as 'ordinary mathematics' in Simpson's sense, as discussed in [41, I.1]. Secondly, filters are studied in the RM of topology (see e.g. [29–31]), and it is well-known that nets and filters provide an equivalent framework (see [3] and Remark 3.6). Thirdly, the weak-$*$-topology is studied in RM (see [41, X.2] for an overview) and this topology has an elegant equivalent formulation in terms of nets by [8, Theorem 3.1].

3   Main Results

We provide a brief introduction to nets in Sect. 3.1. We study the Bolzano-Weierstrass theorem for nets and the unit interval in Sect. 3.2. We study Dini's theorem for nets of functions in Sect. 3.3. In each case, we obtain HBU from Sect. 2.2, and even an equivalence in the latter case. As discussed in Sect. 3.1, we shall only study nets indexed by subsets of Baire space, where the latter are defined in Definition 2.4 in the context of RCA_0^ω.

3.1   Introducing Nets

We introduce the notion of net and associated concepts. We first consider the following standard definition (see e.g. [17, Chap. 2]).

Definition 3.1 [Nets]. A set $D \neq \emptyset$ with a binary relation '$\preceq$' is *directed* if

a. The relation $\preceq$ is transitive, i.e. $(\forall x, y, z \in D)([x \preceq y \wedge y \preceq z] \rightarrow x \preceq z)$.
b. The relation $\preceq$ is reflexive, i.e. $(\forall x \in D)(x \preceq x)$.
c. For $x, y \in D$, there is $z \in D$ such that $x \preceq z \wedge y \preceq z$.

For such $(D, \preceq)$ and topological space X, any $x : D \rightarrow X$ is a *net* in X. To emphasise the similarity with sequences, we write 'x_d' in the stead of $x(d)$.

The relation '$\preceq$' is often not explicitly mentioned; we also write $d_1, \ldots, d_k \succeq d$ as short for $(\forall i \leq k)(d_i \succeq d)$. In this paper, we shall only study nets indexed by subsets of Baire space, as defined in Definition 2.4.

The definitions of convergence and increasing net are of course familiar.

Definition 3.2 [Convergence of nets]. If x_d is a net in X, we say that x_d *converges* to the limit $\lim_d x_d = y \in X$ if for every neighbourhood U of y, there is $d_0 \in D$ such that for all $e \succeq d_0$, $x_e \in U$.

Definition 3.3 [Increasing nets]. A net x_d in $I \equiv [0, 1]$ is *increasing* if $a \preceq b$ implies $x_a \leq_{\mathbb{R}} x_b$ for all $a, b \in D$.

Definition 3.4 [Sub-nets]. A *sub-net* of a net x_d with directed set D, is a net y_b with directed set B such that there is a function $\phi : B \rightarrow D$ such that:

a. the function ϕ satisfies $y_b = x_{\phi(b)}$,
b. $(\forall d \in D)(\exists b_0 \in B)(\forall b \succeq b_0)(\phi(b) \succeq d)$.

Finally, we point out that $\mathbb{N}$ with its usual ordering yields a directed set, i.e. convergence results about nets *always* apply to sequences and *can* even yield the associated convergence results for sequences; the latter are studied in e.g. [41, III.2] as part of classical RM. An example is provided in the proof of Theorem 3.5 where the monotone convergence theorem *for sequences* is shown to follow from the Bolzano-Weierstrass theorem *for nets*.

3.2   The Bolzano-Weierstrass Theorem for Nets

We study the Bolzano-Weierstrass theorem for nets, $\mathsf{BW}_{\mathsf{net}}$ for short, i.e. the statement that every net in the unit interval (indexed by subsets of Baire space) has a convergent sub-net. This theorem is one of the standard results pertaining to nets, and can even be found in mathematical physics (see [37, p. 98]). As discussed in the first paragraph of the proof of Theorem 3.5, $\mathsf{BW}_{\mathsf{net}}$ implies the sequential compactness of the unit interval, but also the Heine-Borel compactness *for uncountable covers* by Theorem 3.5. Hence, nets provide a 'unified' approach to compactness that captures both sequential and (uncountable) open-cover compactness.

Theorem 3.5. *The system* $\mathsf{RCA}_0^\omega + \mathsf{BW}_{\mathsf{net}}$ *proves* HBU.

Proof. First of all, $\mathsf{BW}_{\mathsf{net}}$ implies the monotone convergence theorem *for sequences*, i.e. we have access to ACA_0 by [41, III.2.2]. Indeed, a monotone sequence in $[0, 1]$ is also a net in $[0, 1]$, and has a convergent sub-*net* by $\mathsf{BW}_{\mathsf{net}}$. It is a straightforward checking of definitions that the original monotone sequence also converges (in the sense of sequences) to the limit of the sub-net.

We shall provide two proofs of HBU, one in case $(\exists^2)$ and one in case $\neg(\exists^2)$. The law of excluded middle (as in $(\exists^2) \vee \neg(\exists^2)$) then finishes the proof.

First of all, in case $\neg(\exists^2)$, all functions on $\mathbb{R}$ are continuous by [20, Proposition 3.12], and HBU reduces to WKL by [19, Sect. 4]. Indeed, if $\Psi : I \to \mathbb{R}^+$ is continuous, then the associated canonical cover has a countable sub-cover given by the rationals in the unit interval. By [41, IV.1], WKL is equivalent to the Heine-Borel theorem for *countable* covers of the unit interval.

Secondly, we now prove HBU in case $(\exists^2)$. To this end, suppose $\neg\mathsf{HBU}$ and fix some $\Psi : I \to \mathbb{R}^+$ for which $\cup_{x \in I} I_x^\Psi$ does not have a finite sub-cover. Let D be the set of all finite sequences of reals in the unit interval, and define '$v \preceq w$' for $w, v \in D$ if $\cup_{i < |v|} I_{v(i)}^\Psi \subseteq \cup_{i < |w|} I_{w(i)}^\Psi$, i.e. the cover generated by w is 'bigger' than the cover associated to v. Note that $(\exists^2)$ suffices to define the relation $\preceq$. Clearly, the latter is transitive and reflexive, and our assumption $\neg\mathsf{HBU}$ also implies item (c) in Definition 3.1. To define a net, consider the following formula:

$$(\forall w^{1^*} \in [0, 1])(\exists q \in \mathbb{Q} \cap [0, 1])(\underline{q \notin \cup_{i < |w|} I_{w(i)}^\Psi}), \tag{3.1}$$

which again holds by assumption. Note that the underlined formula in (3.1) is decidable thanks to $(\exists^2)$. Applying $\mathsf{QF\text{-}AC}^{1,0}$ to (3.1), we obtain a net x_w in $[0, 1]$, which has a convergent (say to $y_0 \in I$) sub-net $y_b = x_{\phi(d)}$ for some directed set B and $\phi : B \to D$, by $\mathsf{BW}_{\mathsf{net}}$.

By definition, the neighbourhood $U_0 = I_{y_0}^\Psi$ contains all y_b for $b \succeq b_1$ for some $b_1 \in B$. However, taking $d = \langle y_0 \rangle$ in the second item in Definition 3.4, there is also $b_0 \in B$ such that $(\forall b \succeq b_0)(\phi(b) \succeq \langle y_0 \rangle)$. By the definition of '$\preceq$', $\phi(b)$ is hence such that $\cup_{i < |\phi(b)|} I_{\phi(b)(i)}^\Psi$ contains U_0, for any $b \succeq b_0$. Now use item (c) from Definition 3.1 to find $b_2 \in B$ satisfying $b_2 \succeq b_0$ and $b_2 \succeq b_1$. Hence, $y_{b_2} = x_{\phi(b_2)}$ is in U_0, but $\cup_{i < |\phi(b_2)|} I_{\phi(b_2)(i)}^\Psi$ also contains U_0, i.e. $x_{\phi(b_2)}$ must be *outside* of U_0 by the definition of x_w, a contradiction. In this way, we obtain HBU in case $(\exists^2)$, and we are done. □

We cannot expect a reversal in the previous theorem, as $\mathsf{BW}_{\mathsf{net}}$ implies ACA_0 by [41, III.2.2], while $\mathsf{RCA}_0^\omega + \mathsf{HBU}$ is conservative over WKL_0, which readily follows from applying the ECF-translation from Remark 2.5. We finish this section with a conceptual remark on nets.

Remark 3.6 (Filters versus nets). For completeness, we discuss the connection between filters and nets. By [3, Proposition 3.4], a topological space X is compact if and only if every *filter base* has a *refinement* that converges to a point of X.

Whatever the meaning of the italicised notions, the similarity with the Bolzano-Weierstrass theorem for nets is obvious, and not a coincidence: for every net $\mathfrak{r}$, there is an associated filter base $\mathfrak{B}(\mathfrak{r})$ such that if the erstwhile converges, so does the latter to the same point; one similarly associates a net to a given filter base with the same convergence properties (see [3, Sect. 2]). Hence, one can reformulate $\mathsf{BW}_{\mathsf{net}}$ using filters and obtain the same result as in Theorem 3.5.

3.3 Dini's Theorem for Nets

We study Dini's theorem *for nets*, which is found in [1, 4, 17, 21, 27, 32, 45, 46, 49].

First of all, we introduce some notation: a net $f_d : (D \times I) \to \mathbb{R}$ is *increasing* if $a \preceq b$ implies $f_a(x) \leq_{\mathbb{R}} f_b(x)$ for all $x \in I$ and $a, b \in D$. It goes without saying that properties of $f_d(x)$ like continuity pertain to the variable x, while the net is indexed by $d \in D$. We (do) say that the net $f_d : (D \times I) \to \mathbb{R}$ converges *uniformly* if the net $\lambda d.f_d(x)$ converges *and* $d_0 \in D$ as in Definition 3.2 is independent of the choice of $x \in I$, i.e. the usual definition.

Secondly, by Corollary 3.9, the following version of Dini's theorem for nets is equivalent to HBU. Note the verbatim replacement of 'sequence' by 'net'.

Definition 3.7 [$\mathsf{DIN}_{\mathsf{net}}$]. For continuous $f_d : (D \times I) \to \mathbb{R}$ forming an increasing net and converging to continuous $f : I \to \mathbb{R}$, the convergence is uniform.

Recall that we only consider nets indexed by subsets of Baire space.

Theorem 3.8. *The system* $\mathsf{RCA}_0^\omega + \mathsf{DIN}_{\mathsf{net}}$ *proves* HBU.

Proof. The 'classical' Dini theorem (for sequences) is equivalent to WKL by [7, Theorem 21], and thus have access to the latter lemma. We proceed as in the proof of Theorem 3.5, i.e. using $(\exists^2) \vee \neg(\exists^2)$. First of all, in case $\neg(\exists^2)$, all functions on $\mathbb{R}$ are continuous by [20, Proposition 3.12], and HBU reduces to WKL by [19, Sect. 4], as in the proof of Theorem 3.5

Secondly, we prove HBU in case $(\exists^2)$. Suppose $\neg$HBU and let $\Psi : I \to \mathbb{R}^+$ be such that the canonical cover does not have a finite sub-cover. Now let D be the set of finite sequences of reals in I and define '$v \preceq w$' for $w, v \in D$ if

$$(\forall i < |v|)(\exists j < |w|)(v(i) = w(j)) \wedge \cup_{i<|v|} I_{v(i)}^\psi \subseteq \cup_{i<|w|} I_{w(i)}^\psi.$$

Clearly, $\preceq$ is transitive and our assumption $\neg$HBU yields item (c) in Definition 3.1. Now define $f_w : I \to \mathbb{R}$ as follows: If $w = \langle x \rangle$ for some $x \in I$, then f_w is 0 outside of I_x^Ψ, while inside the latter, $f_w(x)$ is the piecewise linear function that is 1 at x, and 0 in $x \pm \Psi(x)$. If w is not a singleton, then $f_w(x) = \max_{i<|w|} f_{\langle w(i) \rangle}(x)$. Moreover, f_w is also increasing (in the sense of nets) and converges to the constant one function (in the sense of nets), as for any $v \succeq \langle x \rangle$, we have $f_v(x) = 1$. The crucial property of the net $f_w(x)$ is of course that:

$$(\forall x \in I)(f_w(x) >_{\mathbb{R}} 0 \leftrightarrow x \in \cup_{i<|w|} I_{w(i)}^\Psi). \tag{3.2}$$

Now apply $\mathsf{DIN}_{\mathsf{net}}$ and conclude that the convergence is uniform. Hence, for $\varepsilon = 1/2$, there is w_0 such that for all $x \in I$, we have $f_{w_0}(x) > 0$. However, the latter implies that every $x \in I$ is in $\cup_{i<|w_0|} I_{w_0(i)}^{\Psi}$ by (3.2), i.e. we found a finite sub-cover. This contradiction yields HBU and we are done. $\square$

Since Dini's theorem is equivalent to WKL in classical RM ([7]), we expect the following result in Corollary 3.9.

Corollary 3.9. *The system* $\mathsf{ACA}_0^\omega + \mathsf{QF\text{-}AC}^{1,1}$ *proves* HBU $\leftrightarrow$ DIN$_{\mathsf{net}}$.

Proof. We only have to prove the forward direction. As in the usual proof of Dini's theorem, we may assume that the net f_d is decreasing and converges pointwise to the constant zero function. Fix $\varepsilon_0 > 0$ and apply $\mathsf{QF\text{-}AC}^{1,1}$ to $(\forall z \in I)(\exists d \in D)(0 \le f_d(z) < \varepsilon_0)$, to obtain $\Phi^{1\to 1}$ yielding $d \in D$ from $z \in I$. Since $\lambda x.f_{\Phi(z)}(x)$ is continuous for any fixed z, $(\exists^2)$ yields a modulus of continuity g as in the proof of [19, Proposition 4.7], as follows:

$$(\forall \varepsilon > 0)(\forall x, y \in I)(|x - y| < g(x, \varepsilon, z) \to |f_{\Phi(z)}(x) - f_{\Phi(z)}(y)| < \varepsilon), \qquad (3.3)$$

and this for all for all $z \in I$. Define $\Psi : I \to \mathbb{R}^+$ as $\Psi(x) := g(x, \varepsilon_0, x)$ and note that $(0 \le f_{\Phi(x)}(y) < \varepsilon_0)$ for all $y \in I_x^\Psi$ by (3.3) and the definition of Φ. Now let $y_1, \ldots, y_k$ be the associated finite sub-cover provided by HBU. By item (c) of Definition 3.1, there is $d_0 \in D$ such that $d_0 \succeq \Phi(y_i)$ for all $i \le k$. Since f_d is a decreasing net and $[0,1] \subset \cup_{i \le k} I_{y_i}^\Psi$, we have $(0 \le f_d(y) < \varepsilon_0)$ for all $y \in I$ and $d \succeq d_0$, i.e. uniform convergence as required. $\square$

In the previous proof, using the continuity properties of the functions in the net, one can get by with $\mathsf{QF\text{-}AC}^{0,1}$, but the latter does seem essential. Moreover, using the above 'excluded middle' trick, one could omit $(\exists^2)$, as nets (essentially) reduce to sequences if all functions on Baire space are continuous.

4   Nets and the Gödel Hierarchy

We discuss the foundational implications of our results, esp. as they pertain to the *Gödel hierarchy*. Now, the latter is a collection of logical systems ordered via consistency strength. This hierarchy is claimed to capture most systems that are natural or have foundational import, as follows.

> *It is striking that a great many foundational theories are linearly ordered by <. Of course it is possible to construct pairs of artificial theories which are incomparable under <. However, this is not the case for the "natural" or non-artificial theories which are usually regarded as significant in the foundations of mathematics.* ([42])

Burgess and Koellner corroborate this claim in [9, Sect. 1.5] and [18, Sect. 1.1]. The Gödel hierarchy is a central object of study in mathematical logic, as e.g. argued by Simpson in [42, p. 112] or Burgess in [9, p. 40]. Precursors to the Gödel hierarchy may be found in the work of Wang ([48]) and Bernays (see [5,6]). Friedman ([13]) studies the linear nature of the Gödel hierarchy in detail. Moreover, the Gödel hierarchy exhibits some remarkable robustness: we can perform the following modifications and the hierarchy remains essentially unchanged:

1. Instead of the ordering via consistency strength, we can order via inclusion: Simpson claims that inclusion and consistency strength yield the same[2] Gödel hierarchy as depicted in [42, Table 1]. Some exceptional (semi-natural) statements[3] do fall outside of the Gödel hierarchy based on inclusion.
2. We can replace the systems with their higher-order (eponymous but for the 'ω') counterparts. The higher-order systems are generally conservative over their second-order counterpart for (large parts of) the second-order language.

Now, *if* one accepts the modifications (inclusion ordering and higher types) described in the previous two items, *then* an obvious question is where HBU fits into the (inclusion-based) Gödel hierarchy. Indeed, the Heine-Borel theorem has a central place in analysis and a rich history predating set theory (see e.g. [22]).

The answer to this question may come as a surprise: starting with the results in [35, 36], Dag Normann and the author have identified a *large* number of *natural* theorems of third-order arithmetic, including HBU, forming a branch *independent* of the medium range of the Gödel hierarchy based on inclusion. Indeed, none of the systems Π_k^1-CA_0^ω + QF-$\mathsf{AC}^{0,1}$ can prove HBU, while Z_2^Ω can. We stress that both Π_k^1-CA_0^ω + QF-$\mathsf{AC}^{0,1}$ and HBU are part of the language of *third-order arithmetic*, i.e. expressible in the same language.

Results pertaining to 'local-global' theorems are obtained in [36], while the results pertaining to HBU and the gauge integral may be found in [35]. In this paper, we have shown that a number of basic theorems about nets similarly fall outside of the Gödel hierarchy. The Bolzano-Weierstrass and Dini theorems for nets are old and well-established, starting with [26, 27].

Finally, the aforementioned results highlight a fundamental difference between second-order and higher-order arithmetic. Such differences are discussed in detail in [39, Sect. 4], based on helpful discussion with Steve Simpson, Denis Hirschfeldt, and Anil Nerode.

References

1. Aliprantis, C.D., Border, K.C.: Infinite Dimensional Analysis: A Hitchhiker's Guide, 3rd edn. Springer, Berlin (2006). https://doi.org/10.1007/3-540-29587-9
2. Avigad, J., Feferman, S.: Gödel's functional ("Dialectica") interpretation. In: Handbook of Proof Theory. Stud. Logic Found. Math. **137**, 1998, pp. 337–405
3. Bartle, R.G.: Nets and filters in topology. Am. Math. Mon. **62**, 551–557 (1955)
4. Bartle, R.G.: On compactness in functional analysis. Trans. Am. Math. Soc. **79**, 35–57 (1955)
5. Benacerraf, P., Putnam, H.: Philosophy of Mathematics: Selected Readings, 2nd edn. Cambridge University Press, Cambridge (1984)

---

[2] Simpson mentions in [42] the caveat that e.g. PRA and WKL_0 have the same first-order strength, but the latter is strictly stronger than the former.

[3] There are some examples of theorems (predating HBU and [35]) that fall outside of the Gödel hierarchy *based on inclusion*, like *special cases* of Ramsey's theorem and the axiom of determinacy from set theory ([15, 23]). These are far less natural than e.g. Heine-Borel compactness, in our opinion.

6. Bernays, P.: Sur le Platonisme Dans les Mathématiques. L'EnseignementMathématique **34**, 52–69 (1935)
7. Berger, J., Schuster, P.: Dini's theorem in the light of reverse mathematics. In: Lindström, S., Palmgren, E., Segerberg, K., Stoltenberg-Hansen, V. (eds.) Logicism, Intuitionism, and Formalism. Synthese Library, vol. 341, pp. 153–166. Springer, Dordrecht (2009). https://doi.org/10.1007/978-1-4020-8926-8_7
8. Brace, W.J.: Almost uniform convergence. Portugal. Math. **14**, 99–104 (1956)
9. Burgess, P.J.: Fixing Frege: Princeton Monographs in Philosophy. Princeton University Press, Princeton (2005)
10. Cousin, P.: Sur les fonctions de n variables complexes. Acta Math. **19**, 1–61 (1895)
11. Friedman, H.: Some systems of second order arithmetic and their use. In: Proceedings of the International Congress of Mathematicians (Vancouver, B.C. 1974), vol. 1, pp. 235–242 (1975)
12. Friedman, H.M.: Systems of second order arithmetic with restricted induction, I & II (Abstracts). J. Symb. Log. **41**, 557–559 (1976)
13. Friedman, H.M.: Interpretations, according to Tarski. In: The Nineteenth Annual Tarski Lectures on Interpretations of Set Theory in Discrete Mathematics and Informal Thinking, vol. 1, pp. 42 (2007). http://u.osu.edu/friedman.8/files/2014/01/Tarski1052407-13do0b2.pdf
14. Gandy, R.: General recursive functionals of finite type and hierarchies of functions. Ann. Fac. Sci. Univ. Clermont-Ferrand **35**, 5–24 (1967)
15. Hirschfeldt, D.R.: Slicing the Truth. Lecture Notes Series, vol. 28, Institute for Mathematical Sciences, National University of Singapore, World Scientific Publishing (2015)
16. Hunter, J.: Higher-order reverse topology, ProQuest LLC, Ann Arbor, MI. Ph.D. thesis - The University of Wisconsin - Madison (2008)
17. Kelley, J.L.: General Topology: Graduate Texts in Mathematics, vol. 27. Springer, New York (1975). Reprint of the 1955 edition
18. Koellner, P.: Large Cardinals and Determinacy, The Stanford Encyclopedia of Philosophy (2014). https://plato.stanford.edu/archives/spr2014/entries/large-cardinals-determinacy/
19. Kohlenbach, U.: Foundational and mathematical uses of higher types. In: Reflections on the Foundations of Mathematics 2002. Lecture Notes in Logic, vol. 15, ASL 2002, pp. 92–116 (2002)
20. Kohlenbach, U.: Higher order reverse mathematics. In: Reverse Mathematics 2001. Lecture Notes in Logic, vol. 21, ASL 2005, pp. 281-295 (2005)
21. Kupka, I.: A generalised uniform convergence and Dini's theorem. N. Z. J. Math. **27**(1), 67–72 (1998)
22. Medvedev, F.A.: Scenes from the History of Real Functions. Science Networks. Historical Studies, vol. 7. Birkhäuser Verlag, Basel (1991). https://doi.org/10.1007/978-3-0348-8660-4
23. Montalbán, A., Shore, R.A.: The limits of determinacy in secondorder arithmetic. Proc. Lond. Math. Soc. **104**(2), 223–252 (2012)
24. Moore, E.H.: On a form of general analysis with application to linear differential and integral equations. Atti IV Cong. Inter. Mat. (Roma 1908) **2**, 98–114 (1909)
25. Moore, E.H.: Introduction to a Form of General Analysis. Yale University Press (1910)
26. Moore, E.H.: Definition of limit in general integral analysis. Proc. Natl. Acad. Sci. U. S. A. **1**(12), 628–632 (1915)
27. Moore, E.H., Smith, H.L.: A general theory of limits. Am. J. Math. **44**(2), 102–121 (1922)

28. Muldowney, P.: A General Theory of Integration in Function Spaces, Including Wiener and Feynman Integration, vol. 153. Longman Scientific & Technical/Wiley, Harlow (1987)

29. Mummert, C., Simpson, S.G.: Reverse mathematics and Π_2^1 comprehension. Bull. Symb. Log. **11**(4), 526–533 (2005)

30. Mummert, C.: On the reverse mathematics of general topology, ProQuest LLC, Ann Arbor, MI. Ph.D. thesis - The Pennsylvania State University (2005)

31. Mummert, C.: Reverse mathematics of MF spaces. J. Math. Log. **6**(2), 203–232 (2006)

32. Naimpally, S.A., Peters, J.F.: Preservation of continuity. Sci. Math. Jpn. **76**(2), 305–311 (2013)

33. Normann, D., Sanders, S.: Nonstandard analysis, computability theory, and their connections. arXiv:abs/1702.06556 (2017)

34. Normann, D., Sanders, S.: The strength of compactness in computability theory and nonstandard analysis. arXiv:1801.08172 (2018)

35. Normann, D., Sanders, S.: On the mathematical and foundational significance of the uncountable. J. Math. Log. (2018). https://doi.org/10.1142/S0219061319500016

36. Normann, D., Sanders, S.: Pincherle's theorem in Reverse Mathematics and computability theory. arXiv:1808.09783 (2018)

37. Reed, M., Simon, B.: Analysis of Operators. Methods of Modern Mathematical Physics, vol. 4. Academic Press, Cambridge (1978)

38. Sakamoto, N., Yamazaki, T.: Uniform versions of some axioms of second order arithmetic. MLQ Math. Log. Q. **50**(6), 587–593 (2004)

39. Sanders, S.: Splittings and disjunctions in reverse mathematics. NDJFL, p. 18. arXiv:1805.11342 (2018)

40. Simpson, S.G. (ed.): Reverse Mathematics 2001. Lecture Notes in Logic, vol. 21, ASL, La Jolla, CA (2005)

41. Simpson, S.G.: Subsystems of Second Order Arithmetic. Perspectives in Logic, 2nd edn. Cambridge University Press, Cambridge (2009)

42. Simpson, S.G.: The Gödel Hierarchy and Reverse Mathematics. Kurt Gödel. Essays for his centennial, pp. 109–127 (2010)

43. Stillwell, J.: Reverse Mathematics, Proofs from the Inside Out. Princeton University Press, Princeton (2018)

44. Swartz, C.: Introduction to Gauge Integrals, World Scientific (2001)

45. Timofte, V., Timofte, A.: Generalized Dini theorems for nets of functions on arbitrary sets. Positivity **20**(1), 171–185 (2016)

46. Toma, V.: Strong convergence and Dini theorems for non-uniform spaces. Ann. Math. Blaise Pascal **4**(2), 97–102 (1997)

47. Troelstra, A.S.: Meta Mathematical Investigation of Intuitionistic Arithmetic and Analysis. Lecture Notes in Mathematics, vol. 344. Springer, Berlin (1973). https://doi.org/10.1007/BFb0066739

48. Wang, H.: Eighty years of foundational studies. Dialectica **12**, 466–497 (1958)

49. Wolk, E.S.: Continuous convergence in partially ordered sets. Gen. Topol. Appl. **5**(3), 221–234 (1975)

Kalmár's Argument for the Independence of Computer Science

Máté Szabó[1,2]([⊠]) [ID]

[1] Archives Henri-Poincaré, UMR 7117, Université de Lorraine, Nancy, France
mate.szabo@univ-lorraine.fr
[2] IHPST, UMR 8590, Université Paris 1 Panthéon-Sorbonne, Paris, France

Abstract. Computer Science is a rather young discipline, and as usual with new disciplines, in its early stage there were important discussions about its aim, scope and methodology. Throughout these debates, it was claimed at different times that computer science belongs to the natural sciences, mathematics, or engineering. Questions about the organization of the field were raised as well: is there a need for computer science departments, or for separate computer science majors at the university level? The history of these debates has been documented rather well in recent years. However, the literature focuses mostly on sources from the US and Western Europe. The aim of this paper is to include the stance of eminent Hungarian logician and computer scientist László Kalmár in the history of this discussion. Kalmár's view is reconstructed based on recently found, formerly unpublished archival materials from 1970–1971: a conference abstract and his correspondence about Hungarian computer science education. In this paper, I will also situate Kalmár's view among the positions of other prominent scholars in these debates.

1 Introduction

Computer Science is a rather young discipline, and as usual with new disciplines, in its early stage there were important discussions about its aim, scope and methodology. Many people argued that computer science is an independent branch of science worthy of academic examination on its own. However, throughout these debates it was also claimed at different times that computer science belongs to the natural sciences, (applied) mathematics, or engineering. In many cases the arguments were based on the backgrounds or scientific interests of those who put them forward; researchers of artificial intelligence argued for the natural science interpretation, while mathematicians invested in the field of computing emphasized its mathematical aspects. Besides its methodology, the scope of computer science was called into question as well. Is it, to name a few options, the study of machines and related questions (as the name of its first and largest association, the Association of Computing Machinery, indicates), of information and data processing or of algorithms? Again, scholars were usually

I would like to thank Kendra Chilson for her help in writing this paper.

F. Manea et al. (Eds.): CiE 2019, LNCS 11558, pp. 265–276, 2019.
https://doi.org/10.1007/978-3-030-22996-2_23

arguing for one or another view based on their own research interests. During these early, identity-forming years even the name of the field was called into question and generated debates.

These questions about the identity of computer science were not merely intellectual or philosophical questions–they also had practical consequences. For example, if computer science is an independent science, then it should have independent institutions within academia, such as departments at universities. Although today we take the independence of computer science for granted, it was not obvious in the beginning. Indeed, even in 1966, President of the ACM Anthony Oettinger stated, "I personally believe, and still believe that I am right, that departments of computer science have no place in the eternal scheme of things. [...] I am forced to split my mind and say that I believe that it is an intellectual mistake to have departments of computer science, while I believe there is no real tactical alternative to having them" ([16], pp. 27–28). He believed that computer science is not part of either mathematics or engineering, but that it is anchored to both. As a consequence, he was worried that separate computer science departments might become isolated within universities. Another practical question was whether there was a need for separate computer science majors at the university level: if computer science is simply viewed as a tool for natural sciences and engineering, then a couple courses should suffice for the experts of those fields, possibly even only on the graduate level. These questions dominated the discussions about computer science education throughout the 1960s.

Scholars have taken an interest in the history of computer science's quest for its identity as an independent scientific discipline. As early as 1976, Wegner wrote about the different research paradigms in the field [22]. Many of these debates were thematized and further analyzed more recently in [3] and [6]. The most complete historical overview of these debates can be found in Tedre's excellent book [21] from 2014. However, the literature covers almost exclusively Western sources (mainly for language reasons), even though computer science as a discipline clearly had to go through a similar process to gain independence, acceptance, and prestige outside the US and Western Europe.

The aim of this paper is to include a scholar in this discourse from the Eastern Block as well.[1] The short argument presented below comes from prominent Hungarian logician and later computer scientist László Kalmár[2] from 1970–1971. He was at the vanguard of Hungarian research in computer science and automata theory as well as building computing devices. He was also indispensable in the start of computer science education in Hungary ([18,20]). As a consequence, he was involved in many similar discussions about computer science as a discipline, and faced many challenges while fighting for its institutional independence in Hungary.

---

[1] See [7], Sect. 6 and especially page 183, for examples depicting similar struggles in the Soviet Union.

[2] For bibliographical information and description of his work in the field of computer science see ([20], Sect. 3) and [14].

Kalmár, although coming from a mathematical background, argued for the independence of computer science from mathematics based on methodological differences. He used his argument to support certain institutional changes in the Hungarian academic world of computing. I will also situate Kalmár's view among the positions of other prominent scholars in these debates. However, due to lack of space only those with similar views will be indicated.

Sources. While looking through the correspondence between Kalmár and Patrick Suppes[3] in the Kalmár Nachlass at the Klebelsberg Library at the University of Szeged, I accidentally found an interesting acceptance letter from Suppes. The letter, dated the 3rd of May 1971, announces that Kalmár's contributed paper, entitled *Is Computing Science an Independent Science?*, is accepted for presentation at the Fourth International Congress on Logic, Methodology and Philosophy of Science.[4] It seems, however, that Kalmár never did deliver his talk. The proceedings [19] do not mention Kalmár, nor does the list of his official travels [9] mention this congress, and the list of his collected papers in [1] does not contain anything similar. Fortunately, the one-page long abstract can be found in the Nachlass under 'Folder 311' [10]. In addition, again by pure accident, I stumbled upon another exposition of the same argument by Kalmár. Folder 'Lev-12' [11] contains a 24-page long letter from April 10th, 1971 detailing his comments and recommendations about the national computer science education for Hungary's unified computer science initiative. The letter was sent to György Aczél, the secretary for cultural affairs of the Central Committee of the Hungarian Socialist Workers' Party, upon Aczél's request. In this letter, Kalmár gives detailed recommendations for computer science education from elementary school through high school to university, and even postgraduate courses and trainings. The context in which the question of the independence of computer science comes up is the education and training of future academic scholars. I will use these two sources to present Kalmár's argument that computer science is independent from mathematics and the implications he thinks this has for the organization of academic institutions.

Remark. Before turning to Kalmár's writings, I must explain his choice of words. His abstract is entitled *Is Computing Science an Independent Science?*: he uses 'comput*ing* science' instead of the now customary 'computer science.' First it should be remarked that the field itself did not yet have a generally accepted, singular term for the discipline of computing (for examples see Chap. 7 of [21] and p. 324 of [13]). Furthermore, it appears to be a deliberate choice on Kalmár's part, as the typewritten abstract I found originally used the expression "computer scientist," which was later changed to "computing scientist" by hand.

---

[3] Between 1963 and 1965, both Kalmár and Suppes served in the governance of the DLMPS, Kalmár as Vice-President and Suppes as Secretary General, and as members of the Committee on the Teaching of Logic and Philosophy of Science from 1964 until 1968 as well.

[4] The Congress took place in Bucharest, Romania from August 29 to September 4 in the same year.

His letter [11] written in Hungarian provides some clarification about his choice of words. There (pp. 16–17), Kalmár makes a distinction between two commonly used Hungarian terms, 'számítástechnika' and 'számítástudomány,' which can be translated as 'computing technology (or technique)' and 'computing science,' respectively. Kalmár briefly indicates that he uses these different phrases such that 'computing technology' covers hardware-related issues, while questions concerning software design and engineering belong under 'computing science.' For the remaining part of this paper, I will use 'computing science' wherever I discuss Kalmár's view. The reader should keep in mind that Kalmár understands it to mean what we would today call software design and engineering, and that it does not cover the entirety of computer science, broadly understood. Thus, in Kalmár's terminology 'computing science' is part of computer science.[5]

2   Kalmár on the Independence of Computing Science

As mentioned above, there are two sources that contain Kalmár's argument for the independence of computing science. First, I will use his letter [11] on the unified computer science education initiative to provide the context in which Kalmár used the argument, then display his conference abstract that contains the argument in its entirety [10], and finally, explain some of his points in detail and position him among the opinions of others at the time.

Section 6 of Kalmár's letter ([11], pp. 16–19) is devoted to the question of the "education of academic scholars" in computer science, that is, those who received scientific degrees, engaged in research and possibly stayed in academia. They also clearly were to have a serious impact on the education of the subject, as they would be the ones to teach it at the university level. The unified computer science initiative contained a directive of funding ten computer science departments in the five year period between 1971 and 1975.[6] Kalmár emphasized that, in order to provide quality education, computer science departments needed highly trained faculty conducting research in both hardware- and software-related issues. However, as Kalmár pointed out, there were only two PhD[7] holders in the field of software engineering at the time in Hungary, and "even the number of those

---

[5] To make things precise, but possibly even worse, computer science departments in Hungary are usually called 'számítástechnika' departments, thus the word Kalmár uses for hardware-related issues was also used in Hungary as an umbrella term that can be translated as 'computer science' broadly understood.

[6] This period coincides with the Fourth Five Year Plan of Hungary. (Five year plans were overarching, nationwide centralized economic plans in the socialist countries.).

[7] In Hungary, and many other countries in the Eastern Block, this scientific degree was called 'candidate of sciences' ('kandidátusi fokozat' in Hungarian). As it is a PhD-equivalent degree, I decided to use 'PhD' throughout the paper to avoid confusion and cumbersome phrasing. (Indeed, many 'candidate of sciences' degrees were actually converted to PhDs in the 1990s, after the collapse of the Eastern Block).

PhDs is rather low that were defended in computing science broadly understood" (p. 17).[8] This obviously posed a problem for the founding of new departments, as the faculty of university departments had to hold a certain number of scientific degrees in order to be accredited.[9] Kalmár's explanation of the low number of software engineering and design PhDs, and his recommended solution, were tied to his argument for the independence of computing science. To understand the context appropriately, we have to go into more detail about the process of obtaining a PhD in Hungary during this period.

Hungary, among other Eastern Block countries, adopted many features of the Soviet academic system. These changes were put into effect in Hungary in 1949, including the creation of the 'Scientific Qualification Committee'[10] of the Hungarian Academy of Sciences. This committee was a centralized organization responsible for the selection of PhD candidates and approval of dissertation topics, as well as approval of faculty members as supervisors, etc. (these decisions were, in many cases, also not free from political considerations). Thus, in this new system, departments and universities lost their freedom and autonomy to award scientific degrees [15].[11]

Although the Scientific Qualification Committee claimed to assign high priority to software-related topics, the number of applicants remained quite low. Kalmár explained the low number of software-related PhDs by the organizational structure of the committee and its approved dissertation categories. Software-related dissertation topics fit only under the *Mathematical machines and programming* category offered by the Mathematical subcommittee. However, according to Kalmár, the subcommittee contained only mathematicians, who understood "programming" as it was customary in operation research at the time, i.e. as linear programming, convex programming, etc.[12] As a consequence, applicants with software-related research interests were often either rejected,

---

[8] It is well known that the Eastern Block lagged behind the West in computing technologies in general. The gap was even larger in the case of software development and maintenance than in the case of hardware ([4] pp. 98–100, and [8]).

[9] In addition, according to Kalmár, most of the PhD holders had already reached well-paid, high ranks in the industry and were unlikely to leave their jobs for academia.

[10] 'Tudományos Minősítő Bizottság' in Hungarian.

[11] For the sake of completeness, it has to be mentioned that from the 1950s, universities were allowed to award a title, colloquially referred to as 'little doctorate' ('kisdoktori' in Hungarian), but it did not count as a scientific degree and in most cases they were not allowed to be converted into PhD degrees in the 1990s.

[12] On p. 17 Kalmár makes a claim, the accuracy of which is hard to judge today, that this understanding was facilitated by a typo. According to Kalmár, the category was supposed to be called 'Matematikai gépek és programozásuk' which translates as 'Mathematical machines and their programming'. However, the official description read 'Matematikai gépek és programozások,' which differs only in one letter (the second from last), and means 'Mathematical machines and programming,' where programming is actually in plural (which is grammatically correct in Hungarian). Thus, programming wasn't necessarily linked to the mathematical machines anymore, and required multiple kinds of programming, leading to the preference of operation research themed dissertation topics.

as their topic was "not mathematical enough," or directed towards operation research. In addition, Kalmár noted that some members of the committee "even a couple years ago during committee meetings openly proclaimed their opinion that programmers are trained at universities but no one should apply for a PhD with such a topic, as 'programming is not a scientific research topic'." This attitude kept many worthy candidates from even applying.

Kalmár saw these issues as part of a "natural process" in which new branches have to fight for their acceptance and approval. Thus, that computing science faced these difficulties was not surprising–quite the opposite, it was to be expected. He even mentioned operation research itself and probability theory as recent examples of new branches that had to fight for their acceptance as legitimate branches of academic mathematical research.

However, argued Kalmár further, the case of computing science was somewhat different from the acceptance of those branches. While he acknowledged that computing science had its origins in mathematics, he also argued that its methodology was so different from mathematics that it should be considered "an independent science", i.e. independent from mathematics. This difference in methodology explained, according to Kalmár, the rejected dissertation topics as well, since mathematicians did not understand what counts as an (intellectual) achievement in software design and engineering, and thus could not judge which topic was worthy of a PhD degree. The solution Kalmár proposed was, of course, to create a new computing science subcommittee within the Scientific Qualification Committee where the members were computing experts instead of mathematicians. In January of 1970, he submitted a request for such a subcommittee to be created.

Although today it is widely accepted without much argumentation that computing science should be considered independent from mathematics, it was not so at the time.[13] This is why, at this point in the letter, Kalmár put forward his argument for the independence of computing science based on its different methodology from mathematics. The same argument was accepted (without the aforementioned context) to the Fourth International Congress on Logic, Methodology and Philosophy of Science of 1971. As the argument provided in the Hungarian letter [11] for computing science being an independent science is very similar to the English abstract [10], I display the entire abstract below to show Kalmár's argumentation in his own phrasing, instead of providing a summary of it. This is a formerly unpublished abstract of a presentation that was most likely never delivered. To retain its original appearance, I used a typewriter font and kept its original typesetting. However, I silently corrected typos and clear grammatical mistakes, and changed the parentheses from "/" and "/" to "(" and ")". The two references listed in the Bibliography are not referred to in the abstract text by Kalmár. The three footnotes (14, 15 and 16) are added by me.

---

[13] For example Knuth in the preface of his [12] from 1968 wrote that "computers are widely regarded as belonging to the domain of 'applied mathematics'" (p. ix). Interestingly, Knuth uses the term 'computer,' not even '(theoretical) computer science' belonging to applied mathematics.

Is Computing Science an independent science?

Computing Science obviously has its origin in Mathematics. The question is, whether it is a branch of Mathematics or it can be considered as an independent science.

Beside its special subject-field, Computing Science diverges from Mathematics by its method. Indeed, while Mathematics is a proof-oriented science, Computing Science is more algorithm-oriented. In any case, a computing scientist puts generally as much ingenuity into his algorithms as a mathematician into the proofs of his theorems.

True enough, algorithms play some role in Mathematics as well. However, even the most sophisticated mathematical algorithms (e.g. Kronecker's algorithm for decision of the reducibility of a polynomial in the field of rationals, say, or Galois' algorithm, using the latter, for decision of solvability, by means of radicals, of an algebraic equation, with rational coefficients, say) are very short relative to a compiling algorithm or to an operational system.

Also, the computing scientist has to prove his propositions, e.g. the correctness of his programs. However, in most cases, the proof has a verificative character. The name "debugging" given to such verifications shows that the computing scientist does not esteem this activity, though important, so high as the mathematician his proofs. In most cases, the errors found in the course of debugging are easily corrected (at least if the programming idea is sound), while errors in mathematical proofs are in general fatal.[14]

A mathematical problem, asking if some statement is true or not, is finally solved by a proof (or disproof) of the statement in question. On the contrary, if one has a computational algorithm for the solution of a given problem of Computing Science, the problem is not yet finally settled, for one is asking for a better algorithm for the same goal (from the point of view of computing time or memory place).[15] Well, a mathematician can also look for a simpler proof of some theorem. However, to find one is not as great an achievement as to find the first proof. On the other hand, the improvement of a computational algorithm is sometimes as (or more) valuable as producing the first algorithm for the same purpose.

These arguments show that Computing Science requires a way of thinking that is different from that of a traditional mathematician. Hence, Computing Science is appropriately considered an independent science rather than a branch of Mathematics.

László Kalmár

---

[14] This comparison of Kalmár's is not clear without further arguments. For, if the idea behind a mathematical proof is sound, it can be "easily corrected" as well. What he might have meant is that judging an idea to be sound in programming is easier than in mathematics.

[15] On a similar note in the letter (p. 18), Kalmár remarks that a proof of the optimality of a particular algorithm belongs to mathematics.

Bibliography

C.B. Jones, P. Lucas: Proving correctness of implementation techniques, IBM Laboratory Vienna, Technical Report TR 25.110, 12 august 1970.[16]

C.D. Allen: The application of formal logic to programs and programming, IBM Systems Journal, 10:1, 1971.

We see that in arguing for the independence of computing science, Kalmár tried to differentiate it only from mathematics, and not from engineering. Most likely he did not address its relation to engineering for two reasons. First, and most importantly, arguments for computer science and programming being an engineering discipline (i.e. "software engineering") became widespread only later, from the 1970s onward. Second, in Hungary technical universities offered programming majors rather late and did not dominate the field ([18,20]).

Kalmár emphasized the differing methodologies of computing science and mathematics to set computing science aside from applied mathematics. Indeed, it was customary at the time to categorize computing science as applied mathematics. Of course, the subjects of applied mathematics differ from pure mathematics, but it is still considered to be a branch of mathematics. Thus, Kalmár had to argue that the difference between mathematics and computing science is not a mere difference in their subjects, but a difference in their methodologies.

At the beginning of the abstract, Kalmár declared computing science to be an "algorithm-oriented" science. The most famous advocate of this point of view is most likely Donald Knuth, who was originally trained as a mathematician just like Kalmár. Indeed, in his [13], Knuth wrote that his "favorite way to describe computer science is to stay that it is the study of *algorithms* [...] because they are really the central core of the subject, the common denominator which underlies and unifies the different branches." (pp. 323–324) However, Knuth emphasized the mathematical aspect of algorithms and compared programming to creating mathematical proofs: "The construction of a computer program from a set of basic instructions is very similar to the construction of a mathematical proof from a set of axioms." ([12], p. ix) He did so to stress the strong interconnectedness of programming and mathematics, not only with applied, but with pure mathematics as well. Kalmár, on the contrary, downplayed the role of algorithms in mathematics in order to separate it from computing science.

Furthermore, Kalmár distanced the notion of mathematical proof from the social practice of "proving" programs to be correct, i.e. from "debugging." In addition to pointing out the different practices in "verification" in these fields, he also claimed that verification is not considered to be the intellectually challenging part of programming, quite in contrast to the appreciation of proofs in mathematics. This part of the argument can be considered Kalmár's response to

---

[16] Also published as Jones C.B., Lucas P. (1971) Proving correctness of implementation techniques. In: Engeler E. (ed) Symposium on Semantics of Algorithmic Languages. Lecture Notes in Mathematics, vol 188. Springer, Berlin, Heidelberg. DOI: https://doi.org/10.1007/BFb0059698.

the so-called "verificationists."[17] This is the view that programs (algorithms) are mathematical entities, and if all their specifications are fully formally described, their correctness could and should be verified by formal mathematical proofs instead of "debugging." ([3,21]) According to Tedre, "Although the formal verification movement was, from its start in the early 1960s, light years away from the reality of actual programming practice in the industry, many believed in its intellectual superiority." ([21], p. 60) This "intellectual superiority" is inherited from the practices of formal mathematics, which is an accepted and well respected science. Clearly, Kalmár was advocating instead for the examination and acceptance of the practices being used in computing science, such as debugging.

In the last step Kalmár compared where the intellectual effort was invested in these fields. He claimed that the intellectual effort on display in computing science was on par with mathematicians' efforts to provide proofs, but it was used to design ever more sophisticated and complex algorithms. As a consequence, connecting back his argument to the academic and institutional context, mathematicians should not be the ones assessing the intellectual merits of achievements in computing science, simply because they are not acquainted with its methodology and practice.

Fig. 1. Kalmár during a lecture. (Picture is from [2]).

---

[17] Indeed, the two entries in the Bibliography attached to the abstract are proponents of the verificationist view.

3   A Strong Parallel: Kalmár and Perlis

As a final thought for this paper, I would like to put Kalmár's argument and the context in which he gave it in parallel with Alan Perlis' *Computer Science is Neither Mathematics nor Electrical Engineering* [17] from 1968. Although today this similarity might not appear to be surprising, as their stance turned out to be the well accepted one, I find it quite striking just how much their views, background in mathematics and even their positions in the academic institutions lined up with each other despite being on two opposite sides of the world.[18]

Perlis, at Carnegie Mellon University at the time, was instrumental in starting a program in computer science during the mid 1960s, which led to the funding of a separate computer science department which he was the first head of [5]. Similarly, Kalmár started the first university-level training in computer science and programming in Hungary at the University of Szeged in 1957 and was the head of a separate computer science department from 1967 [20].[19] Just as Kalmár expressed frustration over mathematicians assessing computing scientists in the Scientific Qualification Committee, Perlis began his paper by describing how, in the US, the allocation of federal funding for computer science research is decided by various mathematics and applied mathematics committees. Perlis believed this was because "Computer Science is, unfortunately a bit too large to be ignored, and yet too new to be properly treated. As a result, computer science is in danger of being mishandled and misinterpreted" (p. 69).

Similarly to Kalmár, Perlis downplayed the importance of algorithms in mathematics: "Before the advent of the computer, algorithms were encountered, but they were rare, simple, and always consigned to the support and background of other investigations." (p. 70) Then he pointed out features in the practice of computer science that are not shared with mathematics: "Still, there are aspects of computer science's preoccupation with algorithms which are less directly related to mathematics. This is true, for example, of computer programming. The algorithms of computer programming are enormously complex and more specialized than it is the custom of mathematics to treat." (p. 71) Finally, he claimed that since computer science is "preoccupied with design and process" while "mathematics is oriented to abstract analysis" (p. 71), they have different methodological approaches, and thus computer science should be institutionally independent from mathematics.[20]

---

[18] Again, for lack of space, no one else holding this general position is mentioned from among the many. As just one example, see George Forsythe's position as described by Tedre ([21], pp. 37–38). Still, I believe, Kalmár and Perlis' positions show a striking resemblance.

[19] The department was called *Foundations of Mathematics and Computer Technology Department* until 1971, when it morphed into the *Computer Science Department*, still headed by Kalmár until his retirement in 1975.

[20] Interestingly, even though Perlis mentions "engineering" in the title explicitly, he does not provide arguments for the independence of computer science from it, just as Kalmár did not.

Thus, Kalmár and Perlis described the methodological differences between computer science and mathematics slight differently. Perlis named the "abstractness" of mathematics and the "design" focus of computer science as distinguishing features, and Kalmár pointed to the different approaches of their verification processes. Nevertheless, their academic pasts and positions, the context in which they argued for an independent computer (or computing) science, and their arguments themselves are astonishingly similar.

References

1. Ádám, A., Dömösi, P.: Kalmár László. In: Pénzes, I. (ed.) Műszaki nagyjaink, vol. 6. Gépipari Tudományos Egyesület 1986, pp. 47–89 (1986)
2. Bohus, M., Muszka, D., Szabó, P.G.: A szegedi informatikai gyűjtemény (The computer collection in szeged). In: Conference Slides, 18 March 2019 (2005). https://www.yumpu.com/hu/document/read/29881933/a-szegedi-informatikai-gyujtemeny-in-memoriam-kalmar-laszlo
3. Colburn, T.: Philosophy and Computer Science. M. E. Sharpe, Armonk (2000)
4. Davis, N.C., Goodman, S.: The Soviet Bloc's unified system of computers. ACM Comput. Surv. 10(2), 93–122 (1978)
5. Denning, P.J.: Alan J. Perlis, 1922–1990: a founding father of computer science as a separate discipline. Commun. ACM 33(5), 604–605 (1990)
6. Eden, A.: Three paradigms of computer science. Minds Mach. 17(2), 135–167 (2007)
7. Ershov, A.P., Shura-Bura, M.R.: The early development of programming in the USSR. In: Metropolis, N., Howlett, J., Rota, G. (eds.) A History of Computing in the Twentieth Century, pp. 137–196. Academic Press, New York (1980)
8. Goodman, S.E.: Software in the soviet union: progress and problems. In: Yovits, M.C. (ed.) Advances in Computers, vol. 18, pp. 231–287 (1979)
9. Kalmár, L.: The Official Travels of Professor Kalmár. A document assembled by Kalmár Nachlass, Klebelsberg Library, University of Szeged (1928–1975)
10. Kalmár, L.: Is computing science and independent science? Abstract. The abstract is not dated, most likely it was prepared during the previous year to the Congress. In: Folder 311. Kalmár Nachlass, Klebelsberg Library, University of Szeged (1970–1971)
11. Kalmár, L.: Számitástechnikai programunk megvalósitása (The implementation of our computer science initiative). In: Folder Lev-12 (Containing Kalmár's Correspondence Related to the Programming Major 1957–1974). Kalmár Nachlass, Klebelsberg Library, University of Szeged (1971)
12. Knuth, D.E.: The Art of Computer Programming: Fundamental Algorithms, vol. 1. Addison-Wesley Publishing, Reading (1968)
13. Knuth, D.E.: Computer science and its relation to mathematics. Am. Math. Mon. 81(4), 323–343 (1974)
14. Makay, Á.: The activities of László Kalmár in the world of information technology. Acta Cybern. 18(1), 9–14 (2007)
15. Nagy, P.T.: A tudományos továbbképzés Kádár-korszakbeli társadalomtörténetéhez. (On the social history of the postgradual scientific education in the Kádár Era). Kultúra és közösség 2(15), 23–34 (2011)
16. Oettinger, A.: President's letter to the ACM membership. Commun. ACM 9(12), 838–839 (1966)

17. Perlis, A.: Computer science is neither mathematic nor electrical engineering. In: Finerman, A. (ed.) University Education in Computing Science, pp. 69–79. Academic Press, New York (1968)

18. Sántáné-Tóth, E.: Computer oriented higher education in Hungary. Studia Universitatis Babes-Bolyai Digitalia **62**(2), 35–62 (2017)

19. Suppes, P., Henkin, L., Joja, A., Moisil, G.C. (eds.): Proceedings of the Fourth International Congress for Logic, Methodology and Philosophy of Science IV, Bucharest, Romnia, 1971. North-Holland Publishing Company/PWN - Polish Scientific Publishers, Amsterdam/Warszawa (1973)

20. Szabó, M.: László Kalmár and the first university level programming and computer science training in Hungary. In: Leslie, C. (ed.) Proceedings of the IFIP World Computer Congress, WG 9.7, Poznan, Poland, 30 p. (2019, forthcoming)

21. Tedre, M.: The Science of Computing: Shaping a Discipline. CRC Press, Boca Raton (2014)

22. Wegner, P.: Research paradigms in computer science. In: ICSE 1976, Proceedings of the 2nd International Conference on Software Engineering, pp. 322–330. IEEE Computer Society Press, Los Alamitos (1976)

The d.r.e wtt-Degrees are Dense

Shaoyi Wang[1], Guohua Wu[1(✉)], and Mars M. Yamaleev[2]

[1] Division of Mathematical Sciences, School of Physical and Mathematical Sciences,
Nanyang Technological University, Singapore 637371, Singapore
WANG0831@e.ntu.edu.sg, guohua@ntu.edu.sg
[2] Institute of Mathematics and Mechanics, Kazan Federal University,
18 Kremlyovskaya Street, 420008 Kazan, Russia
mars.yamaleev@kpfu.ru

Abstract. We prove in this paper that the d.r.e. wtt-degrees are dense, improving a result of Wu and Yamaleev. Our result is a direct generalization of Ladner and Sasso's splitting theorem for r.e. wtt-degrees. One essential feature of our construction is that the Lachlan sets are used as a help to obtain more information of d.r.e. sets.

In this article, we prove that the d.r.e. wtt-degrees are dense, improving a result of Wu and Yamaleev in their paper [19].

Theorem 1. *For any d.r.e. wtt-degrees* $\mathbf{c} < \mathbf{d}$, *there is a d.r.e. wtt-degree* $\mathbf{a}$ *with* $\mathbf{c} < \mathbf{a} < \mathbf{d}$.

We actually prove Theorem 1 by showing that an analogue of Ladner and Sasso's splitting theorem is true for d.r.e. wtt-degree. Recall that Ladner and Sasso's splitting theorem in [13] says that any r.e. wtt-degree splits above less ones. In Theorem 1, if both $\mathbf{c}$ and $\mathbf{d}$ are r.e. wtt-degrees, then our construction gives Ladner and Sasso's splitting, and due to this, Theorem 1 is a direct generalization of Ladner and Sasso's splitting, from r.e. wtt-degrees to d.r.e. wtt-degrees.

A set $A \subseteq \mathbb{N}$ is recursively enumerable, r.e. for short, if it is a domain of some partial recursive function, and a set $D \subseteq \mathbb{N}$ is d.r.e. if D is the difference of two r.e. sets, i.e. $D = A - B$ for some r.e. sets A and B with $B \subseteq A$. A Turing degree is r.e. (d.r.e.) if it contains an r.e. (d.r.e., respectively) set.

For the r.e. degrees, Sacks proved that this structure is dense (Sacks density theorem, [16]) and every nonzero r.e. degree splits (Sacks splitting theorem, [15]). Cooper proved an analogue of Sacks splitting for d.r.e. degrees (in [4]), and the

The authors appreciate the comments and suggestions from the referees. Wang and Wu are partially supported by MOE2016-T2-1-083 (M4020333), M4011274 (RG29/14) and M4011672 (RG32/16) from Ministry of Education of Singapore. Yamaleev is supported by Russian Foundation for Basic Research (project 18-31-00420), by The President Grant of Russian Federation (project SS-5383.2012.1), and by The Ministry of Education and Science of Russian Federation (projects 14.A18.21.0360, 14.A18.21.0368, 14.A18.21.1127).

F. Manea et al. (Eds.): CiE 2019, LNCS 11558, pp. 277–285, 2019.
https://doi.org/10.1007/978-3-030-22996-2_24

density for d.r.e. degrees is true for low$_2$ d.r.e. degrees [3]. In 1975, Lachlan proved in [12] that the density and splitting above cannot be combined, where Lachlan developed the $0'''$ argument for the first time.

Cooper initiated the study of the structure of d.r.e. degrees in his PhD thesis [2] in 1971. Lachlan observed that the d.c.e. degrees are downwards dense, by using the degrees of Lachlan sets. However, the d.r.e. degrees fail to be dense, by the existence of maximal d.r.e. degrees constructed by Cooper, Harrington, Lachlan, Lempp and Soare in their paper [5].

In this paper, we consider the structure of d.r.e. *wtt*-degrees, i.e., weak-truth-table degrees of d.r.e. sets. In [19], Wu and Yamaleev proved that there is no maximal d.r.e. *wtt*-degree, providing a structural difference between d.r.e. *wtt*-degrees and d.r.e. Turing degrees.

Recall that weak-truth-table reduction was proposed by Friedberg and Rogers in 1959 in [9], and is now also called *bounded Turing* reduction, as the uses of computations are bounded by partial recursive functions. Ladner and Sasso's paper [13] gives a systematic study of r.e. *wtt*-degrees, where they showed that the splitting and density can be combined in this degree structure. Technically, we can handle the constructions of *wtt*-degrees in a much easier manner, due the bound of uses given in advance.

For the proof of Theorem 1, as we are dealing with d.r.e. sets, for a given recursive approximations of d.r.e. A, we need to use Lachlan sets, which is defined based on the approximations we take, to catch the information of numbers leaving A. Lachalan used this kind of sets to show that the d.r.e. Turing degrees are downward dense. Ishmukhametov [10,11] and also Wu, Yamaleev and their coauthors [7,8] had considered the distributions of Lachlan degrees. Our proof of Theorem 1 will be the first construction to use Lachlan sets as an essential help for obtaining detailed information of the given d.r.e. set. The notion of Lachlan sets and Lachlan degrees will be given the preliminary part. We believe that such an idea will be useful for other constructions.

In the remainder of this paper, all degrees are d.r.e. *wtt*-degrees. Our notation and terminology are standard and generally follow Soare [17] and Odifreddi [14]. The readers can refer to Ambos-Spies' paper [1] and Stob's paper [18] for general idea on r.e. *wtt*-degrees. Downey [6] provides an extensive survey of splitting theorems.

1   Preliminary and Requirements

Given $c < d$ as required, we take d.r.e sets C, D with $C \in c$, $D \in d$, and let $\{C_s : s \in \omega\}$ and $\{D_s : s \in \omega\}$ be recursive approximations of C and D, respectively.

Without loss of generality, we assume that $C \oplus L(D) <_{wtt} D$, where $L(D)$ is the Lachlan set of D with respect to the approximation $\{D_s : s \in \omega\}$, i.e.,

$$L(D) = \{s : \exists x[(x \in D_s - D_{s-1}) \text{ and } x \notin D]\}.$$

It is easy to see that $L(D)$ is r.e., if D is a d.r.e. set, then $L(D) <_{wtt} D$ ($L(D)$ is empty, if D is r.e.). Note that it is possible that $C \oplus L(D) \equiv_{wtt} D$ (even though

$C <_{wtt} D$), and if so, we choose $C \oplus L(D)$ instead of D from $\boldsymbol{d}$, and consider the Lachlan set of $C \oplus L(D)$:

$$L(C \oplus L(D)) \equiv_{wtt} L(C) \oplus L(L(D)) \equiv_{wtt} L(C) \leq_{wtt} C <_{wtt} D \equiv_{wtt} C \oplus L(D).$$

This shows that we can always choose C and D with $C <_{wtt} D$ with $C \oplus L(D) <_{wtt} D$. We are targeting to apply this idea to the n-r.e. wtt-degrees, to show that the n-r.e. wtt-degrees are dense. Note that this does not work for ω-r.e. wtt-degrees, as Downey, Ng and Solomon proved recently the existence of minimal ω-r.e. wtt-degrees.

We assume that $C \cap D = \emptyset$. Further, we assume that at each stage $s + 1$, exactly one number enters or leaves $C \cup D$.

To prove the theorem, we split D into two d.r.e sets, D_0 and D_1, such that D cannot be wtt-reducible to either $D_0 \oplus C$ or $D_1 \oplus C$. As a consequence, $D_0 \oplus C$ and $D_1 \oplus C$ are not wtt-reducible to each other, and hence their wtt-degrees are above $\boldsymbol{c}$.

D_0 and D_1 are constructed to meet the following requirements:

P: $D_0 \cup D_1 = D$, $D_0 \cap D_1 = \emptyset$;
$N_{\langle e,i \rangle}$: If $D = \Phi_e^{D_i \oplus C}$ and the use function ϕ_e is bounded by ψ_e, then $D \leq_{wtt}$
    $L(D) \oplus C$.

Here $\{\langle \Phi_e, \psi_e \rangle : s \in \omega\}$ is an effective list of pairs $\langle \Phi_e, \psi_e \rangle$, where Φ_e is a partial recursive functional, with use function ϕ_e, and ψ_e is a partial recursive function.

2   Satisfying the P-Requirement

The P-strategy is a standard set-splitting strategy for wtt-reduction, like Ladner-Sasso's splitting, with modification for d.r.e. sets.

In Ladner-Sasso's splitting, also in Sacks splitting, we construct two disjoint r.e. sets D_0 and D_1 whose union is D. That is, to split D, when a number x enters D, it is immediately enumerated into one of D_0 and D_1, but not both. In this case, once a number x is enumerated into D_i, it will be kept in it forever. We have $D \equiv_{wtt} D_0 \oplus D_1$ easily.

In our construction, D is given as a d.r.e. set and we work in a similar way: construct two disjoint d.r.e. sets D_0 and D_1 whose union is D. If a number x enters D, we enumerate x one of D_0 and D_1, D_i say, but not both. If later x leaves D, then we remove x from D_i immediately. Again, in this case, we have $D \equiv_{wtt} D_0 \oplus D_1$.

A crucial point here is that when x leaves D, s, the stage when x enters D, enters $L(D)$, the Lachlan set of D, and this enumeration into $L(D)$ keeps the consistency between N-strategies. We will give detail in the N strategies.

3   Satisfying One N-Requirement

The core of Sacks splitting is the technique of Sacks preservation, where the main point is to preserve the agreement between D and $\Phi_e^{D_i}$, to force a disagreement, as D is nonrecursive. This technique was developed by Robinson to show that r.e. sets split above low r.e. sets. Lachlan's nonsplitting theorem says that Sacks splitting cannot be combined with Sacks density, and the main obstacle is that one strategy can impose restraint with limit infinite. Ladner-Sasso's splitting theorem says that splitting and density can be combined for r.e. wtt-degrees. That is, $D \not\equiv_{wtt} D_i \oplus C$ can be guaranteed by applying Sacks preservation, via threatening $D \leq_{wtt} C$. In particular, to make $D \neq \Phi_e^{D_i \oplus C}$, where Φ_e is a wtt-reduction, we will construct a wtt-reduction $\Delta_{e,i}$ such that once we see D and $\Phi_e^{D_i \oplus C}$ agree on a certain length, ℓ say, we will preserve the D_i-side, and define $\Delta_{e,i}^C$ up to ℓ. Thus, for $n \leq \ell$, either $\Phi_e^{D_i \oplus C}(n)$ is preserved, and $\Delta_{e,i}^C(n)$ is kept, or C changes below the use $\phi_e(n)$, and this change undefines $\Delta_{e,i}^C(n)$. As for a fixed n, $\Phi_e^{D_i \oplus C}(n)$ converges at most finitely many times, and each change of the computation $\Phi_e^{D_i \oplus C}(n)$ is caused by a C-change, $\Delta_{e,i}^C(n)$ can be defined at most finitely many times, and each time when it is defined, we define it as $D(n)$, so if $\Delta_{e,i}^C(n) \neq D(n)$, we will have $D(n) \neq \Phi_e^{D_i \oplus C}(n)$.

We are now ready to describe how to satisfy one N-requirement, $N_{e,i}$ say, where C and D are d.r.e. sets. At a stage s, we define the length of agreement of D and $\Phi_e^{D_i \oplus C}$ as:

$$\ell(e,i;s) = \max\{x < s : \forall y < x[\psi_e(y)[s] \downarrow, \ \Phi_e^{D_{i,s} \oplus C}(y)[s] \downarrow = D_s(y)$$

$$\text{and the use } \varphi_e^{D_i \oplus C}(y)[s] < \psi_e(y)]\},$$

and define the restriction function as:

$$r(e,i;s) = \max\{\psi_e(x) : x \leq \ell(e,i;s)\}.$$

To simplify the expression, we denote $r(e,i;s)$ by $r(s)$. That is, $r(s)$ is the largest value of the use in computing the initial segment of $\Phi_e^{D_i \oplus C}(y)[s]$, which agrees with the initial segment of D_s. As in Sacks preservation strategy, if $x \leq r(s)$ enters D at this stage, then we enumerate x into D_{1-i}, and we are expecting to preserve the length of agreement $\ell(s)$ by ensuring that no enumerations into the D_i-part can change the computations, like the splitting of Ladner-Sasso. However, in our construction, both C and D are d.r.e. sets, n can enter C or D first and then leave at a later stage, and due to this, we cannot guarantee a disagreement between $D(y)$ and $\Phi_e^{D_i \oplus C}(y)$ if y enters D at a stage s with

$$D(y)[s] = 1 \neq 0 = \Phi_e^{D_i \oplus C}(y)[s],$$

as when y leaves D at a stage $s' > s$, we will come back to

$$D(y)[s'] = 0 = \Phi_e^{D_i \oplus C}(y)[s].$$

Therefore, for Sacks preservation, for x between y and $\ell(s)$, we also need to keep $\Phi_e^{D_i \oplus C}(x)$ from the enumeration of D_i (We definitely cannot prevent the removal of elements from D_i), and because of this, we will keep the restraint we had at stage s forever, even when we see y enters D at stage s. This is a feature different from those in Sacks splitting and Ladner-Sasso splitting.

We comment here that this will not make difference at all in Sacks splitting and Ladner-Sasso splitting, as the restraints imposed in this way will be finite in the whole construction. In our construction, this idea is crucial, as we are expecting that all changes of the disagreement being caused by the enumerations into C or D may change back later, as noted and used in Wu and Yamaleev's paper [19]. We will refer to this idea "once forever". That is, once a restraint $r(s)$ is set at a stage s, no number less than $r(s)$ can be put in D_i at a later stage (unless this strategy is initialized), and we will show that, by the assumption that $L(D) \oplus C <_{wtt} D$, this strategy will impose only finite restraint in the whole construction. We will show this now.

Recall that we are constructing a partial recursive functional $\Delta_{e,i}$, such that if there are infinitely many expansionary stages between D and $\Phi_e^{D_i \oplus C}$, where the use function ϕ_e is bounded by ψ_e, then $\Delta_{e,i}^{L(D) \oplus C}$ will be defined infinitely often, and at any expansionary stage, $\Delta_{e,i}^{L(D) \oplus C}$ computes D correctly at arguments in the current domain.

As above, when we have $\ell(s)$, for $n < \ell(s)$, if $\Delta_{e,i}^{L(D) \oplus C}(n)[s]$ is not defined, we define

$$\Delta_{e,i}^{L(D) \oplus C}(n)[s] = D_s(n),$$

with the C-part of the use as $\psi_e(n)$, and the $L(D)$-part of the use as the latest stage $t < s$ at which some number $y \in D_i[s]$ less than $\psi_e(n)$ enters D_i. Once we define it, the C-part and the $L(D)$-part of the use will not be changed in the remainder of the construction, unless the strategy is initialized.

Note that $L(D)$ is r.e., once a number t is enumerated into $L(D)$, t will be in $L(D)$ forever, which means that $\Delta_{e,i}^{L(D) \oplus C}(n)$ never comes back to $\Delta_{e,i}^{L(D) \oplus C}(n)[s]$, and when it is redefined later, t is already in $L(D)$. This just says that if the computation $\Phi_e^{D_i \oplus C}(n)$ changes due to some number less than $\psi_e(n)$ leaves D_i (and hence leaves D), then $\Delta_{e,i}^{L(D) \oplus C}(n)$ will get undefined, and the new definition will be different from the one at stage s. Also note that when $\Delta_{e,i}^{L(D) \oplus C}(n)$ is defined at stage s, a restraint is put on to prevent the enumeration of any number less than $\psi_e(n)$ into D_i. By this restraint, $\Delta_{e,i}^{L(D) \oplus C}(n)$ does not recover to a previous definition. Thus, $\Delta_{e,i}^{L(D) \oplus C}(n)$ can be undefined in this way by the change of D_i at most $\psi_e(n)$ many times, and as no number less than $\psi_e(n)$ can be enumerated into D_i (by the "once forever" rule), once it is defined again at stage $s' > s$, $D_i[s'] \subseteq D_i[s]$, and hence we can still keep the use of the $L(D)$-part the same as before.

Now consider the C-part of the use. By the discussion above, we only need to consider definitions of $\Delta_{e,i}^{L(D) \oplus C}(n)$ between two $L(D)$-changes, if any. In this case, $\Delta_{e,i}^{L(D) \oplus C}(n)$ changes due to C-changes. In particular, these changes are below $\psi_e(n)$.

There are two cases.

Case 1. C changes and the computation $\Phi_e^{D_i \oplus C}(n)$ does not recover to a previous one.
Then $\Delta_{e,i}^{L(D) \oplus C}(n)$ is redefined again at a later stage.

Case 2. C-changes recover $\Phi_e^{D_i \oplus C}(n)$ to a previous one.
In this case, $\Delta_{e,i}^{L(D) \oplus C}(n)$ is also recovered to a previous definition.

In both cases, if $D(n) = \Phi_e^{D_i \oplus C}(n)$, then we can have $D(n) = \Delta_{e,i}^{L(D) \oplus C}(n)$, either by redefining $\Delta_{e,i}^{L(D) \oplus C}(n)$ and letting it be $D(n)$, or because of $D_{s'}(n) = \Phi_e^{D_i \oplus C}(n)[s'] = \Phi_e^{D_i \oplus C}(n)[s] = D_s(n)$, and

$$\Delta_{e,i}^{L(D) \oplus C}(n)[s'] = \Delta_{e,i}^{L(D) \oplus C}(n)[s] = \Phi_e^{D_i \oplus C}(n)[s] = D_s(n) = D_{s'}(n).$$

Therefore, once $\Delta_{e,i}^{L(D) \oplus C}(n)$ is defined, the use will be kept the same, and if for all n, $D(n) = \Phi_e^{D_i \oplus C}(n)$, then $\Delta_{e,i}^{L(D) \oplus C}(n)$ is defined, where the $L(D)$-part and the C-part of the use are fixed, and computes $D(n)$ correctly. This shows that D is wtt-reducible to $L(D) \oplus C$, which is impossible, by our assumption. Therefore, in the construction, after a certain stage, we will not have more agreement between D and $\Phi_e^{D_i \oplus C}$, and this $N_{e,i}$-requirement is satisfied.

4   Construction

We are ready to construct a splitting D_0, D_1, of D satisfying the requirements. Also we assume that C has changes at odd stages and D has changes at even stages.
We list all requirements as follows:

$$P < N_{0,0} < N_{0,1} < N_{1,0} < N_{1,1} < \cdots\cdots < N_{e,0} < N_{e,1} < \cdots\cdots.$$

P is a global requirement and has the highest priority. We will take care of it at every stage. For $N_{e_1,j_1} < N_{e_2,j_2}$, N_{e_1,j_1} has higher priority. Say that a requirement $N_{e,j}$ *requires attention* at a stage s, if the change of $C \cup D$ at number x at stage s affects the way of satisfying $N_{e,j}$, or s is an expansionary for $N_{e,j}$.

Stage $s = 0$. Do nothing and go to Stage 1.

Stage $s + 1$.

Step 1: Check whether stage $s+1$ is an expansionary stage for some requirement among $N_{0,0}, N_{0,1}, \cdots, N_{s,0}, N_{s,1}$.
(*We are trying to find a requirement that requires attention at this stage* via expansionary stages.)

If Yes, let $N_{e,i}$ be the requirement with the highest priority such that $s + 1$ is an expansionary stage for $N_{e,i}$. Extend the definition of $\Delta_{e,i}^{L(D) \oplus C}$ with the C-part of the use of $\delta_{e,i}(n)$ defined as $\psi_e(n)$ and the $L(D)$-part of the use $\delta_{e,i}$ defined as the last stage $t < s + 1$ such that some number $y < \psi_e(n)$ enters D_i and still in D_i. Initialize all the requirements with lower priority and proceed to step 2.

If Not, then do nothing and proceed to step 2.

Step 2: Suppose that x enters or leaves $C \cup D$ at stage $s + 1$.

s **odd:** If x enters D at this stage, then find the requirement with the highest priority, $N_{e,i}$ say, whose restraint is bigger than x, and enumerate x into D_{1-i}. If x leaves D at this stage, then remove x from $D_0 \cup D_1$. Let $N_{e,i}$ be the requirement with the highest priority that is affected by the removal of D. In both cases, initialize all the requirements with priority lower than $N_{e,i}$. (*Again, we are trying to find a requirement that requires attention at this stage* because of the changes of $D(x)$.)

s **even:** If x enters or leaves C at this stage, then find the requirement with the highest priority, $N_{e,i}$ say, whose definition of $\Delta_{e,i}^{D_i \oplus C}(n)$ is affected by the change of C at x, if any, and initialize all the requirements with priority lower than $N_{e,i}$. If there is no such a requirement, then do nothing. In both cases, go to the next stage. (*We are trying to find a requirement that requires attention at this stage* because of the changes of $C(x)$.)

This completes the construction of D_0 and D_1.

5 Verification

Obviously, the constructed D_0 and D_1 form a disjoint splitting of D. The N-requirement is satisfied.

In this section, we verify that D_0 and D_1 satisfies all the N-requirements. Thus, the wtt-degrees of $D_0 \oplus C$ and $D_1 \oplus C$ form a wtt-degree splitting of $deg_{wtt}(D)$ above $deg_{wtt}(C)$.

Lemma 1. *For each e, i,*

(1) $N_{e,i}$ can be initialized at most finitely many times;

(2) $N_{e,i}$ can be affected by the approximations of C and D at most finitely many times and consequently, $D \neq \Phi_e^{D_i \oplus C}$, $N_{e,i}$ is satisfied;

(3) $r(e, i) = \lim\limits_{s \to \infty} r(e, i, s)$ exists and is finite, and $N_{e,i}$ acts and initializes requirements with lower priority at most finitely many times.

Proof. We prove by induction on $\langle e, i \rangle$. Fix $\langle e, i \rangle$ and assume (1), (2), (3) hold for all $\langle k, j \rangle < \langle e, i \rangle$.

According to the construction, $N_{e,i}$ is initialized only by strategies of higher priority, i.e., $N_{k,j}$, with $\langle k, j \rangle < \langle e, i \rangle$. By the induction hypothesis (3), each such $N_{k,j}$ initializes $N_{e,i}$ at most finitely many times, and hence, $N_{e,i}$ is initialized finitely many times. (1) is true for $N_{e,i}$.

Let t be the least stage such that $N_{\langle e,i \rangle}$ cannot be initialized after this stage.

Assume for contradiction that $N_{e,i}$ is affected by approximations of C and D infinitely many times. As for a fixed n, for computation $\Phi_e^{D_i \oplus C}(n)$, the use $\phi_e(n)$ is bounded by $\psi_e(n)$, the affection of approximations of C and D on it is finite. So our assumption implies that $D = \Phi_e^{D_i \oplus C}$. We show below that under this assumption, D is wtt-reducible to $L(D_i) \oplus C$, via $\Lambda_{e,i}$ defined by us.

Fix p, and assume that after stage $t' > t$, further approximations of C and D do not affect computations $\Phi_e^{D_i \oplus C}(q)$, and also $\Lambda_{e,i}^{L(D_i) \oplus C}(q)$, where $q < p$, and $\Lambda_{e,i}^{L(D) \oplus C}(q) = D(q)$. By our assumption $D = \Phi_e^{D_i \oplus C}$, there is a stage $t'' > t$ such that further approximations of C and D do not affect the computation $\Phi_e^{D_i \oplus C}(p)$, and $\Phi_e^{D_i \oplus C}(p) = \Phi_e^{D_i \oplus C}(p)[t'] = D_{t'}(p) = D(p)$.

We now show that at a stage $s > t$, if we define $\Lambda_{e,i}^{L(D) \oplus C}(p)$ at this stage for the first time, we have $\Phi_e^{D_i \oplus C}(p)[s] = D_s(p)$, and we define $\Lambda_{e,i}^{L(D) \oplus C}(p)[s] = D_s(p)$, with use the C-part use as $\psi_e(p)$ and the $L(D)$-part use as the biggest stage at which some number less than $\psi_e(p)$ in $D_{i,s}$ enters D. This use will be fixed since this stage, and because of this, $\Lambda_{e,i}$ is a wtt-reduction. That is, if at a stage $s' > s$, $\Lambda_{e,i}^{L(D) \oplus C}(p)$ is undefined, then we make the definition, and keep the use unchanged. It is obvious for the C-part use, as $\psi_e(p)$ is fixed, and the C-part use in the computation of $\Phi_e^{D_i \oplus C}(p)$ is below $\psi_e(p)$. For the $L(D)$-part, note that from stage s onwards, no number less than $\psi_e(p)$ is allowed to be enumerated into D_i, and hence for any computation $\Phi_e^{D_i \oplus C}(p)$ we see after stage s, numbers in D_i involved in the computation are all in $D_{i,s}$, and hence when we redefine $\Lambda_{e,i}^{L(D) \oplus C}(p)$, the $L(D)$-part use is always less the one we defined at stage s. Thus, $\Lambda_{e,i}^{L(D) \oplus C}(p)$ is defined, with use fixed since stage s.

Now we show that from stage s, if $\Lambda_{e,i}^{L(D) \oplus C}(p)$ has definition at stage s', or is defined at this stage, then $\Phi_e^{D_i \oplus C}(p)[s'] = D_{s'}(p)$, we have $\Lambda_{e,i}^{L(D) \oplus C}(p)[s'] = D_{s'}(p)$. It is true because for any stage s'' with $s \le s'' < s'$, if computations $\Phi_e^{D_i \oplus C}(p)[s']$ and $\Phi_e^{D_i \oplus C}(p)[s'']$ are different, then either C changes below $\psi_e(p)$ or some numbers less than $\psi_e(p)$ leave D_i, and in the latter case, $L(D)$ changes below the use. In both cases, $\Lambda_{e,i}^{L(D) \oplus C}(p)[s'] = D_{s'}(p)$.

This shows that for each p, $\Lambda_{e,i}^{L(D) \oplus C}(p)$ is defined, $\Lambda_{e,i}^{L(D) \oplus C} = D$ and the reduction $\Lambda_{e,i}$ is a wtt-reduction, contradicting our assumption that $D >_{wtt} L(D) \oplus C$.

Therefore, $N_{e,i}$ can be affected by the approximations of C and D at most finitely many times, and we have $D \ne \Phi_e^{D_i \oplus C}$. (2) holds for $N_{e,i}$. As a consequence, $\Lambda_{e,i}^{L(D) \oplus C}$ can have the definition extended at most finitely many times.

For (3), note that $N_{e,i}$ sets a restraint at a stage s only when $\Lambda_{e,i}^{L(D) \oplus C}(n)$ is defined at this stage for some n, or approximations of C and D affect computations involved. By (2), we know that such actions can happen at most finitely many times. Hence we can set restraint for $N_{e,i}$ finitely many times, and hence has a finite limit. $N_{e,i}$ initializes strategies with lower priority only when $N_{e,i}$ is affected during the construction, and can take only finitely many such initializations during the whole construction. (3) holds for $N_{e,i}$.

This completes the proof of Lemma 1.

References

1. Ambos-Spies, K.: Contiguous r.e. degrees. In: Börger, E., Oberschelp, W., Richter, M.M., Schinzel, B., Thomas, W. (eds.) Computation and Proof Theory. LNM, vol. 1104, pp. 1–37. Springer, Heidelberg (1984). https://doi.org/10.1007/BFb0099477
2. Cooper, S.B.: Degrees of unsolvability. Ph.D. thesis, Leicester University (1971)
3. Cooper, S.B.: The density of the low$_2$ n-r.e. degrees. Arch. Math. Logic **31**, 19–24 (1991)
4. Cooper, S.B.: A splitting theorem for the n-r.e. degrees. Proc. Am. Math. Soc. **115**, 461–471 (1992)
5. Cooper, S.B., Harrington, L., Lachlan, A.H., Lempp, S., Soare, R.I.: The d.r.e. degrees are not dense. Ann. Pure Appl. Logic **55**, 125–151 (1991)
6. Downey, R., Stob, M.: Splitting theorems in recursion theory. Ann. Pure Appl. Logic **65**, 1–106 (1993)
7. Fang, C., Wu, G., Yamaleev, M.M.: On a problem of Ishmuhkametov. Arch. Math. Logic **52**, 733–741 (2013)
8. Fang, C., Liu, J., Wu, G., Yamaleev, M.M.: Nonexistence of minimal pairs in $L[\mathbf{d}]$. In: Beckmann, A., Mitrana, V., Soskova, M. (eds.) CiE 2015. LNCS, vol. 9136, pp. 177–185. Springer, Cham (2015). https://doi.org/10.1007/978-3-319-20028-6_18
9. Friedberg, R.M., Rogers Jr., H.: Reducibility and completeness for sets of integers. Z. Math. Logik Grundlag. Math. **5**, 117–125 (1959)
10. Ishmukhametov, Sh.: On the predececcors of d.r.e. degrees. Arch. Math. Logic **5**, 373–386 (1999)
11. Ishmukhametov, S.: On relative enumerability of turing degrees. Arch. Math. Logic **39**, 145–154 (2000)
12. Lachlan, A.H.: A recursively enumerable degree which will not split over all lesser ones. Ann. Math. Logic **9**, 307–365 (1975)
13. Ladner, R.E., Sasso, L.P.: The weak-truth-table degrees of recursively enumerable sets. Ann. Math. Logic **8**, 429–448 (1975)
14. Odifreddi, P.: Classical Recursion Theory, Volume II, vol. 143. Studies in Logic and the Foundations of Mathematics, Elsevier (1999)
15. Sacks, G.E.: On the degrees less than $0'$. Ann. Math. **77**, 211–231 (1963)
16. Sacks, G.E.: The recursively enumerable degrees are dense. Ann. Math. **80**, 300–312 (1964)
17. Soare, R.I.: Recursively Enumerable Sets and Degrees. Springer, Heidelberg (1987)
18. Stob, M.: *wtt*-degrees and T-degrees of recursively enumerable sets. J. Symb. Log. **48**, 921–930 (1983)
19. Wu, G., Yameleev, M.M.: There are no maximal d.c.e. *wtt*-degrees. In: Day, A., Fellows, M., Greenberg, N., Khoussainov, B., Melnikov, A., Rosamond, F. (eds.) Computability and Complexity. LNCS, vol. 10010, pp. 479–486. Springer, Cham (2017). https://doi.org/10.1007/978-3-319-50062-1_28

Finite State Machines with Feedback: An Architecture Supporting Minimal Machine Consciousness

Jiří Wiedermann[1(✉)] and Jan van Leeuwen[2(✉)]

[1] Institute of Computer Science, AS CR, Prague, Czech Republic
jiri.wiedermann@cs.cas.cz
[2] Department of Information and Computing Sciences, Utrecht University,
Utrecht, The Netherlands
J.vanLeeuwen1@uu.nl

Abstract. Finite state machines with feedback present a novel machine model when considered under the scenario of cognitive computations. The model is designed in the spirit of automata theory and presents a mix of Alan Turing's finite state machines and Norbert Wiener's machines with feedback. For the model we define, what we call, minimal machine consciousness and machine qualia. The design of our model is lead by natural engineering requirements. Its properties are justified by the latest findings in neuroscience and by ideas from the classical literature of the philosophy of mind. For the model a test distinguishing minimally conscious machines from unconscious ones ("zombies") on a given cognitive task is proposed. Our modeling supports the claim that consciousness is a computational phenomenon that is not just a matter of suitable software but also requires a dedicated architecture.

"We need to break down the concept of consciousness into different aspects, all of which tend to occur together in humans, but can occur independently, or some subset of these can occur on its own in an artificial intelligence. . . . We can imagine building something that has some aspects of consciousness and lacks others."
Murray Shanahan, The Space of Possible Minds, Edge, July 10, 2018.

1 Introduction

Although the efforts in the computational modeling of intelligence have a long tradition, it continues to be a challenge to design formal computational models that adequately capture all important aspects of the human mind. The respective efforts have started with Turing's foundational paper on machine intelligence [14]. Shortly later, Kleene [10] clarified the intimate connection between neural

The research of the first author was partially supported by the ICS AS CR fund RVO 67985807 and programme Strategy AV21 "Hopes and Risks of the Digital Age".

F. Manea et al. (Eds.): CiE 2019, LNCS 11558, pp. 286–297, 2019.
https://doi.org/10.1007/978-3-030-22996-2_25

modeling as understood at the time and finite automata theory. Since then, the efforts in computational modeling of the brain have permeated all of artificial intelligence and robotics.

For years, research has focused mainly on specific aspects of cognition, like vision, planning, and natural language processing (cf. [11,15]). Until recently, almost no attention was paid to *consciousness* from this perspective, not in the least because there was no clear-cut separation between intelligence and consciousness [4]. Some notable exceptions exist, such as the work by Aleksander [1]. He presented a theory of consciousness based on neural machines, starting from fundamental postulates that a neural machine should satisfy in order to exhibit "artificial consciousness". Aleksander and Dunmall [2] proposed a set of informal axioms, meant as necessary conditions for consciousness in agents. Bringsjord et al. [3] went even further and used both formal expressions and inference schemata in proposing a definition of artificial consciousness, based on formal axioms. These approaches all aim at modeling human-like consciousness.

With the advent of robotics, the deep issues of high-level cognitive processes and of mind as a whole, including consciousness, have entered the center stage of AI. This was also triggered by the current, unprecedented progress in the area of deep neural networks. Recent findings in neuroscience (cf. [5]) and in the philosophy of mind (cf. [18]) support the functional (or computational) theory of mind – namely, that cognitive functions of the brain, consciousness included, are of a computational nature. However, until now there were practically no attempts to investigate this using the classical means of theoretical computer science. What can theory tell us about the feasibility of designing robots with at least minimal consciousness?

Given this challenge, and leaving the current approaches to modeling consciousness in AI and the philosophy of mind aside for the moment, we propose an approach to consciousness in the best tradition and spirit of the theory of computation and, especially, of automata theory. This is quite different from the formal approaches [1,2] and [3] mentioned earlier, since we employ a formal machine rather than a formal theory model. A further difference is that, while the former approaches aim at modeling human consciousness—in some sense a "maximally" developed consciousness, our model has a different, more modest and pragmatic, purpose. Namely, we aim at the design of a computational model of cognitive systems that is as simple as possible while still capturing the minimalist requirements imposed by the intended presence of some important aspects of consciousness.

Our design is lead by natural engineering requirements, asking that cognitive systems should at least have some means for situating themselves in their environment, be able to monitor their functioning, and coordinate the actions of their sensory and motor units in an unpredictable environment. Using this as a guideline, we develop a model for the respective mechanisms. For this model we define, what we call, *minimal machine consciousness* and *machine qualia*. The presence of qualia is generally considered to be the hallmark of consciousness [13]. We point to analogues of our machine qualia with the classical philosophical notion of qualia (or subjective experience) and justify our model by the latest findings of neuroscience, claiming that conscious systems possess a mechanism for the global availability of information and have self-knowledge, self-monitoring and self-awareness ability [5]. Our model seems to be the first computational model utilizing qualia for enabling a correct behavior of cognitive systems in situations when a malicious adversary occasionally prevents machine's sensory and motor units from optimal performance.

The design of our model, giving sufficient conditions for a machine to exhibit minimal machine consciousness, is already the main result of our paper. Notably, the possession of minimal machine consciousness is not a purposeless property, but presents a set of natural engineering conditions as mentioned above which, when satisfied, facilitate a machine's meaningful behavior. The model also points to an important methodological finding, namely, that the presence of (even minimal) machine consciousness is not a matter of the software design of a model's inner machinery, or of the size of this machinery, as some thinkers believe. On the contrary, it is a matter of the organization of the system's interface with the environment, i.e., a matter of the system's architecture.

We stress that our paper is *not* intended to be a paper in the philosophy of mind and it is not aiming at modeling and explaining the human mind, although it has several philosophical connotations along this line. Rather, our model should be seen as a *theoretical* model whose aim is to study the properties of minimal machine consciousness, and of machine qualia, in computational artifacts like robots that can be designed and constructed. It is our contribution to the feasibility of robots with minimal machine consciousness.

To get an idea of what kind of machines can be endowed with minimal machine consciousness, think of computational devices provided with simple sensors and motor units with broadcasting facilities. Examples of such devices are molecular nanorobots operating in a confined space (e.g. the bloodstream) and communicating via signal molecules, mobile phones communicating via radio on various frequencies, or driverless cars fitted with ultrasound sensors, lidars and radars, GPS units, etcetera. More generally, devices which also include facilities for vision, requiring more sophisticated processing of inputs, also count.

2   An Automata-Based Model of Cognitive Machines

The basis of our model is an embodied interactive finite-state system equipped with sensory and motor units. What makes it different from a classical finite automaton is the *feedback* between its sensory units and its finite-state control, and between its motor units and its finite-state control. Moreover, our model works under a different, non-standard scenario than what is common in automata theory, and in fact, in the majority of contemporary computational systems. This so-called *cognitive scenario* captures the natural requirements that any cognitive system should meet in order to act successfully in its environment.

Namely, by its actions the model is continually generating sensory inputs that must be registered and classified. Doing so, inputs caused by the own actions of the system must be distinguished from information about independent external changes. In this way the system gets acquainted with what goes on in its environment and can single itself out from this environment. Last but not least, the system must be knowledgeable about the proper working of its sensory and motor units in order to be able to take corrective measures, if a need for it arises. Integrating and disambiguating all this knowledge, the system's finite control eventually generates new actions. In this way the system becomes a center of action and perception and builds the basis for its self-awareness.

Within this cognitive computational scenario, we consider a non-standard model of a finite-state system with sensor and motor units called a *cognitive automaton*. This automaton survives and operates in a certain environment. The system obtains inputs from its environment through its sensors. It has a fixed, finite number of sensors of several kinds. Sensory and motor units are of two types: external and internal ones. Depending on its type, a sensor sends both a representation of an occurrence of a phenomenon it is specialized to and the *quality* of the corresponding sensation. The quality of a sensation can be graded and depends on the nature of that sensation. It can be, e.g., its magnitude, intensity, frequency, etc. The system also has various motor units to which it sends instructions how to operate and from which it obtains *reports* whether, or to what extent, the proposed operation could be realized. The qualities of the sensations and reports from the motor units provide important information for a system's self-monitoring and self-awareness. The graded responses allow the system to monitor the working of its sensory and motor units. We assume that among the motor units there is one special unit – called the *engine* – that is always on and that powers all system activities. The source of energy of the engine is not a subject of our modeling. Some motor units may serve for the positioning of sensors, others for the manipulation of various of the machine's effectors.

The model operates as follows. In a single move (i.e., in parallel), it reads a tuple of (input) sensations from its sensory units and to each sensation it also reads the quality of that sensation. Moreover, it also reads reports from the motor units indicating whether and how the motor units have realized the instructions sent to the motor units in the previous step. Then, based in its internal state, the automaton makes a transition. A transition is dictated by a so-called *transition function* which assigns, to all signals mentioned above, a new instruction for each motor unit and a new internal state. In this way, the mechanism applies its "causal power" to the current state in order to determine a next state and a next action. A formal definition is as follows:

Definition 1. *A cognitive automaton with $k \geq 1$ sensory units and $n \geq 1$ motor units is a system $\langle A, S_1, \ldots, S_k, Q_1, \ldots, Q_k, R_1, \ldots, R_n, M_1, \ldots, M_n, \delta, q_0 \rangle$, where*

- *A is the finite set of states,*
- *S_i is the finite set of sensations from the i-th sensory unit, for $i = 1, \ldots, k$,*
- *Q_i is the finite set of qualities of sensations from the i-th sensory unit, for $i = 1, \ldots, k$,*

- R_j is the finite set of reports from the j-th motor unit, for $j = 1, \ldots, n$,
- M_j is the finite set of motor instructions for the j-th motor unit, for $j = 1, \ldots, k$,
- δ is a partial transition function from $A \times S_1 \times \ldots \times S_k \times Q_1 \times \ldots \times Q_k \times R_1 \times \ldots \times R_n$ to $A \times M_1 \times \ldots \times M_n$, and
- $q_0 \in A$ is the initial state.

From the definition of δ it follows that if q is a current internal state of the system, $s_1, s_2, \ldots, s_k$ current sensations, $q_1, q_2, \ldots q_k$ current qualities of the previous sensations, $r_1, r_2, \ldots, r_n$ the reports from the motor units, r a new internal state, and $m_1, m_2, \ldots, m_n$ motor instructions, then each transition has the following form: $(q, s_1, \ldots, s_k, q_1, \ldots q_k, r_1, \ldots, r_n) \rightarrow (r, m_1, \ldots, m_n)$. The left-hand part of the transition is called the *configuration of the automaton*.

From the automata-theoretic point of view, our model represents an interactive model of computation—a transducer—with feedback. It is interactive since the next move of the automaton depends on the reactions of the environment as perceived by the automaton's sensory units. Within the model feedback is provided internally—by the qualities of sensations and reports from the motor units which. Technically, these are not a part of the outputs from the system that are intended to go beyond (i.e., out of) the system. Rather, they are routed back to the system without leaving it. This corresponds to a classical (digital, in our case) feedback as described by Wiener [17]. It is a feedback that is different from the external feedback usually considered in purely interactive systems where feedback means that all outputs of a process can be read again by system's sensors as causal inputs to the ongoing process.

From the viewpoint of cognition, and more specifically, in order to enable minimal machine consciousness to occur in a cognitive automaton in a given environment, the automaton must possess elementary, but important self-control abilities. These abilities include self-knowledge, self-monitoring and self-awareness. *Self-knowledge* means that the system has available information about its state and that produced both by its sensors (the percepts and their qualities) and by its motor units (the reports from all of them). *Self-monitoring* is the ability of a system to evaluate the performance of its own sensory and motor units. Finally, *self-awareness* is the capacity of the system for introspection and to recognize itself as an individual object, separate from the environment and other systems (cf. Wikipedia), and the awareness of changes in the outside world.

Proposition 1. *There exist environments for which cognitive automata can be constructed possessing the capacity of self-knowledge, self-monitoring and self-awareness.*

The information needed for self-knowledge includes the information about a system's state and that produced both by its sensors (the percepts and their qualities) and by its motor units (the reports from all of them). Knowing what a system is presently doing is a form of self-knowledge that leads both to a subjective machine sense of presence and the capacity of reportability.

It is almost obvious that the feedback from the sensory and motor units enables the realization of self-monitoring. Namely, from these units the systems gets the data about the working conditions of the system, and based on this information it can either prolong its working without any further special actions or remedy it. All this leads to a subjective machine sense of certainty or error and enables the repair of a system's own mistakes [5].

The definition of self-awareness requires fulfillment of three conditions. First, the capacity for introspection, i.e., the ability to reflect one's own mental state (cf. [5]. In general, mechanisms of introspection seem to be beyond the ability of finite automata. However, in the framework of finite automata we will model introspection by a finite number of dedicated automata states—so-called machine qualia states. We will explain the latter notion in the next section.

Second, the ability to recognize oneself as an individual object separate from the environment and other objects is usually provided by a proper selection of a pair of sender-receiver units whose cooperation provides the required effect. There are several modalities of signals that can work in this way. For instance, reception of a returned specific olfactory (or chemosensory), electric, optical, acoustic or haptic signal indicates the presence of other instances of the respective object. Obviously, absence of such returning signals indicates that no similar objects are around. Using a vision system for a similar purpose is a more demanding process, which may go beyond our modeling possibilities.

Third, the feedback also allows the system to distinguish its own actions as observed by its sensory units from the similarly observed actions performed by other such systems. That is to say, in the latter case the reports from the motor units do not match the sensations from the sensory units.

Self-awareness provides a cognitive automaton with a rudimentary machine concept of the self – the automaton has the information on what goes on in the outside world, what its own actions are and what is their effect. This information is of the form "here and now" – it is pertinent to the present position of the automaton in its environment and to the present moment.

Definition 2. *Self-knowledgeable, self-monitoring and self-aware cognitive automata are called* minimally conscious *automata.*

By the very definition of a cognitive automaton its current state is "globally available" to all its transitions that can be evaluated independently, in parallel.

3   Classifying Cognitive Machines

A cognitive machine can be designed so as to perform an important cognitive activity—and this is *classification*. Classification is the problem of identifying to which of the sets of inputs a new observation belongs. We do not require that each observation belongs to at most one set of inputs. It may belong to several such sets, or to none at all. To this end we extend our model from Definition 1.

Let $\mathcal{C}$ be the set of all possible inputs to an automaton's finite control, and let $C_1, \ldots, C_p$, (the so-called *classification sets*), be subsets of $\mathcal{C}$ such that $\cup_{i=1}^{p} C_i = \mathcal{C}$.

Informally, a *classifying cognitive machine* with classification sets $C_1, \ldots, C_p$, is a cognitive machine that works as follows. At any time, the automaton is in a current internal state. On any input, the classification automaton determines the classification sets to which the input at hand belongs and assigns to it a new, uniquely defined, so-called *classification state*. Then, based on the current input and both the current internal and new classification state, the transition function of the automaton determines the new actions and the new internal state of the automaton.

Note that what the classification automaton does, is the decomposition of the space of all inputs into equivalence classes of classification sets represented by the respective classification states.

Definition 3. *Let* $A, S_i, Q_i, R_j, M_j, q_0$ *for* $i = 1, \ldots, k$, $j = 1, \ldots, n$ *be as in Definition 1. Let* $\mathcal{C} = S_1 \times \ldots \times S_k \times Q_1 \times \ldots \times Q_k \times R_1 \times \ldots \times R_n$ *be the set of all possible inputs to the automaton's finite control, let* $C_1, \ldots, C_p$, *be the* classification sets.

Given an input from $\mathcal{C}$ *the task of the classifying automaton is to identify all classification sets to which the given input belongs. This is done by determining, for each input a Boolean vector whose non-zero components denote the index of a classification set pertinent to that input. Formally, let* $\varkappa : \mathcal{C} \to \{0,1\}^p$, *the so-called* classification function, *be defined for any* $d \in \mathcal{C}$ *as follows:*

$$\varkappa(d) = (b_1, b_2, \ldots, b_p) \text{ where for } i = 1, \ldots, p : b_i = 1 \text{ iff } d \in C_i$$

The set B *of all vectors of form* $(b_1, b_2, \ldots, b_p) = \varkappa(d)$ *for some* $d \in \mathcal{C}$ *is called the set of* classifying states.

Then the classifying cognitive automaton *with* $k \geq 1$ *sensory units and* $n \geq 1$ *motor units and the classification sets* $C_1, \ldots, C_p$ *is the system*

$$\langle A, B, S_1, \ldots, S_k, Q_1, \ldots, Q_k, R_1, \ldots, R_n, M_1, \ldots, M_n, C_1, \ldots, C_p, \delta, \varkappa, q_0, \bot \rangle$$

where $\bot$ *is the initial classification state and* δ *is the partial transition function from* $(A \times B) \times S_1 \times \ldots \times S_k \times Q_1 \times \ldots \times Q_k \times R_1 \times \ldots \times R_n$ *to* $(A \times B) \times M_1 \times \ldots \times M_n$ *defined as follows:* $\delta((q, u), s_1, \ldots, s_k, q_1, \ldots q_k, r_1, \ldots, r_n) = ((r, v), m_1, \ldots, m_n))$, *with* $v = \varkappa(s_1, \ldots, s_k, q_1, \ldots q_k, r_1, \ldots, r_n)$.

Note that for a given input, classification function $\varkappa$ correctly returns in the respective classification state indexes of those classification sets to which the input belongs. Classification states arise by *fusion* of the data from different sources captured by $\mathcal{C}$. On one hand, this produces more consistent, accurate information than that provided by any individual data source. On the other hand, by possibly omitting information from certain sensors that are not important for judging the current situation the resulting information is more comprehensive and more informative for subsequent decision purposes.

Proposition 2. *Classifying cognitive automata possess the capacity of performing the classification task.*

When compared to the automaton from Definition 1, the classifying cognitive automaton has a set of "structured" internal states. Each such state stores in its first component the internal state that controls the computation of the automaton and in its second component the classifying state denoting the subsets of inputs to which the current input belongs. Thus, the transition function δ is the mechanism that performs two tasks at the same time: it realizes the "classical" transition as in Definition 1, and in addition, it also realizes the classification task (with the help of the classifying function $\varkappa$). Note that, as long as the inputs are from the same classification class, the automaton remains in the same classification state.

4   Machine Qualia

There are many definitions of qualia which have changed over time. None of them is supported by a computational model. Therefore, it makes sense to select a simple, broad definition for our purposes. One of the simplest definitions captures the "what-it-is-like" character of sensations. Hence, we will view qualia as characterizations of certain specific sensations to which specific states are assigned. Thus, a quale is characterized by the corresponding state called a *quale state*, and by both qualitative and quantitative aspects of sensations. Qualitative aspects describe what sensations are important to a given quale state, while quantitative aspects describe the required quality of such sensations.

The key to defining qualia are *classification* sets. To this end, the classification sets must be designed so as to capture important circumstances in the operation of a cognitive machine that matter and require a similar response. Such events are defined by the current input from the sensory units and by the conditions that the feedback information from the sensory and motor units must satisfy. That is, classification sets are no longer defined by enumeration of their members (as it was the case with Definition 3), but in a more compact way by computable predicates. These predicates list the sensory and motor units that are active in the circumstance of interest, the expected outputs from the sensory units, their expected quality, and the expected types of reports from the motor units. Thus, these predicates in fact check whether the inputs at hand satisfy a certain set of conditions that indicate for the machine how it should react to its current inputs.

Definition 4. Qualia sets *are specific classifying sets of the classifying cognitive machine, defined in a compact way with the help of computable predicates. Qualia sets differentiate the inputs with respect to the similarity of circumstance and the actions that should be taken by the machine under a given circumstance. Inputs assigned to the same qualia set cause the machine to enter into the respective, uniquely defined classification state called the* quale state *that represents the entire quale set. The members of a qualia set defined by a given quale state are called the* machine qualia *defined by that state.*

Note that the current quale state is made globally available to the entire automaton, as the automaton's transition depends (also) on this state (cf. Definition 3).

Definition 5. *Let $\mathcal{A}$ be a minimally conscious classification automaton, let b be a quale state. Then we say that $\mathcal{A}$ perceives a quale defined by the given input if and only if this input is classified by $\mathcal{A}$ as belonging to the quale set defined by b.*

It is important to note the perception of machine qualia is defined only for minimally conscious machines. This is because without self-knowledge, self-monitoring and self-awareness there would be no subject, no mechanism playing the role of an internal machine "observer" perceiving the machine qualia. An other implicit condition for a machine quale to be perceived is that all inputs to a machine must be registered in parallel, as the machine must "globally" react to the entire input rather than "locally" to its parts only. Also note that perceiving a quale is a primitive form of introspection which is an important aspect of self awareness—in the quale state a remembrance of the characteristic circumstance that lead to activation of the quale at hand is preserved.

Observe that a machine can remain in the same quale state during a sequence of internal states. This is because a change of a quale state is invoked by signals from specified sensory and motor units, and it can take several clock cycles between the initiation and the end of a sensory or motor operation, since these operations may include slow mechanical actions, or for other reasons.

Examples of specific machine qualia are, e.g., for a nanorobot: "achieving the threshold of quorum sensing", for a smartphone: "receiving a blue-tooth signal", for a driverless car: "the sensing of car sliding", etc.

5   Justification of the Model

Although our model has not been intended as a model of human brain, it is interesting to compare our notion of machine consciousness and machine qualia with the ideas of philosophers of mind concerning human consciousness and human qualia.

Our notion of minimal machine consciousness of (artificial) cognitive machines is close in spirit to the ideas of Edelman [8] on primary consciousness and to those of Damasio [6] on core consciousness concerning humans and animals. According to Edelman, primary consciousness refers to being mentally aware of things in the world in the present time without any sense of past and future. Damasio [6] has it that *"core consciousness enables a sense of self about one moment – now, and one place – here. The scope of core consciousness is here and now."*

As far as our notion of machine qualia is concerned Dennett [7] has explored a classical conception of human qualia and gave four conditions that these must satisfy:

(i) Qualia are directly or immediately apprehensible in consciousness, (ii) qualia are intrinsic, (iii) qualia are ineffable, and (iv), qualia are private.

Dennett famously argued that, if these conditions are to hold, human qualia are vacuous. However, we show that in our model of cognitive machines machine qualia satisfying these conditions *do* exist.

Consider a cognitive machine $\mathcal{A}$ in the context of the epistemic theory of computation (cf. [16]). The machine works in an environment $\mathcal{E}$ that is described by a certain epistemic theory that captures all objects in $\mathcal{E}$, their properties and relations among them knowable by the sensorimotor apparatus of $\mathcal{A}$. Let us see whether our machine qualia (cf. Definition 4) satisfy the four conditions imposed by Dennett.

First of all, machine qualia are obviously directly and immediately apprehensible in a minimally consciousness machine by definition. Remember that minimal machine consciousness requires that the machine must be self-knowledgeable, self-monitoring, and self-aware, meaning that the machine must have knowledge of its own self-monitoring and self-knowledge pertaining to given time and space, here and now.

Secondly, machine qualia in minimally conscious machines are intrinsic, because they are designed in this way and belong to essential nature of the machine. They originate by fusion of machine's sensorimotor percepts by a mechanism that is a part of a machine's design isolated from any other influence.

The matter of the ineffability of machine qualia is interesting. They are ineffable from a machine's own perspective: the machine cannot, in principle, express its qualia in terms of its epistemic theory, since qualia represent non-existing objects and data that are knowable by activities of its sensorimotor units (since such objects do not occur in the machine's environment). However, from the viewpoint of the machine's designer, machine qualia are effable. The designer can "measure" the quality of the machine's instantaneous observation by tampering with the sensorimotor units and he can give a complete description of any machine quale. A designer can even record and copy them into an (genuine) copy of the machine at hand and invoke them in this new machine.

Finally, the latter consideration also gives an answer to the question about the privacy of machine qualia. From the machine's perspective, its qualia are private: without having access to the machine's fusion mechanism, nobody or nothing can infer the machine's quale state.

To conclude, if one accepts that our model captures at least some valid aspects of qualia (as considered by Dennett), it seems that Dennett's arguments for the non-existence of human qualia are too feeble. This is because we have a model based on realistic groundings, whereas Dennett has no such a model.

6   On the Power of Minimal Machine Consciousness

In the philosophy of mind, a (philosophical) *zombie* is a hypothetical being that from the outside its behavior is indistinguishable from a normal human being, but lacks conscious experience, qualia, or sentience (cf. [9]). We argue that a minimally machine conscious machine would be able to perform under certain circumstance in ways a *zombie machine* lacking minimal machine consciousness would never be able to act.

Proposition 3. *Let* $\mathcal{M}$ *be an arbitrary minimally conscious (non-zombie) machine and* $\mathcal{Z}$ *an arbitrary zombie machine. Then there exist situations in which* $\mathcal{Z}$ *cannot behave in the same way as* $\mathcal{M}$ *does.*

Consider an arbitrary zombie machine $\mathcal{Z}$ which is, by definition, not minimally machine conscious. That is, $\mathcal{Z}$ is either not self-knowledgeable, or not self-monitoring, or not self-aware. If $\mathcal{Z}$ is not self-knowledgeable, then in some situation it will not be registering some information vital for the behavior of $\mathcal{M}$ and therefore, in this particular situation, $\mathcal{Z}$ cannot behave in a way as $\mathcal{M}$ does. If $\mathcal{Z}$ is not self-monitoring then again, in situations where an adversary prevents some non-monitored motor unit of $\mathcal{Z}$ from performing as expected, then $\mathcal{Z}$, not being informed about the failure of its action, cannot behave in a way as $\mathcal{M}$ does. Last but not least, if $\mathcal{Z}$ is not self-aware, it must err sometimes in the classification of situations that have been evaluated by $\mathcal{M}$ as a consequence of its own previous actions.

Based on the previous argumentation we can design an experimental test for minimal machine consciousness. The test is based on the assumption that minimally conscious machines are designed so as to behave purposefully under all external conditions that a machine at hand can encounter. The idea of the test of an unknown machine is to check whether this assumption is always fulfilled. If the tested machine happens to be a zombie machine then it cannot pass the test since it will miss information which is important to its behavior due to the lack of self-knowledge, or self-monitoring, or self-awareness ability.

To this end, we test the machine at hand whether it can adjust its behavior to variable conditions in its environment that would complicate the machine's mission. That is, we check whether the machine will operate meaningfully under the conditions of temporarily preventing its sensors or motor units in their free operation. If under such conditions the machine starts to behave erratically or nonsensically, we conclude that it was a zombie machine.

Comparing our experimental test with the recently proposed test by Schneider and Turner [12], which is in fact a variant of the Turing test focused on discovering subjective experience during a verbal interaction with the subject, we immediately see the advantages of our test. Namely, our test is suitable for consciousness testing in living and non-living entities, and it is a behavioral test not dependent upon understanding a natural language.

The example of zombies illustrates the advantages of possessing minimal machine consciousness: without it, a cognitive machine would at best be a zombie machine, unable to deal with non-standard situations.

7 Conclusions

We have presented a relatively simple computational model of cognitive machines that makes use of machine qualia to implement the basic ingredients of the minimal machine consciousness: global availability of information, self-knowledge, self-monitoring and self-awareness. The emergence of minimal machine consciousness has been enabled by a specific architecture of cognitive machine with internal feedback between its sensory units and its finite-state control, and between its motor units and its finite-state control. Therefore it is to be expected that in general, a "full" (machine) consciousness will not only be a matter of a suitably programmed "standard" model of computation, but will also require a non-standard, specific architecture such as in our model. This finding seems to be the essential contribution to the theory of machine consciousness.

References

1. Aleksander, I.: Neuroconsciousness: a theoretical framework. Neurocomputing **12**, 91–111 (1996)
2. Aleksander, I., Dunmall, B.: Axioms and tests for the presence of minimal consciousness in agents I: preamble. J. Consc. Stud. **10**(4–5), 7–18 (2003)
3. Bringsjord, S., Bello, P., Govindarajulu, N.S.: Toward axiomatizing consciousness. In: Jacquette, D. (ed.) The Bloomsbury Companion to the Philosophy of Consciousness, pp. 289–324. Bloomsbury Academic, London (2018)
4. Chella, A., Manzotti, R.: Artificial consciousness. In: Cutsuridis, V., Hussain, A., Taylor, J. (eds.) Perception-Action Cycle: Models, Architectures, and Hardware. SSCNS, pp. 637–671. Springer, New York (2011). https://doi.org/10.1007/978-1-4419-1452-1_20
5. Dehaene, S., Lau, W., Kouider, S.: What is consciousness, and could machines have it? Science **358**(6362), 486–492 (2017)
6. Damasio, A.: The Feeling of What Happens: Body and Emotion in the Making of Consciousness. Harcourt, New York (1999)
7. Dennett, D.: Quining qualia. In: Marcel, A., Bisiach, E. (eds.) Consciousness in Modern Science. Oxford University Press, Oxford (1988)
8. Edelman, G.: Naturalizing consciousness: a theoretical framework. Proc. Natl. Acad. Sci. **100**(9), 5520–5524 (2003)
9. Kirk, R.: Zombies. In: Zalta, E.N. (ed.) The Stanford Encyclopedia of Philosophy (Summer 2009 Edition) (2009)
10. Kleene, S.C.: Representation of events in nerve nets and finite automata. Project RAND, Research Memorandum RM-704, The RAND Corporation (1951)
11. Samsonovich, A.V.: Toward a unified catalog of implemented cognitive architectures. BICA **221**(2010), 195–244 (2010)
12. Schneider S., Turner E.: Is anyone home? A way to find out if AI has become self-aware. Scientific American, Observations, July 19 2017
13. Tegmark, M.: Life 3.0: Being Human in the Age of Artificial Intelligence. Knopf, New York (2017)
14. Turing, A.: Computing machinery and intelligence. Mind **236**(59), 433–460 (1950)
15. Vernon, D., von Hofsten, C., Fadiga, L.: Computational models of cognition. In: Vernon, D., von Hofsten, C., Fadiga, L. (eds.) A Roadmap for Cognitive Development in Humanoid Robots. COSMOS, vol. 11, pp. 81–99. Springer, Heidelberg (2010). https://doi.org/10.1007/978-3-642-16904-5_5
16. Wiedermann, J., van Leeuwen, J.: Epistemic computation and artificial intelligence. In: Müller, V.C. (ed.) PT-AI 2017. SAPERE, vol. 44, pp. 215–224. Springer, Cham (2018). https://doi.org/10.1007/978-3-319-96448-5_22
17. Wiener, N.: Cybernetics: Or Control and Communication in the Animal and the Machine. Hermann & Cie, MIT Press, Paris, Cambridge (1948)
18. Yampolskiy, R.: Detecting qualia in natural and artificial agents. arXiv:1712.04020 (2017)

Representations of Natural Numbers and Computability of Various Functions

Michał Wrocławski[(✉)]

Institute of Philosophy, University of Warsaw, Warsaw, Poland
michalwro@wp.pl

Abstract. We discuss various ways of representing natural numbers in computations. We are primarily concerned with their computational properties, i.e. which functions each of these representations allows us to compute. We show that basic functions, such as successor, addition, multiplication and exponentiation are largely computationally independent from each other, which means that in most cases computability of one of them in a certain representation does not imply that others will be computable in it as well.

We also examine what difference can be made if we restrict our attention only to those representations in which it is decidable whether two numerals represent the same number. It turns out that the impact of such restriction is huge and that it allows us to rule out representations with certain unusual properties.

Keywords: Representations of numbers · Computable functions · Characteristic functions

1 Introduction

Various authors have been considering the view that the notion of computability applies in the first place to functions on numerals, rather than on numbers themselves. Such position has been suggested by Shapiro in [4] and further discussed, among others, by Rescorla in [3], Copeland and Proudfoot in [1] and Quinon in [2]. I have also considered related issues in [5].

All algorithms are performed on strings of symbols which denote numbers (or other objects)—but a certain number can be represented by different strings. E.g. the number 6 is represented as VI when we use Roman numerals, but by 110 if we want to use binary numerals. Computation of a function such as addition is different in each of these cases.

According to Church's thesis, computable functions are exactly recursive functions. However, if we allow non-standard ways of encoding numbers, this does not have to be true. A set of numerals (satisfying a few additional conditions specified in the next section) together with a function assigning a natural number to each numeral, shall be called a representation.

© Springer Nature Switzerland AG 2019
F. Manea et al. (Eds.): CiE 2019, LNCS 11558, pp. 298–309, 2019.
https://doi.org/10.1007/978-3-030-22996-2_26

In this paper we are going to examine computability of the most important functions on natural numbers: successor, addition, multiplication and exponentiation. While they are all recursive and hence their computability is normally taken for granted, we want to show that this is not always the case (i.e. not in every representation). Furthermore, as it turns out, these functions are computationally largely independent from each other—i.e. the assumption of computability of one of them in most cases does not guarantee computability of the others.

We shall also provide a suggestion of an additional constraint on representations which will allow us to rule out representations with particularly irregular properties. Namely, if for a certain representation there exists an algorithm which for any two numerals determines whether they represent the same number or not, then such representations exhibit properties much more similar to representations usually employed.

2   Defining the Concept of Representation

In this section we are going to define some basic notions regarding representations.

Definition 1. *Let Σ be a finite alphabet. We shall call (S, σ) a representation of $\mathbb{N}$, where $S \subseteq \Sigma^*$ is an infinite computable set and $\sigma : S \to \mathbb{N}$ is a surjection.*

Definition 2. *Let (S, σ) be a representation of $\mathbb{N}$. We shall say that this representation is unambiguous iff for every $n \in \mathbb{N}$ there exists exactly one numeral $\alpha \in S$ such that $\sigma(\alpha) = n$. Otherwise we shall call the representation ambiguous.*

The basic example of a representation is the unary representation defined as follows:

Let $\Sigma = \{\bar{1}\}$. S is the set of all finite sequences comprised of $\bar{1}$ and the empty word ε, and the function σ is defined in the following way:

$$\sigma(\varepsilon) = 0,$$

$$\text{if } \sigma(\alpha) = n, \text{then } \sigma(\alpha \frown \bar{1}) = n + 1.$$

Another representation, which we shall refer to throughout this paper as the standard representation, is the decimal representation, defined as follows:

Let $\Sigma = \{\bar{0}, \bar{1}, ..., \bar{9}\}$. S is the set of all standard decimal numerals (i.e. the set consisting of the numeral $\bar{0}$ and of all finite sequences of digit from Σ which do not begin with $\bar{0}$), and the function σ is defined in the following way:

$$\sigma(\overline{a_n}...\overline{a_0}) = \sum_{i=0}^{n} a_i \cdot 10^i,$$

Both these representations are unambiguous.

In unambiguous representations, the concept of computability is simple. A function is computable if there exists an algorithm which for every numeral (representing a certain number) supplied on the input, returns the numeral representing the value of the function on the output. The issue gets more complicated when it comes to ambiguous representations. This is how we define computability in general case:

Definition 3. *Let (S, σ) be a representation of $\mathbb{N}$. Then for any function, $f : \mathbb{N}^n \to \mathbb{N}$, by $f^\sigma : S^n \to S$ we shall denote a function such that for any $\alpha_1, ..., \alpha_n, \beta \in S$ the following condition is satisfied:*

$$f^\sigma(\alpha_1, ..., \alpha_n) = \beta \Rightarrow f(\sigma(\alpha_1), ..., \sigma(\alpha_n)) = \sigma(\beta).$$

If there exists a computable function f^σ satisfying the above condition, than we shall say that f is computable in (S, σ).

Note that in case of ambiguous representations, many such functions f^σ can exist. It is possible that some of them are computable, and some are not. We adopt a convention that "to compute the function f in (S, σ)" and "to compute f^σ" are both going to mean "to compute any function f^σ which satisfies the above condition".

We will also want to be able to compute Boolean functions, i.e. functions whose values are $TRUE$ and $FALSE$.

Definition 4. *Let $R \subseteq \mathbb{N}^n$. The characteristic function of the relation R is the function χ_R such that for any $a_1, ..., a_n \in \mathbb{N}$ the following holds:*

$$\chi_R(a_1, ..., a_n) = TRUE \Leftrightarrow R(a_1, ...a_n).$$

$$\chi_R(a_1, ..., a_n) = FALSE \Leftrightarrow \neg R(a_1, ...a_n).$$

In this paper we are going to be particularly concerned with the characteristic function of identity:

$$\chi_=(a_1, a_2) = TRUE \Leftrightarrow a_1 = a_2,$$

$$\chi_=(a_1, a_2) = FALSE \Leftrightarrow a_1 \neq a_2.$$

The computability of characteristic functions is defined in a similar way as in the case of numerical functions.

Definition 5. *Let (S, σ) be a representation of $\mathbb{N}$. Then for any relation $R \subseteq \mathbb{N}^n$ we shall define $R^\sigma \subseteq S^n$ in the following way:*

$$(\alpha_1, ..., \alpha_n) \in R^\sigma \Leftrightarrow (\sigma(\alpha_1), ..., \sigma(\alpha_n)) \in R,$$

for all $\alpha_1, ..., \alpha_n \in S$. We shall say that χ_R is computable (or simply that R is computable) in (S, σ) if and only if R^σ is computable.

Note that $TRUE$ and $FALSE$ are neither numerals, nor numbers, but they are entirely different symbols.

3   Computability of Successor, Addition, Multiplication and Exponentiation in Representations of Natural Numbers

In this section we are going to show what are the relations between computability of some basic functions. In particular, we want to emphasise the role of computability of characteristic function of identity $\chi_=$.

The proofs of Theorems 6 (in a modified form) and 7 come from my paper [5]. The former theorem is a generalised version of Shapiro's result included in his paper [4]. Shapiro considered only unambiguous representations (in his terminology—notations), which is a very common approach among authors dealing with this subject. I have generalised his result to include all types of representations.

Theorem 6. *Let (S, σ) be a representation of $\mathbb{N}$ in which successor and $\chi_=$ are computable. Then all functions computable in the standard representation, including addition, multiplication and exponentiation, are also computable in (S, σ).*

Proof. Let (S, σ) be a representation of $\mathbb{N}$ in which the successor function (denoted as $Succ$) and $\chi_=$ are computable. In this representation there is a numeral representing number 0. Let us denote such a numeral as α, i.e. let $\alpha \in S$ be such that $\sigma(\alpha) = 0$. Note that for the purpose of this proof we only need to know that such α exists, not which numeral it is. This is because it is our aim here only to prove the existence of an algorithm, not to state which exactly algorithm it is.

We shall first show how to translate numerals from (S, σ) to the standard representation.

Let $\bar{n}$ be a numeral representing n in the standard representation, for every natural number n. The purpose of this convention is to clearly distinguish between standard numerals and numbers which they denote.

Let λ be a numeral of (S, σ). For every natural number n, let us denote $\lambda_n = Succ^\sigma(Succ^\sigma(...(\alpha)...))$, where the successor is iterated n times in λ_n. We compare one by one each λ_n with λ until we find such n that $\chi_{\underline{=}}^\sigma(\lambda, \lambda_n) = TRUE$. Then $\sigma(\lambda) = n$, so the numeral $\bar{n}$ represents the same number in the standard representation as the numeral λ in (S, σ).

Let $\bar{n}$ be a numeral of the standard representation. To find its counterpart in (S, σ), we calculate λ_n defined as above.

Now suppose that f is computable in the standard representation. We want to compute this function in (S, σ) on some given input. In order to do so, we translate this input to the standard representation, perform an algorithm in the standard representation and then translate the output back to (S, σ).

Theorem 7. *There exists a representation (S, σ) of $\mathbb{N}$ in which the successor function is computable, but addition, multiplication and exponentiation are not computable.*

Proof. We construct (S, σ) as follows:

The alphabet consists of symbols: $\bar{0}, \bar{1}, a$.

The set of numerals S consists of all finite non-empty sequences of symbols from the alphabet which contain at most one occurrence of a.

Let $A \subseteq \mathbb{N}$ be uncomputable in the standard representation.

We construct σ in the following way:

$$\sigma(\bar{0}) = 0,$$
$$\sigma(\bar{1}) = 1,$$
$$\sigma(a) = 0 \Leftrightarrow 1 \notin A,$$
$$\sigma(a) = 1 \Leftrightarrow 1 \in A.$$

Also, for any $\alpha \in S$:

$$\sigma(\alpha \frown \bar{0}) = \sigma(\alpha),$$
$$\sigma(\alpha \frown \bar{1}) = \sigma(\alpha) + 1,$$
$$\sigma(\alpha \frown a) = \sigma(\alpha) \Leftrightarrow lh(\alpha) + 1 \notin A,$$
$$\sigma(\alpha \frown a) = \sigma(\alpha) + 1 \Leftrightarrow lh(\alpha) + 1 \in A,$$

where $\frown$ is a concatenation and $lh(\alpha)$ is the length of the sequence α.

This is a correct representation because every natural number n is represented by at least one numeral, namely $\bar{1}...\bar{1}$ consisting of n digits $\bar{1}$, with the exception of number 0, which is represented by the numeral $\bar{0}$.

For any $\alpha \in S$, let $\#_{\bar{1}}(\alpha)$ denote the number of occurrences of symbol $\bar{1}$ in the numeral α.

The successor function in (S, σ) can be computed as follows:

$$Succ^{\sigma}(\alpha) = \alpha \frown \bar{1}.$$

We shall show that addition is not computable in this representation. Suppose to the contrary that it is.

For any natural number $n \geq 1$ let us denote:

$$\lambda_n = \bar{0}...\bar{0}a,$$

where λ_n consists of $n - 1$ digits $\bar{0}$ followed by one occurrence of a.

We want to find out whether $n \in A$. We compute $\lambda_n + \lambda_n$ in (S, σ). We know that $\sigma(\lambda_n)$ is equal to 0 or 1. Thus $\sigma(\lambda_n +^{\sigma} \lambda_n)$ is equal to 0 or 2.

If $n \in A$, then $\sigma(\lambda_n) = 1$ and $\sigma(\lambda_n +^{\sigma} \lambda_n) = 2$. Then $\#_{\bar{1}}(\lambda_n +^{\sigma} \lambda_n) \geq 1$. If, however, $n \notin A$, then $\sigma(\lambda_n) = \sigma(\lambda_n +^{\sigma} \lambda_n) = 0$ and then $\#_{\bar{1}}(\lambda_n +^{\sigma} \lambda_n) = 0$.

It is easy to find out which of these cases occurs and thus—whether $n \in A$. It follows that A is computable in the standard representation, which contradicts our assumption. Therefore, addition is not computable in (S, σ).

Similarly we show that multiplication and exponentiation are not computable in (S, σ). Let us denote:

$$\delta_n = \bar{1}...\bar{1}a,$$

where λ_n consists of $n-1$ digits $\bar{1}$ followed by one occurrence of a. Then we compute respectively $\delta_n \cdot \delta_n$ or $\delta_n^{\bar{1}\bar{1}}$ in (S, σ) (note that they both return the same result, we shall only provide a proof for the case with multiplication).

Suppose that multiplication is computable in (S, σ). We shall prove that A is also computable. Let $n \in \mathbb{N}$. We want to find out whether $n \in A$. Without loss of generality we can assume that $n \geq 2$.[1] Let $\alpha \in S$ be the result of multiplication $\delta_n \cdot \delta_n$ in (S, σ). We know that $\sigma(\delta_n)$ is equal to either $n-1$ or n. Therefore:

1. If $\sigma(\delta_n) = n - 1$, then $\sigma(\delta_n \cdot^\sigma \delta_n) = (n-1)^2 = n^2 - 2n + 1$. Therefore $\#_{\bar{1}}(\alpha) = n^2 - 2n$ or $\#_{\bar{1}}(\alpha) = n^2 - 2n + 1$.
2. If $\sigma(\delta_n) = n$, then $\sigma(\delta_n \cdot^\sigma \delta_n) = n^2$. Therefore $\#_{\bar{1}}(\alpha) = n^2 - 1$ or $\#_{\bar{1}}(\alpha) = n^2$.

Note that for $n \geq 2$ we can find out which of these cases occurs. If the first case occurs, then $n \notin A$, otherwise $n \in A$. Thus we have obtained contradiction with the assumption that A is not computable. Therefore multiplication (and similarly exponentiation) is not computable in (S, σ).

Theorem 8. *Let (S, σ) be a representation of $\mathbb{N}$ in which addition is computable. Then the successor function is also computable in this representation.*

Proof. Let (S, σ) be a representation of $\mathbb{N}$ in which addition is computable. In (S, σ) there must be a numeral representing number 1. Let us denote this numeral as β. Then we can calculate the successor function in (N, σ) as follows:

$$Succ(\alpha) = \alpha +^\sigma \beta.$$

Theorem 9. *There exists a representation (S, σ) of $\mathbb{N}$ in which addition (and thus also successor) is computable, but multiplication and exponentiation are not computable.*

Proof. For any natural number n, let $\bar{n}$ denote the numeral which represents n in the standard representation of $\mathbb{N}$.

We construct the following representation (S, σ):

The alphabet Σ consists of digits $\bar{0}, \ldots, \bar{9}$, of symbols (,) and the comma.

We construct the set S of numerals as follows:

For any standard numerals $\overline{a_0}, \ldots, \overline{a_n}$, the sequence $(\overline{a_0}, \ldots, \overline{a_n})$ is a numeral of the representation (S, σ) if $a_0 \geq \sum_{i=1}^{n} a_i$.

Let $A \subseteq \mathbb{N}$ be uncomputable in the standard representation such that $0 \in A$. For any $(\overline{a_0}, \ldots, \overline{a_n}) \in S$ the function σ is defined as follows:

$$\sigma((\overline{a_0}, \ldots, \overline{a_n})) = \sum_{i=0}^{n}(a_i \cdot \chi_A(i)),$$

where for any natural number i: $\chi_A(i) = 1$ if $i \in A$, and $\chi_A(i) = 0$ if $i \notin A$.

---

[1] The algorithm which is supposed to find out whether $n \in A$ will have answers for $n \in \{0, 1\}$ explicitly given as special cases.

This representation is well-defined because every natural number is represented by at least one numeral, in particular n is represented by $(\overline{n})$.

For any numerals $(\overline{a_0}, ..., \overline{a_m})$ and $(\overline{b_0}, ..., \overline{b_n})$ (without loss of generality we assume that $m \leq n$), we define addition in (S, σ) in the following way:

$$(\overline{a_0}, ..., \overline{a_m}) +^{\sigma} (\overline{b_0}, ..., \overline{b_n}) = (\overline{a_0 + b_0}, ..., \overline{a_m + b_m}, \overline{b_{m+1}}, ...\overline{b_n}),$$

where $+_S$ is interpreted as addition of numbers represented by respective numerals in the standard representation. It is obviously computable.

We shall prove that multiplication is not computable in this representation. Suppose that it is computable. We shall show that then A is computable in the standard representation which leads to a contradiction.

We want to find out whether $n \in A$.

For any natural number n we define the following numeral:

$$\lambda_n = (\overline{1}, \overline{0}, ..., \overline{0}, \overline{1}),$$

where λ_n has $\overline{1}$ on the zeroth and n-th position and $\overline{0}$ on all the other positions.

We compute the multiplication $\lambda_n \cdot \lambda_n$ in (S, σ). There are two possible cases:

If $n \in A$, then $\sigma(\lambda_n) = 2$ and $\sigma(\lambda_n \cdot \lambda_n) = 4$. From the condition that $a_0 \geq \sum_{i=1}^{n} a_i$ it follows that $a_0 \geq 2$ for every numeral representing number 4 in this representation.

If $n \notin A$, then $\sigma(\lambda_n) = 1$ and $\sigma(\lambda_n \cdot \lambda_n) = 1$, so $a_0 = 1$ in a numeral representing number 1 in this representation.

We determine which of these cases occurs and thus we can find out whether $n \in A$. Therefore A is a computable set in the standard representation, which leads to a contradiction. It follows that multiplication is not computable in (S, σ).

Similarly, by considering the result of the computation $\lambda_n{}^{\lambda_n}$ we can show that exponentiation is not computable in this representation.

We compute $\lambda_n{}^{\lambda_n}$ in (S, σ). There are two possible cases:

If $n \in A$, then $\sigma(\lambda_n) = 2$ and $\sigma(\lambda_n{}^{\lambda_n}) = 4$. From the condition that $a_0 \geq \sum_{i=1}^{n} a_i$ it follows that $a_0 \geq 2$ for every numeral representing number 4 in this representation.

If $n \notin A$, then $\sigma(\lambda_n) = 1$ and $\sigma(\lambda_n{}^{\lambda_n}) = 1$, so $a_0 = 1$ in a numeral representing number 1 in this representation.

We determine which of the cases occurs and thus we can find out whether $n \in A$. Therefore A is a computable set in the standard representation, which leads to a contradiction. It follows that exponentiation is not computable in (S, σ).

Theorem 10. *There exists a representation (S, σ) of $\mathbb{N}$ in which multiplication and $\chi_=$ are computable, but addition and exponentiation are not computable.*

Proof. Let π be a permutation of $\mathbb{N}$ uncomputable in the standard representation.

We construct the following representation. The alphabet consists of the digits $\bar{0}, \ldots, \bar{9}$, of the symbols $(\ ,\ )$ and the comma.

The admissible numerals in (S, σ) are all finite sequences of the form $(\overline{a_0}, \ldots, \overline{a_n})$, where each a_i is a natural number. Additionally, the numeral $\bar{0}$ belongs to S.

We construct σ as follows:

$$\sigma(\bar{0}) = 0,$$

$$\sigma((\overline{a_0}, \ldots, \overline{a_n})) = p_{\pi(0)}^{a_0} \cdot \ldots \cdot p_{\pi(n)}^{a_n},$$

where $\overline{a_i}$ is the numeral representing a_i in the standard representation and p_i is the i-th prime number.

It is a correct representation because each natural number is represented by a certain numeral, which results from the fundamental theorem of arithmetic.

For any numerals $(\overline{a_0}, \ldots, \overline{a_m})$ and $(\overline{b_0}, \ldots, \overline{b_n})$ (without loss of generality we assume that $k \leq l$), we define multiplication in (S, σ) in the following way:

$$(\overline{a_0}, \ldots, \overline{a_m}) \cdot^{\sigma} (\overline{b_0}, \ldots, \overline{b_n}) = (\overline{a_0 + b_0}, \ldots, \overline{a_m + b_m}, \overline{b_{m+1}}, \ldots \overline{b_n}),$$

where $+$ is interpreted as addition of numbers in the standard representation.

Additionally, for any $\alpha \in S$, let $\alpha \cdot^{\sigma} \bar{0} = \bar{0} \cdot^{\sigma} \alpha = \bar{0}$.

Hence, multiplication is computable in (S, σ). The function $\chi_=$ is also computable, as a consequence of the fundamental theorem of arithmetic. We shall show that addition and exponentiation are not computable in this representation.

Let us assume that addition is computable in this representation. We shall show that then the permutation π must be computable in the standard representation, which leads to a contradiction.

Let n be any natural number. We want to find the value of $\pi^{-1}(n)$. We take any non-zero numeral $\lambda \in S$ and we calculate $\underbrace{\lambda + \ldots + \lambda}_{p_n \text{ times}}$ in (S, σ). Then

we check on which position of λ the number has increased by 1 (note that it can also be a new position on which $\bar{1}$ has appeared). The number of this position is equal to $\pi^{-1}(n)$. Thus we can compute the permutation π^{-1} in the standard representations. However, if π^{-1} is computable, then obviously π is also computable.

Now suppose that exponentiation is computable in this representation. We shall prove that then the permutation π must be computable in the standard representation.

For any natural number n we shall find $\pi(n)$ using the following method:

Let λ_n be a numeral of the form $(\bar{0}, \ldots, \bar{0}, \bar{1})$, where the digit $\bar{1}$ is proceeded by n occurrences of the digit $\bar{0}$. Then $\sigma(\lambda_n) = p_{\pi(n)}$. We compute the result of $(\bar{1})^{\lambda_n}$ in (S, σ). Obviously:

$$\sigma((\bar{1})^{\lambda_n}) = p_{\pi(0)}^{p_{\pi(n)}}.$$

When we calculate this exponentiation, we will get the numeral $(\overline{p_{\pi(n)}})$ as a result. Thus we find out the value of the $\pi(n)$-th prime number, so we can easily compute $\pi(n)$.

Theorem 11. *Let (S, σ) be a representation of $\mathbb{N}$ in which exponentiation and $\chi_=$ are computable. Then multiplication and addition are also computable in this representation.*

Proof. Let $\alpha, \beta \in S$. We want to calculate $\alpha \cdot^\sigma \beta$ and $\alpha +^\sigma \beta$. Let λ be a numeral representing number 2 in (S, σ) and let $(\zeta_n)_{n \in \mathbb{N}}$ be a recursive enumeration of all numerals from S. For each ζ_n we check if the following equality holds:

$$(\lambda^\alpha)^\beta = \lambda^{\zeta_n}$$

until we find a numeral for which it is true. Such ζ_n shall be the result of calculating $\alpha \cdot^\sigma \beta$ in (S, σ).

To calculate $\alpha +^\sigma \beta$ in (S, σ), for each ζ_n we check whether the following equality holds:

$$\lambda^\alpha \cdot^\sigma \lambda^\beta = \lambda^{\zeta_n}.$$

until we find a numeral for which it is true. Such ζ_n shall be the result of calculating $\alpha +^\sigma \beta$ in (S, σ).

We conclude that addition and multiplication are computable in (S, σ).

Theorem 12. *There exists a representation (S, σ) of $\mathbb{N}$ in which exponentiation is computable, but successor, addition and multiplication are not computable.*

Proof. We construct such a representation as follows:

The alphabet consists of digits $\bar{0}, \ldots, \bar{9}$, symbols $\overline{\pi}$, E, $($, $)$ and the comma.

The set of numerals S is the smallest set satisfying the following conditions:

Every numeral of the standard representation belongs to S.

If $\alpha, \beta \in S \setminus \{\bar{0}, \bar{1}\}$, then $E(\alpha, \beta) \in S$.

If $\alpha \in S$ and α represents a prime number in the standard representation, then $\overline{\pi}(\alpha) \in S$.

We construct the function σ in the following way:

Let π be a permutation of prime numbers (i.e. a bijection from prime numbers onto prime numbers) uncomputable in the standard representation such that $\pi(2) = 2$.

For any standard numeral $\overline{n}$, let $\sigma(\overline{n}) = n$

For any $\alpha, \beta \in S \setminus \{\bar{0}, \bar{1}\}$, let $\sigma(E(\alpha, \beta)) = \sigma(\alpha)^{\sigma(\beta)}$.

For any prime number p, let $\sigma(\overline{\pi}(\bar{p})) = \pi(p)$.

This representation is well-defined because every natural number is represented by a certain numeral, in particular by the same numeral as in the standard representation.

We define exponentiation in (S, σ) as follows:

$\alpha^\beta = E(\alpha, \beta)$, for $\alpha, \beta \in S \setminus \{\bar{0}, \bar{1}\}$,

$\alpha^{\bar{0}} = \bar{1}$, $\alpha^{\bar{1}} = \alpha$, $\bar{1}^\alpha = \bar{1}$, for any $\alpha \in S$,

$\bar{0}^\alpha = \bar{0}$, for any $\alpha \in S \setminus \{\bar{0}, \bar{1}\}$.

Exponentiation is computable in this representation.

We shall prove that successor is not computable in (S, σ). Suppose to the contrary that it is computable. We shall show that π is then computable in the standard representation, which leads to a contradiction.

Let $T = (\alpha_{ij})_{i,j \in \mathbb{N}}$ be defined as follows:

$$\alpha_{ij} = Succ^\sigma(Succ^\sigma(...(\overline{\pi}(\overline{p_i}))...)),$$

where the successor is iterated j times, and p_i is the i-th prime number.

Since we assumed that the successor function is computable, it follows that T is a computable family of numerals indexed by pairs of natural numbers.

Note that each prime number p is represented by exactly two numerals in (S, σ), namely $\overline{p}$ and $\overline{\pi}(\overline{q})$, for a certain prime number q. Let us consider the following cases:

Case 1. Suppose that there exists a prime number p and that there exist natural numbers i, j such that $\alpha_{ij} = \overline{p}$ (by this we understand the equality of numerals, not just the equality of numbers represented by them) and for every prime number $p\prime > p$ and for any natural numbers $i\prime, j\prime$ the following holds: $\alpha_{i\prime j\prime} \neq \overline{p\prime}$. Then we consider the infinite sequence of results of the following computations (which is a row of T, possibly with the exception of a certain initial segment):

$$\overline{p}, Succ^\sigma(\overline{p}), Succ^\sigma(Succ^\sigma(\overline{p})), ...$$

It is a sequence of numerals representing consecutive natural numbers, starting with $\sigma(\overline{p})$. For any natural numbers i, j, if α_{ij} is a numeral representing a certain prime number p, then $\alpha_{ij} = \overline{p}$ or $\alpha_{ij} = \overline{\pi}(\overline{q})$, for a certain prime number q. According to our assumption, there are only finitely many prime numbers represented in the first of these two ways. In each row of T nearly all prime numbers are represented by numerals of the second type. Since T is computable, by calculating consecutive numerals from any row of T and choosing only those of them which represent prime numbers, we obtain an infinite sequence representing consecutive prime numbers:

$$\overline{\pi}(\overline{p_{i_0}}), \overline{\pi}(\overline{p_{i_1}}), \overline{\pi}(\overline{p_{i_2}}), ...$$

Then we compute π in the following way: nearly all of its elements can be obtained from the above sequence, the rest of them (which are finite in number) can be enumerated as special cases.

Case 2. Suppose that there exists a natural number i such that for any natural number j and any prime number p: $\alpha_{i,j} \neq \overline{p}$. Then from the i-th row of T, like in the previous case, we calculate nearly all values of π. Since there are only finitely many values outside of this row, it follows that π is computable.

Case 3. Suppose that there is no natural number which satisfies either of the conditions from cases 1 and 2. Therefore, for every natural number i there exist a natural number j and a prime number p such that $\alpha_{i,j} = \overline{p}$. Let us take any natural number i. We shall show how to compute $\pi(p_i)$, where p_i is the i-th prime number. Let j, p be such that $\alpha_{i,j} = \overline{p}$, where p is a prime number. Also:

$$\alpha_{i,j} = Succ^\sigma(Succ^\sigma(...(\overline{\pi}(\overline{p_i}))...)),$$

where the successor function is iterated j times.

Therefore $\pi(p_i) + j = p$. It follows that $\pi(p_i) = p - j$. We have obtained a contradiction with the assumption that π is not computable in the standard representation. Therefore the successor function is not computable in (S, σ).

From this and Theorem 8 it follows that addition is not computable in (S, σ) either.

We shall show that multiplication is not computable in (S, σ). Assume to the contrary that it is. We shall show that π is then computable in the standard representation and thus we shall obtain a contradiction. Let $p > 2$ be a prime number. We shall show how to compute $\pi(p)$.

Let us calculate $\overline{2} \cdot^{\sigma} \overline{\pi}(\overline{p})$. From the definition of (S, σ) it follows that the result of this calculation cannot be of the form $E(\alpha, \beta)$ for any numerals α, β. It cannot be of the form $\overline{\pi}(\overline{q})$, for any prime number q because the result of this multiplication is not a prime number. Therefore it must be a certain standard numeral $\overline{n}$. However, all such numerals are interpreted in (S, σ) just like in the standard representation. Therefore $\pi(p) = \frac{n}{2}$.

It follows that π is computable in the standard representation, which contradicts our assumption. Therefore, multiplication is not computable in (S, σ).

4   Conclusions

In this paper we have considered computability of the most important functions on natural numbers. We believe that we have managed to establish in what ways their computability depends on each other.

Based on these results it seems that except for some trivial cases, we can usually construct a representation in which one function is computable and the other is not. The example of such a trivial case was that computability of addition implies computability of successor—which is not surprising because successor function can be obtained by addition by substituting a constant for one of the arguments.

It is our purpose to find general rules governing such dependencies. Suppose that for a function f, we define computational closure of f as the set of all functions computable in every representation in which f is computable. Certainly f will be closed under such operations as substitution, composition of functions, etc. But is it possible to give a complete description of what functions belong to such closure? This is a question we are currently investigating.

Another important conclusion is that the computational landscape dramatically changes as soon as Boolean functions are included. The assumption of computability of $\chi_=$ ensures that we are already able to say a lot more about properties of various representations. The question arises whether there are other equivalence relations of similar importance.

References

1. Copeland, B.J., Proudfoot, D.: Deviant encodings and Turing's analysis of computability. Stud. Hist. Philos. Sci. Part A **41**(3), 247–252 (2010)
2. Quinon, P.: A taxonomy of deviant encodings. In: Manea, F., Miller, R.G., Nowotka, D. (eds.) CiE 2018. LNCS, vol. 10936, pp. 338–348. Springer, Cham (2018). https://doi.org/10.1007/978-3-319-94418-0_34
3. Rescorla, M.: Church's thesis and the conceptual analysis of computability. Notre Dame J. Formal Logic **48**(2), 253–280 (2007)
4. Shapiro, S.: Acceptable notation. Notre Dame J. Formal Logic **23**(1), 14–20 (1982)
5. Wrocławski, M.: Representing Numbers. Filozofia Nauki **26**(4), 57–73 (2018)

On the Differences and Sums of Strongly Computably Enumerable Real Numbers

Klaus Ambos-Spies[1] and Xizhong Zheng[2(✉)]

[1] Department of Mathematics and Computer Science, University of Heidelberg,
69120 Heidelberg, Germany
ambos@math.uni-heidelberg.de
[2] Department of Computer Science and Mathematics, Arcadia University,
Glenside, PA 19038, USA
zhengx@arcadia.edu

Abstract. A real number is called left c.e. (right c.e.) if it is the limit of an increasing (decreasing) computable sequence of rational numbers. In particular, if a left c.e. real has a c.e. binary expansion, then it is called strongly c.e. While the strongly c.e. reals have nice computational properties, the class of strongly c.e. reals does not have good mathematical properties. In this paper, we show that, for any non-computable strongly c.e. real x, there are strongly c.e. reals y_1 and y_2 such that the difference $x - y_1$ is neither left c.e., nor right c.e., and the sum $x + y_2$ is not strongly c.e. Thus, the class of strongly c.e. reals is not closed under addition and subtraction in an extremely strong sense.

Keywords: C.e. reals · Strongly c.e. reals · Semi-computable reals

1 Introduction

A precise definition of the "computable real numbers" was the main motivation of Alan Turing's seminal paper [16]. According to Turing, *"the computable (real) numbers are those whose decimals are calculable by finite means"*, where the "finite means" were defined in the same paper as the "automatic machines" which are now widely called "Turing machines". In other words, the decimal expansion of a computable real can be represented by a computable function. As Turing also mentioned, the computability of the reals can be defined equivalently based on binary expansions. Actually, the notion of "computable reals" can be defined equivalently based on any classic definitions of reals (see [12]).

One important property of the binary (as well as decimal) representations of reals is that every real has at most two representations. This representation is not "admissible" (see [13] for more details), i.e. it does not transfer the topological

This research was done when the first author visited Arcadia University in the fall 2018. We appreciate very much the support of the Department of Mathematics and Computer Science, Arcadia University.

© Springer Nature Switzerland AG 2019
F. Manea et al. (Eds.): CiE 2019, LNCS 11558, pp. 310–322, 2019.
https://doi.org/10.1007/978-3-030-22996-2_27

properties well from the Cantor space (i.e., $\{0,1\}^{\omega}$) to the Euclidean space of reals. Topological properties are especially important when the computable real functions are investigated, because the computable real functions are essentially the effectivization of uniform continuity. Therefore, binary expansion is not a suitable representation for computable analysis in general.

However, binary expansion does closely relate the computable analysis with the classic computability and randomness theory (cf. [8,17]), because a set of natural numbers corresponds naturally to a real in terms of binary expansion. For example, if $A \subseteq \omega$ and if the characteristic sequence of A is also denoted by A, then the set A corresponds to a real x in the unit interval with the binary expansion $0.A$ $(= 0.A(0)A(1)A(2)\cdots)$, or equivalently, $x = \sum_{i \in A} 2^{-(i+1)}$. In this case, we call A the *(binary) representation* of x. We also write $x = x_A$ if A represents x. Throughout this paper we consider only reals in the unit interval $[0,1]$. If x has a finite or co-finite representation then x is a *dyadic rational*.

By means of binary representation almost all results of classic computability theory about sets of natural numbers can be "translated" directly into the corresponding results about reals. In particular, a real x is *computable or computably approximable* (*c.a.* for short) if it has a computable or Δ_2^0 representation, respectively. Most of this kind of translations are quite reasonable and make perfect sense. For example, we can define the Turing reducibility between two reals by $x \leq_T y$ iff there are sets $A, B \subseteq \omega$ such that $A \leq_T B$ with $x = x_A$ and $y = x_B$.

Unfortunately, however, some of these translations do not seem reasonable. For example, computable enumerability (c.e. for short), one of the most important notions in classical computability theory, can not be well translated into reals in such a straightforward way. As it was first observed by Jockusch and Soare ([15]), for a non-computable c.e. set A, the real $x_{A \oplus \overline{A}}$ does not have a c.e. binary expansion, but its left Dedekind's cut is a c.e. set of rational numbers and, equivalently, it is the limit of an increasing computable sequence of rational numbers. It seems that these reals are better counterparts of the c.e. sets. Therefore, we call a real x *left c.e.* or simply *c.e.* if it is the limit of an increasing computable sequence of rational numbers. The example of Jockusch and Soare shows that not every c.e. real has a c.e. binary representation. So, we call a real x *strongly c.e.* if it has a c.e. binary representation, i.e. $x = x_A$ for a c.e. set A. Moreover, we call a real *right c.e.* or *co-c.e.* if it is the limit of a decreasing computable sequence of rational numbers, and we call a real *semi-computable* if it is either left c.e. or right c.e.

The class of c.e. reals has been widely investigated (see [6–8,10,21]) and it has very rich and nice properties with respect to computability as well as randomness. Unfortunately, however, the class of c.e. reals does not have nice algebraic properties. In particularly, it is not closed under subtraction, because $-x$ is not c.e. if x is c.e. but not computable. The class of strongly c.e. reals is even worse with respect to the algebraic properties, and it is not closed under addition. For example, the sum $x_{A \oplus \overline{A}} = x_{2\omega+1} + x_{2A+1}$ of two strongly c.e. reals $x_{2\omega+1}$ and x_{2A+1} is not strongly c.e. if A is a non-computable c.e. set. This observation is extended by Wu [18] as follows. He calls a real *n-strongly c.e.* if it is the sum of up to n strongly c.e. reals, and he calls a real *regular* if it is the sum of finitely many strongly c.e. reals. Then he shows that these classes form a proper hierarchy and they do not exhaust all c.e. reals. In this paper, we strengthen the observation that the class of the strongly c.e. reals is not closed under sums in a different direction. We show that, for any strongly c.e. real x, if it is not a dyadic rational, then there exists a strongly c.e. real y such that $x + y$ is not strongly c.e. (Theorem 2).

Since $(x_{2A} + x_{2\omega+1}) - x_{2A+1} = x_{A \oplus \overline{A}}$ we know that the class of strongly c.e. reals is also not closed under subtraction. Furthermore, as it is shown in [2], the real $x_{A \oplus \overline{B}}$ is neither c.e. nor co-c.e. if A and B are Turing incomparable c.e. sets. This immediately implies that there are strongly c.e. reals such that their difference is not even semi-computable. Furthermore, the second author shows in [19] that, there are strongly c.e. reals x and y such that their difference $x - y$ does not have an ω-c.e. Turing degree. Moreover, for the c.e. reals Barmpalias and Lewis-Pye [4] show that, for any non-computable c.e. real x, there is a c.e. real y such that the difference $x - y$ is not semi-computable. Here we will show the corresponding result for strongly c.e. reals (Theorem 1) namely that, for any non-computable strongly c.e. real x, there is a strongly c.e. real y such that the difference $x - y$ is not semi-computable. Therefore, the class of strongly c.e. reals is not closed under both addition and subtraction in an extremely strong sense. For some more recent work on differences of c.e. reals we refer the reader to Barmpalias and Lewis-Pye [5] and Miller [11].

We close this section by explaining some useful notions and notations. In the following the term set refers to sets of nonnegative integers, i.e., subsets of ω. $\overline{A}$ is the complement of A. $A \oplus B = \{2n : n \in A\} \cup \{2n+1 : n \in B\}$ is the effective disjoint union of A and B. We call a set A *almost c.e.* if there is a computable sequence (A_s) of finite sets such that $\lim A_s = A$ and, for any n and s, if $n \in A_s \setminus A_{s+1}$ then $m \in A_{s+1} \setminus A_s$ for some $m < n$. Such a sequence (A_s) is called a *computable almost-enumeration* of A. Notice that we have $A_s \leq_{\text{lex}} A_{s+1}$ if (A_s) is a computable almost-enumeration. A real x is c.e. iff its binary representation is almost c.e. (see [6]). In the literature, c.e. and co-c.e. reals are also called *left computable* and *right computable*, respectively. Since the reals in the unit interval are not closed under sums and differences, we tacitly use the following convention in this paper. If $x = x_A + x_B$ is greater than 1 then we replace x by $x - 1$ and if $x = x_A - x_B$ is less than 0 then we replace x by $x + 1$. By these shifts we can keep the sums and differences of the reals from the unit interval still in the unit interval.

2   Differences of Strongly C.E. Reals

This section will explore the differences of strongly c.e. reals. We will show

Theorem 1. *For any noncomputable strongly c.e. real x there is a strongly c.e. real $y < x$ such that $x - y$ is not semi-computable.*

Before we turn to the proof, we first give some of the underlying ideas. In [2] Ambos-Spies, Weihrauch and Zheng observed that, for any c.e. sets A and B such that A and B are Turing incomparable, the real $x_{A \oplus \overline{B}}$ is not semi-computable. Since $x_{A \oplus \overline{B}} = (x_{2A} + x_{2w+1}) - x_{2B+1}$ this immediately gives an example of two strongly c.e. reals such that their difference is not semi-computable. This argument can be easily modified to show that $x_A - x_B$ is not semi-computable if $B \subseteq A$ and A and B are Turing incomparable c.e. sets. Moreover, by Sacks' Splitting Theorem [14], one can easily show that any noncomputable Turing incomplete set A contains a c.e. subset B as above. This proves Theorem 1 for strongly c.e. reals $x = x_A$ represented by Turing incomplete sets A. In order to obtain the theorem for *all* noncomputable c.e. sets *including* the Turing complete sets, we first observe that for the above argument, it suffices to assume that the sets A and B are ibT incomparable, where ibT denotes identity bounded Turing reducibility. The ibT reducibility admits only such Turing reductions $\Phi^X(x)$ where any oracle query y is bounded by the input x, i.e., $y \leq x$ (see e.g. Downey and Hirschfeldt [8]). Moreover, as shown in [8], there are no ibT complete sets. So, in order to complete the proof of Theorem 1, it suffices to extend the latter result by showing that, for any noncomputable c.e. set A, there is a c.e. *subset* B of A such that A and B are ibT incomparable.

We now turn to the proof of Theorem 1. By the above observations, the following two lemmas are the keys to the proof.

Lemma 1. *Let A and B be c.e. sets such that $B \subseteq A$, $A \not\leq_{\mathrm{ibT}} B$ and $B \not\leq_{\mathrm{ibT}} A$. Then neither $A \setminus B$ nor $\overline{A \setminus B}$ is almost-c.e.*

Proof. We first show that $C = A \setminus B$ is not almost-c.e. The proof is indirect. For a contradiction assume that $A \setminus B$ is almost-c.e. We show that $B \leq_{\mathrm{ibT}} A$ contrary to assumption. Given n, it suffices to show that $B(n)$ can be computed from $A \restriction (n+1)$ and $B \restriction n$ uniformly in n. Fix a computable enumeration $(B_s)_{s \geq 0}$ of B and a computable almost-enumeration $(C_s)_{s \geq 0}$ of $A \setminus B$. Now if $n \notin A$ then, by $B \subseteq A$, $B(n) = 0$. So w.l.o.g. we may assume that $A(n) = 1$. Then there is a least stage s such that

$$C_s \restriction (n+1) = ((A \setminus B) \restriction n)(1 - B_s(n)), \tag{1}$$

and s can be computed from $A \restriction (n+1)$ and $B \restriction n$. So it suffices to show that $B(n) = B_s(n)$. If $B_s(n) = 1$ then this is immediate since $(B_s)_{s \geq 0}$ is a computable enumeration of B. So, for the remainder of the argument we may assume that $B_s(n) = 0$ hence $C_s(n) = 1$. Since $(C_s)_{s \geq 0}$ is a computable almost-enumeration of $C = A \setminus B$,

$$C_s \restriction (n+1) \leq_{\mathrm{lex}} C_t \restriction (n+1) \leq_{\mathrm{lex}} ((A \setminus B) \restriction n)C(n)$$

for all numbers $t \geq s$. But, by (1) and by $C_s(n) = 1$, this implies that $C(n) = 1$ hence $B(n) = 0$.

The proof that $\overline{C} = \overline{A \setminus B}$ is not almost-c.e. is similar and left to the reader. $\qquad \square$

Lemma 2. *Let A be a noncomputable c.e. set. There is a c.e. set $B \subseteq A$ such that $A \not\leq_{\text{ibT}} B$ and $B \not\leq_{\text{ibT}} A$.*

Proof. It suffices to give a c.e. set B such that

$$B \subseteq A \text{ and } B \not\leq_{\text{ibT}} A. \tag{2}$$

Then, by Sacks' Splitting Theorem, we can split B into two c.e. subsets $\hat{B}_0$ and $\hat{B}_1$ such that $A \not\leq_T \hat{B}_0, \hat{B}_1$ hence $A \not\leq_{\text{ibT}} \hat{B}_0, \hat{B}_1$. Moreover, since $\hat{B}_0 \cup \hat{B}_1 = B \not\leq_{\text{ibT}} A$ it follows that $\hat{B}_0 \not\leq_{\text{ibT}} A$ or $\hat{B}_1 \not\leq_{\text{ibT}} A$. So, for some $i \leq 1$, $\hat{B}_i \subseteq A$, $A \not\leq_{\text{ibT}} \hat{B}_i$ and $\hat{B}_i \not\leq_{\text{ibT}} A$.

The c.e. set B satisfying (2) is constructed in stages. Let B_s denote the finite part of B enumerated by the end of stage s.

By the noncomputability, the set A contains a computable ascending sequence $(k_e)_{e \geq 0}$ such that the interval $I_e = [k_e, k_{e+1})$ contains at least $k_e + 2$ elements of A which can be effectively given as $k_e^0 = k_e < k_e^1 < \cdots < k_e^{k_e+1}$. Moreover, we may fix a computable enumeration $(A_s)_{s \geq 0}$ of A such that, for $e < s$, all numbers k_e^m ($m \leq k_e + 1$) have been enumerated in A_s.

In order to ensure that $B \not\leq_{\text{ibT}} A$, fix a computable numbering $(\hat{\Phi}_e)_{e \geq 0}$ of the ibT-functionals, and let $\hat{\Phi}_{e,s}^X(m) = \hat{\Phi}_e^X(m)$ if the computation on the right hand side converges in less than s steps, and let $\hat{\Phi}_{e,s}^X(m) \uparrow$ otherwise. Then it suffices to meet the requirements

$$\mathcal{R}_e : B \neq \hat{\Phi}_e^A$$

for $e \geq 0$. The strategies for meeting these requirements are finitary and do not interfere with each other. The strategy for meeting the requirement $\mathcal{R}_e$ will enumerate only elements of A from the interval I_e into B. So in order to define the computable enumeration $(B_s)_{s \geq 0}$, given e and s it suffices to determine which numbers from I_e are enumerated into B by the $\mathcal{R}_e$-strategy at stage $s + 1$. (B_0 will be empty.)

Requirement $\mathcal{R}_e$ requires attention at stage $s + 1$ if the following hold.

(i) $s > e$.
(ii) For all $m \in I_e$, $B_s(m) = \hat{\Phi}_{e,s}^{A_s}(m)$.
(iii) There is an $m \in I_e$ such that $m \in A_s \setminus B_s$.

If $\mathcal{R}_e$ requires attention then enumerate the greatest number m as in (iii) into B_{s+1}. (Moreover say that $\mathcal{R}_e$ requires attention via any number m as in (iii) and that it becomes active via the greatest such number.)

It remains to show that B has the required properties. Obviously, $B \subseteq A$ and B is computably enumerable. So it suffices to show that the requirements $\mathcal{R}_e$ are met. Assume by contradiction that $\mathcal{R}_e$ is not met for some e. Then $\hat{\Phi}_e^A$ is total and $B(m) = \hat{\Phi}_e^A(m)$ for all numbers m. So we may fix a stage $s_0 > e$ such that $B_{s_0} \upharpoonright k_{e+1} = B \upharpoonright k_{e+1}$, $A_{s_0} \upharpoonright k_{e+1} = A \upharpoonright k_{e+1}$, and $B_{s_0}(m) = \hat{\Phi}_{e,s_0}^A(m)$ for all $m \in I_e$. Then conditions (i) and (ii) in the definition of requiring attention hold for all $s \geq s_0$. On the other hand, however, $\mathcal{R}_e$ does not require attention after stage s_0 since no number from I_e becomes enumerated into B after stage s_0. So (iii) fails for $s \geq s_0$ whence $A \cap I_e \subseteq B_{s_0}$. Hence all elements of A in I_e have been enumerated into B by the end of stage s_0.

Now let $m_0, \ldots, m_p$ be the numbers from I_e entering B in order of their enumeration, say m_q is enumerated into B at stage $t_q + 1$. We claim that, for $q < p$,

$$\left|(A_{t_q} \setminus (I_e \cap B_{t_q})) \restriction m_q\right| \leq \left|(A_{t_{q+1}} \setminus (I_e \cap B_{t_{q+1}})) \restriction m_{q+1}\right|. \tag{3}$$

Note that this gives the desired contradiction. Namely, the first number from I_e put into B is $k_e^{k_e+1}$ or a greater number hence $m_0 \geq k_e^{k_e+1}$ and the $k_e + 1$ numbers $k_e^0, k_e^1, \ldots, k_e^{k_e+1}$ less than m_0 are in I_e and A_{t_0} but not in B_{t_0}. So

$$\left|(A_{t_0} \setminus (I_e \cap B_{t_0})) \restriction m_0\right| \geq k(e) + 1. \tag{4}$$

On the other hand, since $k_e = k_e^0$ is enumerated into B too, there is a number $q \geq 0$ such that $m_{q+1} = k_e$. Then, obviously,

$$\left|(A_{t_{q+1}} \setminus (I_e \cap B_{t_{q+1}})) \restriction m_{q+1}\right| \leq m_{q+1} = k(e).$$

By (4) this contradicts (3).

In the rest of the proof we show (3). So fix $q < p$. If the set $(A_{t_q} \setminus (I_e \cap B_{t_q})) \restriction m_q$ is empty or does not intersect I_e (hence is contained in $\omega \restriction k_e$) then the claim is trivial. So w.l.o.g. assume that $n_q \in I_e$ is the greatest element of this set. Now, by assumption, $\mathcal{R}_e$ requires attention via m_q and m_{q+1} at stages $t_q + 1$ and $t_{q+1} + 1$, respectively, and does not require attention at any stage s with $t_q + 1 < s \leq t_{q+1}$. So m_q is the only number from I_e which is in $B_{t_{q+1}}$ but not in B_{t_q} whence

$$(I_e \cap B_{t_q}) \restriction m_q = (I_e \cap B_{t_{q+1}}) \restriction m_q.$$

It follows by choice of n_q that

$$(A_{t_q} \setminus (I_e \cap B_{t_q})) \restriction m_q \subseteq (A_{t_q} \setminus (I_e \cap B_{t_{q+1}})) \restriction (n_q + 1). \tag{5}$$

and $n_q \in A_{t_q} \setminus B_{t_{q+1}}$. So $\mathcal{R}_e$ requires attention via n_q at stage $t_{q+1} + 1$. Since $\mathcal{R}_e$ acts via the greatest number via which it requires attention it follows that $n_q \leq m_{q+1}$. Moreover, by (5), (3) holds if $n_q < m_{q+1}$.

So, for the remainder of the proof, we may assume that $m_{q+1} = n_q$. Now, since $\mathcal{R}_e$ acts via m_q at stage $t_q + 1$ and requires attention at stage $t_{q+1} + 1$ again, it follows by clause (ii) in the definition of requiring attention that

$$\hat{\Phi}_e^{A_{t_{q+1}}}(m_q) = B_{t_{q+1}}(m_q) = B_{t_q+1}(m_q) = 1 \neq 0 = B_{t_q}(m_q) = \hat{\Phi}_e^{A_{t_q}}(m_q)$$

whence

$$A_{t_{q+1}} \restriction m_q \neq A_{t_q} \restriction m_q.$$

(Recall that $\hat{\Phi}_e^A$ is an ibT-functional. So the computation $\hat{\Phi}_e^{A_{t_q}}(m_q)$ can only change after stage s if a number $\leq m_q$ becomes enumerated into A later, and

m_q is in A at stage t_q already.) So, by $m_q \in I_e$, there is a greatest number $n' \leq \max I_e$ such that $n' \in A_{t_{q+1}} \setminus A_{t_q}$. Since

$$I_e \cap B_{t_{q+1}} = I_e \cap (B_{t_q} \cup \{m_q\}) \subseteq A_{t_q}$$

and $n_q \in A_{q_t} \cap I_e$, it follows that $n' \in A_{t_{q+1}} \setminus (I_e \cap B_{t_{q+1}})$ and $n' \neq n_q$. By the former, $n' < \min I_e$ or $\mathcal{R}_e$ requires attention via n' but, by $n' \neq n_q = m_{q+1}$, does not become active via n'. In either case this implies $n' < n_q$. Hence n' is witnessing the following (first) inequality (the second inequality is trivial).

$$\left| (A_{t_{q+1}} \setminus (I_e \cap B_{t_{q+1}})) \upharpoonright n_q \right| \geq \left| (A_{t_q} \setminus (I_e \cap B_{t_{q+1}})) \upharpoonright n_q \right| + 1$$
$$\geq \left| (A_{t_q} \setminus (I_e \cap B_{t_{q+1}})) \upharpoonright (n_q + 1) \right|.$$

By (5) this implies (3) which completes the proof of (3) and the proof of the lemma. □

Remark. The complexity of the proof of Lemma 2 is due to the fact that the desired c.e. set B which is ibT-incomparable with the given noncomputable c.e. set A has to be a subset of A. Without this condition, the proof easily follows from Sacks's Splitting Theorem by using the simple observation (see e.g. Lemma 11 in [1]) that, by noncomputability of A, $A <_{\text{ibT}} A_{-1}$ for the left shift $A_{-1} = \{n : n + 1 \in A\}$ of A. Namely, if we split the c.e. set A_{-1} into c.e. sets B_0 and B_1 such that $A \not\leq_T B_0$ and $A \not\leq_T B_1$ (hence $A \not\leq_{\text{ibT}} B_0$ and $A \not\leq_{\text{ibT}} B_1$) then B_0 or B_1 will be ibT incomparable with A. The sets B_0 and B_1, however, might not be subsets of A.

It remains to show how the above two lemmas imply Theorem 1.

Proof (of Theorem 1). Given a noncomputable c.e. set A it suffices to give a c.e. set B such that $x_A - x_B$ is not semi-computable. By Lemma 2 let B be a c.e. set such that $B \subseteq A$ and A and B are ibT incomparable. Then, by $B \subseteq A$, $x_{A \setminus B} = x_A - x_B$ and $x_B \leq x_A$ while, by Lemma 1, neither $A \setminus B$ nor $\overline{A \setminus B}$ is almost-c.e. So $x_{A \setminus B}$ is neither left computable nor right computable, hence not semi-computable.[1] □

Note that in Theorem 1 the assumption that x is noncomputable is necessary since, for any computable real x and any (strongly) c.e. real y, the difference $x - y$ is right computable hence semi-computable. Nevertheless, for any computable real x the difference $x - y$ is not c.e., hence not strongly c.e., as long as y is c.e. but not computable.

---

[1] One of the referees pointed out an alternative proof of Theorem 1 which follows more easily from results in the literature. It is based on the observation that the difference of two left computable reals which are Solovay incomparable is not semi-computable. So, since Downey, Hirschfeldt and LaForte [9] have shown that Solovay reducibility coincides with computable Lipschitz reducibility on the strongly c.e. reals, it suffices to show that for any noncomputable c.e. set A there is a c.e. set B which is cl-incomparable with A. But, since Barmpalias [3] has shown that there are no maximal c.e. cl-degrees, this can be deduced from Sacks' Splitting Theorem by duplicating the argument for cl-reducibility which we have given for ibT-reducibility in the remark following the proof of Lemma 2.

3   Sums of Strongly C.E. Reals

We now consider sums of strongly c.e. reals. Since the class of the c.e. reals is closed under addition, the sum of any two strongly c.e. reals is c.e. As we mentioned in Sect. 1, however, the class of the strongly c.e. reals is not closed under addition. Here we will show a very strong version of this non-closure property:

Theorem 2. *Let x be any strongly c.e. real such that x is not a dyadic rational. There is a strongly c.e. real y such that $x + y$ is not strongly c.e.*

Here the assumption that x is not a dyadic rational is necessary: As one can easily show, for any dyadic rational x and any strongly c.e. real y, the sum $x + y$ is strongly c.e.

Proof. (of Theorem 2). The outline of the proof is as follows. Fix a c.e. set A such that $x = x_A$. Since x is not a dyadic rational, A is infinite and coinfinite. We define a c.e. subset B of A such that, for the strongly c.e. real $y = x_B$, the sum $x + y$ is not strongly c.e. So, for the unique set C such that $x_C = x_A + x_B$, we have to ensure that C is not c.e. For this sake, we first show how the set C can be described in terms of A and B (using that B is a subset of A and that A is infinite and coinfinite).

Claim 1. Let $\{I_e\}_{e\geq 0}$ be the unique sequence of nonempty finite intervals $I_e = (p_e, q_e)$ such that p_0 is the least number which is not in A, $q_e \leq p_{e+1}$, $I_e \subseteq A$, $p_e, q_e \notin A$ and $\bigcup_{e\geq 0} I_e = A \setminus \{0,\dots,p_0 - 1\}$. Moreover, assume that B is a subset of $A \setminus \{0,\dots,p_0 - 1\}$. Then $x_A + x_B = x_C$ for the set $C \subseteq \{0,\dots,p_0 - 1\} \cup \left(\bigcup_{e\geq 0}[p_e, q_e)\right)$ defined as follows (where $e \geq 0$).

(i) $\{0,\dots,p_0 - 1\} \subseteq C$.
(ii) If $B \cap I_e = \emptyset$ then $C \cap [p_e, q_e) = I_e$.
(iii) If $B \cap I_e \neq \emptyset$ and b_e is the greatest element of $B \cap I_e$ then $C \cap [p_e, q_e)$ is defined by

$$
C(n) = \begin{cases}
1 & \text{if } b_e < n < q_e \\
0 & \text{if } n = b_e \\
B(n) & \text{if } p_e < n < b_e \\
1 & \text{if } n = p_e.
\end{cases}
$$

We omit the straightforward proof of Claim 1. Note that the intervals I_e are the maximal (i.e. nonextendible) subintervals of A. Now, the actual construction of B is nonuniform. It depends on whether the lengths of the intervals I_e are bounded or not. For this sake we say that A is *k-scattered* ($k \geq 1$) if k is maximal such that there are infinitely many intervals I of length k such that $I \subseteq A$, i.e., if $|I_e| = k$ for infinitely many e and $|I_e| \leq k$ for almost all e. Moreover, we say that A is *scattered* if A is k-scattered for some $k \geq 1$, and that A is *nonscattered* otherwise.

Now if A is scattered, then the construction of B is rather straightforward. If A is nonscattered then the construction of B is more involved and requires a finite injury argument. We treat the two cases in the following two claims.

Claim 2. Assume that A is scattered. There is a c.e. set $B \subseteq A$ such that $x_A + x_B$ is not strongly c.e.

Proof. Fix $k \geq 1$ such that A is k-scattered, and fix $I_e = (p_e, q_e)$ as in Claim 1. Then there are infinitely many numbers e such that $|I_e| = k$ whereas there are only finitely many numbers e such that I_e has length greater than k. So, since A is c.e., we may effectively enumerate an infinite computable subsequence $\{\hat{I}_e\}_{e \geq 0}$ of $\{I_e\}_{e \geq 0}$ such that $|\hat{I}_e| = k$ and such that, for the numbers $\hat{p}_e$ and $\hat{q}_e$ such that $\hat{I}_e = (\hat{p}_e, \hat{q}_e)$, $\hat{q}_e \leq \hat{p}_{e+1}$. Then the desired set B is defined by

$$B = \{\hat{q}_e - 1 : e \geq 0 \ \& \ \hat{q}_e - 1 \in W_e\}.$$

Obviously, B is c.e. and B is contained in $A \setminus \{0, \dots, p_0 - 1\}$. So, by Claim 1, for the unique set C such that $x_A + x_B = x_C$, it holds that $\hat{q}_e - 1 \in C$ iff $\hat{q}_e - 1 \notin W_e$. So C is not c.e. whence $x_A + x_B$ is not strongly c.e. $\qquad \square$

Claim 3. Assume that A is nonscattered. There is a c.e. set $B \subseteq A$ such that $x_A + x_B$ is not strongly c.e.

Proof (sketch). The proof is by a finite injury argument. We enumerate the desired set B in stages where we let B_s be the finite part of B enumerated by the end of stage s. Let $\{A_s\}_{s \geq 0}$ be a computable enumeration of A, and let $I_e = (p_e, q_e)$ be the nonextendible subintervals of A (to the right of p_0) as defined in Claim 1. Since we will ensure that B is a subset of $A \setminus \{0, \dots, p_0 - 1\}$, it suffices to ensure that the set C (depending on B) defined in Claim 1 is not c.e., i.e., for $n \geq 0$, the requirements

$$\mathcal{R}_n : \exists m \ (C(m) \neq W_n(m))$$

are met.

The basic strategy for meeting $\mathcal{R}_n$ is as follows. We appoint a follower $m \in A \setminus \{0, \dots, p_0 - 1\}$ to $\mathcal{R}_n$ for which we attempt to ensure $C(m) \neq W_n(m)$. We put the follower m into B if and only if $m \in W_n$. Now, by clauses (ii) and (iii) in the definition of C, this guarantees that $\mathcal{R}_n$ is met unless there is a number $m' > m$ in I_{e_m} which is put into B where e_m is the unique index such that $m \in I_{e_m}$. Namely, if $W_n(m) = 0$ (whence m is not put into B), then either $B \cap I_{e_m} = \emptyset$ or b_{e_m} (as defined in Claim 1) is less than m. So, in either case, $C(m) = 1$ by definition of C. On the other hand, if $W_n(m) = 1$ then $b_{e_m} = m$ hence $C(m) = 0$.

The problem with this strategy is how to ensure that no number $m' > m$ in I_{e_m} is put into B. As usual, we can ensure that, for followers m and m' of requirements $\mathcal{R}_n$ and $\mathcal{R}_{n'}$, respectively, which exist at the same stage, $m < m'$ iff $n < n'$ and that m' is appointed later than m. So it suffices to ensure that, once m is appointed at stage $s+1$, then, for any follower m' of a lower priority requirement $\mathcal{R}_{n'}$ ($n' > n$) which is appointed later, either m' is not in I_{e_m} or m' is prevented from entering B (by cancellation). Now if we would know the interval I_{e_m} then this problem could be overcome by choosing m' to be bigger than $\max I_{e_m}$. Since A is nonscattered, however, in general we cannot compute the borders of an interval I_e from an element of the interval. So if we appoint m' at stage $s'+1 > s+1$ then we do so only if there is at least one number k separating m from m' which is not yet in A, i.e., $m < k < m'$ and $k \notin A_{s'+1}$. Now, if one of these numbers k is never enumerated into A, then $e_m < e_{m'}$ whence enumerating m' into B does not affect the $\mathcal{R}_n$-strategy. If, however, it eventually turns out that m and m' are in the same interval, i.e., $e_m = e_{m'}$ then the enumeration of m' in B will kill the $\mathcal{R}_n$-strategy. So, if we see that m and m' are in the same interval at a stage where m' has not (yet) been enumerated in B then the $\mathcal{R}_n$-strategy cancels the lower priority follower m'. (In this case $\mathcal{R}_{n'}$ has to get a new greater follower. Since the interval I_{e_m} is finite, $\mathcal{R}_{n'}$ will be injured by $\mathcal{R}_n$ via m in this way only finitely often.) If m' has been put into B, however, before it became apparent that m and m' are in the same interval, then the attack on $\mathcal{R}_n$ via m fails and has to be abandoned.

The above problem caused by the fact that A is nonscattered can be resolved by using this property of A too. Since A is nonscattered there are arbitrarily long intervals contained in A. So we may appoint the follower m' of $\mathcal{R}_{n'}$ only when m' *and* the following n' numbers $m'+1, \ldots, m'+n'$ have been enumerated into A already, and we reserve these numbers in decreasing order as potential replacements of the current followers of the higher priority requirements. Note that all of these numbers are in the same interval. So if m' is enumerated into B and later we see that m and m' are in the same interval then we replace m (and its entourage $m+n, m+(n-1), \ldots, m+1$ of candidates for the higher priority requirements $\mathcal{R}_0, \ldots, \mathcal{R}_{n-1}$) by $m'+n'-n$ (and $m+n', m+n'-1, \ldots, m+n'-(n+1)$, respectively). Then the new $\mathcal{R}_n$-follower $m'+n'-n$ is greater than m', and the numbers m, m' and $m'+n'-n$ are in the same interval. So the enumeration of m' into B which took place earlier does not affect the $\mathcal{R}_n$-strategy for the new follower $m'+n'-(n+1)$. Moreover, since a follower is replaced by a greater follower from the same interval, eventually the follower stabilizes and, for the final follower, the strategy works. So we may argue that the requirements are finitary and will be eventually met.

This completes the intuitive description of the strategy for meeting a requirement $\mathcal{R}_n$. In order to formally describe the construction, we start with some notation. Two numbers k, k' are called *equivalent* if the interval $[\min(k, k'), \max(k, k')]$ is contained in A, i.e., if $I_{e_k} = I_{e_{k'}}$, and we say that k, k' are *equivalent at stage* s or *s-equivalent* for short if $[\min(k, k'), \max(k, k')] \subseteq A_s$. Moreover, we say that a number k is *unused at stage* s if, for any follower k' appointed by the end of stage s, k is greater than k' and any number in the entourage of k'.

At any stage s of the construction, a requirement $\mathcal{R}_n$ may or may not have followers. If $\mathcal{R}_n$ has some followers at stage s then either all of these followers or all but one are *frozen*. (Once frozen, a follower remains frozen forever.) So there is at most one follower which is not frozen and which is called *active*. At stage 0 no requirement has any follower. Requirement $\mathcal{R}_n$ requires attention at stage $s+1$ if $n < s+1$ and one of the following holds.

(i) There is no active follower of $\mathcal{R}_n$ at the end of stage s and there is an unused number $m > p_0$ such that $[m, m+n] \subseteq A_s$.

(ii) $\mathcal{R}_n$ has active follower m at the end of stage s and there is a number $m' > m$ such that, at the end of stage s, m' is a (frozen or active) follower of a lower priority requirement $\mathcal{R}_{n'}$, $n < n'$, and m and m' are s-equivalent.

(iii) $\mathcal{R}_n$ has active follower m at the end of stage s, (ii) does not hold, $m \notin B_s$ and $m \in W_{n,s}$.

Using these definitions the construction is as follows where stage 0 is vacuous.

Stage $s+1$. Fix $n \leq s$ minimal such that $\mathcal{R}_n$ requires attention. Declare that $\mathcal{R}_n$ *becomes active* at stage $s+1$, and distinguish the following cases according to the clause via which $\mathcal{R}_n$ requires attention. (If there is no such n, stage $s+1$ is vacuous.)

(i) For the least unused number m such that $[m, m+n] \subseteq A_s$, appoint m as active follower of $\mathcal{R}_n$ and let $\{m+1, \ldots, m+n\}$ be the entourage of m where the number $m+n-n''$ becomes reserved for requirement $\mathcal{R}_{n''}$ ($n'' < n$).

(ii) Fix the greatest $m' > m$ such that m' is follower of a lower priority requirement and m and m' are s-equivalent. Replace the active follower m of $\mathcal{R}_n$ by the number $m' + n' - n$ in the entourage of m' reserved for $\mathcal{R}_n$ (i.e., freeze m and declare $m' + n' - n$ to be active follower of $\mathcal{R}_n$) and let $m'+n'-n+1, \ldots, m'+n'$ be the entourage of $m'+n'-n$ where $m'+n'-n''$ becomes reserved for requirement $\mathcal{R}_{n''}$ ($n'' < n$).

(iii) Enumerate the active follower m of $\mathcal{R}_n$ into B.

In any case *initialize* all lower priority requirements $\mathcal{R}_{n'}$, $n < n'$, i.e., *freeze* the active follower of $\mathcal{R}_{n'}$ (if there is any).

This completes the construction. It remains to show that the set B has the required properties. Obviously, the construction is effective, hence B is computably enumerable. Moreover, B is a subset of $A \setminus \{0, \ldots, p_0 - 1\}$. Hence the set C corresponding to B according to Claim 1 is well defined. So it only remains to show that the requirements $\mathcal{R}_n$ ($n \geq 0$) are met. Using the above given intuition behind the construction, this is done in the standard way. For lack of space, the formal verification is left to the reader.   □

Note that Theorem 2 is immediate by Claims 2 and 3.   □

Note that the proofs of our two main results (Theorems 1 and 2) are nonuniform. We do not know whether there are uniform proofs.

References

1. Ambos-Spies, K., Ding, D., Fan, Y., Merkle, W.: Maximal pairs of computably enumerable sets in the computably Lipschitz degrees. Theory Comput. Syst. **52**(1), 2–27 (2013)
2. Ambos-Spies, K., Weihrauch, K., Zheng, X.: Weakly computable real numbers. J. Complex. **16**(4), 676–690 (2000)
3. Barmpalias, G.: Computably enumerable sets in the solovay and the strong weak truth table degrees. In: Cooper, S.B., Löwe, B., Torenvliet, L. (eds.) CiE 2005. LNCS, vol. 3526, pp. 8–17. Springer, Heidelberg (2005). https://doi.org/10.1007/11494645_2
4. Barmpalias, G., Lewis-Pye, A.: A note on the differences of computably enumerable reals. In: Day, A., Fellows, M., Greenberg, N., Khoussainov, B., Melnikov, A., Rosamond, F. (eds.) Computability and Complexity. LNCS, vol. 10010, pp. 623–632. Springer, Cham (2017). https://doi.org/10.1007/978-3-319-50062-1_37
5. Barmpalias, G., Lewis-Pye, A.: Differences of halting probabilities. J. Comput. Syst. Sci. **89**, 349–360 (2017)
6. Calude, C.S., Hertling, P.H., Khoussainov, B., Wang, Y.: Recursively enumerable reals and Chaitin Ω numbers. Theor. Comput. Sci. **255**, 125–149 (2001)
7. Downey, R., Terwijn, S.A.: Computably enumerable reals and uniformly presentable ideals. MLQ Math. Log. Q. **48**(suppl. 1), 29–40 (2002)
8. Downey, R.G., Hirschfeldt, D.R.: Algorithmic Randomness and Complexity Theory and Application of Computability. Springer, Heidelberg (2010). https://doi.org/10.1007/978-0-387-68441-3
9. Downey, R.G., Hirschfeldt, D.R., LaForte, L.G.: Randomness and reducibility. J. Comput. Syst. Sci. **68**(1), 96–114 (2004)
10. Downey, R.G., LaForte, G.L.: Presentations of computably enumerable reals. Theor. Comput. Sci. **284**(2), 539–555 (2002)
11. Miller, J.S.: On work of Barmpalias and Lewis-Pye: a derivation on the D.C.E. reals. In: Day, A., Fellows, M., Greenberg, N., Khoussainov, B., Melnikov, A., Rosamond, F. (eds.) Computability and Complexity. LNCS, vol. 10010, pp. 644–659. Springer, Cham (2017). https://doi.org/10.1007/978-3-319-50062-1_39
12. Robinson, R.M.: Review of Peter, R. Rekursive Funktionen. J. Symbol. Log. **16**, 280–282 (1951)
13. Schröder, M.: Admissible representations in computable analysis. In: Beckmann, A., Berger, U., Löwe, B., Tucker, J.V. (eds.) CiE 2006. LNCS, vol. 3988, pp. 471–480. Springer, Heidelberg (2006). https://doi.org/10.1007/11780342_48
14. Sacks, G.E.: Degrees of Unsolvability. Princeton University Press, Princeton (1963)
15. Robert Irving Soare: Recursion theory and Dedekind cuts. Trans. Am. Math. Soc. **140**, 271–294 (1969)
16. Turing, A.M.: On computable numbers, with an application to the "Entscheidungsproblem". Proc. Lond. Math. Soc. **42**(2), 230–265 (1936)
17. Weihrauch, K.: An introduction. In: Weihrauch, K. (ed.) Computable Analysis. Springer, Heidelberg (2000). https://doi.org/10.1007/978-3-642-56999-9_1
18. Guohua, W.: Regular reals. Math. Log. Q. **51**(2), 111–119 (2005)
19. Zheng, X.: On the turing degrees of weakly computable real numbers. J. Log. Comput. **13**(2), 159–172 (2003)
20. Zheng, X., Rettinger, R.: Weak computability and representation of reals. Math. Log. Q. **50**(4/5), 431–442 (2004)

21. Zheng, X., Rettinger, R.: Computability of real numbers. In: Brattka, V., Hertling, P. (eds.) Handbook on Computability and Complexity in Analysis, Theory and Applications of Computability. Springer-Verlag (to appear)

Author Index

Printed in the United States
By Bookmasters